Comp

Computer Busses

William Buchanan, BSc (Hons), CEng, PhD
Napier University, Edinburgh, UK

A member of the Hodder Headline Group
LONDON

Co-published in North America by CRC Press
Boca Raton • New York • Washington, D.C.

First published in Great Britain in 2000 by
Arnold, a member of the Hodder Headline Group,
338 Euston Road, London NW1 3BH

http://www.arnoldpublishers.com

Co-published in North America by CRC Press LLC,
2000 N.W. Corporate Blvd.,
Boca Raton,
Florida 33431
USA

British Library Cataloguing in Publication Data
A catalogue record for this book is available from the British Library

Library of Congress Cataloging-in-Publication Data
A catalog record for this book is available from the Library of Congress

ISBN 0 340 74076 0
ISBN 0 8493 0825 9 (CRC Press)

1 2 3 4 5 6 7 8 9 10

Printed and bound in Great Britain by J W Arrowsmith Ltd, Bristol

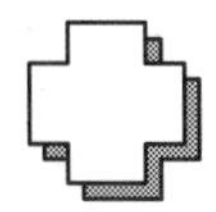

Table of Contents

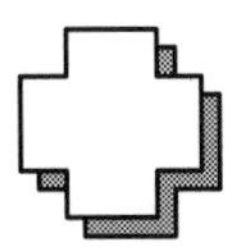

Preface

What is it that really determines the performance of a computer? Is it the processor? No, not really. It is the amount of memory that it has? No, not really. Is it the speed of the disk drives? No, not really. This is because computers can have a fast processor, and lots of memory, and a fast disk drive, but they do not count for much if the busses that connect them to each other do not operate efficiently. The performance of a computer thus directly relates to the busses that connect it. The computer bus is thus the foundation of the modern computer. Without them, a computer would just be a bundle of components.

Busses provide the mechanism for the orderly flow of data over the required channel. They range vastly in their specification. From busses that transmit hundreds of millions of bytes every second (such as with the PCI bus) to busses which transmit only a few thousand bytes per second (such as with the RS-232 bus). They vary in their specification as no one bus can provide the required specification for all applications. For example, graphics adaptors and electronic memory require high data throughputs, and must thus be closely coupled to the processor (known as a local bus connection), whereas modems and printers require relatively slow transfer rates, and must be coupled to a bus which does not try and hog the processor for long periods.

The perfect bus system would use a single connector for every device that connects to it, would be able to sense and configure whichever devices connected to it, would be able to use any type of cable, and devices which connect to it would simply require a tap from one connection onto the next (a daisy-chain connection). It would support high data transfer devices, alongside low data transfer devices, but the low data transfer devices would not hog the bus in favour of the high data transfer devices. It would support real-time data (such as speech and audio) and non-real-time data (such as computer data) in an integrated way, so that the non-real-time data would not swamp the real-time data. This bus, of course, does not exist, or if it does exist, it will be too expensive, and would be incompatible with all the existing busses. Thus, we have many different types of busses, each with their own application. It is impossible to immediately change computer systems every time a new application comes along. We do not immediately knock down our house every time we want to upgrade it. This would be expensive, and we probably would be able to sell it after we had done it. We thus try to use our existing framework and integrate with it.

Internal busses connect the processor to its memory and its interface busses (such as the PCI and the ISA busses). The external busses allow the connection the external devices to the computer, in an orderly manner.

The book splits into five main areas, these are:

1. **PC Interfaces.**
    - Introduction
    - PC Interfacing.
    - Interfacing Standards
2. **Local busses.**
    - PC/ISA.

- PCI/AGP.
- Motherboard Design
- USB.
- Games Port, Keyboard and Mouse.
- Fibre Channel.
- RS-232/RS-422/Modems.
- Parallel Port.

3. **Instrumentation busses.**
   - Modbus.
   - Fieldbus.
   - WorldFIP.
   - CAN bus.
   - IEEE-488.
   - VME/VXI.
4. **Network busses.**
   - Ethernet.
   - ISDN/HDLC.
   - Protocols (TCP/IP).
5. **Bus programming/protocols**
   - TCP/IP.
   - RS-232.
   - Parallel port.

Slides and backup information can be found on my WWW site at:

`http://www.dcs.napier.ac.uk/~bill/books.html`

Questions and any feedback that you have on the book should be sent to:

`w.buchanan@napier.ac.uk` or `bill@dcs.napier.ac.uk`

I have included some notes at the end of most of the chapters which are much lighter in content than the main text. These are my own options, and, of course, should not be taken as fact. In fact they are there for debate, and in some cases your may disagree with some of my comments. For example, I think that the TCP and IP protocols have done more for the freedom of speech, and world peace than all of the diplomats around the world, put together. They have no respect for borders, they do not favour any language, and they do not mind what the data is, and on what computer it came from. They are truly making the world into a village.

Before I start on this book, I must reveal a little secret. My favourite bus, apart from the Number 45 bus which takes me to work every day, is the RS-232 bus. It's not because it is the most technological advanced bus, or that it is easy to interface. Its because I grew an excellent consultancy company by writing program for it. So, I've got a soft spot for RS-232. Long may it reign.

Introduction

1.1 Pre-PC Development

One of the first occurrences of computer technology occurred in the USA in the 1880s. It was due to the American Constitution demanding that a survey is undertaken every 10 years. As the population in the USA increased, it took an increasing amount of time to produce the statistics. By the 1880s, it looked likely that the 1880 survey would not be complete until 1890. To overcome this, Herman Hollerith (who worked for the Government) devised a machine which accepted punch cards with information on them. These cards allowed a current to pass through a hole when there was a hole present.

Hollerith's electromechanical machine was extremely successful and used in the 1890 and 1900 Censuses. He even founded the company that would later become International Business Machines (IBM): CTR (Computer Tabulating Recording). Unfortunately, Hollerith's business fell into financial difficulties and was saved by a young salesman at CTR, named Tom Watson, who recognized the potential of selling punch card-based calculating machines to American business. He eventually took over the company Watson, and, in the 1920s, he renamed it International Business Machines Corporation (IBM). After this, electromechanical machines were speeded up and improved. Electromechnical computers would soon lead to electronic computers, using valves.

The first electronic computers were developed, independently, in 1943; these were the 'Harvard Mk I' and Colossus. Colossus was developed in the UK and was used to crack the German coding system (Lorenz cipher), whereas 'Harvard Mk I' was developed at Harvard University and was a general-purpose electromechanical programmable computer. These led to the first generation of computers which used electronic valves and used punched cards for their main, non-volatile storage.

The world's first large electronic computer (1946), containing 19 000 values was built at the University of Pennsylvania by John Eckert during World War II. It was called ENIAC (Electronic Numerical Integrator and Computer) and it ceased operation in 1957. By today's standards, it was a lumbering dinosaur and by the time it was dismantled it weighed over 30 tons and spread itself over 1500 square feet. Amazingly, it also consumed over 25 kW of electrical power (equivalent to the power of over 400, 60 W light bulbs), but could perform over 100 000 calculations per second (which is reasonable, even by today's standards). Unfortunately, it was unreliable, and would only work for a few hours, on average, before a valve needed to be replaced. Faultfinding, though, was easier in those days, as a valve, which was working, would not glow, and would be cold to touch.

Valves were fine and were used in many applications, such as in TV sets and radios, but they were unreliable and consumed great amounts of electrical power, mainly to the heating element on the cathode. By the 1940s, several scientists at the Bell Laboratories were investigating materials called semiconductors, such as silicon and germanium. These substances only conducted electricity moderately well, but when they where doped with impurities their resistance changed. From this work, they made a crystal called a diode, which worked like a valve, but had many advantages, including the fact that it did not require a vacuum and was much smaller. It also worked well at room temperatures, required little electrical current and had no warm-up time. This was the start of microelectronics.

One of the great revolutions of all time occurred on December 1948 when William Shockley, Walter Brattain, and John Bardeen at the Bell Labs produced a transistor that could act as a triode. It was made from a germanium crystal with a thin p-type section sandwiched between two n-type materials. Rather than release its details to the world, Bell Laboratories kept its invention secret for over seven months so that they could fully understand its operation. They soon applied for a patent for the transistor and, on 30 June 1948, they finally revealed the transistor to the world. Unfortunately, as with many other great inventions, it received little public attention and even less press coverage (the *New York Times* gave it 4½ inches on page 46). It must be said that few men have made such a profound change on the world, and Shockley, Brattain, and Bardeen were deservedly awarded the Nobel Prize in 1956. To commercialize on his success, Shockley, in 1955, founded Shockley Semiconductor. Then in 1957, eight engineers decided they could not work within Shockley Semiconductor and formed Fairchild Semiconductors, which would become one of the most inventive companies in Silicon Valley. Unfortunately, most of the time Fairchild Semiconductors did not fully exploit its developments, and was more of an incubator for many of the innovators in the electronics industry. Around the same time, Kenneth Olsen founded the Digital Equipment Corporation (DEC), who would go on to become one of the key companies in the computer industry, along with IBM.

Previously, in 1952, GW Dummer, a radar expert from Britain's Royal Radar Establishment had presented a paper proposing that a solid block of materials could be used to connect electronic components, without connecting wires. This would lay the foundation of the integrated circuit.

Transistors were initially made from germanium, which is not a robust material and cannot withstand high temperatures. The first company to propose the use of silicon transistors was a geological research company named Texas Instruments (which had diversified into transistors). Then, in May 1954, Texas Instruments started commercial production of silicon transistors. Soon many companies were producing silicon transistors and, by 1955, the electronic valve market had peaked, while the market for transistors was rocketing. The larger electronic valve manufacturers, such as Western Electric, CBS, Raytheon and Westinghouse failed to adapt to the changing market and quickly lost their market share to the new transistor manufacturing companies, such as Texas Instruments, Motorola, Hughes and RCA.

In July 1958, at Texas Instruments, Jack St. Clair Kilby proposed the creation of a monolithic device (an integrated circuit) on a single piece of silicon. Then, in September, he produced the first integrated circuit, containing five components on a piece of germanium that was half an inch long and was thinner than a toothpick.

The following year, Fairchild Semiconductor filed for a patent for the planar process of manufacturing transistors. This process made commercial production of transistors possible and led to Fairchild's introduction, in two years, of the first commercial integrated circuit. Within a few years, transistors were small enough to make hearing aids that fitted into the ear, and soon within pacemakers. Companies, such as Sony, started to make transistors operate over higher frequencies and within larger temperature ranges. Eventually they became so small that many of them could be placed on a single piece of silicon. These were referred to as microchips and they started the microelectronics industry. The first two companies who developed the integrated circuit, were Texas Instruments and Fairchild Semiconductor. At Fairchild Semiconductor, Robert Noyce constructed an integrated circuit with components connected by aluminium lines on a silicon-oxide surface layer on a plane of silicon. He then went on to lead one of the most innovate companies in the world, the Intel Corporation.

After ENIAC, progress was fast in the computer industry and, by 1948, small electronic computers were being produced in quantity within five years (2000 were in use), in 1961 it was 10 000, 1970 100 000. IBM, at the time, had a considerable share of the computer market, so much so that a complaint was filed against them alleging monopolistic practices in its computer business, in violation of the Sherman Act. By January 1954, the US District Court made a final judgment on the complaint against IBM. For this, a 'consent decree' was then signed by IBM, which placed limitations on how IBM conducts business with respect to 'electronic data processing machines'.

In 1954, the IBM 650 was built and was considered the workhorse of the industry at the time (which sold about 1000 machines, and used valves). In November 1956, IBM showed how innovative they were by developing the first hard disk, the RAMAC 305. It was towering by today's standards, with 50 two-foot diameter platters, giving a total capacity of 5 MB. Around the same time, the Massachusetts Institute of Technology produced the first transistorised computer: the TX-O (Transistorized Experimental computer). Seeing the potential of the transistor, IBM quickly switched from valves to transistors and, in 1959, they produced the first commercial transistorised computer. This was the IBM 7090/7094 series, and it dominated the computer market for years.

Programs written on these mainframe computers were typically either machine code (using the actual binary language that the computer understood) or using one of the new compiled languages, such as COBOL and FORTRAN. FORTRAN was well suited to engineering and science as it is based around mathematical formulas. COBOL was more suited to business applications. FORTRAN was developed in 1957 (typically known as FORTRAN 57) and considerably enhanced the development of computer programs, as the program could be writing in a near-English form, rather than using a binary language. With FORTRAN, the compiler converts the FORTRAN statements into a form that the computer can understand. At the time, FORTRAN programs were stored on punch cards, and loaded into a punch-card reader to be read into the computer. Each punch card had holes punched into them to represent ASCII characters. Any changes to a program would require a new set of punch cards.

In 1959, IBM built the first commercial transistorised computer named the IBM 7090/7094 series, which dominated the computer market for many years. In 1960, in New York, IBM went on to develop the first automatic mass-production facility for transistors. In 1963, the Digital Equipment Company (DEC) sold their first minicomputer, to Atomic Energy of Canada. DEC would become the main competitor to IBM, but eventually fail as they dismissed the growth in the personal computer market.

The second generation of computers started in 1961 when the great innovator, Fairchild Semiconductor, released the first commercial integrated circuit. In the next two years, significant advances were made in the interfaces to computer systems. The first was by Teletype who produced the Model 33 keyboard and punched-tape terminal. It was a classic design and was on many of the available systems. The other advance was by Douglas Engelbart who received a patent for the mouse-pointing device for computers.

The production of transistors increased, and each year brought a significant decrease in their size. Gordon Moore, in 1964, plotted the growth in the number of transistors that could be fitted onto a single microchip, and found that the number of transistors that can be fitted onto an integrated circuit approximately doubles every 18 months. This is now known as Moore's law, and has been surprisingly accurate ever since. In 1964, Texas Instruments also received a patent for the integrated circuit.

At the time, there were only three main ways of writing computer programs: machine code, FORTRAN or COBOL. These languages were often difficult for inexperienced users to use. So, in 1964, John Kemeny and Thomas Kurtz at Dartmouth College developed the BASIC (Beginners All-purpose Symbolic Instruction Code) programming language. It was a great success, although has never been used much in 'serious' applications, until Microsoft developed Visual BASIC, which used BASIC as a foundation language, but enhanced it with an excellent development system. Many of the first personal computers used BASIC as a standard programming language.

The third generation of computers started in 1965 with the use of integrated circuits rather than discrete transistors. IBM again was innovative and created the System/360 mainframe. In the course of history, it was a true classic computer. Then, in 1970, IBM introduced the System/370, which included semiconductor memories. All of the computers were very expensive (approx. $1 000 000), and were the great computing workhorses of the time. Unfortunately, they were extremely expensive to purchase and maintain. Most companies had to lease their computer systems, as they could not afford to purchase them. As IBM happily clung to their mainframe market, several new companies were working away to erode their share. DEC would be the first, with their minicomputer, but it would be the PC companies of the future who would finally overtake them. The beginning of their loss of market share can be traced to the development of the microprocessor, and to one company: Intel. In 1967, though, IBM again showed their leadership in the computer industry by developing the first floppy disk. The growing electronics industry started to entice new companies to specialize in key areas, such as International Research who applied for a patent for a method of constructing double-sided magnetic tape utilizing a Mumetal foil inter layer.

The beginning of the slide for IBM occurred in 1968, when Robert Noyce and Gordon Moore left Fairchild Semiconductors and met up with Andy Grove to found Intel Corporation. To raise the required finance they went to a venture capitalist named Arthur Rock. He quickly found the required start-up finance, as Robert Noyce was well known for being the person who first put more than one transistor of a piece of silicon.

At the same time, IBM scientist John Cocke and others completed a prototype scientific computer called the ACS, which used some RISC (Reduced Instruction Set Computer) concepts. Unfortunately, the project was cancelled because it was not compatible with the IBM's System/360 computers.

Several people were proposing the idea of a computer-on-a-chip, and International Research Corp. were the first to develop the required architecture, modelled on an enhanced DEC PDP-8/S concept. Wayne Pickette, at the time, proposed to Fairchild Semiconductor that they should develop a computer-on-a-chip, but was turned down. So, he went to work with IBM and went on to design the controller for Project Winchester, which had an enclosed flying-head disk drive.

In the same year, Douglas C. Engelbart, of the Stanford Research Institute, demonstrated the concept of computer systems using a keyboard, a keypad, a mouse, and windows at the Joint Computer Conference in San Francisco's Civic Center. He also demonstrated the use of a word processor, a hypertext system, and remote collaboration. His keyboard, mouse and windows concept has since become the standard user interface to computer systems.

In 1969, Hewlett-Packard branched into the world of digital electronics with the world's first desktop scientific calculator: the HP 9100A. At the time, the electronics industry was producing cheap pocket calculators, which led to the development of affordable computers, when the Japanese company Busicom commissioned Intel to produce a set of between eight and 12 ICs for a calculator. Then instead of designing a complete set of ICs, Ted Hoff, at Intel, designed an integrated circuit chip that could receive instructions, and perform simple integrated functions on data. The design became the 4004 microprocessor. Intel produced a set of ICs, which could be programmed to perform different tasks. These were the first ever microprocessors and soon Intel (short for *Int*egrated *El*ectronics) produced a general-purpose 4-bit microprocessor, named the 4004.

In April 1970, Wayne Pickette proposed to Intel that they use the computer-on-a-chip for the Busicom project. Then, in December, Gilbert Hyatt filed a patent application entitled 'Single Chip Integrated Circuit Computer Architecture', the first basic patent on the microprocessor.

The 4004 caused a revolution in the electronics industry as previous electronic systems had a fixed functionality. With this processor, the functionality could be programmed by software. Amazingly, by today's standards, it could only handle four bits of data at a time (a nibble), contained 2000 transistors, had 46 instructions and allowed 4 KB of program code and 1 KB of data. From this humble start, the PC has since evolved using Intel microprocessors. Intel had previously been an innovative company, and had produced the first memory device (static RAM, which uses six transistors for each bit stored in memory), the first DRAM (dynamic memory, which uses only one transistor for each bit stored in memory) and the first EPROM (which allows data to be downloaded to a device, which is then permanently stored).

In the same year, Intel announced the 1 KB RAM chip, which was a significant increase over previously produced memory chip. Around the same time, one of Intel's major partners, and also, as history has shown, competitors, Advanced Micro Devices (AMD) Incorporated was founded. It was started when Jerry Sanders and seven others left – yes, you've guessed it, Fairchild Semiconductor. The incubator for the electronics industry was producing many spin-off companies.

At the same time, the Xerox Corporation gathered a team at the Palo Alto Research Center (PARC) and gave them the objective of creating 'the architecture of information.' It would lead to many of the great developments of computing, including personal distributed computing, graphical user interfaces, the first commercial mouse, bit-mapped displays, Ethernet, client/server architecture, object-oriented programming, laser printing and many of the basic protocols of the Internet. Few research centers have ever been as creative, and forward thinking as PARC was over those years.

In 1971, Gary Boone, of Texas Instruments, filed a patent application relating to a single-chip computer and the microprocessor was released in November. Also in the same year, Intel copied the 4004 microprocessor to Busicom. When released the basic specification of the 4004 was:

- Data bus: 4-bit
- Clock speed: 108 kHz
- Price: $200
- Speed: 60 000 operations per second
- Transistors: 2300

Computer Generations

1st Valves (ENIAC)
2nd Transistors (PDP-1)
3rd Integrated Circuits/ Time sharing (IBM System/ 360)
4th Large Scale Integration (ZX81)
5th Systems-on-a-Chip (Pentium)

Intel 4000-series

4001 PROM (4096×8 bits)
4002 RAM (5120 bits)
4003 Registers
4004 Processor

2nd Generation Computer Companies (Transistorized)

1. IBM
2. Univac
3. Burroughs
4. NCR
5. Honeywell
6. Control Data Corporation
7. Siemens
8. Fuji
9. Bendix
10. Librascope

Computer message:
'Error, no keyboard, press F1 to continue.'

All computers wait at the same.

- Silicon: 10-micron technology, 3×4 mm^2
- Addressable memory: 640 bytes

Intel then developed an EPROM, which integrated into the 4004 to enhance development cycles of microprocessor products.

Another significant event occurred when Bill Gates and Paul Allen, calling themselves the 'Lakeside Programming Group' signed an agreement with Computer Center Corporation to report bugs in PDP-10 software, in exchange for computer time.

Other significant effects at the time were:

- Ken Thompson, at AT&T's Bell Laboratories, wrote the first version of the Unix operating system.
- Gary Starkweather, at Xerox, used a laser beam along with the standard photocopying processor to produce a laser printer.
- The National Radio Institute introduced the first computer kit, for $503.
- Texas Instruments develops the first microcomputer-on-a-chip, containing over 15 000 transistors.
- IBM introduced the memory disk, or floppy disk, which was an 8-inch floppy plastic disk coated with iron oxide.
- Wang Laboratories introduced the Wang 1200 word processor system.
- Niklaus Wirth invented the Pascal programming language. BASIC and FORTRAN had long been known for producing unstructured programs, with lots of GOTOs and RETURNs. Pascal was intended to teach good, modular programming practices, but was quickly accepted for its clean, pseudocode-like language. Today it still survives, but has struggled against C/C++ (mainly because of the popularity of Unix) and Java (because of its integration with the Internet), but lives with Borland Delphi, an excellent Microsoft Windows development system.

1.2 8008/8080/8085

In 1974, Intel was a truly innovative company, and was the first to develop an 8-bit microprocessor. These devices could handle eight bits (a byte) of data at a time and were:

- 8008 (0.2 MHz, 0.06 MIPS, 3500 transistors, 10-micron technology, 16 KB memory).
- 8080 (2 MHz, 0.64 MIPS, 6000 transistors, 6-micron technology, 64 KB memory).
- 8085 (5 MHz, 0.37 MIPS, 6500 transistors, 3-micron technology, 64 KB memory).

These were much more powerful than the previous 4-bit devices and were used in many early microcomputers and in applications such as electronic instruments and printers. The 8008 had a 14-bit address bus and could thus address up to 16 KB of memory, and the 8080 and 8085 had 16-bit address busses, giving them limit of 64 KB. Table 1.1 outlines the basic specification for the main 8-bit microprocessors. At the time, Intel's main product area was memory, and microprocessors seemed like a good way of increasing sales for other product lines, especially memory.

Table 1.1 Popular 8-bit microprocessors

Processor	*Release date (manufacturer)*	*Computer used in*	*Example computers*
8008	April 1972 (Intel)	Mark-8	
8080	April 1974 (Intel)	Sol-20 MITS Altair 8800 IMSAI 8080	
8085	March 1976 (Intel)		
Z80 Z80A	July 1976 (Zilog)	Radio Shack TRS-80 Exidy Sorcerer Sinclair ZX81 Osborne 1 Xerox 820 DEC Rainbow 100 Sord M5/ M23P Sharp X1 Sony SMC-70	1. TRS-80 microcomputer, 4 KB RAM, 4 KB ROM, keyboard, black-and-white video display, and tape cassette, $600, Aug. 1977. 2. ZX81 (1 KB), $200, March 1981. ZX81 (2KB), $200. March 1981. 3. Osborne 1, 5-inch display, 64 KB RAM, keyboard, keypad, modem, and two 5.25-inch 100 KB disk drives, $17, April 1981.
6502/ 6502A	June 1976 (MOS Technologies)	Franklin Ace 1000 Atari 400/800 Commodore PET Apple II/III	1. Atari 400/800, 8 KB, $550/1000, Oct 1979. 2. PET 2001,4 KB RAM, 14 KB ROM, keyboard, display, and tape drive, $600. 3. Apple II, 4 KB RAM, 16 KB ROM, keyboard, 8-slot motherboard, game paddles, graphics/text interface to colour display (first ever), and built-in BASIC, $1300, April 1977. 4. Apple II Plus, 48 KB, June 1979. 5. Apple III, 5.25-inch floppy drive, $4500–$8000, May 1980. 6. BBC Microcomputer System. 48 KB RAM, 73-key keyboard, and 16-colour graphics, Sept 1981.
6800/ 6809	1974 (Motorola)	MITS Altair 680	1. TRS-80 Colour Computer, 4 KB RAM, $400.
780-1	NEC		1. ZX80, 1 KB RAM and 4 KB ROM, $200, Feb. 1980.

Excited by the new 8-bit microprocessors, two kids from a private high school, Bill Gates and Paul Allen, rushed out to buy the new 8008 device. This they believed would be the beginning of the end of the large, and expensive, mainframes (such as the IBM range) and minicomputers (such as the DEC PDP range). They bought the processors for the high price of $360 (possibly, a joke at the expense of the IBM System/360 mainframe), but even they could not make it support BASIC programming. Instead, they formed the Traf-O-Data company and used the 8008 to analyse tickertape read-outs of cars passing in a street. The company would close down in the following year (1973) after it had made $20 000, but from this enterprising start, one of the leading computer companies in the world would grow: Microsoft (although it would initially be called Micro-soft).

Intel knew that providing a processor alone would have very little impact on the market. It required a development system, which would allow industrial developers an easy method of developing hardware and software around the new processor. Thus, Intel introduced the Intellec 4 development system.

The main competitors to the 8080 were: the Motorola 6800, the Zilog Z80 and the MOS Technology 6502. The Z80 had the advantage that it could run any programs written for the 8080, and, because it was also pin compatible, it could be easily swapped with the 8080 processor, without a change of socket. It also had many other advantages over the 8080, such as direct memory access, serial I/O technology, and full use of the 'reserved' op-codes (Intel had used only 246 out of the 256 available op-codes). The Z80 was also much cheaper than the 8080 and had a 2.5 MHz clock speed. After the release of the Z80, Intel produced a quick response: the 8085. This device fully used all the op-codes, but it was too late to stop the tide towards Zilog. Many personal computers started to appear that were based on the Z80 processor, including the Radio Shack TRS-80, Osborne 1 and the Sinclair/Timex ZX81. The ZX81 caused a great revolution because of its cheapness, but unfortunately, most home users had to wait for many months to receive their kit, or for their prebuilt computer. However, as the computer was so original and cost effective, users were willing to wait for their prized system. Another great challenger was the 6502, which was released in June 1975 and cost $25. This compared well with the 8080, which cost $150. It was used in many of the great personal computer systems, such as the Apple II and Atari 400.

For the first time, home users could actually build their own computer, and were available from Altair and Mistral. With the success of the Z80, many companies were demanding to produce a second-source supply for the Z80 processors. The Motorola processor was also more powerful than the 8080. It was simpler in its design and only required a single 5 V supply, whereas the 8080 required three different power supplies.

"Sinclair ZX80: with an unusable keyboard and a quirky BASIC, this machine discouraged millions of people from ever buying another computer.'
PC World Magazine, October 1985

"I don't think it's that significant",
Tandy president John Roach, on the IBM PC.

"... the 'irresistible tide' of AT&T's Unix now threatens to engulf the current microcomputer operating system standard, MS-DOS",
Datamation, 1984

"The previous stars – Digital Research and Microsoft – may soon find themselves playing cameo roles as AT&T and IBM take center stage"
ComputerWorld, 1984

"I have traveled the length and breadth of this country and talked with the best people, and I can assure you that data processing is a fad that won't last out the year"
Editor, Prentice Hall, 1957

"Welcome IBM. Seriously.", Headline, produced by Apple, for the full page advert in the Wall Street Journal

"DOS addresses only 1 Megabyte of RAM because we cannot imagine any applications needing more."
Microsoft, 1980, on the development of DOS

"640K ought to be enough for anybody", Bill Gates, 1981

"The PCjr is bound to be around for a while", Ken Williams, Serria On-Line founder.

"We're just sitting here trying to put our PCjrs in a pile and burn them. And the damn things won't burn. That's the only thing IBM did right with it - they made it flameproof.", William Boman, Spinnaker Software

At the end of the 1970s, IBM's virtual monopoly on computer systems started to erode from the high-powered end as DEC developed their range of minicomputers and from the low-powered-end by companies developing computers based around the newly available 8-bit microprocessors, such as the 6502 and the Z80. IBM's main contenders, other than DEC, were Apple and Commodore who introduced a new type of computer – the personal computer (PC). The leading systems, at the time, were the Apple I and the Commodore PET. These captured the interest of the home user and for the first time individuals had access to cheap computing power. These flagship computers spawned many others, such as the Sinclair ZX80/ZX81, the BBC microcomputer, the Sinclair Spectrum, the Commodore Vic-20 and the classic Apple II (all of which where based on the 6502 or Z80). Most of these computers were aimed at the lower end of the market and were mainly used for playing games and not for business applications. IBM finally decided, with the advice of Bill Gates, to use the 8088 for its version of the PC, and not, as they had first thought, to use the 8080 device. Microsoft also persuaded IBM to introduce the IBM PC with a minimum of 64 KB RAM, instead of the 16 KB that IBM planned.

Also, in 1972, at XEROX PARC, Alan Kay proposed that XEROX should build a portable personal computer, called the Dynabook, which would be the size of an ordinary notebook; unfortunately, the PARC management did not support it. In future years, companies such as Toshiba and Compaq would fully exploit the idea. PARC eventually choose to develop the Alto personal computer.

At the time, most people thought that personal computers would be used mainly as games computers. One of the major innovators in this was Atari, who were founded by Nolan Bushnell. They produced the first ever commercial game based on tennis, named Pong. By today's standards, Pong used simple graphics. It had just two paddle lines, which could be moved left and right, and a square ball, which moved back and forward between the paddles. Atari and other companies would release many other classic games, such as Space Invaders, Asteroids and Frogger.

At the time, Texas Instruments was well advanced in microprocessor development and introduced the TMS1000 one-chip microcomputer. It had 1 KB ROM, 32 bytes of RAM with a simple 4-bit processor. In the following year (1973), Intel filed a patent application for a memory system for a multichip digital computer.

In 1973, the model for future computer systems occurred at Xerox's PARC, when the Alto workstation was demonstrated with a bit mapped screen (showing the Cookie Monster, from Sesame Street). The following year, at Xerox, Bob Metcalfe demonstrated the Ethernet networking technology, which was destined to become the standard local area networking technique. It was far from perfect, as computers contended with each other for access to the network, but it was cheap and simple, and it worked relatively well.

Also in 1973, before the widespread acceptance of PC-DOS, the future for personal computer operating systems looked to be CP/M (Control Program/Monitor), which was written by Gary Kildall of Digital Research. One of his first applications of CP/M was on the Intel 8008, and then on the Intel 8080. At the time, computers based on the 8008 started to appear, such as the Scelbi-8H, which cost $565 and had 1 KB of memory.

IBM was also innovating at the time, creating a cheap floppy disk drive. They also produced the IBM 3340 hard disk unit (a Winchester disk) which had a recording head which sat on a cushion of air, 18 millionths of an inch above the platter. The disk was made with four platters, each was 8-inches in diameter, giving a total capacity of 70 MB.

A year later (1974), at IBM, John Cocke produced a high-reliability, low-maintenance computer called the ServiceFree. It was one of the first computers in the world to use RISC technology and it operated at the unbelievable speed of 80 MIPS. Most computers at the time were measured in a small fraction of a MIP, and, at the time, were over 50 times faster than IBM's fastest mainframe. The project was eventually cancelled as a competing project named 'Future Systems' was consuming much of IBM's resources.

In the next year (1974), several personal computers began to appear, including the MITS-built (Micro Instrumentation and Telemetry Systems) computer based on Intel's new 8080 device, at the cheap price of $500. It was released as the Altair 8800 microcomputer. One of the first prototypes for the Altair computer was lost, en-route, to New York, as it was to be reviewed and photographed for Popular Electronics. Eventually they did receive a new version and at a selling price of $439, it received great reviews.

At PARC, the Bravo was developed for the Xerox Alto computer and demonstrated the first WYSIWYG (What You See Is What You Get) program for a personal computer. The Alto computer was then released onto the market. The following year Xerox demonstrated the Gypsy word-processing system, which was fully WYSIWYG. At Motorola, Chuck Peddle and Charlie Melear developed the 6800 microprocessor, which was never really successful in the personal computer market, but was used in many industrial and automotive applications.

While many of the processors at the time ran at 1 MHz or, at the most, 5 MHz, RCA released the RISC-based 1802 processor, which ran at 6.4 MHz. It was used on a variety of systems, from video games to NASA space probes.

Up to 1974, most programming languages had been produced either as a teaching language, such as Pascal or BASIC, or had been developed in the early days of computers, such as FORTRAN and COBOL. No software language had been developed that would properly interface with the operating system, and used both high-level commands, and supported low-level commands (such as AND, OR and NOT bitwise operations). To overcome these problems, Brian Kernighan and Dennis Ritchie developed the C programming language. Its main advantage was that it was supported in the Unix operating system. C has since led a charmed existence by software developers for many proven (and unproven) reasons, and quickly took off in a way that Pascal had failed to do. Its main advantages were stated as: being both a high- and a low-level language, it produced small and efficient code, and that it was portable on different systems. The main advantage was probably that it was a standard software language that was supported on most operating systems, and the ANSI C standard helped its adoption. For this, a program written on one computer system would compile on another system, as long as both compilers conformed to a given standard (typically ANSI C). Pascal always struggled because many compiler developments used non-standard additions to the basic language, and thus Pascal programs were difficult to port from one system to another. FORTRAN never really had this problem, as it only had a few standards, mainly FORTRAN 57 and FORTRAN 77. BASIC also had few problems because of the lack of additional facilities. Most BASIC programs did not port well from one system to another, as they tended to use different methods to access the hardware. Typically, BASIC accessed the hardware directly, whereas C has tended to use the operating system to access the hardware. The non-direct method had many advantages over direct access. Non-direct accesses allow for multi-access to hardware, hardware independence, time-sharing, smoother running programs and better error control. C moved from the Unix operating system down to the PCs, as they become more advanced. It normally requires a relatively large amount of storage space (for all of its standardised libraries), whereas BASIC requires very little storage space.

In 1975, Micro-soft (as it was known before the hyphen was dropped) realized the potential of BASIC for the newly developed 8-bit computers and use it to produce the first programming language for the PC. Their first product was BASIC for the Altair, and licensed it to MITS, their first customer. The MITS, Altair 8800 was a truly innovative system and sold for $375 and has 1 KB memory. Soon Microsoft BASIC 2.0, for the Altair 8800, was available in 4 K and 8 K editions. The Altair was an instant success, and MITS begin work on a Motorola 6800-based system. Even its bus become a standard: the S-100 bus.

> "But what ... is it good for?", Engineer at the Advanced Computing Systems Division of IBM, 1968, commenting on the microchip.

At Xerox, work began on the Alto II, which would be easier to produce, more reliable, and more easily maintained, whereas IBM segmented their mainframe market and moved down-market, with their first briefcase-sized portable computer: the IBM 5100. It cost $9000, used BASIC, had 16 KB RAM, tape storage, and a built-in 5-inch screen. Also at IBM, after the rejection of the ServiceFree computer, John Cocke began working on the 801 project, which would develop scaleable chip designs that could be used in small computers, as well as large ones.

In 1976, the personal computer industry started to evolve around a few companies. For software development two companies stood out:

- **Microsoft**. The development of BASIC on the Altair allowed Microsoft to concentrate on the development of software (while many other companies concentrated on the cutthroat hardware market). Its core team of Paul Allen (ex-MITS) and Bill Gates (ex-Harvard) left their job/study to devote their efforts, full-time, to Microsoft. They even employed their first employee: Marc McDonald. The Microsoft trademark was also registered.
- **Digital Research**. Microsoft's biggest competitor for PC software was Digital Research who had copyrighted CP/M, which it hoped would become the industry-standard microcomputer operating system. Soon CP/M was licensed to GNAT Computers and IMSAI. But for a bad business decision at Digital Research, CP/M would have become the standard operating system for the PC, and the world may never have heard about MS-DOS.

> "Computers in the future may weigh no more than 1.5 tons." Popular

For personal computer systems, five computers were leading the way:

- **Apple**. Steve Wozniak and Steve Jobs completed work on the Apple I computer, and on April Fool's Day, 1976, the Apple Computer Company was formed. It was initially available in kit form and cost $666.66 (hopefully nothing to do with it being a beast to construct). With the success of the Apple I computer, Steve Wozniak began working on the Apple II, and he soon left Hewlett-Packard to devote more time to this development. Steve Wozniak and Steve Jobs proposed that Hewlett-Packard and Atari create a personal computer. Both proposals were turned down.
- **Commodore**. Things were looking very good at Commodore, as Chuck Peddle designed the Commodore PET. To ensure a good supply of the 6502, Commodore International bought MOS Technology.
- **Xerox**. The innovation continued at great pace at Xerox with the Display Word Processing Task Force recommending that Xerox produce an office information system, like the Alto (the Janus project). On the negative side, Xerox management had always been slightly suspicious about the change of business area, and rejected two proposals to market the Alto computer as part of an advanced word processing system.

> "I think there is a world market for maybe five computers", Thomas Watson, chairman of IBM, 1943

- **Cray Research**. Cray Research developed one of the first supercomputers with the Cray-1. It used vector-processing computers and was a direct attack on IBM's traditional computer market. This caused major rumbles in IBM which was seeing its market attacked from three sides: the personal computers (which started to show potential in lower-end applications), the minicomputer (which were cheaper and easier to use than the mainframes) and from the supercomputers (at the upper end). Processing power became the key factor for supercomputers, whereas connectivity was the main feature for mainframe computers. As DEC has done, Cray concentrated on the scientific and technical areas of high-performance computers.
- **Wang Laboratories**. Wang emerged in the computing industry with its innovative word-processing system which used computer technology, instead of traditional electronic typewriters. It initially cost $30 000.
- **MITS**. After the success of the Altair 8800, MITS released the Altair 680, which was based on the Motorola 6800 microprocessor.

And for microprocessors there were five major competitors:

- **Zilog**. Zilog released the 2.5 MHz Z80; an 8-bit microprocessor whose instruction set was a superset of the Intel 8080.
- **AMD**. Intel realized that they must create alliances with key companies, in order to increase the acceptance of the 8080 processor. Thus, they signed a patent cross-license agreement with AMD, which gave AMD the right to copy Intel's processor microcode and instruction codes.
- **MOS Technology**. MOS Technology released the 1 MHz 6502 microprocessor to a great reception, and started a wave of classic computers, such as the Apple II. The 6502A processor would increase the clock speed.
- **National Semiconductor**. Released the SC/MP microprocessor, which used advanced multiprocessing.
- **Texas Instruments**. After years of innovation at Intel in producing the first 4-bit (4004) and the first 8-bit processor (8008), it was TI who developed the first 16-bit microprocessor: the TMS9900. Its first implementation was within the TI 990 minicomputer. The processor was extremely advanced for the time, but, unfortunately, TI failed to provide proper support for the processor. Its main failing was that there was no usable development system (something that Intel and Motorola always made sure was available for their systems).

"There is no reason anyone would want a computer in their home.", Ken Olson, president, chairman and founder of Digital Equipment Corp., 1977.

The following year belonged to Apple, Commodore and Radio Shack, who released the excellent Apple II, the Commodore PET and the TRS-80, respectively, to an eager market. In 1977, the Apple Computer Company was incorporated, and the employees moved to California. The Apple II computer sold initially for $1300 and used the 6502 CPU, had 4 KB RAM, 16 KB ROM, a QWERTY keyboard, 8-slot motherboard, game paddles, graphics/text interface to colour display and came with the Applesoft system (built-in BASIC provided by Microsoft). Soon, Steve Wozniak was working on software for a floppy disk controller.

In has been shown that a killer software application, or game, is required for the widespread adoption of a new computer system. This killer application occurred for the Apple II when Dan Bricklin developed the VisiCalc spreadsheet program. Unfortunately, for him, and fortunately for others, such as Lotus and Microsoft, he never patented his technology. If he had done this, he would have become a multibillionaire. Dan got the idea of the electronic spreadsheet while he sat in a class at Harvard Business School. He designed the interface, while his partner, Bob Frankston, wrote the code. The VisiCalc software ran on the Apple II computer, and had a significant effect on the sales of the computer. It has since been the father of all other spreadsheet programs, such as Lotus 123 and Microsoft Excel (Lotus eventually bought the rights to VisiCalc for $800 000 in 1985), and was released in 1979.

The Commodore PET 2001 was also based around the 6502 CPU, and had a simpler specification (4 KB RAM, 14 KB ROM, keyboard, display, and tape drive), but it only cost $600. In competition, and at the same price, Radio Shack developed the TRS-80 microcomputer. It was based around the Z80 processor and had 4KB RAM, 4KB ROM, keyboard, black-and-white video display, and tape cassette, and sold well beyond expectations.

Microsoft expanded their market by developing Microsoft FORTRAN for CP/M-based computers, and granted Apple Computer a license to Microsoft's BASIC.

1.3 8086/8088

The third generation of microprocessors began, in June 1976, with the launch of the 16-bit processors, when Texas Instruments introduced the TMS9900. It initially used the TI 990 minicomputer. The processor never took-off as it lacked peripheral devices, and it was on May 1978 that Intel released the 8086 microprocessor. This processor was mainly an extension to the original 8080 processor and thus retained a degree of software compatibility. Intel first introduced the 4.77 MHz 8086 microprocessor, which had 16-bit registers, a 16-bit data bus, and 29 000 transistors, using three-micron technology. It had a 20-bit address bus and could thus access 1 MB of memory. It had good performance at 0.33 MIPS and initially sold for $360 (maybe a joke at the expense on the IBM System/360). Later speeds included 8 MHz (0.66 MIPS) and 10 MHz (0.75 MIPS).

IBM's designers, after discussions with Bill Gates, realized the power of the 8086 and used it in the original IBM PC and IBM XT (eXtended Technology). It had a 16-bit data bus and a 20-bit address bus, and thus has a maximum addressable capacity of 1 MB, and could handle either 8 or 16 bits of data at a time (although in a messy way). Its main competitors were the Motorola 68000 and the Zilog Z8000.

It was important for Intel to keep compatibility with 8080. The difficulty was that the 8080 used a 16-bit address (64 KB or 65,536 locations), whereas the 8086 would use a 20-bit address bus, allowing up to 1 MB of memory to be addressed. Thus, the 8086 was designed with a segmented memory, where the memory was segmented in 64 KB chunks. The 20-bit address was then made up of a segment address, and an offset address.

In February 1979, Intel released the 8086 processor as follows:

The Intel 8086, a new microcomputer, extends the midrange 8080 family into the 16-bit arena. The chip has attributes of both 8- and 16-bit processors. By executing the full set of 8080A/8085 8-bit instructions plus a powerful new set of 16-bit instructions, it enables a system designer familiar with existing 8080 devices to boost performance by a factor of as much as 10 while using essentially the same 8080 software package and development tools.

The goals of the 8086 architectural design were to extend existing 8080 features symmetrically, across the board, and to add processing capabilities not to be found in the 8080. The added features include 16-bit arithmetic, signed 8- and 16-bit arithmetic (including multiply and divide), efficient interruptible byte-string operations, and improved bit manipulation. Significantly, they also include mechanisms for such minicomputer-type operations as reentrant code, position-independent code, and dynamically relocatable programs. In addition, the processor may directly address up to 1 megabyte of memory and has been designed to support multiple-processor configurations.

The 8086 and 8088 were binary compatible with each other, but not pin compatible. Binary compatibility means that either microprocessor could execute the same program. Pin incompatibility means that you cannot plug the 8086 into the 8088, and vice-versa, and expect the chips to work. The new 'x86' devices implemented a CISC (Complex Instruction Set Computer design methodology). At the time, many companies were promoting RISC as the fasting processor technology. Intel would eventually win the CISC battle with the release of the Pentium processor, many years in the future.

At the time, Intel Corporation struggled to supply enough chips to feed the hungry assembly lines of the expanding PC industry. Therefore, to ensure sufficient supply to the personal computer industry, they subcontracted the fabrication rights of these chips to AMD, Harris, Hitachi, IBM, Siemens, and possibly others. Amongst Intel and their cohorts, the 8086 line of processors ran at speeds ranging from 4 MHz to 16 MHz.

The Z80 processor, which had beaten the 8080 processor in many ways, led the way for its new 16-bit processor: the Z8000. Zilog had intended that it was to be compatible with the previous processor. Unfortunately, the designer decided to redesign the processor, so that it had an improved architecture, but was not compatible with the Z80. From that time on, Zilog lost their market share, and this gives an excellent example of compatibility winning over superior technology. The 8086 design was difficult to work with and was constrained by compatibility, but it allowed easy migration for system designers.

IBM realized the potential of the PC and microprocessor. Unlike many of their previous computer systems, they developed their version of the PC using standard components, such as Intel's 16-bit 8086 microprocessor. They released it as a business computer, which could run word processors, spread sheets and databases and was named the IBM PC. It has since become the parent of all the PCs ever produced. To increase the production of this software for the PC they made information on the hardware freely available. This resulted in many software packages being developed and helped clone manufacturers to copy the original design. So the term 'IBM compatible' was born and it quickly became an industry standard by sheer market dominance.

On previous computers, IBM had written most of their programs for their systems. For the PC they had a strict time limit, so they first went to Digital Research who was responsible for developing CP/M, which was proposed as a new standardised operating system for microprocessors. Unfortunately, for Digital Research, they were unable to reach a final deal because they could not sign a strict confidentiality agreement. They then went to a small computer company called Microsoft. For this Bill Gates bought a program called Q-DOS (often called the Quick and Dirty Operating System) from Seattle Computer Products. Q-DOS was similar to CP/M, but totally incompatible. Microsoft paid less than $100 000 for the rights to the software. It was released on the PC as PC-DOS, and Microsoft released their own version called MS-DOS, which has since become the best selling software in history, and IBM increased the market for Intel processors, a thousand times over.

To give users some choice in their operating system, the IBM PC was initially distributed with three operating systems: PC-DOS (provided by Microsoft), Digital Research's CP/M-86 and UCSD Pascal P-System. Microsoft understood that to make their operating system the standard, that they must provide IBM with a good deal. Thus, Microsoft offered IBM the royalty-free rights to use Microsoft's operating system forever, for $80 000. This made PC-DOS much cheaper than the other two (such as $450 for P-System, $175 for CP/M and $60 for PC-DOS). Microsoft was smart in that they allowed IBM to use PC-DOS for free, but they held the control of the licensing of the software. This was one of the greatest pieces of business ever conducted. Eventually CP/M and P-System died off, while PC-DOS become the standard operating system for the PC.

Top 20 Computer People

1. DAN BRICKLIN (VisiCalc)
2. BILL GATES (Microsoft)
3. STEVE JOBS (Apple)
4. ROBERT NOYCE (Intel)
5. DENNIS RITCHIE (C Programming)
6. MARC ANDREESSEN (Netscape Communications)
7. BILL ATKINSON (Apple Mac GUI)
8. TIM BERNERS-LEE (CERN/WWW)
9. DOUG ENGELBART (Mouse/Windows/etc)
10. GRACE MURRAY HOPPER (COBOL)
11. PHILIPPE KAHN (Turbo Pascal)
12. MITCH KAPOR (Lotus 123)
13. DONALD KNUTH (TEX)
14. THOMAS KURTZ
15. DREW MAJOR (NetWare)
16. ROBERT METCALFE (Ethernet)
17. BJARNE STROUSTRUP (C++)
18. JOHN WARNOCK (Adobe)
19. NIKLAUS WIRTH (Pascal)
20. STEVE WOZNIAK (Apple)

-- Byte, Sept 1995

The developed program was hardly earth shattering, but has since gone on to make billions of dollars. It was named the Disk Operating System (DOS) because of its original purpose of controlling the disk drives. Compared with some of the work that was going on at Apple and at Xerox, it was a very basic system. It had no graphical user interface and accepted commands from the keyboard and displayed them to the monitor. These commands were interpreted by the system to perform file management tasks, program execution and system configuration. Its function was to run programs, copy and remove files, create directories, move within a directory structure and to list files. To most people this was their first introduction to computing, but for many, DOS made using the computer too difficult, and it would not be until proper graphical user interfaces, such as Windows 95, that PCs would truly be accepted and used by the majority.

It did not take long for the computer industry to start 'cloning' the IBM PC. Many companies tried; but most of them failed because their BIOS were not compatible with IBM PC BIOS. Columbia, Kayro and others went by the wayside because they were not totally PC compatible. Compaq eventually broke though the compatibility barrier with the introduction of the Compaq portable computer. Compaq's success created the turning point that enabled today's modern computer industry. They produced sales of $111 million in the first year of their operation, making it the fastest growing company in history.

In Japan, NEC bought a license on the 8086/8088. They improved the design and produced two Intel 'clones', called the V20 (8088-compatible) and V30 (8086-compatible). The V-series ran approximately 20% faster than the Intel chips when running at the same clock speed. Therefore, the V-series chips provided a cheap upgrade to owners of the IBM-PC and other clones computers. Although these chips were pin compatible with the 8086 and 8088, they also had some extensions to the architecture. They featured all of the 'new' instructions on the 80186/80188, and were capable of running in Z80 mode (directly running programs written for the Z80 microprocessor). Much to Intel's embarrassment, NEC refused to pay royalties to NEC on the sale of their processors. Intel found that it was difficult to copyright the actual silicon design, and have since copyrighted the microcode, which runs on the processor. The microcode for the 8086/8088 consisted of 90 different mini-programs. However, in a courtroom, NEC showed that they had not copied these mini-programs and had designed their own.

At this time, Intel was loosing a great deal of their memory product to Japanese companies. Their focus, from now on, would be the PC-processor market. If they could always keep one step ahead of the cloners they would have a virtual monopoly. Eventually they would become so powerful as a market leader that they would overcome the basic rule that you always need a second source of processors for new processors to be accepted in the market. IBM had developed a system that would end up reducing their market share, and create a quasi-monopoly at the end of the 1990s and the beginning of the millennium for Intel (with processors and support devices) and Microsoft (for operating systems, and eventually application software). IBM would eventually fail in its introduction of new industry standards, such as MXA bus technology, whereas Intel would gain acceptance of new standards, such as the PCI bus, and Microsoft would develop new standard in operating systems, such as Windows NT.

At the same time as Intel was developing the 8086 they were developing the 8800 processor, which would not be compatible with the 8080, and would be a great technological break-though (as it would not have to be compatible with the older 8080 device). When the 8800 was finally released in 1981 as the iAPX432 (Intel Advanced Processor Architecture), it reached the market just as the IBM PC took off, and died a quick death, as everyone wanted the lower-powered 8086 device. The iAPX lives on as the 'x86' architecture.

Apple was growing fast in 1978 and released a BASIC version of VisiCalc spreadsheet. They also produced their first Apple II disk drive and Disk II, which was a 5.25-inch floppy disk drive linked to the computer by a cable ($495). At the end of 1978, Apple Computer began work on an enhanced Apple II with custom chips, code-named Annie, a supercomputer with a bit-sliced architecture, code-named Lisa, and also on Sara (the Apple III). Atari released the Atari 400 and 800 personal computers, which used the 6502 processor. Microsoft was quick to spot the potential of the 8086 processor and developed Microsoft COBOL and Microsoft BASIC for it.

Computer systems also started to find their way into social pursuits when Atari developed the Asteroids computer game and Taito developed the Space Invaders arcade game. They were classics of their time, but hardly powerful by today's bit-mapped, 3D graphics.

Epson, who had had a successful market in typewriters, started to produce low-price, high-performance dot matrix printers (the MX-80), and at the same time, Commodore released the CBM 2020 dot-matrix printer (as well as a dual 5.25-inch floppy disk drive unit).

In 1979, Xerox finally lost its foothold on the computer industry when the Alto was advertised on TV, but then the president decides to drop its development. Microsoft, on the other hand, was going from strength to strength. Microsoft 8080 BASIC eventually broke the one million-dollar barrier, the first microprocessor product to do this. Soon, Microsoft had developed BASIC, and FORTRAN for the 8086. They had also released Assembler language system for 8080/Z80 microprocessors.

Apple Computer released DOS 3.2, and the Apple II Plus computer, which had a 48 KB memory, and cost $1195. They also highlighted their growing strength by introduces their first printer, the Apple Silentype ($600). At PARC, Xerox was the leader in developing a graphical user interface with their Alto computer. As a learning process, a group of engineers and executives from Apple were given a demonstration of the Alton, and its associated software, in exchange for Xerox spending $1 million buying 100 000 Apple Computer shares. The investment would pay off many times over for Apple as it helped in their development of the Apple Mac computer.

1979 produced mixed fortunes for two of Intel's competitors: Zilog and Motorola. It was a bad year for Zilog when it distributed its new 16-bit processor, the Z8000. It main drawback was its incompatibility with its 8-bit predecessor, the classic Z80. For Motorola, it was one of success as they released the excellent 68000, 16-bit microprocessor. It used 68 000 transistors (thus, its derived name).

Radio Shack continued development of their TRS-80 computer, with the TRS-80 Model II, and Texas Instruments introduced the TI-99/4 personal computer ($1500). Atari also started to distribute Atari 400 (8 KB memory, $550) and Atari 800 ($1000) personal computers.

In the UK, Clive Sinclair created Sinclair Research, and was distended to develop classic computers, such as the ZX81 and the Sinclair Spectrum. He had already been a major innovator in the 1960s and the 1970s, with watches, audio amplifiers and pocket calculators. In the main these were extremely successful however, he was also destined to develop an electric car (Sinclair C5), which had the opposite effect on sales as he had had with his computer systems.

In 1983, Time magazine selected the microcomputer as its "Man" of the Year.

"The 32-bit machine would be 'overkill for a personal computer", Sol Libes, ByteLines

"Since human beings themselves are not fully debugged yet, there will be bugs in the code no matter what you do",

"We could conceivably put a company out of business with a bug in a spreadsheet, database, or word processor",
both by Chris Mason, Microsoft

A key to the acceptance, and the sales of a computer was its software. This was in terms of its operating systems and its applications. Initially it was games that were used with the PCs, but three important application packages were released, these were:

"I'm glad to be out of that bag", "Hello, I am Macintosh. Never trust a computer you cannot lift.". Quotes from the Macintosh computer when it, introduced itself.

By the middle of the 1990s, Intel and Microsoft were so profitable that they accounted for nearly half of the entire profits in the worldwide PC industry (which was worth over $100 billion, each year).

"Microcomputers are the tool of the 80's. BASIC is the language that all of them use. So the sooner you learn BASIC, the sooner you will understand the microcomputer revolution.", *30 Hour BASIC Standard, 1981*

- **Spreadsheet**. The VisiCalc software was released for the Apple II at a cost of $100. Apple Computer eventually tried to buy the company, which developed VisiCalc, for $1 million in Apple stock, but Apple's president refuses to approve the deal. Its eventual rights would have been worth much more than this small figure.
- **Wordprocessor**. MicroPro released the WordStar word processor (written by Rob Barnaby). It is available for Intel 8080A and Zilog Z80-based CP/M-80 systems. Apple Computer also released AppleWriter 1.0. The following year (1980) would see the release of the popular WordPerfect (from Satellite Software International).
- **Database**. The Vulcan database program, which become known at dBase II.

Two new companies were created in 1979, which would become important industry leaders in peripherals. These were Seagate Technologies (founded by Alan Shugart founded in Scotts Valley, California), and Hayes Microcomputer Products who produced the 110/300-baud Micromodem II for the Apple II ($380).

The following year (1979) saw Radio Shack (with their TRS-80 range), Commodore (with the PET range), Apple Computer (with their Apple II/III) and Microsoft at the forefront of the personal computer market. Two new companies joined the growing personal computer market, at different ends of technology. At the bottom end, which covered the games and hobby market, Sinclair Research appeared, and at the top end of the market, the workstation end, which was aimed at serious applications, came Apollo. Clive Sinclair in the UK had started Sinclair Research. He had already had a significant effect on the electronics industry. In the 1960s, he had developed hi-fi, amplifier and radio kits for hobbyists, and then in the 1970s he had further developed into calculators, multimeters and, even, pocket TVs. His main market in the 1980s would be personal computers, and it was on price that his company would gain the most on his competitors.

The major developments of the year were:

Top 20 Computers of All Time

1. MITS Altair8800
2. Apple II
3. Commodore PET
4. Radio Shack TRS-80
5. Osborne I Portable
6. Xerox Star
7. IBM PC
8. Compaq Portable
9. Radio Shack TRS-80 Model 100
10. Apple Macintosh
11. IBM AT
12. Commodore Amiga 1000
13. Compaq Deskpro 386
14. Apple Macintosh II
15. Next Nextstation
16. NEC UltraLite
17. Sun SparcStation 1
18. IBM RS/6000
19. Apple Power Macintosh
20. IBM ThinkPad 701C

Byte, Sept 1995

- **Radio Shack**. In 1980, Radio Shack followed up their success of the TRS-80, with the TRS-80 Model III. It was based around the Zilog Z80 processor and was priced between $700 and $2500. They also released the TRS-80 Color Computer, which was based on the Motorola 6809E processor and had 4 KB RAM. It was priced well below the Model III and cost $400. Radio Shack at the time were innovating in other areas, and produced the TRS-80 Pocket Computer, which had a 24-character display, and sold for $230.
- **Apple Computer**. Apple Computer accelerated their development work and released the Apple III computer. It was based on the 2 MHz 6502A microprocessor, and included a 5.25-inch floppy drive. It initially cost between $4500 and $8000. Work also began on the Diana project, which would eventually become the Apple IIe. The company was also floated on the stock market, where 4.6 million shares were sold at $22 a share. This made many Apple employees instant millionaires.
- **Sinclair Research**. Sinclair Research burst on the computer market place with the ZX80 computer. It was based on the 3.25 MHz NEC Technologies 780-1 processor and came with 1 KB RAM and 4 KB ROM. It was priced at a cut-down rate of $200, but it was far from perfect. Its main drawback was its membrane type keyboard.
- **Intel**. Along with development of the 8086 processor, Intel released a number of support devices, including the 8087math coprocessor.
- **Microsoft**. Microsoft released a Unix operating system, Microsoft XENIX OS, for the Intel 8086, Zilog Z8000, Motorola M68000, and Digital Equipment PDP-11.
- **Hewlett-Packard**. HP had developed a good market in powerful calculators, and produced a mixture of a computer and a calculator, with the HP-85. It cost $3250, had a 32-character wide CRT display, a built-in printer, a cassette tape recorder, and a keyboard.
- **Commodore**. Commodore Business Machines enhanced their product range with the CBM 8032 computer, which had 32 KB RAM and an 80-column monochrome display. They also developed a dual 5.25-inch floppy disk drive unit (the CBM 8050). In Japan, Commodore released the VIC-1001, which would later become the VIC-20. It had 5 KB RAM, and a 22-column colour video output capability.
- **Apollo**. Apollo burst onto the computer market with high-end workstations based on the Motorola 68000 processor. They were aimed at the serious user, and their main application area was in computer-aided design. One of the first to be introduced was the DN300, which was based around the excellent Motorola 68000 processor. It had a built-in mono

monitor, an external 60 MB hard disk drive, an 8-inch floppy drive, built-in ATR (Apollo Token Ring) network card, and 1.5 MB RAM. It even had its own multiuser, networked operating system called Aegis. Unfortunately, for all its power and usability, Aegis never really took off, and when the market demanded standardized operating systems, Apollo switched to Domain/IX (which was a Unix clone). It is likely that Apollo would have captured an even larger market if they had had changed to Unix at an earlier time, as Sun (the other large workstation manufacturer) had done. The Token Ring network was excellent in its performance, but suffered from several problems, such as the difficulty in tracing faults, and the difficulty in adding and deleting nodes from the ring. Over time, Ethernet eventually became the standard networking technology, as it was relatively cheap and easy to maintain and install. Apollo attacked directly at the IBM/DEC mainframe/minicomputer market, and soon developed a large market share of the workstation market. The advantage that workstations had over mainframes is that each workstation had its own local resources, including a graphical display, and typically, windows/graphics-based packages. Mainframes and minicomputers tended to be based on a central server with a number of text terminals. Apollo were successful in developing the workstation market and their only real competitor was Sun. Hewlett-Packard eventually took Apollo over. However, Apollo computers, as with the classic computers, such as the Apple II and the Apple Macintosh, were well loved by their owners and some would say that they were many years ahead of their time. There are many occurrences of Apollo computers working continuously for five years, with only short breaks for Xmas holidays, and so on. After a skilled network manager set them up, they tended to cause few problems. No crashes, no hardware problems, no network problems, no software incompatibilities. Nothing. Aegis, as Unix does, supported a networked file system, where a global file system could be built up with local disk resources. Thus, a network of 10 workstations, each with 50 MB hard disks allowed for a global file system of 500 MB.

- **Seagate Technology**. Seagate become a market leader for hard disk drives when they developed a 5.25-inch Winchester disk, with four platters and a capacity of 5 MB.
- **Philips/Sony**. These companies developed the CD–Audio standard for optical disk storage of digital audio. At the same time, Sony Electronics introduced a 3.5-inch floppy disk and drive, double-sided, double-density, which had a capacity of 875 KB (but less, when formatted).
- **Texas Instruments**. TI were busy adding peripherals to their TI 99/4 computer, including a thermal printer (30 cps on a 5×7 character matrix), a command module ($45), a modem, RS-232 interface ($225), a 5.25-inch mini-floppy disk drive which could store up to 90 KB on each disk. The floppy disk controller cost $300, and the disk drive cost $500.
- **Digital Research**. DR released CP/M-86 for Intel 8086- and 8088-based systems. Digital Research could have easily become the Microsoft of the future, but for a misunderstanding with IBM.

1.4 80186/80188

Intel continued the evolution of the 8086 and 8088 by, in 1982, introducing the 80186 and 80188. These processors featured new instructions, new fault tolerance protection, and were Intel's first of many failed attempts at the x86 chip integration game.

The new instructions and fault tolerance additions were logical evolutions of the 8086 and 8088. Intel added instructions that made programming much more convenient for low-level (assembly language) programmers. Intel also added some fault tolerance protection. The original 8086 and 8088 would hang when they encountered an invalid computer instruction, whereas the 80186 and 80188 added the ability to trap this condition and attempt a recovery method.

Intel integrated this processor with many of the peripheral chips already employed in the IBM PC. The 80186/80188 integrated interrupt controllers, interval timers, DMA controllers, clock generators, and other core support logic. In many ways, the device was produced a decade ahead of its time. Unfortunately, this device did not catch on with many hardware manufacturers; this spelled the end of Intel's first attempt at CPU integration. However, this device has enjoyed a tremendous success in the world of embedded processors. If you look on your high performance disk driver or disk controller, you might still see an 80186 being used.

Eventually, many embedded processor vendors began manufacturing these devices as a second source to Intel, or in clones of their own. Between the various vendors, the 80186/80188 was available in speeds ranging from 6 MHz to 40 MHz.

1.5 80286

In 1982, Intel introduced the 80286. For the first time, Intel did not simultaneously introduce an 8-bit bus version of this processor (such as the 80288). The 80286 introduced some significant microprocessor extensions. Intel continued to extend the instruction set; more significantly, Intel added four more address lines and a new operating mode called 'protected mode'. The 8086, 8088, 80186 and 80188 all contained 20 address lines, giving these processors one megabyte of addressibility (2^{20} = 1 MB). The 80286, with its 24 address lines, gives 16 megabytes of addressibility (2^{24} = 16 MB).

For the most part, the new instructions of the 80286 were introduced to support the new protected mode. Real mode was still limited to the one megabyte program addressing of the 8086, et al. Essentially, a program could not take advantage of the 16-megabyte address space without using protected mode. Unfortunately, protected mode could not run real-mode (DOS) programs. These limitations thwarted attempts to adopt the 80286 programming extensions for mainstream consumer use.

During the reign of the 80286, the first 'chipsets' were introduced. These were nothing more than a set of devices that replaced dozens of other peripheral devices, while maintaining identical functionality. Chips and Technologies became one of the first popular chipset companies.

IBM was spurred by the huge success of the IBM PC and decided to use the 80286 in their next generation computer, the IBM PC-AT. However, the PC-AT was not introduced until 1985, which was three years after introduction of the 80286. IBM, it seems, were actually frightened by the thought of the 32-bit processors as they allowed PCs to challenge their thriving minicomputer market. A new threat to the PC emerged from Apple, who used the Motorola 68000 processor, with an excellent operating system, to produce the Apple Mac computer. It had a full graphical user interface, which was based around windows and icon, and had a mouse pointer to allow users to easily move around the computer system.

Like the IBM PC, the PC-AT was hugely successful for home and business use. Intel continued to second-source the device to ensure an adequate supply of chips to the computer industry. Intel, AMD, IBM and Harris were now producing 80286 chips as OEM products, whereas Siemens, Fujitsu, and Kruger either cloned it, or were also second sources. Between these various manufacturers, the 80286 was offered in speeds ranging from 6 MHz to 25 MHz.

Intel had had considerable trouble providing enough 8086/80186 processors, and had created technology-sharing agreements with companies such as AMD. This also allowed companies to have a second source for processors, as many organizations (especially military-based organizations) did not trust a single-source supply for a product. In 1984, it was estimated that Intel could only supply between one-fifth and one-third of the current demand for the 80186 device. For the coming 80386 design, Intel decided to break the industry practice of second sourcing and go on their own.

1.6 Post-PC development

IBM dominated the computer industry in the 1950s and 1960s, and it was only in the 1970s that their quasimonopoly started to erode but, at the time, most of their competitors feared their power. If a competitor released a new product, they would often sit back and wait for IBM to trump them, with a better product that had the magical IBM badge. Few companies had the sales turnover to match IBM in research and development. This was shown to great effect with the development of the System/360 range, which had one of the largest ever research and development budgets ($5 billion). After initial development setbacks, the System/360 range was a great success and paid off the initial investment, many times over. IBM sold over 50 000 System/360 computers in a period of six years, and then replaced it with the System/370 series, which was one of the first computers with memory made from integrated circuits.

In 1981, IBM started the long slide from front-runner to also-ran, and within ten years, their own child (the IBM PC) would match the power of their own mainframes. For example, when the Pentium was released, in 1989, it had a processing power of 250 MIPS, while the IBM System/370 mainframe had, at the time, a processing power of 400 MIPS. IBM even, in the development of the IBM AT computer, tried to slug the power of the PC so that it would not impinge on their lucrative mainframe market. As will be seen, IBM, after the overwhelming success of the IBM PC, made two major mistakes:

- The PCjr. The PCjr quickly sank without trace, as it was not compatible with the IBM PC. The time and money spent on the PCjr was completely wasted and gave other manufacturers an opportunity to clone, and improve on the original IBM PC.
- Missing the portable market. IBM missed the IBM PC portable market, and when they did realize its potential, their attempt was inferior to the market leader (Compaq Computers). Later, though, they would produce an excellent portable, called the ThinkPad, but, by that time, they had lost a large market share to companies such as Toshiba, Compaq and Dell.

After making these mistakes, other factors continued to affect their loss of market share. These included:

- Initially missing the market for systems based on 32-bit microprocessors (80386). IBM missed the 32-bit processor when they developed their AT and PS/2 ranges of computers, as, initially, they used the 16-bit 80286. This had been intentional, as IBM did not want to make their new computer too powerful, as they would start to compete with their lucrative mainframe market.

- Trying to move the market towards MCA. After IBM realized that they had lost the battle against the cloners, they developed their own architecture: MicroChannel Architecture (MCA), which would force manufacturers to license the technology from them. Unfortunately, for IBM, Compaq took over the standard as they introduced a computer, which used standard IBM PC architecture, but improved on it as they used the new Intel 80386 in their DeskPro range. IBM would, in time, come back into the fold and follow the rest in their architecture. From then on, IBM became a follower rather than a leader.

After loosing a large market share, IBM soon realized, after the failure of MCA, that they had also lost the market leadership for hardware development. They then decided to try to turn the market for operating system software, with OS/2. It was becoming obvious that the operating system held the key to the hardware architectures, and application software. In a perfect world, an operating system can hide the hardware from the application software, so the hardware becomes less important. Thus, if the software runs fast enough, the hardware can be of any type and of any architecture, allowing application programmers to write their software for the operating system and not for the specific hardware. Whichever company developed the standards for the operating system would hold the key to hardware architecture, and also the range of other packages, such as office tools, networking applications, and so on. OS/2 would eventually fail, and it would be left to one company to lead in this area: Microsoft. Not even the mighty Intel could hold the standards, as Microsoft holds the key link between the software and the hardware. Their operating system would eventually decouple the software from the hardware. With the Microsoft Windows NT operating system, they produced an operating system that could run on different architectures.

Unfortunately, for IBM, OS/2 was a compromised operating system, which was developed for all their computers, whether they be mainframes or low-level PCs. Unlike the development of the PC, many of the organizational units within IBM, including the powerful main-frame divisions, had a say about what went into OS/2 and what was left out. For the IBM PC, the PC team at Boca Raton was given almost independence from the rest of the organization, but the development of OS/2 was riddled with compromises, reviews and specification changes. At the time, mainframes differed from PCs in many different ways. One of the most noticeable ways was the way that they were booted, and the regularity of system crashes. Most users of PCs demanded fast boot times (less than a minute, if possible), but had no great problems when it crashes at a few times a day. These crashes were typically due to incorrectly functioning and configured hardware, and incorrectly installed software. In the mainframe market, an operating system performs a great deal of system checks and tries to properly configure the hardware. This causes long boot-up times, and is not a problem with a mainframe, which will typically run for many weeks, months, or years without requiring a re-boot. However, for the PC, a boot time of anything more than a few minutes is a big problem. In the end, OS/2 had too long a boot time, and was too slow (possibly due to its complexity) to compete in the marketplace. In total, IBM spent over $2 billion on OS/2 with very little in return. It is perhaps ironical that new versions of the Microsoft operating system perform a great deal of system checks and try to configure the system each time it is booted. Now, though, this can now be done in a relatively short time, as the hardware is a great deal faster than it was when graphical user interfaces first reached the market.

Another casualty of the rise of the IBM PC was DEC. As IBM had done with the System/360 range, DEC invested billions of dollars in their VAX range, which became an unbelievable success. As Compaq Computers would do in the 1980s, DEC achieved unbelievable growth, going from its foundation in 1957 to a sales turnover of $8 billion in 1986 (the peak year for DEC, before the PC destroyed the market for minicomputers).

The introduction of the PC would see the end of computer manufacturing for Osborne, Altair, Texas Instruments and Xerox. Going in the opposite direction were the new companies such as Compaq Computers, Sun Microsystems, Apollo (for a while), Cray and Microsoft. Compaq Computers, in 1981, generated $110 million in their first year, a further two years on it was $503.9 million, and two years after that it was $1 billion. The following year it was $2 billion. From zero to $2 billion, in six years (a world record, at the time). Microsoft was another high-growth company going from $16 million in 1981 to $1.8 billion in 1991. In most years, Microsoft doubled its size. Consistently Microsoft was also highly profitable with at least 30% of sales resulting in profit, and at least 10% invested in research and development. The next 20 years would also see the creation of many computer-related multibillionaires, such as Bill Gates who, within in twenty years, would be worth almost $100 billion.

Before the introduction of the IBM PC, the biggest threats to IBM came, at the top end from DEC and at the bottom end from Apple. Both companies could do little wrong. DEC released their classic PDP-11, and then followed it up with the VAX range. Apple quickly developed their range of computers, and moved from a mainly game-playing computer, to one which could be used for game playing and also for business applications. For Apple, the key to the move into the business environment was the introduction of VisiCalc. From the 1980s, software would become the dominant driving force, and the best hardware in the world could not make up for a lack of application software.

1981 would become a pivotal year for the development of computers. Before this year, different computer standards thrived, and incompatibility reigned. After it, there would only be one main standard, which would be a truly open standard, which would be driven not by IBM, but by Intel and Microsoft.

At the time, the computer industry split itself into two main areas:

- **Serious/commercial computers**. Mainly IBM and DEC with their range of mainframe computers and minicomputers. Within 10 years, both IBM and DEC would change to be different companies. IBM would end up loosing their quasi-monopoly on computers systems, and DEC would end up being taken over by Compaq, who would evolve from the new market created by IBM.
- **Hobby/home/game-playing computers**. These computers had grown from the basic 8-bit processors, such as the 6502 and the Z80. The main product leaders were Commodore, Sinclair, Apple, Osborne, Altair, Acorn, Radio Shack and Xerox.

Few of these computers, at the time, were compatible with each other, and it was a great advantage to a manufacturer that their computers were incompatible with others, as software written for one would not work on another. For example, the Apple II and the Commodore PET were based on the same processor, but had incompatible hardware, especially with the graphics system.

It was in 1981 that IBM released, ahead of schedule, the IBM 5150 PC Personal Computer. It featured the 4.77 MHz Intel 8088 processor, 64 KB RAM, 40 KB ROM, one 5.25-inch floppy drive (160 KB capacity), and PC-DOS 1.0 (Microsoft's MS-DOS). It cost $3000, and could be installed with Microsoft BASIC, VisiCalc, UCSD Pascal, CP/M-86, and Easywriter 1.0. Another version used a CGA graphics card, which gave 640×200 resolution with

16 colours.

At the time, many of the other computer companies were following up the success of their previous products, and few had any great worries of the business-oriented IBM PC. The main developments were:

* **Commodore**. After its release in Japan, Commodore eventually released the VIC-20 to an eager world market. It has a full-size 61-key keyboard, 5 KB RAM (expandable to 32 KB), 6502A CPU, a 22×23 line text display, and colour graphics. It initially sold for $299, and at its peak, it was being produced at 9 000 units per day.
* **Sinclair**. Sinclair followed up the success of the ZX80 with the ZX81, which was released for $150 (in the USA it was released as the TS1000), and was based on the Z80A processor. Within 10 months, over 250 000 were sold.
* **Apple**. Apple was very much a market leader, and would eventually be the only real competitor to the IBM PC. In 1981, they reintroduced the Apple III, which was their first with a hard disk. In 1981, Apple Computer got into a little bit of trouble over the Apple name, as The Beatles used it for their record company (Apple Corps Limited). Eventually, Apple Computer signed an agreement allowing them to use the Apple name for their business, but they were not allowed to market audio/video products with recording or playback capabilities.
* **Osborne**. The Osborne Computer Corporation was going from strength to strength, and if not for the release of the IBM PC would have been a major computer manufactures. In 1981, they released the Osborne 1 PC, which was based on the Z80A processor and included a 5-inch display, 64 KB RAM, keyboard, a keypad, modem, and two 5.25-inch 100 KB disk drives. It sold for $1795, but included CP/M, BASIC, WordStar, and SuperCalc. Sales were much great than expected, in fact they sold as many in a single month as they expected for their total sales (up to 10 000 per month).
* **Xerox**. Xerox continued to innovate and released the Star 8010, which contained many of the features that were used with the Alto, such as a bitmapped screen, WYSIWYG word processor, mouse, laser printer, Smalltalk language, Ethernet, and software for combining text and graphics in the same document. It sold for the unbelievably high price of $16 000. This price, especially up against the IBM PC, was too great for the market, and it quickly failed. At the same time, Xerox was planning the Xerox 820 (code named The Worm), which would be based on the 8-bit Z80 processor, whereas the new IBM PC was based on the 16-bit 8088. It, like the Star 8010, was doomed to fail. These were classic cases of releasing the products at the wrong time, for the wrong price.
* **Acorn**. In the UK, Acorn Computer released an excellent computer named the BBC Microcomputer System. It was quickly adopted for a UK TV program, which the BBC was running to introduce microcomputers. Against the ZX81, it had an excellent specification, such as being based on the 6502A processor, addressing up to 48 KB RAM, and a 16-color graphics display. Its great advantage, though, was that it had a real keyboard (and not a horrible membrane keyboard, like the ZX81). The BBC TV program was a great success in the UK, and so was the BBC Microcomputer.

Two other companies that became industry leaders, developed products in 1981. These were Novell Data Systems and Aston-Tate. Novell created a simple networking operating system that allowed two computers to share a single hard disk drive. Soon Novell would develop their Novell NetWare operating system, which allowed computers to share resources over a network. Ashton-Tate released the dBase II package which was the standard database package for many years.

For Intel, the adoption of the 8088 in the IBM PC was a godsend, and they had great difficulties keeping up with the supply of the processor. Unlike the 8080, though, they did not actively seek AMD for a second source for the processor. Intel had learnt that some second - source rights caused problems when the second source company actually moved ahead of them in their technology. Typically, second- source companies are able to charge a lower rate, as they do not have to recoup the initial research and development investment. Intel would eventually seek other companies, and AMD sought out Zilog for second source rights for their up-and-coming Z8000 device. It seemed to AMD that Zilog would have greatest potential for their new device, as they had shown with their Z80 device.

Intel was starting to realise that the processor market was a winner as it had a great deal of intellectual effort added to it. It differed from the memory market where designs could be easily copied by competitors. With microprocessors, they could set new standards and protect their designs with copyrights. If they established a lead in the processor market, and kept one step ahead of the copiers, they could make a great deal of profit in releasing new products and producing support devices for their processors, especially for the 8086/8088. For this, Intel released the 8087 math coprocessor, which greatly speeded up mathematical calculations, especially floating point ones. The use of floating point long division would eventually come back to haunt Intel, when a college tutor discovered a bug in their Pentium processor.

Intel were an innovative company, and had produced the first 4-bit and the first 8-bit processor, but with the 16-bit market they were beaten by Texas Instruments (TI). Unfortunately, for TI, the TMS 9900 was a rehash of an earlier product, and was generally underpowered. Intel, though, had the great strength in their 8088 processor of releasing a whole series of support devices which made it easier for designers to integrate the new processor. Anyway, no one could have guessed the impact that the IBM PC would have on the market. Intel was also beaten by National Semiconductor for the first 32-bit processor (the 32000).

The year 1982 would see IBM throw open the market for computers, with the IBM PC, and also through two great mistakes. Apart from IBM, five other companies would dominate the year: Commodore, Sinclair, Compaq, Apple and Sun. Three of them, Commodore, Apple and Sinclair, were from the old school, and the other two, Compaq and Sun Microsystems, were from the new school, and would learn to adapt to the new 'serious' market in computing that the IBM PC had created. In the same year, the US Justice Department threw out an antitrust lawsuit filed against IBM 13 years earlier. Within 15 years, it would be Microsoft who was facing similar action.

At IBM, the PC was taking off in ways that could never have been imagined. The IBM PC was a work of genius in which everything had been planned with perfection. It would sell over 200 000 computers within 12 months of its introduction, but the following year would see two major mistakes by IBM. The first was the introduction of the PCjr, which was intentionally incompatible with the IBM PC (because IBM did not want it to effect the IBM PC market) and the IBM AT. The PCjr failed because of its incompatibility, whereas the AT failed as it used a 16-bit processor (the 80286), while other computers were released using the new Intel 32-bit processor (the 80386). IBM could have easily have overcome these drawbacks, but, as these developments involved a much wider team than the IBM PC, they were held back by the interests of other parties. For example, the mainframe division was keen for the AT to use 16-bit processors, rather than the more powerful 32-bit processors, as this could further erode their market. These two decisions would open the door to the new kid on the block – Compaq.

Three former Texas Instruments managers founded Compaq Computer Corporation in

1982: Rod Canion, Jim Harris, and Bill Murto. Their first product was Compaq Portable PC. It was released in the following year (1983), and cost $3000. The Compaq Portable was totally compatible with the IBM PC and used the Intel 8088 (4.77 MHz), had 128 KB RAM, a 9-inch monochrome monitor and had a 320 KB 5.25-inch disk drive (Sony Electronics in the same year demonstrated the 3.5-inch microfloppy disk system). A large part of the start-up finance was used to create a version of the ROM BIOS which was IBM compatible, but did not violate IBM's copyright – a stroke of genius that many failed to follow. Compaq would soon become the fastest growing company ever. Only in the computer industry could a company grow from zero to hundreds of millions of dollars within 12 months. Compaq created a new market, which was based on IBM PC compatibility. They then waited for the great IBM to come along and sink their product, but when IBM did produce a portable, it was too late, too heavy, and failed to match the Compaq in its specification. Compaq were not in fact the first company to clone the IBM PC as they finally released it in 1983 – that was achieved by Columbia Data Products, with their MPC.

Two companies who would battle against the PC for market share were Sun and Apple. Sun Microsystems would quickly become a major computer company, and derived its name from an acronym from the Stanford University Network. Their first product was the Sun 1 workstation computer. They, like Apple, fought the IBM in terms of architecture and operating system. Sun, almost single-handedly, made the Unix operating system popular. Their computers succeed in the market, not because they were compatible with any other computer, but because they were technically superior to anything that the IBM PC could offer. The software that ran on the system fully used the processing power of the processor, and the Unix operating system provided an excellent robust and reliable operating system. Compatibility can often lead to a great deal of problems, especially if the compatibility involves the 8088 processor.

At Apple, champagne corks were popping, as they became the first PC company to generate $1 billion in annual sales. The Apple II Plus and Apple II had sold over 750 000 units. After toying with the Lisa computer and new versions of the Apple II, Apple would have one more trump card up their sleeve: the Apple Macintosh. Microsoft was keen to work with Apple, in case the relationship with IBM did not work out, and signed an agreement to develop applications for the forthcoming Macintosh (of which Microsoft were given an initial prototype to work on). IBM had become slightly annoyed with the success of Microsoft, especially from the success of their own creation. For Microsoft, it was a no-loose situation. They were, in the main, sharing code across the two architectures, which would quickly become industry standards. One would become an open standard (the IBM PC), and the other would be a closed standard (the Apple Mac).

The year 1982 saw a fantastic growth at Intel, and the only way that they could keep up with demand was to license their products to other silicon design companies. For this, they signed a 10-year technology exchange agreement with Advanced Micro Devices (AMD) that focused on the x86 microprocessor architecture. This agreement would be later regretted as AMD started to overtake them in the 80486 market. Intel, in the same year, released an update to the 8086 processor, called the 80286. The processor was destined for the IBM AT computer and it ran initially at 6 MHz, which improved on the 4.77 MHz of the 8088 processor. It had a 16-bit data bus, like the 8086, but had an extended 24-bit address bus that gave it an addressing range of 16 MB, rather than the 1 MB addressing range of the 8086/8088, or 1 GB of virtual memory. It outperformed the 8086 with a throughput of 0.9 MIPS, but this increased to 1.5 MIPS with a 10 MHz clock and 2.66 MIPS with a 12 MHz clock.

Commodore was never slow at developing their products. After the success of the Vic-20, in 1981, they released the Commodore 64 in the following year. It sold for $600 and had an excellent specification based around the new 6510 processor, and was released with 64 KB RAM, 20 KB ROM, sound chip (the first PC to have integrated sound), eight sprites, 16-colour graphics, and a 40-column screen. It was the first personal computer with an integrated sound synthesizer chip. They then released a whole range of peripherals, such as the VIC Modem ($110). Commodore also moved into the business market, with the BX256 and B128 computers for $3000 and $1700, respectively. The BX256 was a 16-bit multiprocessor computer. It included 256 KB RAM, Intel 8088 for CP/M-86, 6509 CPU, 80-column B/W monitor, built-in dual disk drives, and three-voice sound. The B128 computer featured 128 KB RAM, 40 KB ROM, 6509 CPU, 5.25-inch floppy drive, three-voice sound chip, cartridge slot, and an 80-column green screen.

At Sinclair, the ZX81 had been an unbelievable success and, knowing that alone they could not succeed in the USA market, they signed an agreement with the Timex Corporation to license Sinclair computers in the USA. By the end of 1982, Sinclair Research had sold over 500 000 ZX81s in over 30 countries. Atari also become a major computer company with the Atari 800. Its main feature was an advanced graphics display. Radio Shack also released the powerful TRS-80 Model 16. It used a 16-bit Motorola MC68000 microprocessor, a Z80 microprocessor, had 8-inch floppy drives, and an optional 8 MB hard drive. At the same time as Compaq were releasing their portable, Radio Shack produced the TRS-80 Pocket Computer; unfortunately, it was relatively slow as it used a 1.3 MHz 8-bit microprocessor, with a 26-character display.

DEC also finally decided to enter the personal computer market with the dual-processor Rainbow 100. It had an excellent specification with both a Z80 and an 8088 microprocessor, and could run CP/M, CP/M-86 or MS-DOS. Unfortunately, at $3000, it was too expensive for the market, which was already hot for the IBM PC.

1983 was a mixed year for IBM. They continued their success with the released of the IBM PC XT. It sold for $5000 and had a 10 MB hard drive, three extra expansion slots, and a serial interface. In its basic form it had 128 KB RAM, and a 360 KB floppy drive. With the success of PC-DOS 1.0, IBM followed it up with PC-DOS 2.1. On the downside, IBM released the IBM PCjr, which cost $700.

The greatest winners in 1983 were the newly created Compaq Computers, and Microsoft. In their first year, Compaq sold 47 000 computers, with a turnover of $111 million (and raised $67 million on their first public stock offering). They would eventually reach the $1 billion within five years of their creation.

The other winner was Microsoft who knew that they had to completely rewrite the MS-DOS operating system, so that it coped better with current and future systems. For this they introduced MS-DOS 2.0, which supported 10 MB hard drives, a tree-structured file system, and 360 KB floppy disks. They had quickly released the potential of the IBM PC, and released XENIX 3.0 (a PC version of Unix), Multi-Tool Word for DOS (which would eventually become Microsoft Word 1.0), as well as producing the Microsoft mouse (which sold for $200, with interface card and mouse). Microsoft also announced, in 1983, that it would be developing Microsoft Windows (initially known as Interface Manager), which would eventually be released in 1985. At the same time as Microsoft announced Windows, IBM was developing a program called TopView, and Digital Research was developing GEM (Graphics Environment Manager). These programs would use DOS as the basic operating system, but would allow the user to run multiple programs. The great problem with TopView was that it was text based and not a graphical user interface (GUI, or 'gooey'). Even allowing for this, most predicted, because of IBM's strength, that TopView would become the standard user interface. If IBM had won the battle for the user interface, they would have probably taken over the standard for both the user interface and the operating system, and then eventually the standard for the architecture. IBM, though, did agree to work with Microsoft on OS/2. Microsoft would eventually invest hundreds of millions of dollars on OS/2, with little in return. Businesses must learn from their mistakes, and Microsoft has always done this. The expertise gained in developing OS/2 was used in the development of Microsoft Windows.

In the same year as Microsoft released their new version of MS-DOS, AT&T was releasing the version of Unix that would become a standard: Unix System V. It was the first attempt at bringing together the different versions of Unix, including XENIX, SunOS and Unix 4.3 BSD. The two main families of Unix have become Unix System V and BSD (Berkeley Software Distribution) Version 4.4. System V would eventually be sold to SCO (Santa Cruz Operation). Currently available Unix systems include AIX (on IBM workstations and mainframes), HP-UX (on HP workstations), Linux (on PC-based systems), OSF/1 (on DEC workstations) and Solaris (on Sun workstations).

Other attempts at standardising Unix occurred with X/Open, OSF, and COSE, but have mainly failed. The great strength of Unix is its communications and networking protocols (such as TCP/IP, SMTP, SNMP, and so on), which provide the foundation for the Internet. Many organizations have tried to create a new operating system, such as VMS (from DEC) and Aegis (Apollo), but only Unix has become a serious competitor to Microsoft in operating systems. In the PC market, they would totally dominate the market, although Linux (a Unix clone) created a small market share for the technical experts. Unix-based systems used the standardised networking software that was built-into Unix, but the PC still lacked any proper form of networking. So, in 1983, Novell create one of the standards of the PC networking market: the Novell NetWare network operating system. The only other operating system which could have competed again Microsoft's DOS and Windows, was the up-and-coming Mac OS from Apple, which was at least 10 years ahead of its competitors. However, Apple refused to license their system to other vendors, or to other computer manufacturers.

Another significant event in software development occurred at AT&T, when Bjarn Stroustrup designed the new object-oriented language C++. Its great strength, and also one of its weaknesses, was that it was based on the popular C programming language. Its usage is now widespread and most current applications have been written using C++, whether they be for microcomputers, minicomputers and mainframe computers. The main drawback of C++ was that programmers could still use the C programming language, which, because of its looseness and simplicity, allowed the programmer to produce programs that would compile, but could crash because of a run-time error which was due to badly designed software. Typical errors were running off the end of an array, bad parameter passing into modules, or using memory that was not reserved for other purposes. Object-oriented programming languages are much tighter in their syntax, and the things that are allowed to be done. Thus, the compiler will typically catch more errors, whether they are run-time or syntax errors, before the program is run. Java has since overcome the problems of C++, as it is totally object-oriented, and much tighter in the rules of software coding.

The great strength for the adoption of the PC was IBM's intention to allow software companies to quickly develop application software. They thus released information on the hardware of the computer so that software companies could write compatible applications. Like VisiCalc for the Apple II, in 1983, the two killer applications to help boost the acceptance, and sales, of the IBM PC were:

- **Lotus 1-2-3**. This was a spreadsheet designed and developed by Jonathan Sachs and Mitch Kapor at Lotus Development. It initially required an extremely large amount of memory, 256 KB. Over $1 million was spent on its initial promotion but it paid back its original investment a thousand times over. Its sales hit Microsoft's Multiplan spreadsheet, which had sold over 1 million copies. Microsoft learnt from this, and in the coming years would release Excel, which would become the standard spreadsheet.

- **WordPerfect**. This was a word processing package developed by Satellite Software International (who would eventually change their name to the WordPerfect Corporation.). It initially cost $500, and was an instant success. Many believed that WordPerfect 5.1 was the classic touch-type program, as it used keystrokes instead of long-winded menu options. Many typists have since had real troubles moving from WordPerfect to WIMPs-based packages such as Microsoft Word (so much so that many current word processors support all of the WordPerfect keystrokes).

The year 1983 was to be bleak for non-IBM PC compatible computers and saw prices falling month upon month. It also spelt the end of the line, in different ways, for three great innovators in the personal computer industry: Zilog, Osborne and Texas Instruments. It was the beginning of the end for Zilog when they released their 32-bit microprocessor: the Z8000. It was an advanced device that had a 256-byte on-chip cache, instruction pipelining, memory management, and 10–25 MHz clock speed. Unfortunately, for Zilog, it was incompatible with the great Z80 processor. It thus failed in the market against the strength of the Intel 8086, and the up-and-coming 80386 processor. Of the many computer manufacturers who rushed to the market and used the 8086/8088, only one, Commodore, introduced a Z8000-based system (Commodore Z8000). Apart from the failings at IBM and DEC, the release of the Z8000 processor must rank amongst the poorest decisions in computing history. No one could predict the effect that a Z80-compatiable 32-bit processor would have had on the market. Certainly a 32-bit processor, which was functionally compatible with the 8086/8088 (as the Z80 had been with the 8080) would have blown the market wide open, and would have possibly stopped the slide to quasimonopoly of the Intel processors. Another failure in the processor market was the extremely powerful 6 MHz, 32-bit NS32032 microprocessor from National Semiconductor.

Commodore Business Machines were becoming dominant in the home computers market, and highlighted their dominance with the release of the Commodore 64, for $400, which quickly fell to $200 and dropped the prices of the VIC-20 to below $100 (breaking it for the first time). In 1983, the sales of the VIC-20 reached 1 000000.

Commodore was also keen to develop the business market, and released the Commodore Executive 64. It cost $1000 and had 64 KB RAM, a detachable keyboard, a 5-inch colour monitor, and a 170 KB floppy drive. In 1983, Commodore became the first personal computer to sell over $1 billion worth of computers.

Many companies in the home computer market had made large profits, but one failure in a product range could spell disaster for a company. The high profits for all would not last long as Commodore, Atari and Sinclair started slashing prices. Sinclair, through Timex, introduced the Timex/Sinclair 2000 in the USA (which was called the Sinclair Spectrum in other countries). It cost $149 for a 16 KB model, while the ZX81 price was reduced to $49. The squeeze was on, as prices tumbled.

Atari released their 600XL for $199, and ceased production of the Atari 5200. The 600XL was based on the 1.79 MHz 6502C processor, had 16 KB RAM, 24 KB ROM, and an optional CP/M module. As the push was on from other manufacturers to reduce prices, they also did the same and reduced the Atari 800 to $400. Atari also released the 1200XL home computer, which had 64 KB RAM, and 256 colour capability, and cost $900. Production eventually ended for the 1200XL, mainly because of compatibility problems.

At the time, Japanese companies had been making great advances in the electronics industry, and many, such as NEC and Fujitsu were starting to overtake USA silicon companies, such as Texas Instruments, Intel and National Semiconductor, in their product of integrated circuits (although Intel had the x86 series of processors as their trump card). They were also winning in producing peripherals and accessories for computer systems, such as:

- Fujitsu producing the first 256 Kbit memory chips.
- Sony developed a new standard for 3.5-inch floppy disks, with the Microfloppy Industry Committee, and thus created the first double-sided, double-density, holding floppy disk system that could store up to 1 MB. Sony was also working with Philips in creating the CD-ROM, which was an extension of audio CD technology.

However, in computer manufacturing, Japanese companies struggled as the USA companies, such as IBM and Apple, were setting the standards. The IBM PC was relatively easy to clone, but the Apple computer required a license to manufacture, which, at the time, was almost impossible to gain. Compared with many USA-based companies, the Japanese companies were efficient and produced reliable electronics, but as long as they were one step behind the US-based computer companies, they could not gain a serious share of the home computer market. To overcome this, 14 Japanese companies and Microsoft joined an alliance to create the MSX standard It used the Zilog Z80, TI TMS9918A video processor, General Instruments AY-8910 sound processor, NEC cassette interface chip, Atari joystick interface, 64 KB RAM, and 32 KB ROM-based extended Microsoft BASIC. This was one of the first attempts to standardize computer architecture, but was doomed to failure with the release of the IBM PC, and that it was based on 8-bit technology. Several MSX computers did reach the market, but quickly failed. It was a great idea, and one that should have worked. The key to its failure was that there was a better, more defined standard: the IBM PC.

In a classic case of releasing the right product at the wrong time, Osborne Computer released their own portable computer. Unlike the IBM PC, or Compaq's portable, it was based on the Z80A processor. The computer quickly failed in the market and Osborne eventually filed for bankruptcy. Around the same time, Radio Shack also produced a non-IBM compatible portable: the TRS-80 Model 100. They were also following the tried and tested technique of improving their product line by releasing the TRS-80 Model 4. It would fail as it cost $2000, and was non-IBM PC compatible (as it was based on the 4 MHz Zilog Z80A microprocessor). Against the IBM PC, and the lower-end computers, such as the VIC-64 and the Sinclair Spectrum, it was vastly overpriced.

Another casualty of the success of the IBM PC was Texas Instruments who eventually withdrew from the personal computer market. The TI 99/1 had sold well over the years (over 1 million), but was now struggling against the new, cheaper computers.

Apple took a big gamble with the LISA[1] (local integrated software architecture) computer, as it cost $50 million, and its software cost $100 million (showing that the costs of developing hardware were reducing, while software development costs were increasing). It was released in 1984 and was expensive ($10 000), it was underpowered, but it was the first personal computer to have a graphical user interface (GUI). Rather than going with the 8086, as most of the market was doing, they based it on the excellent 5 MHz 68000 microprocessor. It had 1 MB RAM, 2 MB ROM, a 12-inch black/white monitor, 720×364 graphics, dual 5.25-inch 860 KB floppy drives, and a 5 MB hard disk drive. LISA would sell over 100 000 units. Apple was keen to develop the LISA computer, but it would be the new Mac, which would

---

[1] LISA was actually named after Steve Job's young daughter.

become the focus for their operation.

Apple was investing a great deal of effort in the Mac, and gave the Mac developers the best environment possible. This caused considerable friction with the Apple II division, as all the finances for the Mac facilities were generated from sales of the Apple II. Apple intentionally kept the two divisions apart, which only helped to increase the friction. In the year, the Apple II highlighted its success by selling its one-millionth unit. They continued its development with the Apple IIe, which had 64 KB RAM, Applesoft BASIC, upper/lower case keyboard, seven expansion slots, 40×24 and 80×24 text, 1 MHz 6502 processor, up to 560×192 graphics, and a 140 KB 5.25-inch floppy drive.

The software market, especially related to the IBM PC was growing fast. Satellite Software International released WordPerfect 3.0 for $500, and Borland International, founded by Philippe Kahn, created the first version of their excellent Turbo Pascal compiler. Borland, single-handily, saved Pascal from an early exit. Borland were for years the main company involved in producing software development tools for the PC, with Borland C++, Borland Delphi and Borland JBuilder. Unfortunately, they would eventually struggle against the might of Microsoft (who were able to invest a great deal of money into their development tools, especially in Visual Basic and Visual C++). Microsoft has the privileged position of being able to invest money in other areas of development, but redirecting them from profits made from other areas. For example, they used profits from the DOS system to invest in Windows, and profits from Windows to invest into office applications (Word, Excel and PowerPoint), and profits from office applications to invest into software development tools (Visual Basic, Visual C++ and Visual Java). Obviously, it is to Microsoft's advantage that they keep the tools up-to-date, as this is the same development system that they use to generate their own applications.

1984 obviously had futuristic connotations to it. However, it was less of a futuristic year for IBM and more of a nightmare, when IBM released the IBM PCjr. It used the 8088 CPU, includes 64 KB RAM, a 'Freeboard' keyboard (IBM would eventually release a new keyboard, which was a free upgrade to those who wanted it), and one 5.25-inch disk drive, and no monitor, for $1300. A year later, in 1985, the PCjr was dropped. As the market became more competitive, IBM started to show their teeth as the number of cloners increased. The unfortunate companies who were the first to be taken to court were Corona Data Systems and Eagle Computer. IBM sued them over a copyright violation of the IBM PC's BIOS, and easily won the case. It was clear that, to avoid litigation, that companies required rewriting the BIOS. This would not give a technical advantage, but would keep IBM's lawyers away.

The next step for IBM was important in the development of the PC. For this they learned from their mistakes with the PCjr, and made their new computer, the PC AT, compatible with the IBM PC. It used the new Intel 6MHz 80286 processor, and had a 5.25-inch 1.2 MB floppy drive, with 256 KB or 512 KB RAM, optional 20 MB hard drive, monochrome or colour monitor. The initial cost was $4000.

As the demand for IBM PCs increased, there was also an increase for demand for graphics adaptors. For this IBM released the Enhanced Color Display (EGA) monitor with 640×350 resolution, 16 colours, at a cost of $850. They also released TopView which failed in the market because it was text-based, and not a GUI. If they had done, they may have captured the market which Microsoft Windows gained.

The battle for the processor market started to heat up when Intel released the 80188, which was an integrated version of the 8086. They also allowed IBM the legal rights to use microlithography masks to make x86 processor chips. Intel, having survived the new 32-bit processors, from Zilog and National Semiconductor, faced their biggest threat from NEC and

Motorola. NEC was the first to clone the 8088, with the 8 MHz V20 microprocessor and a clone of the 8086 processor, with the 8 MHz V30 microprocessor. Another threat came from Motorola who added the 68010 and 68020 32-bit processors to their range. Many non-PC-based developers adopted the Motorola processor in favour of the 8086, as it was typically easier to develop hardware for it, and much easier to write software (as the 8086 had a horrible segmented memory architecture). For most, it was the only way for a computer manufacturer to differentiate themselves from the clone market. Some, such as Radio Shack, followed the IBM PC market with the Tandy 1000/1200 HD, but there was little to differentiate their clone from any other clone.

New entries for the year included Silicon Graphics, who would go on to produce excellent workstations, which had state-of-the-art graphics power. In 1984, they produced the first 3D graphics workstation. They were also involved, in the 1990s, in the development of the graphics for *Jurassic Park*.

It was to be the year of Compaq Computer and Apple Computers. Compaq introduced the Compaq Deskpro. Apple Computer created the ultimate entry for their Macintosh computer, by running their *1984* advert once, during the NFL SuperBowl. The advert had cost $1.5 million, but soon became one of the most talked about adverts, ever. The Macintosh was as brilliant a computer as anyone could have conceived. It was designed by creative people, and not just by technocrats. It was a fully integrated unit, which could be easily ported from place to place. The Mac used the 8 MHz 32-bit Motorola 68000 processor, along with a 9-inch B/W screen, 512×342 graphics, 400 KB 3.5-inch floppy disk drive, mouse, and 128 KB RAM. It cost $2500. Microsoft knew that they could not just rely on the IBM PC market, so they worked closely with Apple and released Microsoft BASIC (MacBASIC) and Microsoft Multiplan for the Macintosh. After just 74 days of its introduction, over 50000 Macs had been sold, and after 100 days they had sold 70 000 units. After six months, it was 100 000 units, and within the year, 250 000 units. This, to Apple, was a great disappointment as they estimated that they would sell over 2 million units by the end of 1985. The main problem is that it lacked resources, especially memory. Apple Computer overcame this by releasing the Macintosh 512K for $3200.

The Macintosh had everything going for it. It was a totally integrated system, where the IBM PC felt like a basic system, which required lots of extra bits and pieces to make it work properly. A great confusion at the time was the number of application packages which were entering the market. Apple eased this problem with the release of AppleWorks, which integrated a word processor, database management program, and a spreadsheet.

Apple also continued developing the Lisa computer with Lisa 2, and also with the Apple II, with the Apple IIc computer (the Apple III computer had not sold well, and production of it soon stopped). The Apple IIc computer cost $1300 and was based on the 6502A processor, had 128 KB RAM and a 3.5-inch floppy disk drive. On the first day of its release, Apple received 52 000 orders. By the end of the year, over 2 million Apple II computers had been sold. The Lisa 2 computer came with 512 KB RAM, and a 10 MB hard disk. Apple was also innovating in the printer market, with the colour Apple Scribe printer and the LaserWriter. At the same time, Hewlett-Packard introduced the LaserJet laser printer, for $3600, with 300 dpi resolution.

As Apple had done, Commodore would release, in the following year, a computer based on the 68000 processor (the Amiga, from newly purchased Amiga Corporation). In 1984, they introduced the Commodore Plus/4 which used the 7501 microprocessor, had 64 KB RAM, 320×200 pixel graphics with 128 colours, and also released the Commodore 16 with 16 KB of RAM, at a selling price of $100.

At Microsoft, development was continuing on both Apple and IBM PC systems. No one at the time could predict that the IBM PC market would eventually dwarf the Apple market. The Macintosh looked to be the system of the future, thus Microsoft stopped working on Excel, their new PC-based spreadsheet package, and switched their resources to developing software for the Macintosh. This included Excel for the Macintosh. From now on Microsoft would concentrate of GUI applications, for Microsoft Windows and for the Macintosh. They released MS-DOS 3.0/3.1 which supported larger hard disks, networks and high capacity floppy disks. After IBM lost out on the DOS operating system, Microsoft held out an olive branch to them by demonstrating Microsoft Windows. IBM refused to become involved, mainly because it competed with its newly developed interface, TopView. Microsoft and Lotus Development also nearly agreed to merge their companies, but Jim Manzi at Lotus Development convinced Mitch Kapor to back out of it. Microsoft's Windows was superior to TopView as it used a graphical user interface. The only other real competitor to Microsoft was Digital Research, who had missed out on the IBM PC market. In 1984, they released the Graphics Environment Manager (GEM) icon/desktop user interface for the IBM PC computer.

In the Unix market, in 1984, the Massachusetts Institute of Technology (MIT) began developing the X Window System. Their main objective was to create a good windows system for Unix machines. Many versions evolved from this and, by 1985, it was decided that X would be available to anyone who wanted it for a nominal cost. X, itself, is a portable user interface and can be used to run programs remotely over a network. It has since become a de facto standard because of its manufacturer independence, its portability, its versatility and its ability to operate transparently across most network technologies and operating systems. The main features of X-Windows are that:

- It is network transparent. The output from a program can either be sent to the local graphics screen or to a remote node on the network. Application programs can output simultaneously to displays on the network. The communication mechanism used is machine-independent and operating system independent.
- Many different styles of user interface can be supported. The management of the user interface, such as the placing, sizing and stacking of windows is not embedded in the system, but is controlled by an application program which can easily be changed.
- As X is not embedded into an operating system, it can be easily transported to a wide range of computer systems.
- Calls are made from application programs to the X-windows libraries which control WIMPs. The application program thus does not have to create any of these functions.

1985 was the year that Microsoft released their first version of Windows, at a price of $100. It was hardly startling, and would take another two versions before it completely dominated the market. It could not multitask, and still used DOS. Another major failing was that it did not use the full capabilities of the new 32-bit processor (80386) or the enhanced 16-bit processor (80286), and could thus only access up to 1 MB of memory.

Just as IBM were releasing their AT computer with the 80286, Intel released their new 32-bit 16 MHz 80386DX microprocessor, and the 80287 math coprocessor. The 80386 used 32-bit registers and a 32-bit data bus, and incorporated 275 000 transistors (1.5 microns). The initial price was $299. It could access up to 4 GB of physical memory, or up to 64 TB of virtual memory. A worrying development for Intel came from the new start up company, Chips & Technologies, which developed a set of five chips that were equivalent to 63 smaller chips that were found on the IBM PC AT motherboard. This development meant that many of the support devices produced by Intel could be replaced by many less devices, thus cutting production costs. At Motorola, the success of the 68000 brought out the 68008 processor.

After Apple had released their 68000-based Macintosh in the previous year, Commodore released their new flagship computer: the Amiga 1000. Unlike the IBM PC, it was fully multitasking and used a WIMPs (windows, icons, menus and pointers) system. In its basic form it cost $1300 and had 256 KB RAM, and 880 KB 3.5-inch disk drive. They also released the Commodore 128 computer, which was an upgrade of the Commodore 64. Along with the Amiga, Commodore were trying to get into the PC market with the PC10 and PC20 computers, and tried to stop production of the Commodore 64 (but public demand restarted production several times).

At Apple Computer the success of the Macintosh continued. The battle was now on for the PC market, and they had the IBM PC in their sights. During the SuperBowl, Apple ran a TV advert for Macintosh Office, which showed blindfolded business executives walking off a cliff, like lemmings. Things were becoming turbulent in Apple, after years of growth had produced a grown-up company with formal business methods. This type of environment did not suit Steve Jobs (the co-founder of Apple Computer), and he left, along with five senior managers, to form NeXT Incorporated. In fact, John Sculley, the former Pepsi-Cola president who, in 1984, had been brought in to train Steve to become the CEO, forced Steve Jobs out. From then on, John Sculley was the man in charge of Apple.

The future for Apple looked difficult, but the key to future growth would be the Macintosh, and not Lisa or the Apple II. The software for the Macintosh was being produced as quickly as the market was buying it. Microsoft released Microsoft Word 1.0 and the Microsoft Excel spreadsheet ($95). Apple were not really impressed with the first version of Excel, and reckoned that Lotus Development's equivalent was better (named Jazz). Another key package for the Macintosh was Aldus PageMaker from Aldus, which created a new industry, which for the first time, integrated text and graphics with a design package: desktop publishing. For years, PageMaker was the de facto standard package for graphics design and desktop publishing. Microsoft obviously had a foot in both the IBM PC and the Macintosh market, as they released Microsoft Word 2.0 for DOS, and QuickBASIC 1.0.

The year produced many good deals for Microsoft, including:

- Microsoft signed a deal with IBM for a joint-development agreement to work together on future operating systems and environments.
- Microsoft signed a deal with Apple to cover Apple's copyrights on the visual display of the Macintosh.
- Microsoft purchased all rights to DOS from Seattle Computer Products for $925000. The deal of the century!

Atari struggled on, in face of the competition from Apple, Compaq, Commodore and the IBM PC. With the might of Microsoft added to the equation, they had little chance in the profitable business market. They continued their previous success in the home market with the 65XE, the 130XE, and the 520ST, for $120, $400, and $600, respectively. Radio Shack also continued to swim against the tide with the release of the Tandy 6000 multi-user system (with up to nine users). It was extremely powerful and used the both a Z80A and a 68000 processor. It had 512 KB RAM, 80×24 text, graphics, 1.2 MB 8-inch disk, an optional 15 MB hard drive, TRS-DOS, or XENIX 3.0. Another struggler with an excellent product was Acorn who released The Advanced RISC Machine (ARM), which used a powerful 32-bit processor.

At IBM, there was despondency as they stopped production of their PCjr and released their first version of TopView for $150. One of the successes of the previous year, Compaq Computer, was jubilant as they reported second year revenues of $329 million. They quickly followed up the success of the Compaq portable with the Compaq Deskpro 286 and Portable 286, which was similar in specification to the IBM AT. IBM also moved into networking with IBM Token Ring; unfortunately, even though Token Ring was an excellent networking technology, the future would be Ethernet.

Each year in the computer industry had seen a new significant company being born. The previous years had seen the birth of Compaq Computers, Sun Microsystems and Apollo. In the 1985, it was Nintendo, and Chip and Technologies. Nintendo would become one of the leading computer companies in the lucrative computer games market. They again highlighted the strength of the USA in generating new and innovative computer companies. Software companies were also being created, such the Corel Corporation (by Michael Cowpland), and Quarterdeck Office Systems. On of the successes of the previous years was Sun Microsystems who had started work on their SPARC processor.

On the PC, new software versions were coming thick and fast. Lotus 1-2-3 has moved to 2.0, WordPerfect moved to Version 4.1, Novell NetWare was now at Version 2.0 and dBase was at Version 3. 1985 also saw the first CD-ROM drives for computer use.

After a few frantic years, things started to settle down in 1986. The IBM PC and the Apple Macintosh would now dominate the market, especially at the business end. One of the biggest winners was Compaq Computers who had seen their turnover for their third year rise to $503.9 million and, by the middle of the year, they would sell their 50 000th computer. Compaq Computer introduced the Compaq Portable II. Against its excellent quality and specification, IBM would eventually withdraw from the portable computer for a while, as it was obviously inferior to the Compaq portable. It would take many years before IBM would regain some of the portable market (with the ThinkPad).

Compaq blasted the PC market wide open with the first 16 MHz Intel 80386-based PC: the Compaq Deskpro 386. The best that IBM could manage was the IBM AT which had an 8 MHz Intel 80286. The Deskpro 386 was thus running at twice the clock speed, and had the potential, with 32-bit software, to run twice as fast again. The 80386 also had significant improvements in the number of clock cycles that it took for an operation to be performed. Thus, the Deskpro 386 sprinted, while all the other PCs dawdled, and its full potential was yet to be realised.

IBM knew that the PC was a compromised system, and released the IBM RT Personal Computer. This was based on a 32-bit RISC-based processor, with 1 MB RAM, a 1.2 MB floppy, and 40 MB hard drive, and cost $11 700. Even with the RISC processor, it only had a performance of 2 MIPS, and thus its price/performance ratio was too great for it to be adopted in the market.

Apple was starting to suffer against the growing power of the IBM PC developers. They still had a closed system, where it was up to them to develop the software and hardware for the Macintosh, whereas the IBM had hundreds, if not thousands, working on it, and improving it. The Apple Mac was now looking underpowered and lacking other facilities, especially in networking on IBM PC-based networks. Apple overcame part of this with the release of the Macintosh Plus, which was based around the 8 MHz 68000 processor, had 1 MB RAM, SCSI-based hard disk connector (the first ever computer to have integrated SCSI interfaces) and an 800 KB 3.5-inch floppy drive. It cost $2600 (while a 512 KB version cost $2000). Unfortunately, it was still not possible to connect an Apple Mac onto an IBM PC-based network, unless a telephone connection was used. This held it back from wider adoption in the commercial market. Apple, though, was starting to make great inroads into the publishing industry with the release of the innovative LaserWriter Plus printer.

Microsoft had over the past few years initiated many new products for both the IBM PC, and the Apple Macintosh. In 1985, they consolidated their market with new versions of the successful software, such as MS-DOS 3.2 and Microsoft Word 3.0. In MS-DOS 3.2, support was added for 3.5-inch 720 KB floppy disk drives (these disks were much more reliable than the older, 'floppy', 5-inch floppy disk, as they had a hard case to protect them). The initial investment of time, and energy, for those involved in Microsoft was rewarded when, for the first time, Microsoft sold its shares to the public. When floated, each share was worth $21, which raised $61 million for Microsoft, and made Bill Gates the world's youngest billionaire.

The UK also showed that they could innovate in market niches with the release of the Inmos T800 Transputer, which was a powerful RISC processor that could be used in parallel processing applications.

Several computer manufacturers, such as Silicon Graphics, started to move towards the new range of RISC processors produced by MIP Technologies, such as the 8 MHz, 32-bit, R2000 processor. This used 110 000 transistors and gave a speed of 5 MIPS. At Motorola, they were working on the 68030 processor, which would have over 300 000 transistors. They also began work on the 88000 processor.

At IBM, work had begun on a computer range which would become a classic: the IBM RS/6000 series. The newcomer of the year was Gateway 2000, which shipped its first PC. In addition, after using the Small Computer System Interface (SCSI) on Apple's Macintosh (SCSI-1), it was standardised with the ANSI X3.131-1986 standard.

1.7 Exercises

The following questions are multiple choice. Please select from a–d.

1.7.1 Who solved the US Governments Census problems:

(a) Bill Gates (b) Herman Hollerith
(c) William Shockley (d) Lee De Forest

1.7.2 Which computer helped aid the British Government to crack codes in World War II:

(a) ENIAC (b) Harvard Mk I
(c) IBM System/360 (d) Colossus

1.7.3 What is ENIAC an acronym for:

(a) Electronic Numerical Integrator and Computer
(b) Electronic Number Interface Analysis Computer
(c) Electronic Number Interface and Computer
(d) Electronic Numerical Interchange Computer

1.7.4 Who invented the transistor:

(a) Bill Gates (b) Herman Hollerith
(c) William Shockley (d) Lee De Forest

1.7.5 Which company did William Shockley form:

(a) Shockley Semiconductor (b) Shockley Devices
(c) Shockley Electronics (d) Shockley Electrics

1.7.6 Which company proposed that silicon could be used for transistors:

(a) IBM (b) Texas Instruments
(c) Motorola (d) Fairchild Semiconductors

1.7.7 Which company first proposed the integrated circuit:

(a) IBM (b) Texas Instruments
(c) Motorola (d) Fairchild Semiconductors

1.7.8 Who first produced an integrated circuit:

(a) John Cocke (b) Robert Noyce
(c) Gordon Moore (d) William Shockley

1.7.9 Who proposed that the number of transistors that can be fitted onto an integrated circuit doubles each year:

(a) John Cocke (b) Robert Noyce
(c) Gordon Moore (d) William Shockley

1.7.10 Which computer was the first to use integrated circuits:

(a) Apple I (b) IBM System/360
(c) IBM PC (d) DEC PDP-11

1.7.11 Which one of the following formed Intel:

(a) Bill Gates and Paul Allen
(b) Robert Noyce, Gordon Moore and Andy Grove
(c) Jerry Sanders
(d) Steve Wozniak and Steve Jobs

1.7.12 Which one of the following formed Microsoft:

(a) Bill Gates and Paul Allen
(b) Robert Noyce, Gordon Moore and Andy Grove
(c) Jerry Sanders
(d) Steve Wozniak and Steve Jobs

1.7.13 Which one of the following formed Apple Computers:

(a) Bill Gates and Paul Allen
(b) Robert Noyce, Gordon Moore and Andy Grove
(c) Jerry Sanders
(d) Steve Wozniak and Steve Jobs

1.7.14 Which company did Kenneth Olsen help form:

(a) Compaq (b) DEC
(c) Microsoft (d) IBM

1.7.15 Which company developed the first microprocessor:

(a) Texas Instruments (b) Motorola
(c) Zilog (d) Intel

1.7.16 Which company developed the first 8-bit microprocessor:

(a) NEC (780-1) (b) Motorola (6800)
(c) Zilog (Z80) (d) Intel (8008)

1.7.17 Which company developed the first 16-bit microprocessor:

(a) Texas Instruments (9900) (b) Motorola (68000)
(c) Zilog (Z8000) (d) Intel (8086)

1.7.18 Which company was the first to demonstrate the usage of windows, mouse and keyboard:

(a) IBM (b) Xerox
(c) Microsoft (d) DEC

1.7.19 Which company was the first to demonstrate the WYSIWYG concept:

(a) IBM (b) Xerox
(c) Microsoft (d) DEC

1.7.20 What was the name of the Xerox research center:

(a) PARC (b) XRES
(c) PERC (d) RESP

1.7.21 Which company did Bill Gates and Paul Allen initially create:

(a) Micro-Traffic (b) Traf-O-Data
(c) Traffic Software (d) Gates & Allen

1.7.22 Who developed the C programming language:

(a) Bill Gates and Paul Allen
(b) Brian Kernighan and Dennis Ritchie
(c) Niklaus Wirth
(d) Steve Wozniak and Steve Jobs

1.7.23 Who developed the Pascal programming language:

(a) Bill Gates and Paul Allen
(b) Brian Kernighan and Dennis Ritchie
(c) Niklaus Wirth
(d) Steve Wozniak and Steve Jobs

1.7.24 Which was the first ever commercial microprocessor:

(a) 4000 (b) 4004
(c) 8080 (d) 1000

1.7.25 The 8008 device could address up to 1 KB. Thus how much address lines did it have:

(a) 6 (b) 10
(c) 20 (d) 1000

1.7.26 Which processor did Zilog produce:

(a) Z80 (b) 6502
(c) 8080 (d) 6800

1.7.27 Which processor did MOS Technology produce:

(a) Z80 (b) 6502
(c) 8080 (d) 6800

1.7.28 Which processor did Motorola produce:

(a) Z80 (b) 6502
(c) 8080 (d) 6800

1.7.29 Which processor did the Apple II use:

(a) Zilog Z80 (b) MOS Technology 6502
(c) Intel 8080 (d) NEC 780-1

1.7.30 Which processor did the Commodore PET use:

(a) Zilog Z80 (b) MOS Technology 6502
(c) Intel 8080 (d) NEC 780-1

1.7.31 Which processor did the TRS-80 use:

(a)	Zilog Z80	(b)	MOS Technology 6502
(c)	Intel 8080	(d)	NEC 780-1

1.7.32 Which processor did the ZX80 use:

(a)	Zilog Z80	(b)	MOS Technology 6502
(c)	Intel 8080	(d)	NEC 780-1

1.7.33 Which company distributed CP/M:

(a)	Microsoft	(b)	Digital Research
(c)	Xerox	(d)	Applesoft

1.7.34 Which software language was standard on most early PCs:

(a)	C	(b)	BASIC
(c)	Pascal	(d)	Assembly Language

1.7.35 How did the Motorola 68000 gain its name:

(a)	No reason	(b)	It was sold for $680.00
(c)	It sounded like the 8008	(d)	It had 68 000 transistors

1.7.36 Which company produced the VAX range of computers:

(a)	IBM	(b)	DEC
(c)	Compaq	(d)	Apple

1.8 Notes from the author

The history of the PC is an unbelievable story, full of successes and failures. Many people who used some of the computer systems before the IBM PC was developed, wipe a tear from their eyes, for various reasons, when they remember their first introduction to computers, typically with the Sinclair Spectrum or the Apple II. In those days, all your programs could be saved to a single floppy disk, 128KB of memory was more than enough to run any program, and the nearest you got to a GUI was at the adhesives shelf at your local DIY store. It must be said that computers were more interesting in those days. Open one up, and it was filled with processor chips, memory chips, sound chips, and so on. You could almost see the thing working (a bit like it was in the days of valves). These days, computers lack any soul; one computer is much like the next. There's the processor, there's the memory, that's a bridge chip, and, oh, there's the busses, that's it.

As we move to computers on a chip, they will, in terms of hardware, become even more boring to look at. But, maybe I'm just biased. Oh, and before the IBM PC, it was people who made things happen in the computer industry, such as William Shockley, Steve Jobs, Kenneth Olson, Sir Clive Sinclair, Bill Gates, and so on. These days it is large teams of software and hardware engineers who move the industry. Well, enough of this negative stuff. The PC is an extremely exciting development, which has changed the form of modern life. Without its flexibility, its compatibility, and, especially, its introduction into the home, we would not have seen the fast growth of the Internet.

Here are my Top 15 successes in the computer industry:

1. ***IBM PC** (for most), which was a triumph of design and creativity. One of the few computer systems to ever to be released on time, within budget, and within specification. Bill Gates must take some credit in getting IBM to adopt the 8088 processor, rather than 8080. After its success, every man and his dog had a say in what went into it. The rise of the bland IBM PC was a great success of an open system over closed systems. Companies who have quasimonopolies are keen on keeping their systems closed, while companies against other competitors prefer open systems. The market, and thus, the user, prefers open systems.*
2. ***TCP/IP**, which is the standard protocol used by computers communicating over the Internet. It has been designed to be computer independent to any type of computer, can talk to any other type. It has withstood the growth of the Internet with great success. Its only problem is that we are now running out of IP addresses to grant to all the computers that connect to the Internet. It is thus a victim of its own success.*
3. ***Electronic mail**, which has taken the paperless office one step nearer. Many mourned the death of the letter writing. Before email, TV and the telephone had suppressed the art of letter writing, but with email it is back again, stronger than ever. It is not without its faults, though. Many people have sent emails in anger, or ignorance, and then regretted them later. It is just too quick, and does not allow for a cooling off period. My motto is: 'If you are annoyed about something sleep on it, and send the email in the morning'. Also, because email is not a face-to-face communication, or a voice-to-voice communication, it is easy to take something out of context. So another motto is: 'Carefully read everything that you have written, and make sure there is nothing'. Only on the Internet could email address format be accepted, worldwide, in such a short time.*
4. ***Microsoft**, who made sure that they could not lose in the growth of the PC, by teaming up with the main computer manufacturers, such as IBM (for DOS and OS/2), Apple (for Macintosh application software) and for their own operating system: Windows. Luckily for them it was their own operating system which became the industry standard. With the might of having the industry-standard operating system, they captured a large market for industry-standard application programs, such as Word and Excel.*
5. ***Intel**, who was presented with an enormous market with the development of the IBM PC, but have since invested money in enhancing their processors, but still keeping compatibility with their earlier ones. This has caused a great deal of hassle for software developers, but is a dream for users. With processors, the larger the market you have, the more money you can invest in new ones, which leads to a larger market, and so on. Unfortunately, the problem with this is that other processor companies can simply copy their designs, and change them a little so that they are still compatible. This is something that Intel has fought against, and, in most cases has succeeded in regaining their market share, either with improved technology or through legal action. The Pentium processor was a great success, as it was technologically superior to many other processors in the market, even the enhanced RISC devices. It has since become faster and faster.*
6. ***6502** and **Z80** processors, the classic 16-bit processors which became a standard part in most of the PCs available before the IBM PC. The 6502 competed against the Motorola*

6800, while the Z80 competed directly with the Intel 8080.

7. ***Apple II**, which brought computing into the classroom, the laboratory and even the home.*
8. ***Ethernet**, which has become the standard networking technology. It is not the best networking technology, but has survived because of its upgradeabliity, its ease-of-use, and its cheapness. Ethernet does not cope well with high capacity network traffic. This is because it is based on contention, where nodes must contend with each other to get access to a network segment. If two nodes try to get access at the same time, a collision results, and no data is transmitted. Thus, the more traffic there is on the network, the more collisions there are. This reduces the overall network capacity. However, Ethernet had two more trump cards up its sleeve. When faced with network capacity problems, it increased its bit rate from the standard 10 Mbps (10 BASE) to 100 Mbps (100 BASE). So there was 10 times the capacity, which reduced contention problems. For networks backbones it also suffered because it could not transmit data fast enough. So, it played its next card: 1000 BASE. This increased the data rate to 1 Gbps (1000 Mbps). Against this type of player, no other networking technology had a chance.*
9. ***WWW**, which is often confused with the Internet, and is becoming the largest, database ever created (okay, 99% of it is rubbish, but even if 1% is good then its all worthwhile). The WWW is one of the uses of the Internet (others include file transfer, remote login, electronic mail, and so on).*
10. ***Apple Macintosh**, which was one of few PC systems which competed with the IBM PC. It succeeded mainly because of its excellent operating system (MAC OS), which was approximately 10 years ahead of its time. Possibly if Apple had spent as much time in developing application software rather than for their operating system it would have considerably helped the adoption of the Mac. Apple refusing to license it to other manufacturers also held its adoption back. For a long time it thus stayed a closed system.*
11. ***Compaq DeskPro 386**. Against all the odds, Compaq stole the IBM PC standard from the creators, who had tried to lead the rest of the industry up a dark alley, with MCA.*
12. ***Sun SPARC**, which succeed against of the growth of the IBM PC, because of its excellent technology, its reliable Unix operating system, and its graphical user interface (X-Windows). Sun did not make the mistakes that Apple made, and allowed other companies to license their technology. They also supported open systems in terms of both the hardware and software.*
13. ***Commodore**, who bravely fought on against the IBM PC. They released mainly great computers, such as the Vic range and the Amiga. Commodore was responsible for forcing the price of computers down.*
14. ***Sinclair**, who, more than any other company, made computing acceptable to the masses. Okay, most of them had terrible membrane keyboards, and memory adaptor that wobbled, and it took three fingers to get the required command (Shift-2nd Function-Alt etc). and it required a cassette recorder to download programs, and it would typically crash after you had entered one thousand lines of code. However, all of this aside, in the Sinclair Spectrum they found the right computer, for the right time, at the right price. Sometimes success can breed complacency, and so it turned out with the Sinclair QL and the Sinclair C-5 (the electric slipper).*
15. ***Compaq**, for startling growth, that is unlikely to be ever repeated. From zero to one billion dollars in five years. They achieved this growth, not by luck, but by shear superior technology, and with the idea of sharing their developments.*

Other contenders include Hewlett-Packard (for their range of printers), CISCO (for their networking products), Java (for ignoring the make of the computer, and its network, and, well, everything), the Power PC (for trying to head off the PC, at the pass), Dell notebooks (because I've got one), the Intel 80386, the Intel Pentium, Microsoft Visual Basic (for bringing programming to the masses), Microsoft Windows 95, Microsoft Windows NT, and so on. Okay, Windows 95, Windows NT, the 80386 and the Pentium would normally be in the Top 10, but, as Microsoft and Intel are already there, I've left them out. Here's to the Wintel Corporation. We are in their hands. One false move and they will bring their world around themselves. Up to now, Wintel have made all the correct decisions.

When it comes to failures, there are no failures really, and it is easy to be wise after the event. Who really knows what would have happened if the industry had taken another route. So instead of the Top 15 failures, I've listed the following as the Top 15 under-achievers (please forgive me for adding a few of my own, such as DOS and the Intel 8088):

1. ***DOS**, which became the best selling, standard operating systems for IBM PC systems. Unfortunately, it held the computer industry back for at least 10 years. It was text-based, command-oriented, had no graphical user interface, could only access up to 640 KB, it could only use 16 bits at a time, and so on, ... Many with a short memory will say that the PC is easy to use, and intuitive, but they are maybe forgetting how it used to be. With Windows 95 (and to a lesser extent with Windows 3.x), Microsoft made computers much easier to use. From then on, users could actually switch their computer on without have to register for a high degree in Computer Engineering. DOS would have been fine, as it was compatible with all its previous parents, but the problem was MAC OS, which really showed everyone how a user interface should operate. Against this competition, it was no contest. So, what was it? Application software. The PC had application software coming out of its ears.*
2. ***Intel 8088**, which became the standard processor, and thus the standard machine code for PC applications. So why is it in the failures list? Well, like DOS, it's because it was so difficult to use, and was a compromised system. While Amiga and Apple programmers were writing proper programs which used the processor to its maximum extent, PC programs were still using their processor in 'sleepy-mode' (8088-compatiable mode), and could only access a maximum of 1 MB of memory (because of the 20-bit address bus limit for 8088 code). The big problem with the 8088 was that it kept compatibility with its father: the 8080. For this Intel decided to use a segmented memory access, which is fine for small programs, but a nightmare for large programs (basically anything over 64 KB).*
3. ***Alpha** processor, which was DEC's attack on the processor market. It had blistering performance, which blew every other processor out of the water (and still does). It has never been properly exploited, as there is a lack of development tools for it. The Intel Pentium proved that it was a great all-comer and did many things well, and was willing to improve the bits that it was not so good at.*
4. ***Z8000** processor, which was a classic case of being technically superior, but was not compatible with its father, the mighty Z80, and its kissing cousin, the 8080. Few companies have given away such an advantage with a single product. Where are Zilog now? Head buried in the sand, probably.*
5. ***DEC**, who was one of the most innovative companies in the computer industry. They developed a completely new market niche with their minicomputers, but they refused to see, until it was too late, that the microcomputer would have an impact on the computer market. DEC went from a company that made a profit of $1.31 billion in 1988, to a*

company which, in one quarter of 1992, lost $2 billion. Their founder, Ken Olsen, eventually left the company in 1992, and his successor brought sweeping changes. Eventually, though, in 1998 it was one of the new PC companies, Compaq, who would buy DEC. For Compaq, DEC seemed a good match, as DEC had never really created much of a market for PCs, and had concentrated on high-end products, such as Alpha-based workstations and network servers.

6. ***Fairchild Semiconductor**. Few companies have ever generated so many ideas and incubated so many innovative companies, and got so little in return.*
7. ***Xerox**. Many of the ideas in modern computing, such as GUIs and networking, were initiated at Xerox's research facility. Unfortunately, Xerox lacked force to develop them into products, maybe because they reduced Xerox's main market, which was, and still is, very much based on paper.*
8. ***PCjr**, which was another case of incompatibility. IBM lost a whole year in releasing the PCjr, and lost a lot of credibility with their suppliers (many of whom were left with unsold systems) and their competitors (who were given a whole year to catch up with IBM).*
9. ***OS/2**, IBM's attempt to regain the operating system market from Microsoft. It was a compromised operating system, and their development team lacked the freedom of the original IBM PC development. Too many people and too many committees were involved in its development. It thus lacked the freedom, and independence that the Boca Raton development team had. IBM's mainframe divisions were, at the time, a powerful force in IBM, and could easily stall, or veto a product if it had an effect on their profitable market.*
10. ***CP/M**, which many believed would become the standard operating system for microcomputers. Digital Research had an excellent opportunity to make it the standard operating system for the PC, but Microsoft overcame them by making their DOS system much cheaper.*
11. ***MCA**, which was the architecture that IBM tried to move the market with. It failed because Compaq, and several others, went against it, and kept developing the existing architecture.*
12. ***RISC processors**, which were seen as the answer to increased computing power. As Intel has shown, one of the best ways to increase computing speed is to simply ramp up the clock speed, and make the busses faster.*
13. ***Sinclair Research**, who after the success of the ZX81 and the Spectrum, threw it all away by releasing a whole range of under-achievers, such as the QL, and the C5.*
14. ***MSX**, which was meant to be the technology that would standardize computer software on PCs. Unfortunately, it hadn't heard of the new 16-bit processors, and most of all, the IBM PC.*
15. ***Lotus Development**, who totally misjudged the market, by not initially developing their Lotus 1-2-3 spreadsheet for Microsoft Windows. They instead developed it for OS/2, and eventually lost the market leadership to Microsoft Excel. Lotus also missed an excellent opportunity to purchase a large part of Microsoft when they were still a small company. The profits on that purchase would have been gigantic.*

So was/is the IBM PC a success? Of course it was/is. But, for IBM it has been a double-edged sword. It opened up a new and exciting market, and made the company operate in ways that would have never been possible before. Before the IBM PC, their systems sold by themselves, because they were made by IBM. It also considerably reduced their market share. Many questions remained unanswered: 'Would it have been accepted in the same way if it had been a closed system, which had to be licensed from IBM?' 'Would it have been accepted if it had used IBM components rather than other standard components, especially the Intel processors?', 'Would they have succeeded in the operating system market if they had written DOS by themselves?', and so on. Who knows? But, from now on we will refer to those computers based on the x86 architecture as PCs.

Oh, and as an academic I would like to give a special mention to the C programming language, which has given me great heartaches over the years. Oh, yeah, it's used extensively in industry and is extremely useful. It is the programming language that I would automatically use for consultancy work. C is well supported by the major language package developers, and there is a great deal of code available for it. But for teaching programming, it is a complete non-starter. Without going into too much detail, the problems with C are not to do with the basic syntax of the language. It's to do with a thing called pointers. They are the most horrible things imaginable when it comes to teaching programming languages, and they basically 'point' to a location in memory. This is fine, but in most cases, you do not really have to bother about where in memory things are stored. But, C forces you to use them, rather than hiding them away. So, in a C programming class, things go very well until about the 8th week, when pointers are introduced, and then that's it. Oh, and don't get me started on C++.

1.9 DEC

The main rival to IBM before the advent of the PC was DEC (Digital Equipment Corporation). They were formed in 1957, and grew to become the second largest computer company in the world. Their unbelievable growth, and fall, is a lesson for any industry. Brothers Kenneth Olson and Stanley Olson, and Harlan Anderson started DEC on a start-up capital of $70 000 (which was 70% owned by American Research and Development Corporation). This should compare this with the start-up capital of Compaq, which was $10 million). DEC had an initial clear strategy, which was to make cheap computers, which appealed to the specialist scientific and technical market. At the time, IBM had a quasimonopoly, and DEC did not have a chance to compete with them on a like-for-like product range. DEC eventually thrived because they attacked a small market niche with technically superior products. At the time, they could not possibly compete with IBM in the larger commercial market, where IBM had made a considerable investment. So, DEC turned to the scientific and technical market, which required relatively small and configurable products. DEC could not compete with the mighty IBM, who had a solid foundation of great marketing and sales teams. DEC was basically a company of engineers, and they were proud of it. Their main product was the minicomputer, which was much cheaper than mainframes, but had a great deal of power, and could be easily configured and managed by a small group.

The big winner for DEC was the PDP (Programmed Data Processor) series, which become the foundation of many scientific and engineering groups. No research group or industrial company was complete without a PDP computer. By today's standards, there was more power in a pocket calculator, as there was in the PDP-8. It was also relatively large, weighing 250 pounds, and came in a rack-mounted unit which was over 6 feet tall. However, the PDP range was much cheaper than IBM mainframes. For example, the PDP-1 sold for $120 000, while the comparable IBM computer cost millions. The PDP range also introduced computing to many young minds. Two exceptional minds, Bill Gates and Paul Allen, cut their teeth on a DEC PDP-8, where they wrote programs to support the BASIC programming language.

The next great winner for DEC was the VAX (Virtual Address eXtension) computer which cost billions to develop, but was a great technical and commercial success. It covered the complete range of computer hardware from basic terminal interface up to large mainframe computers. For the first time, DEC produced every part of the computer system: the operating system, the hardware and the software. One of the great successes of the VAX range was the VMS operating system (produced by David Culter). It allowed computer programmers to create programs which had more memory than the computer actually had (a virtual memory), and allowed several programs to run at the same time (multitasking). After the success of VMS, David Culter eventually went on develop a RISC operating system, but DEC management cancelled the project. After this, he left DEC in disgust and went to Microsoft to lead the development of the Windows NT operating system. Microsoft and Intel have strong recruitment policies, and often hire the best brains in the computer industry.

In these days of networked computers, it is difficult to believe, but, at the time, the VAX range was a radical concept. Before VAX computers, DEC, with their PDP range, was never touched $7.6 billion. Unfortunately, DEC's bubble burst for two reasons. The first was the really a threat to IBM's core market in mainframes. However, the VAX range was. The future looked destined to be DEC's, and not IBM's. In 1986, their sales reached $2 billion, and soon recession of the 1990s. It was a situation that many companies had difficulty coping with, but it could not be avoided. The only reason was one that could have been avoided if DEC had realized the changing market, and the power of the new 16-bit microprocessors. It was basically the IBM PC which eventually beat IBM's mainframes and DEC's minicomputers on performance, at a fraction of the price, from whichever company you wanted. DEC actually, in 1979, had the opportunity to enter into the PC market when they allowed Heath-Kit to sell the PDP-11 minicomputer in kit form. At the time, DEC believed there was more profit to be made with corporate clients, thus didn't really believe there was a great market for PCs. Ken Olsen believed that PCs were a passing fad that would never really evolve into proper computers. Many computers at the time were bought, played with, and then put in the cupboard, never to be used again. The great advantages with personal computers were that they were designed for individuals, whereas minicomputers where designed for businesses.

DEC struggled though the 1990s and could never regain their dominance. As with IBM's mainframe business, they relied on their existing customer base buying their new products. A well-known brand name, with its associated image is extremely important for corporate companies when they buy computers. Most companies believe that brand names such as DEC (as they were), IBM, Compaq and Dell are associated with reliable and well-built products. Companies buying the brand name kept DEC's brand alive in many cases. As many companies used DEC equipment, DEC in the 1990s was still a well-respected brand name. They showed that they could innovate and lead the market with one of the most respected RISC processors ever made: the Alpha. This had a blistering performance and is still used in many workstations. It would take several years before Intel could even match the power of the Alpha device. Unfortunately, DEC failed to support the processor with the required software. DEC, as IBM had, had always seen itself as a computer hardware company, and not a software one.

So from the 1980s to the 1990s, DEC had gone from being a fast-moving, innovative and enterprising company, to one which was entrenched in its existing product lines. As PCs grew in strength, DEC kept developing their minicomputers (as IBM was doing with their mainframes). DEC's other main problem was that, like IBM, they did everything, from writing software, design and making the processors, developing hard disk drives, and so on. This made them vulnerable from specialist companies who could beat DEC in each of the areas. A focused, specialist company will typically innovate faster than a large, generalized company. They also failed to become involved in alliances. This was because DEC felt that they could turn the market in whichever way they wanted, thus they did not need alliances. At present, only Microsoft and Intel can claim to not requiring alliance pacts. All other companies typically need to become involved in alliances to get their non-Intel and non-Microsoft products accepted in the market.

DEC went from a company that made a profit of $1.31 billion in 1988, to a company that, in one quarter of 1992, lost $2 billion. Olsen eventually left the company in 1992, and his successor brought sweeping changes. Eventually, though, in 1998 it was one of the new PC companies, Compaq, who would buy DEC. For Compaq, DEC seemed a good match, as DEC had never really created much of a market for PCs, and had concentrated on high-end products, such as Alpha-based workstations and network servers.

Unlike IBM, DEC did not pull the walls down around themselves. They had found an excellent market share and were coping well. If not for the advent of the PC, DEC would probably be the market leader by now. Their VAX range would have probably evolved to include a closed-system personal computer in which DEC could have held control of (as IBM would have done). However, the open-system approach of the PC spelt disaster for both IBM and DEC.

1.10 Open .v. closed system

In 1985, Apple was having difficult times. The sales of the Macintosh were not as great as expected, and the Apple II was facing a great deal of competition from other manufacturers. Many people at the time, including Bill Gates, were advising Apple to open-up the market for Macintosh computers by allowing other manufacturers build their own systems, under strict license arrangements. Bill Gates had advised them that they should tie up with companies such as HP and AT&T. However, Apple held onto both their Mac operating system, and their hardware, which they believed were totally intertwined. A Mac could not exist without both its operating system and its hardware. Rather than open the market up, Apple decided to trample cloners, especially in software cloners. Apple's first target was Digital Research, who had developed GEM for the PC. Digital Research believed that they had borrowed the look-and-feel of the Mac operating system, but not the actual technology. Apple immediately shot GEM out of the water when Apple's lawyers, in 1985, visited Digital Research and threatened them with court action. At the time, IBM had been keen to license GEM for their own products, but they IBM were frightened away over the fear of litigation, and that was the end of GEM.

Apple then turned to Microsoft to head off their attempt at producing a GUI. Bill Gates, though, had much greater strength than Digital Research against Apple. His main point was that the true originator of the GUI was Xerox. Thus, for its Microsoft Windows, it was Xerox's ideas that were being used, and not Apple's. Bill Gates, though, had another trump card: If Apple were going to stop Microsoft from producing Windows then Microsoft would stop producing application software for the Macintosh. Apple knew that they need Microsoft more than Microsoft needed Apple. In the face of a lack of investment in their application software, Apple signed a contact with Microsoft which stated that Microsoft would:

'have a non-exclusive, worldwide, royalty-free, perpetual, nontransferable license to use derivate works in ***present and future software programs****, and to license them to and through third parties for*

use in their software programs'

which basically gave Microsoft carte blanche for all future versions of their software, and were quite free to borrow which ever features they wanted. John Scully at Apple signed it, and gave away one of the most lucrative markets in history. Basically, Apple were buying peace with Microsoft, but it was peace with a long-term cost.

> "It is practically impossible to teach good programming style to students that have had prior exposure to BASIC; as potential programmers they are mentally mutilated beyond hope of regeneration."
> "The use of COBOL cripples the mind; its teaching should, therefore, be regarded as a criminal offense."
>
> *Diikstra*

1.11 RIP, Sinclair Research

The biggest failure of 1984 was Sinclair Research, after many years of success. They released the 16/32-bit QL microcomputer, which was their first attempt at the business market. It cost $500, and used the Motorola 68008 microprocessor, had 128KB RAM, two built-in tape drives, and multitasking ROM-based operating system. It was their first attempt at the business marked, but unfortunately, the IBM PC now dominated that market.

In the following year, Sinclair Research, was struggling with disappointing sales of the QL, but still buoyed by the Sinclair Spectrum. The biggest disaster for Sinclair, though, was the C-5, which was to be a revolutionary electric vehicle. Unfortunately, it looked less like a motor vehicle, and more like a large plastic slipper. It was also powered by a washing machine motor, giving it a top speed of 15mph. Along with this it was heavy and short-ranged. Within a few short months, the production of the C-5 was stopped, and Sinclair faced large financial losses (£7 million). In the face of these losses, Sinclair sold his name and the rights to his computers to Amstrad. Clive Sinclair then went on to form a new company, Cambridge Computers. which created a laptop, which was not PC-compatible and based on the Z80-based machine and had an LCD screen. It was reasonability successful, but no-where near as successful as the ZX-series of computers.

The C-5 was not Clive Sinclair's worse failure; it was the Zike (ultra-light electric bicycle), which only sold 2,000 units (as opposed to 17,000 C-5s). The Zeta (Zero-Emission Transport Accessory) has sold better with more than 15,000 units being sold. However, Clive Sinclair will always be remembered for classic computers, which were so popular, you had to wait months to get one.

1.12 How to miss a market opportunity

"This 'telephone' has too many shortcomings to be seriously considered as a means of communication. The device is inherently of no value to us."
– Western Union internal memo, 1876.

"So we went to Atari and said, 'Hey, we've got this amazing thing, even built with some of your parts, and what do you think about funding us? Or we'll give it to you. We just want to do it. Pay our salary, we'll come work for you.' And they said, 'No.' So then we went to Hewlett-Packard, and they said, 'Hey, we don't need you. You haven't got through college yet.'"
– Apple Computer Inc. founder Steve Jobs on attempts to get Atari and H-P interested in his and Steve Wozniak's personal computer.

2 Busses, Interrupts and PC Systems

2.1 Busses

The part that makes computers operate and allows devices to be easily plugged in is the computer bus, which allows the orderly flow of data between one device and another. The PC, and other computer systems, has an amazing number of different types of interfaces and bus systems, these include the PC bus, ISA bus, PCI bus, CAN bus, AGP bus, games port, parallel port, serial port, and so on.

The main elements of a basic computer system are a central processing unit (or microprocessor), memory, and I/O interfacing circuitry. These connect by means of three main buses: the address bus, the control bus and the data bus. A bus is a collection of common electrical connections grouped by a single name. Figure 2.1 shows a basic system. External devices such as a keyboard, display, disk drives can connect directly onto the data, address and control buses or through the I/O interface circuitry.

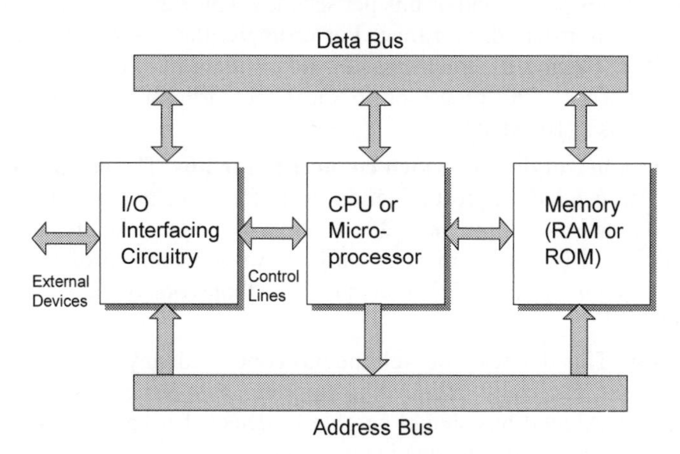

Figure 2.1 Block diagram of a simple computer system

Electronic memory consists of RAM (random access memory) and ROM (read only memory). ROM stores permanent binary information, whereas RAM is a non-permanent memory and loses its contents on a loss of power. Applications of this type of memory include running programs and storing temporary information. RAM is normally made up of either DRAM (Dynamic RAM) or SRAM (Static RAM). DRAM uses a single capacitor and a transistor to store a single bit of data, whereas SRAM uses six transistors, arranged as a flip-flop device, to store a single bit of data. DRAM has the advantage that more memory can be crammed onto a microchip (as only one transistor is required for each bit stored). DRAM, though, has two major disadvantages: it is relatively slow (because of the charging and discharging of the storage capacitors) and it requires that the complete contents of its memory be refreshed with power many times a second (because the tiny capacitors loose their charge over a short time). This power refresh is thus wasteful of electrical power and leads to heat dissipation.

The microprocessor is the main controller of the computer. It only understands binary information and operates on a series of binary commands known as machine code. It fetches binary instructions from memory, decodes these instructions into a series of simple actions and carries out the actions in a sequence of steps. A system clock synchronises these steps.

To access a location in memory the microprocessor puts the address of the location on the address bus. The contents at this address are then placed on the data bus and the microprocessor reads the data from the data bus. To store data in memory the microprocessor places the data on the data bus. The address of the location in memory is then put on the address bus and data is read from the data bus into the memory address location.

The classification of a microprocessor relates to the maximum number of bits it can process at a time, that is their word length. The evolution has gone from 4-bit, 8-bit, 16-bit, 32-bit and to 64-bit architectures.

2.1.1 Bus specification

The basic specification of a computer can be determined by analysing the performance of the busses within the system. Each bus performs a specific function and is suited to the devices that connect to it. The basic specifications for busses include:

- **Data rate** (in bytes per second or bits per second). This defines the maximum amount of data that can be transferred, at a time. For example, the ISA bus has a maximum data rate of 16MB/s, Gigabit Ethernet has a maximum data rate of 125MB/s, and the local bus which connects a PC processor to local memory can have a data rate of over 800MB/s (64 bits at 100MHz).
- **Maximum number of devices which connect to the bus**. The number of devices which connect to a bus can have a great effect on its performance as they all provide an electrical loading on the bus and the more that connect to the bus, the greater the overhead of bus arbitration will be. Standard SCSI only allows a maximum of seven devices to be connected to the bus, whereas Ethernet can allow thousands of devices to connect to the bus.
- **Bus reliability**. This defines how well the bus copes with any errors which occur on the bus. Some busses, especially in industrial environments, can be susceptible to externally generated noise. A good bus should be able to detect if it has received data which has been corrupted by noise (or was sent incorrectly).
- **Data robustness**. This is the ability of the bus to react to faults within the bus or from the malfunctioning of connected devices. Busses such as the CAN bus can isolate incorrectly operating devices.
- **Electrical/physical robustness**. This is the ability of the bus to cope with electrical faults, especially due to short-circuits and power surges. Problems can also be caused by open circuit electrical connections, although these tend not to cause long term damage to the bus. The physical robustness of a bus is also important, especially in industrial or safety critical situations.
- **Electrical characteristics**. This involves the basic electrical parameters of the bus, such as the range of voltage levels used, electrical current ranges, short-circuit protection sys-

tem, capacitance and impedance of cables, cross-talk (the amount of interference between local signal transmissions), and so on.

- **Ease-of-connection**. This includes the availability of cables and connectors, and how easy it is to add and remove devices from the bus. Some busses allow devices to be added or removed while the bus is in operation (hot pluggable). A good example of a hot-pluggable bus, which is easy to connect to, is the USB.
- **Communications overhead**. This is a measure of the amount of data that is added to the original data, so that it can be sent in a reliable way. Local, fast busses normally have a minimum of overhead, whereas remote, networked busses have a relatively large overhead on the transmitted data.
- **Bus controller topology**. This relates to the method that is used to control the flow of data around the bus. Some busses, such as SCSI, require a dedicated bus controller which is involved in all of the data transfers, whereas the PCI bus can operate with one or more bus controller devices taking control of the bus. Other busses, such as Ethernet, have a distributed topology where any device can take control of the bus.
- **Software interfacing**. This defines how easy it is to interface to the bus with software, especially when using standard interface protocols, such as TCP/IP or MODBUS.
- **Cable and connectors**. This defines the range of cables and connectors that can be used with the bus. There is a wide range of cables available, such as ribbon cables (which are light and are useful inside computer systems), twisted-pair cables (which are easy to connect to and are useful in minimising cross-talk between transmitted signals) and fibre optic cables (which provide a high capacity communications link and minimise cross talk between transmitted signals). For example, Ethernet can use BNC connectors with coaxial cables, RJ-45 connectors with twisted-pair cables and SNA connectors with fibre optic cables.
- **Standardisation of the bus**. Most busses must comply with a given international standard, which allows hardware and software to interconnect in a standard form. There are normally standards for the electrical/mechanical interface, the logical operation of the bus, and its interface to software. For example, the IEEE has defined most of the Ethernet standard (especially IEEE 802.3), and the EIA have defined the RS-232 standard. International standard agencies, such as the IEEE, ISO, ANSI and EIA, provide a more secure standard than a vendor-led standard.
- **Power supply modes**. Some busses allow power saving modes, where devices can power themselves down and be powered up by an event on the bus. This is particularly useful with devices that have a limited power supply, such as being battery supplied.

2.1.2 Bus components

Devices connect to each other within a computer using a bus. The bus can either be an internal bus (such as the IDE bus which connects to hard disks and CD-ROM drives within a PC) or an external bus (such as the USB which can connect to a number of external devices, typically to scanners, joypads and printers). Busses typically have a number of basic components: a data bus, an optional address bus, control lines and handshaking lines, as illustrated in Figure 2.2. Other lines, such as clock rates and power supply lines are not normally displayed when discussing the logical operation of the bus. If there is no address bus, or no control and handshaking lines, then the data bus can be used to provide addressing, control and handshaking. This is typical in serial communications, and helps to reduce the number of connections in the bus, although will generally slow down the communications.

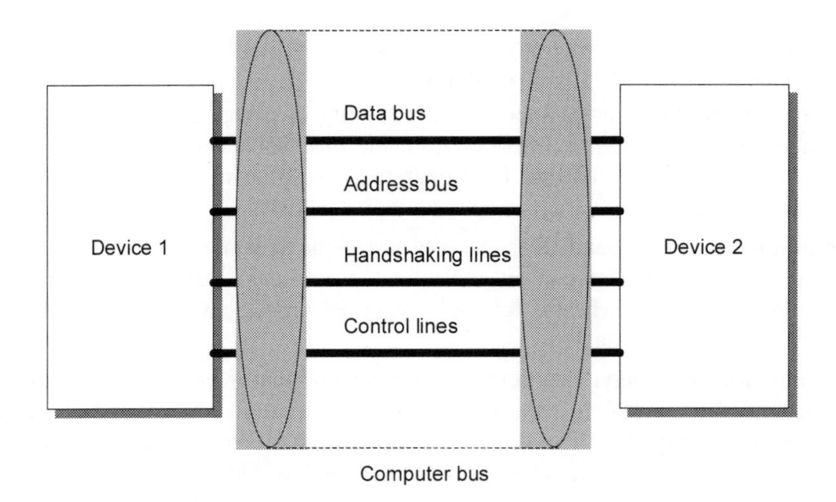

Figure 2.2 Model of a computer bus

Data bus

The data bus is responsible for passing data from one device and another. This data is either passed in a serial manner (one bit at a time) or in parallel (several bits at a time). In a parallel data bus, the bits are normally passed in a multiple of eight bits at a time. Typical parallel data busses are 8 bits, 16 bits, 32 bits, 64 bits or 128 bits wide.

The bus size defines the maximum size of the bus, but the bus can be used to transmit any number of bits which is less than the maximum size. For example, a 32-bit bus can be used to transmit eight bits, 16 bits or 32 bits at a time. Most modern computer systems use a 64-bit address bus, although the software which runs on the computer only uses a maximum of 32 bits at a time (known as 32-bit software).

Parallel busses are normally faster than serial busses (as they can transmit more bits in a single operation), but require many more lines (thus requiring more wires in the cable). A parallel data bus normally requires extra data handshaking lines to synchronise the flow of data between devices. Serial data transmission normally uses a start and end bit sequence to define the start and end of transmission. Figure 2.3 illustrates the differences between serial and parallel data busses. Parallel busses are typically used for local busses, or where there are no problems with cables with a relatively large number of wires. Typically, parallel busses are SCSI and IDE which are used to connect to hard disk drives, and typical serial busses are RS-232, and the USB.

Serial communications can operate at very high transmission rates; the main limiting factor is the transmission channel and the transmitter/receiver electronics. Gigabit Ethernet, for example, uses a transmission rate of 1 Gbps (125 MB/s) over high-quality twisted-pair copper cables, or over fibre optic cables (although this is a theoretical rate as more than one bit is sent at a time). For a 32-bit parallel bus, this would require a clocking rate of only 31.25 MHz (which requires much lower quality connectors and cables than the equivalent serial interface).

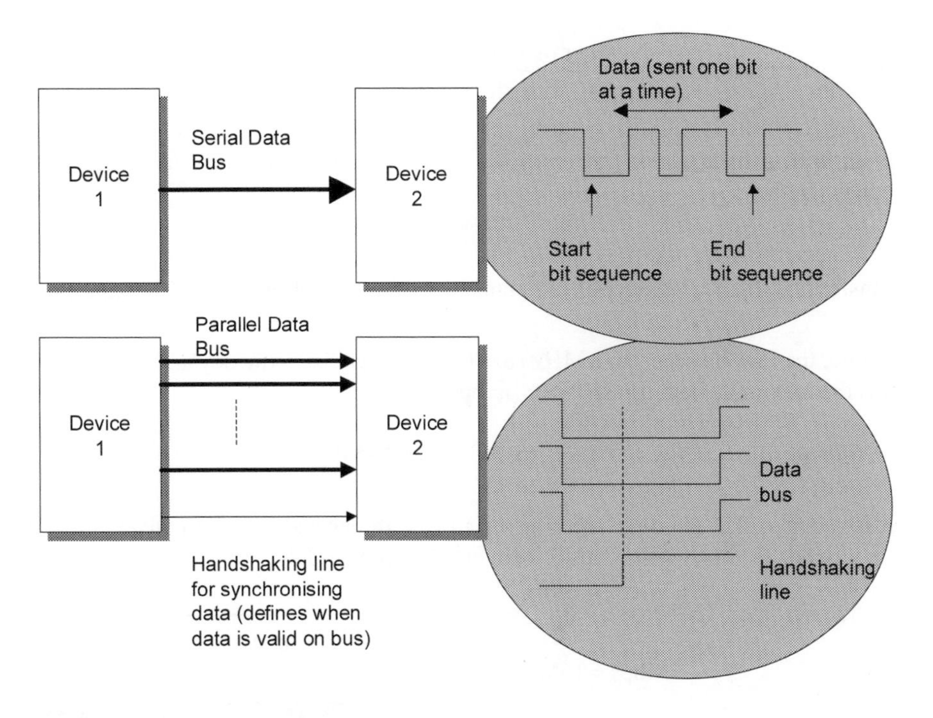

Figure 2.3 Serial/parallel data busses

Data transfer rates

The amount of data that a system can transfer at a time is normally defined either in bits per second (bps) or bytes per second (B/s). The more bytes (or bits) that can be transferred the faster the transfer will be. Typically serial busses are defined in bps, whereas parallel busses use B/s.

The transfer of the data occurs are regular intervals, which is defined by the period of the transfer clock. This period is either defined as a time interval (in seconds), or as a frequency (in Hz). For example, if a clock operates at a rate of 1 000 000 cycles per second, its frequency is 1 MHz, and its time interval will be one millionth of a second (1×10^{-6} s).

In general, if f is the clock frequency (in Hz), then the clock period (in seconds) will be

$$T = \frac{1}{f}\text{sec}$$

Conversion from clock frequency to clock time interval

For example, if the clock frequency is 8 MHz, then the clock period will be:

$$T = \frac{1}{8\times10^{6}} = 0.000000125\,\text{sec}$$
$$= 0.125\mu\text{s}$$

Example of a calculation of clock time interval from clock frequency

The data transfer rate (in bits/second) is defined as:

$$\text{Data transfer rate (bps)} = \frac{\text{Number of bits transmitted per operation (bits)}}{\text{Transfer time per operation (s)}}$$

If operated with a fixed clock frequency for each operation then the data transfer rate (in bits/second) will be

$$\text{Data transfer rate (bps)} = \text{Number of bits transmitted per operation (bits)} \times \text{Clocking rate (Hz)}$$

For example, the ISA bus uses an 8 MHz (8×10^6 Hz) clocking frequency and has a 16-bit data bus. Thus the maximum data transfer rate (in bps) will be:

$$\text{Data transfer rate} = 16\times8\times10^6 = 128\times10^6 \text{ b/s} = 128\text{Mbps}$$

Often it is required that the data rate is given in B/s, rather and bps. To convert from bps to B/s, eight divides the bps value. Thus to convert 128Mbps to B/s

$$\text{Data transfer rate} = 128\text{Mbps}$$
$$= \frac{128}{8}\text{Mbps} = 16\text{MB/s}$$

Example conversion from bps to B/s

For serial communication, if the time to transmit a single bit is 104.167 μs then the maximum data rate will be

$$\text{Data transfer rate} = \frac{1}{104.167\times10^{-6}} = 9600 \text{ bps}$$

Example conversion to bps for a serial transmission with a given transfer time interval

2.1.3 Address bus

The address bus is responsible for identifying the location into which the data is to be passed into. Each location in memory typically contains a single byte (8 bits), but could also be arranged as words (16 bits), or long words (32 bits). Byte-oriented memory is the most flexible as it also enables access to any multiple of eight bits. The size of the address bus thus indicates the maximum addressable number of bytes. Table 2.3 shows the size of addressable memory for a given address bus size. The number of addressable bytes is given by:

$$\text{Addressable locations} = 2^n \ bytes$$

Addressable locations for a given address bus size

where n is the number of bits in the address bus. For example:

- A 1-bit address bus can address up to two locations (that is 0 and 1).
- A 2-bit address bus can address 2^2 or 4 locations (that is 00, 01, 10 and 11).
- A 20-bit address bus can address up to 2^{20} addresses (1 MB).
- A 32-bit address bus can address up to 2^{32} addresses (4 GB).

The units used for computers for defining memory are B (Bytes), kB (kiloBytes), MB (megaBytes) and GB (gigabytes). These are defined as:

- **KB** (kiloByte). This is defined as 2^{10} bytes, which is 1024 B.
- **MB** (megaByte). This is defined as 2^{20} bytes, which is 1024 kB, or 1 048 576 bytes.
- **GB** (gigaByte). This is defined as 2^{30} bytes, which is 1024 MB, or 1 048 576 kB, or 1 073 741 824 B.

Table 2.1 gives a table with addressable space for given address bus sizes.

Table 2.1 Addressable memory (in bytes) related to address bus size

Address bus size	*Addressable memory (bytes)*	*Address bus size*	*Addressable memory (bytes)*
1	2	15	32 K
2	4	16	64 K
3	8	17	128 K
4	16	18	256 K
5	32	19	512 K
6	64	20	1 M†
7	128	21	2 M
8	256	22	4 M
9	512	23	8 M
10	1K*	24	16 M
11	2K	25	32 M
12	4K	26	64 M
13	8K	32	4G‡
14	16K	64	16GG

* 1 K represents 1024 † 1 M represents 1 048 576 (1024 K)
‡ 1 G represents 1 073 741 824 (1024 M)

Data handshaking

Handshaking lines are also required to allow the orderly flow of data. This is illustrated in Figure 2.4. Normally there are several different types of busses which connect to the system, these different busses are interfaced to with a bridge, which provides for the conversion between one type of bus and another. Sometimes devices connect directly onto the processor's bus; this is called a local bus, and is used to provide a fast interface with direct access without any conversions.

The most basic type of handshaking has two lines:

- Sending identification line – this identifies that a device is ready to send data.
- Receiving identification line – this identifies that device is a device is ready to receive data, or not.

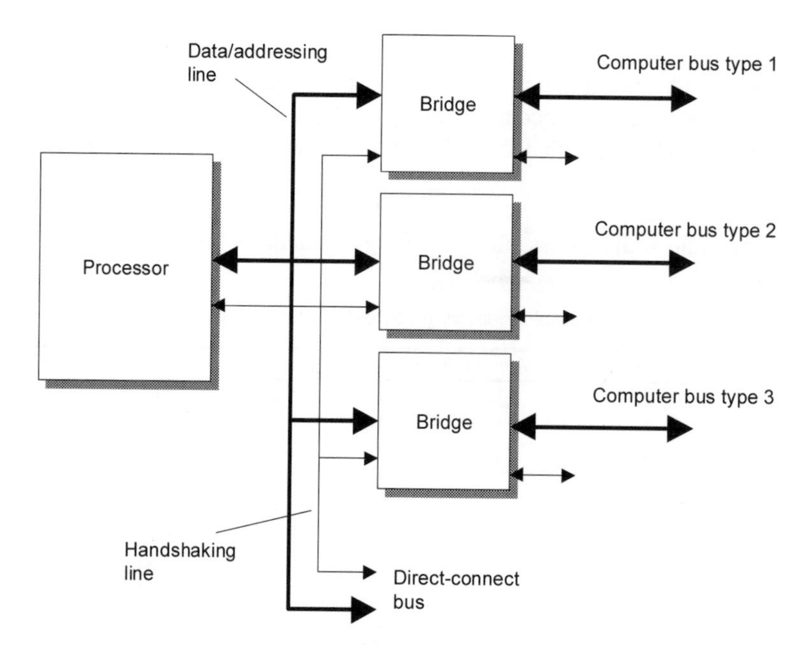

Figure 2.4 Computer bus connections

Figure 2.5 shows a simple form of handshaking of data, from Device 1 to Device 2. The sending status is identified by READY? and the receiving status by STATUS. Normally an event is identified by a signal line moving from one state to another, this is described as edge-triggered (rather than level-triggered where the actual level of the signal identifies its state). In the example in Figure 2.5, initially Device 1 puts data on the data bus, and identifies that it is ready to send data by changing the READY? line from a LOW to a HIGH level. Device 2 then identifies that it is reading the data by changing its STATUS line from a LOW to a HIGH. Next it identifies that it has read the data by changing the STATUS line from a HIGH to a LOW. Device 1 can then put new data on the data bus and start the cycle again by changing the READY? line from a LOW to a HIGH.

This type of communication only allows communication in one direction (from Device 1 to Device 2) and is know as simplex communications. The main types of communication are:

- **Simplex communication**. Only one device can communicate with the other, and thus only requires handshaking lines for one direction.
- **Half-duplex communication**. This allows communications from one device to the other, in any direction, and thus requires handshaking lines for either direction.
- **Full-duplex communications**. This allows communication from one device to another, in either direction, at the same time. A good example of this is in a telephone system, where a caller can send and receive at the same time. This requires separate transmit and receive data lines, and separate handshaking lines for either direction.

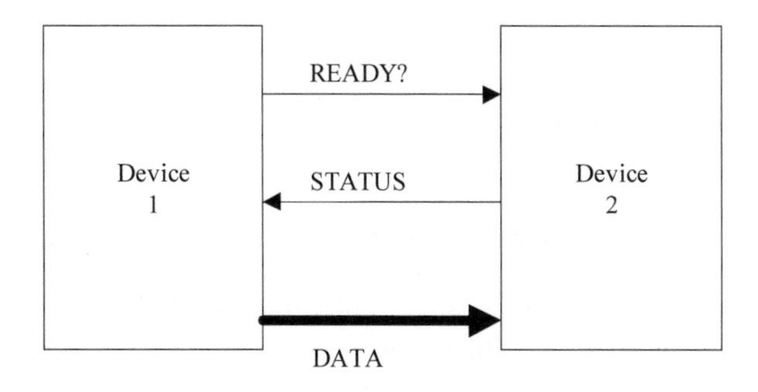

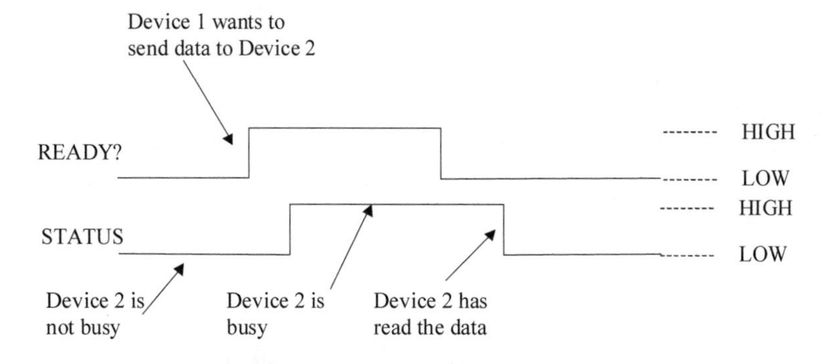

Figure 2.5 Simple handshaking of data

Control lines

Control lines define the operation of the data transaction, such as:

- Data flow direction – this identifies that data is either being read from a device or written to a device.
- Memory addressing type – this is typically either by identifying that the address access is direct memory accessing or indirect memory access. This identifies that the address on the bus is either a real memory location or is an address tag.
- Device arbitration – this identifies which device has control of the bus, and is typically used when there are many devices connected to a common bus, and any of the devices are allowed to communicate with any other of the devices on the bus.

2.1.4 Cables

The cable type used to transmit the data over the bus depends on several parameters, including:

- The signal bandwidth.
- The reliability of the cable.
- The maximum length between nodes.
- The possibility of electrical hazards.
- Power loss in the cables.

- Tolerance to harsh conditions.
- Expense and general availability of the cable.
- Ease of connection and maintenance.
- Ease of running cables, and so on.

The main types of cables used are standard copper cable, unshielded twisted-pair copper (UTP), shielded twisted-pair cable (STP), coaxial and fibre optic. Twisted-pair and coaxial cables transmit electric signals, whereas fibre optic cables transmit light pulses. Twisted-pair cables are not shielded and thus interfere with nearby cables. Public telephone lines generally use twisted-pair cables. In LANs they are generally used up to bit rates of 10 Mbps and with maximum lengths of 100 m.

Coaxial cable has a grounded metal sheath around the signal conductor. This limits the amount of interference between cables and thus allows higher data rates. Typically, they are used at bit rates of 100 Mbps for maximum lengths of 1 km.

The highest specification of the three cables is fibre optic. This type of cable allows extremely high bit rates over long distances. Fibre optic cables do not interfere with nearby cables and give greater security, give more protection from electrical damage by external equipment and greater resistance to harsh environments; they are also safer in hazardous environments.

Cable characteristics

The main characteristics of cables are attenuation, cross-talk and characteristic impedance. Attenuation defines the reduction in the signal strength at a given frequency for a defined distance. It is normally defined in dB/100 m, which is the attenuation (in dB) for 100 m. An attenuation of 3 dB/100 m gives a signal voltage reduction of 0.5 for every 100 m. Table 2.2 lists some attenuation rates and equivalent voltage ratios; they are illustrated in Figure 2.6. Attenuation is given by

$$\text{Attenuation} = 20\log_{10}\left(\frac{V_{in}}{V_{out}}\right) \text{ dB}$$

Calculation of attenuation from input and output voltages

For example if the input voltage to a cable is 10 V and the voltage at the other end is only 7 V, then the attenuation is calculated as

$$\text{Attenuation} = 20\log_{10}\left(\frac{10}{7}\right) = 3.1 \text{ dB}$$

Coaxial cables have an inner core separated from an outer shield by a dielectric. They have an accurate characteristic impedance (which reduces reflections), and because they are shielded they have very low cross-talk levels. They tend also to have very low attenuation, (such as 1.2 dB at 4 MHz), with a relatively flat response. UTPs (unshielded twisted-pair cables) have either solid cores (for long cable runs) or are stranded patch cables (for shorts run, such as connecting to workstations, patch panels, and so on). Solid cables should not be flexed, bent or twisted repeatedly, whereas stranded cable can be flexed without damaging the cable. Coaxial cables use BNC connectors while UTP cables use either the RJ-11 (small connector which is used to connect the handset to the telephone) or the RJ-45 (larger connector which is typically used in networked applications to connect a network adapter to a network hub).

The characteristic impedance of a cable and its connectors are important, as all parts of the transmission system need to be matched to the same impedance. This impedance is normally classified as the characteristic impedance of the cable. Any differences in the matching result in a reduction of signal power and produce signal reflections (or ghosting).

Cross-talk is important as it defines the amount of signal that crosses from one signal path to another. This causes some of the transmitted signal to be received back where it was transmitted. Capacitance (pF/100 m) defines the amount of distortion in the signal caused by each signal pair. The lower the capacitance value, the lower the distortion.

Table 2.2 Attenuation rates as a ratio

dB	*Ratio*	*dB*	*Ratio*	*dB*	*Ratio*
0	1.000	10	0.316	60	0.001
1	0.891	15	0.178	65	0.000 6
2	0.794	20	0.100	70	0.000 3
3	0.708	25	0.056	75	0.000 2
4	0.631	30	0.032	80	0.000 1
5	0.562	35	0.018	85	0.000 06
6	0.501	40	0.010	90	0.000 03
7	0.447	45	0.005 6	95	0.000 02
8	0.398	50	0.003 2	100	0.000 01
9	0.355	55	0.001 8		

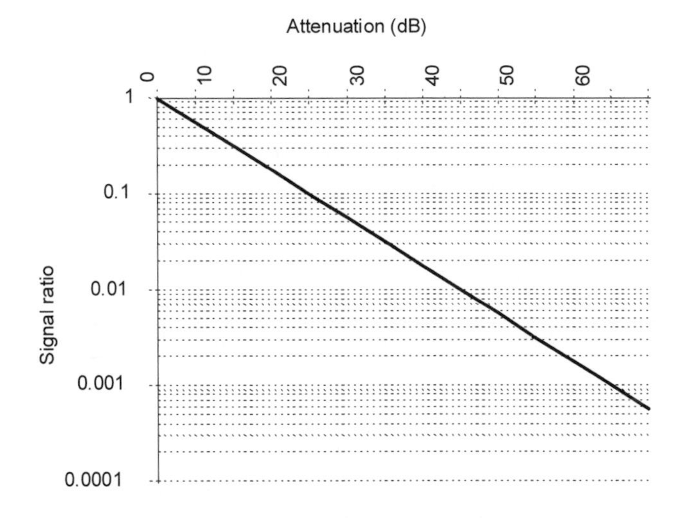

Figure 2.6 Signal ratio related to attenuation

Typical cables used are:

- Coaxial cable – cables with an inner core and a conducting shield having characteristic impedance of either 75 Ω for TV signal or 50 Ω for other types.
- Cat-3 UTP cable – level 3 cables have non-twisted-pair cores with a characteristic impedance of 100 Ω (±15 Ω) and a capacitance of 59 pF/m. Conductor resistance is around 9.2 Ω/100 m.
- Cat-5 UTP cable – level 5 cables have twisted-pair cores with a characteristic impedance of 100 Ω (±15 Ω) and a capacitance of 45.9 pF/m. Conductor resistance is around 9 Ω/100 m.

The Electrical Industries Association (EIA) has defined five main types of cables. Levels 1 and 2 are used for voice and low-speed communications (up to 4 Mbps). Level 3 is designed for LAN data transmission up to 16 Mbps and level 4 is designed for speeds up to 20 Mbps. Level 5 cables, have the highest specification of the UTP cables and allow data speeds of up to 100 Mbps. The main EIA specification on these types of cables is EIA/TIA568 and the ISO standard is ISO/IEC11801.

Table 2.3 gives typical attenuation rates (dB/100 m) for Cat-3, Cat-4 and Cat-5 cables. Notice that the attenuation rates for Cat-4 and Cat-5 are approximately the same. These two types of cable have lower attenuation rates than equivalent Cat-3 cables. Notice that the attenuation of the cable increases as the frequency increases. This is due to several factors, such as the skin effect, where the electrical current in the conductors becomes concentrated around the outside of the conductor, and the fact that the insulation (or dielectric) between the conductors actually starts to conduct as the frequency increases.

The Cat-3 cable produces considerable attenuation over a distance of 100 m. The table shows that the signal ratio of the output to the input at 1 MHz, will be 0.76 (2.39 dB), then, at 4 MHz it is 0.55 (5.24 dB), until at 16 MHz it is 0.26. This differing attenuation at different frequencies produces not just a reduction in the signal strength but also distorts the signal (because each frequency is affected differently by the cable. Cat-4 and Cat-5 cables also produce distortion but their effects will be lessened because attenuation characteristics have flatter shapes.

Table 2.4 gives typical near-end cross-talk rates (dB/100 m) for Cat-3, Cat-4 and Cat-5 cables. The higher the figure, the smaller the cross-talk. Notice that Cat-3 cables have the most cross-talk and Cat-5 have the least, for any given frequency. Notice also that the cross talk increases as the frequency of the signal increases. Thus, high-frequency signals have more cross-talk than lower-frequency signals.

Table 2.3 Attenuation rates (dB/100 m) for Cat-3, Cat-4 and Cat-5 cable

Frequency (MHz)	*Attenuation rate (dB/100 m)*		
	Cat-3	*Cat-4*	*Cat-5*
1	2.39	1.96	2.63
4	5.24	3.93	4.26
10	8.85	6.56	6.56
16	11.8	8.2	8.2

Table 2.4 Near-end cross-talk (dB/100 m) for Cat-3, Cat-4 and Cat-5 cable

Frequency (MHz)	*Near end cross-talk (dB/100 m)*		
	Cat-3	*Cat-4*	*Cat-5*
1	13.45	18.36	21.65
4	10.49	15.41	18.04
10	8.52	13.45	15.41
16	7.54	12.46	14.17

2.2 Interrupts

An interrupt allows a program or an external device to interrupt the execution of a program. The generation of an interrupt can occur by hardware (hardware interrupt) or software (software interrupt). When an interrupt occurs an interrupt service routine (ISR) is called. For a hardware interrupt the ISR then communicates with the device and processes any data. When it has finished the program execution returns to the original program. A software interrupt causes the program to interrupt its execution and goes to an interrupt service routine. Typical software interrupts include reading a key from the keyboard, outputting text to the screen and reading the current date and time. The operating system must respond to interrupts from external devices, as illustrated in Figure 2.7.

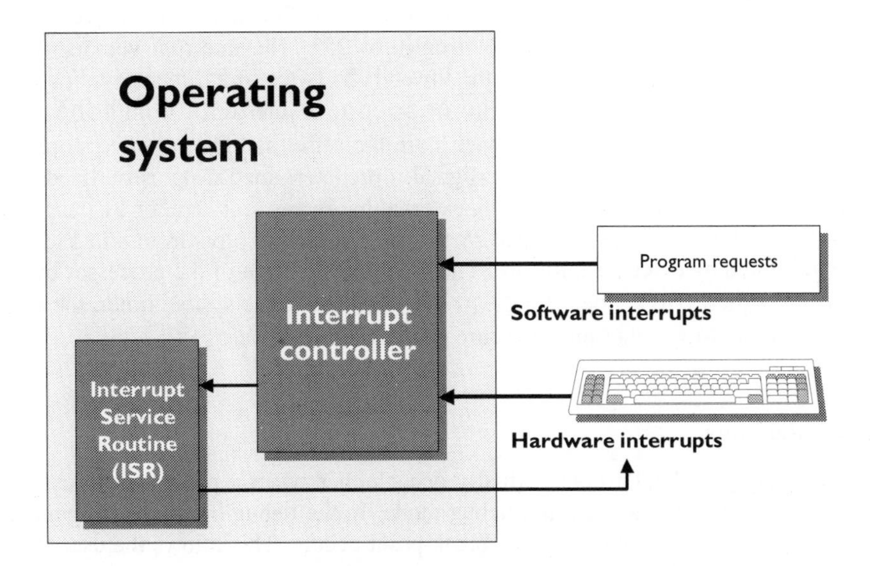

Figure 2.7 Interrupt service routine

2.2.1 Software interrupts

BIOS and the operating system

The Basic Input/Output System (BIOS) communicates directly with the hardware of the computer. It consists of a set of programs which interface with devices such as keyboards, displays, printers, serial ports and disk drives. These programs allow the user to write application programs that contain calls to these functions, without having to worry about controlling them or which type of equipment is being used. Without BIOS, the computer system would simply consist of a bundle of wires and electronic devices.

There are two main parts to BIOS. The first is the part permanently stored in a ROM (the ROM BIOS). It is this part that starts the computer (or bootstap) and contains programs which communicate with resident devices. The second stage is loaded when the operating system is started. This part is non-permanent.

An operating system allows the user to access the hardware in an easy-to-use manner. It accepts commands from the keyboard and displays them to the monitor. The Disk Operating System, or DOS, gained its name from its original purpose of providing a controller for the computer to access its disk drives. The language of DOS consists of a set of commands which are entered directly by the user and are interpreted to perform file management tasks, program execution and system configuration. It makes calls to BIOS to execute these. The main functions of DOS are to run programs, copy and remove files, create directories, move within a directory structure and to list files. Microsoft Windows calls BIOS programs directly.

Interrupt vectors

Interrupt vectors are addresses which inform the interrupt handler as to where to find the ISR. All interrupts are assigned a number from 0 to 255. The interrupt vectors associated with each interrupt number are stored in the lower 1024 bytes of PC memory. For example, interrupt 0 is stored from `0000:0000` to `0000:0003`, interrupt 1 from `0000:0004` to `0000:0007`, and so on. The first two bytes store the offset and the next two store the segment address. Each interrupt number is assigned a predetermined task, as outlined in Table 2.5. An interrupt can be generated either by external hardware, software, or by the processor. Interrupts 0, 1, 3, 4, 6 and 7 are generated by the processor. Interrupts from 8 to 15 and interrupt 2 are generated by external hardware. These get the attention of the processor by activating a interrupt request (IRQ) line. The `IRQ0` line connects to the system timer, the keyboard to `IRQ1`, and so on. Most other interrupts are generated by software.

Processor interrupts

The processor-generated interrupts normally occur either when a program causes a certain type of error or if it is being used in a debug mode. In the debug mode the program can be made to break from its execution when a break-point occurs. This allows the user to test the status of the computer. It can also be forced to step through a program one operation at a time (single-step mode).

Table 2.5 Interrupt handling (codes followed by 'h' are in hexadecimal)

Interrupt	*Name*	*Generated by*
00 (00h)	Divide error	processor
01 (00h)	Single step	processor
02 (02h)	Non-maskable interrupt	external equipment
03 (03h)	Breakpoint	processor
04 (04h)	Overflow	processor
05 (05h)	Print screen	Shift-Print screen key stroke
06 (06h)	Reserved	processor
07 (07h)	Reserved	processor
08 (08h)	System timer	hardware via IRQ0
09 (09h)	Keyboard	hardware via IRQ1
10 (0Ah)	Reserved	hardware via IRQ2
11 (0Bh)	Serial communications (COM2)	hardware via IRQ3
12 (0Ch)	Serial communications (COM1)	hardware via IRQ4
13 (0Dh)	Reserved	hardware via IRQ5
14 (0Eh)	Floppy disk controller	hardware via IRQ6
15 (0Fh)	Parallel printer	hardware via IRQ7
16 (10h)	BIOS – Video access	software
17 (11h)	BIOS – Equipment check	software
18 (12h)	BIOS – Memory size	software
19 (13h)	BIOS – Disk operations	software
20 (14h)	BIOS – Serial communications	software
22 (16h)	BIOS – Keyboard	software
23 (17h)	BIOS – Printer	software
25 (19h)	BIOS – Reboot	software
26 (1Ah)	BIOS – Time of day	software
28 (1Ch)	BIOS – Ticker timer	software
33 (21h)	DOS – DOS services	software
39 (27h)	DOS – Terminate and stay resident	software

2.2.2 *Hardware interrupts*

Computer systems either use polling or interrupt-driven software to service external equipment. With polling the computer continually monitors a status line and waits for it to become active, whereas an interrupt-driven device sends an interrupt request to the computer, which is then serviced by an interrupt service routine (ISR). Interrupt-driven devices are normally better in that the computer is thus free to do other things, whereas polling slows the system down as it must continually monitor the external device. Polling can also cause problems in that a device may be ready to send data and the computer is not watching the status line at that point. Figure 2.8 illustrates polling and interrupt-driven devices.

The generation of an interrupt can occur by hardware or software, as illustrated in Figure 2.9. If a device wishes to interrupt the processor, it informs the programmable interrupt controller (PIC). The PIC then decides whether it should interrupt the processor. If there is a processor interrupt then the processor reads the PIC to determine which device caused the interrupt. Then, depending on the device that caused the interrupt, a call to an ISR is made. The ISR then communicates with the device and processes any data. When it has finished the program execution returns to the original program.

A software interrupt causes the program to interrupt its execution and goes to an interrupt service routine. Typical software interrupts include reading a key from the keyboard, outputting text to the screen and reading the current date and time.

Hardware interrupts allow external devices to gain the attention of the processor. Depending on the type of interrupt the processor leaves the current program and goes to a special program called an interrupt service routine (ISR). This program communicates with the device and processes any data. After it has completed its task then program execution returns to the program that was running before the interrupt occurred. Examples of interrupts include the processing of keys from a keyboard and data from a sound card.

As previously mentioned, a device informs the processor that it wants to interrupt it by setting an interrupt line on the PC. Then, depending on the device that caused the interrupt, a call to an ISR is made. Each PIC allows access to eight interrupt request lines. Most PCs use two PICs which gives access to 16 interrupt lines.

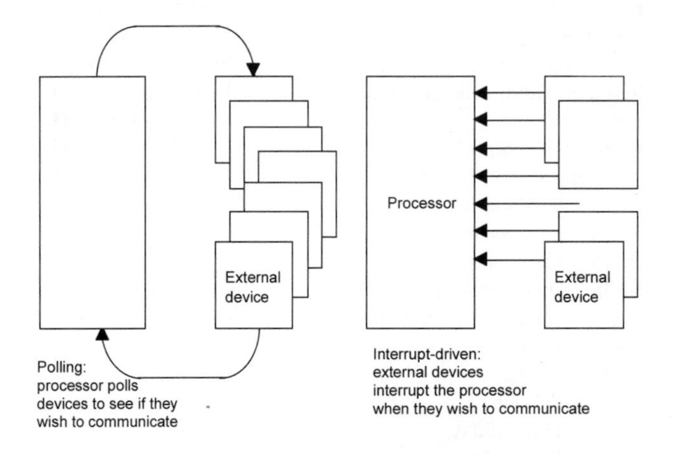

Figure 2.8 Polling or interrupt-driven communications

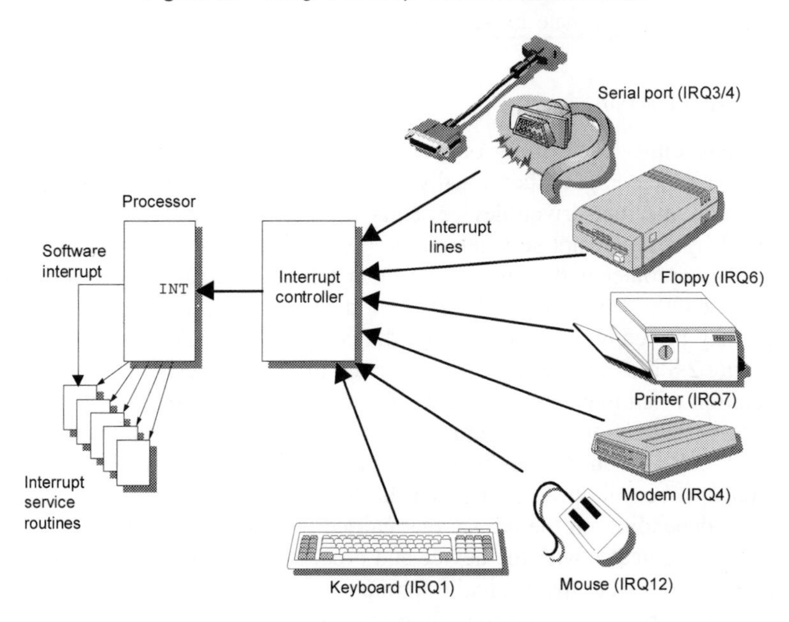

Figure 2.9 Interrupt handling

Interrupt vectors

Each device that requires to be 'interrupt-driven' is assigned an IRQ (interrupt request) line. Each IRQ is active high. The first eight (`IRQ0–IRQ7`) map into interrupts 8 to 15 (08h–0Fh) and the next eight (`IRQ8–IRQ15`) into interrupts 112 to 119 (70h–77h). Table 2.6 outlines the usage of each of these interrupts. When `IRQ0` is made active, the ISR corresponds to interrupt vector 8. `IRQ0` normally connects to the system timer, the keyboard to `IRQ1`, and so on. The standard set-up of these interrupts is illustrated in Figure 2.10. The system timer interrupts the processor 18.2 times per second and is used to update the system time. When the keyboard has data, it interrupts the processor with the `IRQ1` line.

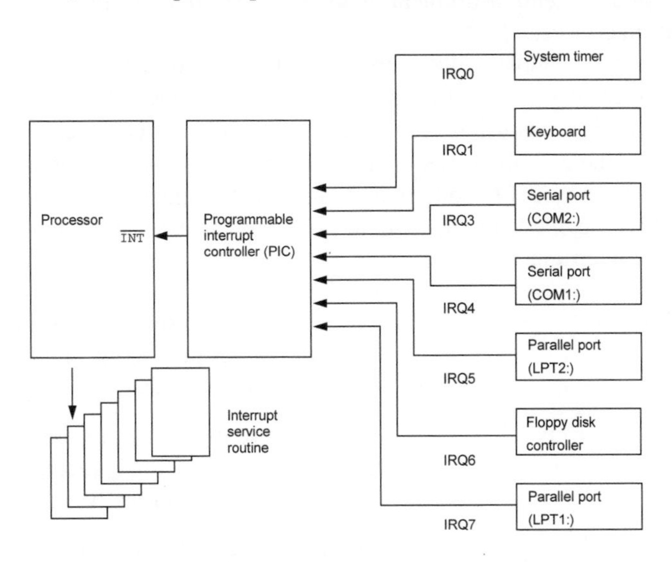

Figure 2.10 Standard usage of IRQ lines

Table 2.6 Interrupt handling

Interrupt	*Name*	*Generated by*
08 (08h)	System timer	IRQ0
09 (09h)	Keyboard	IRQ1
10 (0Ah)	Reserved	IRQ2
11 (0Bh)	Serial communications (COM2:)	IRQ3
12 (0Ch)	Serial communications (COM1:)	IRQ4
13 (0Dh)	Parallel port (LPT2:)	IRQ5
14 (0Eh)	Floppy disk controller	IRQ6
15 (0Fh)	Parallel printer (LPT1:)	IRQ7
112 (70h)	Real-time clock	IRQ8
113 (71h)	Redirection of IRQ2	IRQ9
114 (72h)	Reserved	IRQ10
115 (73h)	Reserved	IRQ11
116 (74h)	Reserved	IRQ12
117 (75h)	Math co-processor	IRQ13
118 (76h)	Hard disk controller	IRQ14
119 (77h)	Reserved	IRQ15

Data received from serial ports interrupts the processor with `IRQ3` and `IRQ4` and the parallel ports use `IRQ5` and `IRQ7`. If one of the parallel, or serial, ports does not exist then the IRQ line normally assigned to it can be used by another device. It is typical for interrupt-driven I/O cards, such as a sound card, to have a programmable IRQ line which is mapped to an IRQ line that is not being used.

Note that several devices can use the same interrupt line. A typical example is `COM1:` and `COM3:` sharing `IRQ4` and `COM2:` and `COM4:` sharing `IRQ3`. If they do share then the ISR must be able to poll the shared devices to determine which of them caused the interrupt. If two different types of device (such as a sound card and a serial port) use the same IRQ line then there may be a contention problem as the ISR may not be able to communicate with different types of interfaces.

Figure 2.11 shows a sample window displaying interrupt usage. In this case it can be seen that the system timer uses `IRQ0`, the keyboard uses `IRQ1`, the PIC uses `IRQ2`, and so on. Notice that a sound blaster is using `IRQ5`. This interrupt is normally reserved for the secondary printer port. If there is no printer connected then `IRQ5` can be used by another device. Some devices can have their I/O address and interrupt line changed. An example is given in Figure 2.12. In this case, the IRQ line is set to IRQ7 and the base address is 378h.

Typical uses of interrupts are:

IRQ0: System timer

The system timer uses `IRQ0` to interrupt the processor 18.2 times per second and is used to keep the time-of-day clock updated.

IRQ1: Keyboard data ready

The keyboard uses `IRQ1` to signal to the processor that data is ready to be received from the keyboard. This data is normally a scan code.

IRQ2: Redirection of IRQ9

The BIOS redirects the interrupt for `IRQ9` back here.

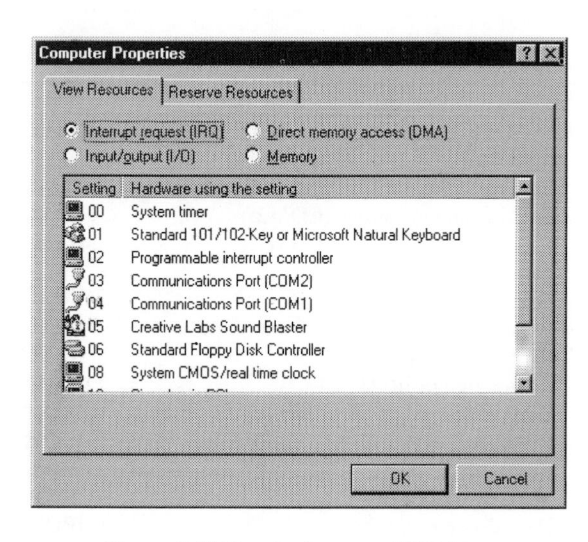

Figure 2.11 Standard usage of IRQ lines

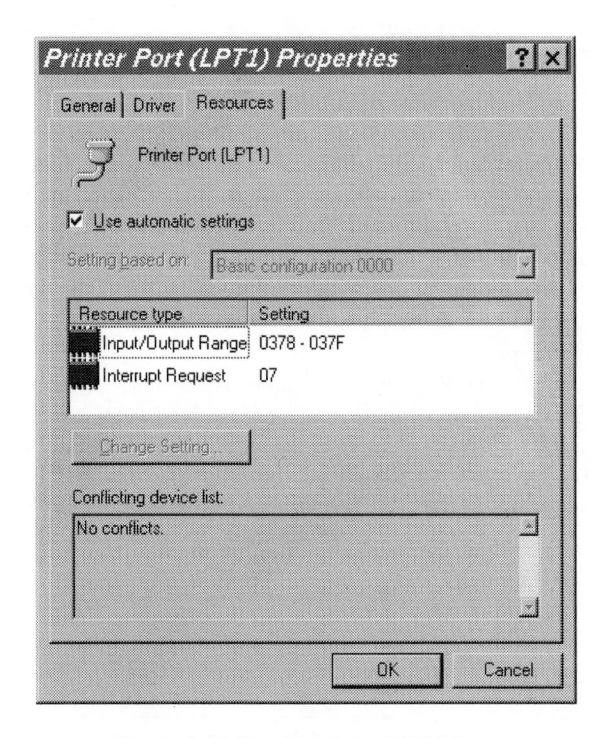

Figure 2.12 Standard set-up of IRQ lines

IRQ3: Secondary serial port (COM2:) — The secondary serial port (`COM2:`) uses `IRQ3` to interrupt the processor. Typically, `COM3:` to `COM8:` also use it, although `COM3:` may use `IRQ4`.

IRQ4: Primary serial port (COM1:) — The primary serial port (`COM1:`) uses `IRQ4` to interrupt the processor. Typically, `COM3:` also uses it.

IRQ5: Secondary parallel port (LPT2:) — On older PCs the `IRQ5` line was used by the fixed disk. On newer systems the secondary parallel port uses it. Typically, it is used by a sound card on PCs which have no secondary parallel port connected.

IRQ6: Floppy disk controller — The floppy disk controller activates the `IRQ6` line on completion of a disk operation.

IRQ7: Primary parallel port (LPT1:) — Printers (or other parallel devices) activate the `IRQ7` line when they become active. As with `IRQ5` it may be used by another device, if there are no other devices connected to this line.

IRQ9 — Redirected to `IRQ2` service routine.

Programmable interrupt controller (PIC)

The PC uses the 8259 PIC to control hardware-generated interrupts. It is known as a programmable interrupt controller and has eight input interrupt request lines and an output line to secondary PIC are then assigned IRQ lines of `IRQ8` to `IRQ15`. This set-up is shown in Figure 2.13. When an interrupt occurs on any of these lines it is sensed by the processor on interrupt the processor. Originally, PCs only had one PIC and eight IRQ lines (`IRQ0-IRQ7`). Modern PCs can use up to 15 IRQ lines which are set up by connecting a secondary PIC interrupt request output line to the `IRQ2` line of the primary PIC. The interrupt lines on the `IRQ2` line. The processor then interrogates the primary and secondary PIC for the interrupt line which caused the interrupt.

The primary and secondary PICs are programmed via port addresses 20h and 21h, as given in Table 2.7. The operation of the PIC is programmed using registers. The IRQ input lines are either configured as level-sensitive or edge-triggered interrupt. With edge-triggered interrupts, a change from a low to a high on the IRQ line causes the interrupt. A level-sensitive interrupt occurs when the IRQ line is high. Most devices generate edge-triggered interrupts.

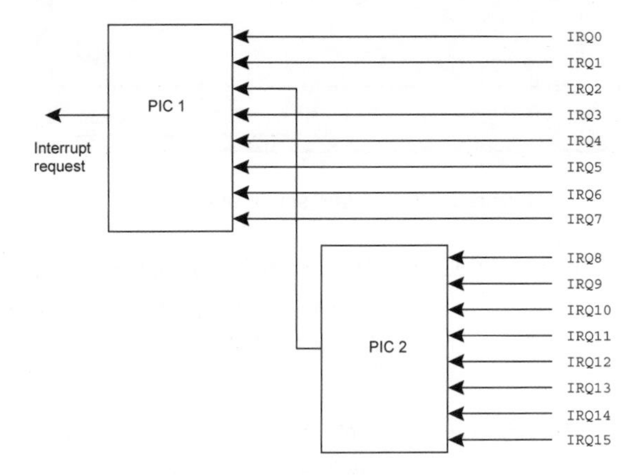

Figure 2.13 PC PIC connections

In the IMR an interrupt line is enabled by setting the assigned bit to a 0 (zero). This allows the interrupt line to interrupt the processor. Figure 2.14 shows the bit definitions of the IMR. For example, if bit 0 is set to a zero then the system timer on `IRQ0` is enabled.

Table 2.7 Interrupt port addresses

Port address	*Name*	*Description*
20h	Interrupt control register (ICR)	Controls interrupts and signifies the end of an interrupt
21h	Interrupt mask register (IMR)	Used to enable and disable interrupt lines

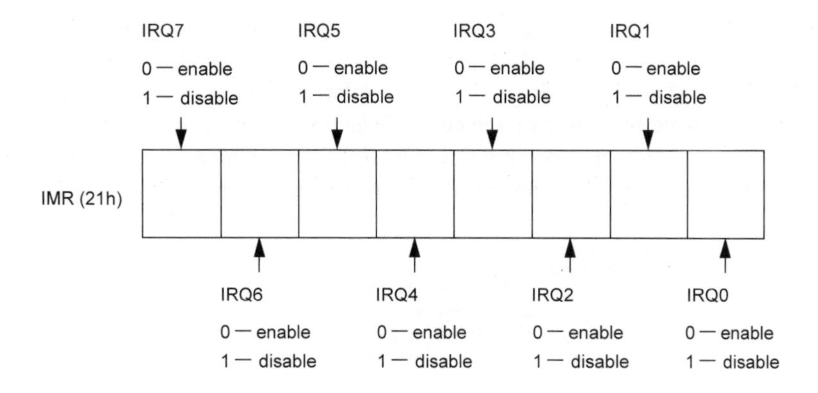

Figure 2.14 Interrupt mask register bit definitions

In the example code given next the lines `IRQ0`, `IRQ1` and `IRQ6` are allowed to interrupt the processor, whereas, `IRQ2`, `IRQ3`, `IRQ4` and `IRQ7` are disabled:

```
_outp(0x21)=0xBC;   /* 1011 1100 enable disk
       (bit 6), keyboard (1) and timer (0) interrupts                */
```

When an interrupt occurs all other interrupts are disabled and no other device can interrupt the processor. Interrupts are enabled again by setting the EOI bit on the interrupt control port, as shown in Figure 2.15.

The following code enables interrupts:

```
_outp(0x20,0x20);  /* EOI command  */
```

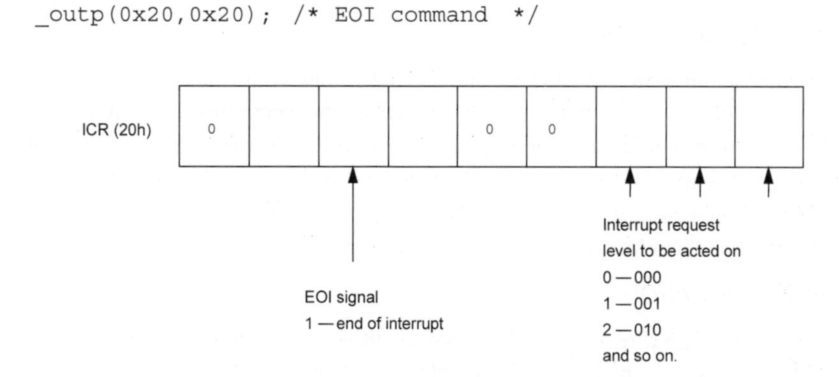

Figure 2.15 Interrupt control register bit definitions

2.3 Interfacing

There are two main methods of communicating with external equipment, either they are mapped into the physical memory and given a real address on the address bus (memory mapped I/O) or they are mapped into a special area of input/output memory (isolated I/O).

Figure 2.16 shows the two methods. Devices mapped into memory are accessed by reading or writing to the physical address. Isolated I/O provides ports which are gateways between the interface device and the processor. They are isolated from the system using a buffering system and are accessed by four machine code instructions. The `IN` instruction inputs a byte, or a word, and the `OUT` instruction outputs a byte, or a word. A high-level compiler interprets the equivalent high-level functions and produces machine code which uses these instructions.

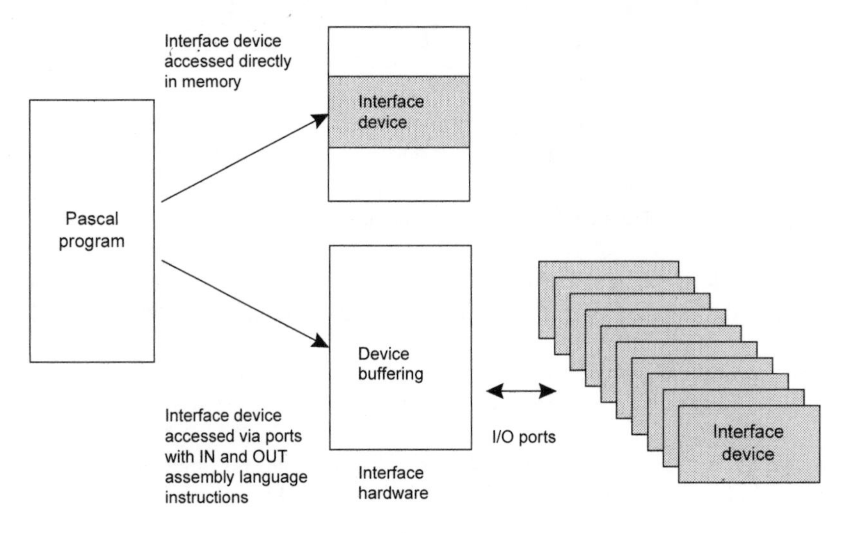

Figure 2.16 Memory mapping or isolated interfacing

2.3.1 Interfacing with memory

The 80x86 processor interfaces with memory through a bus controller, as shown in Figure 2.17. This device interprets the microprocessor signals and generates the required memory signals. Two main output lines differentiate between a read or a write operation ($R/\overline{W}$) and between direct and isolated memory access ($M/\overline{IO}$). The $R/\overline{W}$ line is low when data is being written to memory and high when data is being read. When $M/\overline{IO}$ is high, direct memory access is selected and when low, the isolated memory is selected.

2.3.2 Memory mapped I/O

Interface devices can map directly onto the system address and data bus. In a PC-compatible system the address bus is 20 bits wide, from address `00000h` to `FFFFFh` (1 MB). If the PC is being used in an enhanced mode (such as with Microsoft Windows) it can access the area of memory above 1 MB. If it uses 16-bit software (such as Microsoft Windows 3.1) then it can address up to 16 MB of physical memory, from `000000h` to `FFFFFFh`. If it uses 32-bit software (such as Microsoft Windows 95/98/NT/2000) then the software can address up to 4 GB of physical memory, from `00000000h` to `FFFFFFFFh`. Figure 2.18 gives a typical memory allocation.

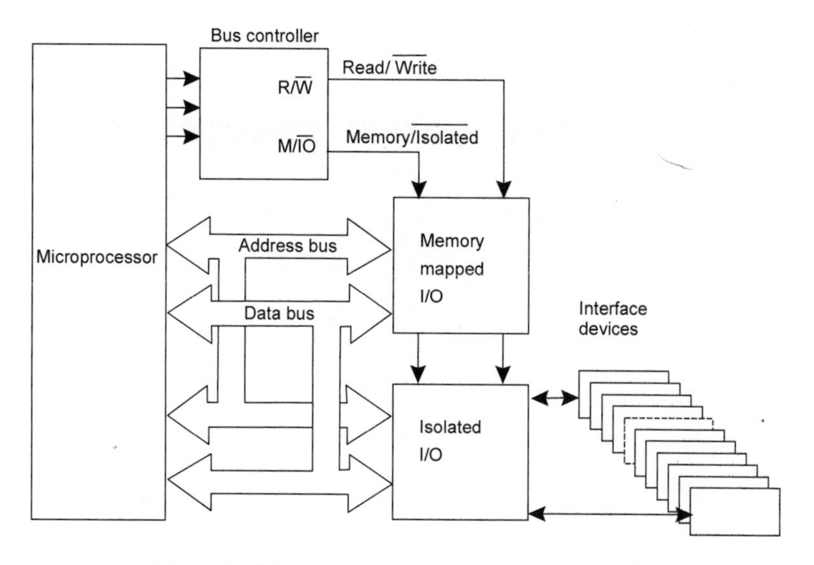

Figure 2.17 Access memory mapped and isolated I/O

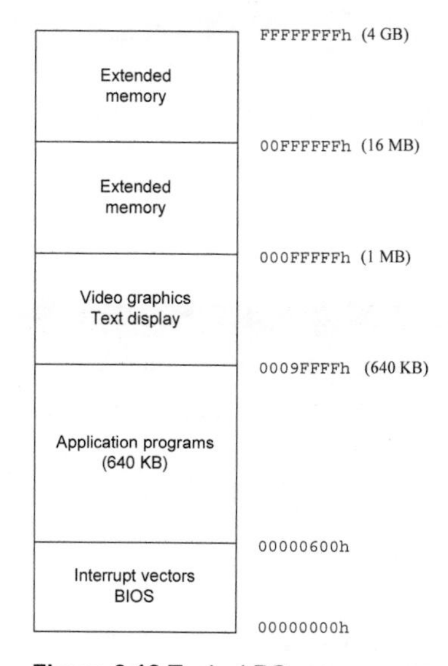

Figure 2.18 Typical PC memory map

2.3.3 Isolated I/O

Devices are not normally connected directly onto the address and data bus of the computer because they may use part of the memory that a program uses or they could cause a hardware fault. On modern PCs only the graphics adaptor is mapped directly into memory, the rest communicate through a specially reserved area of memory, known as isolated I/O memory.

Isolated I/O uses 16-bit addressing from `0000h` to `FFFFh`, thus up to 64 KB of memory can be mapped. Figure 2.19 shows an example for a computer in the range from `0000h` to `0064h` and Figure 2.20 shows from `0378h` to `03FFh`. It can be seen from Figure 2.19 that the keyboard maps into addresses `0060h` and `0064h`, the speaker maps to address `0061h` and the system timer between `0040h` and `0043h`. Table 2.8 shows the typical uses of the isolated memory area.

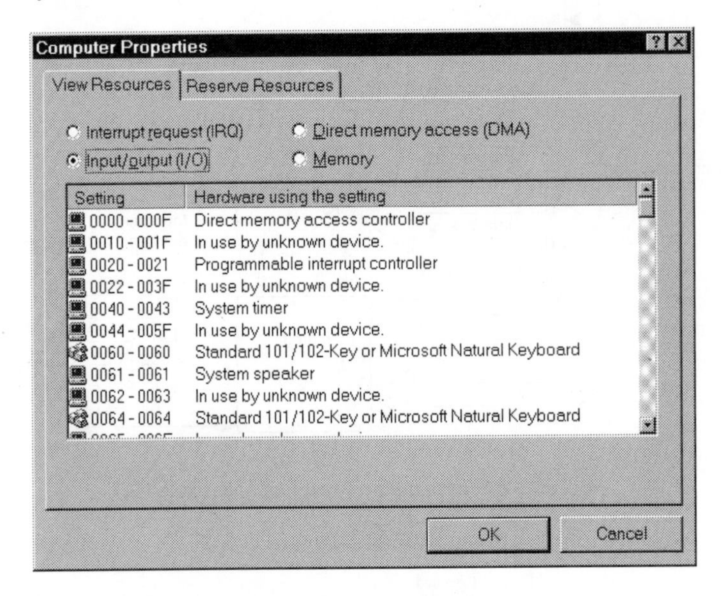

Figure 2.19 Example I/O memory map from `0000h` to `0064h`

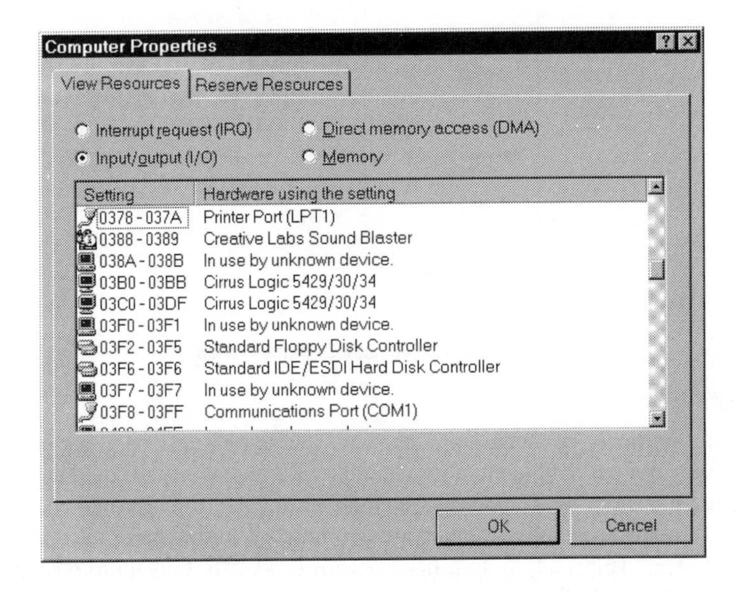

Figure 2.20 Example I/O memory map from `0378h` to `03FFh`

Table 2.8 Typical isolated I/O memory map

Address	*Device*
000h-01Fh	DMA controller
020h-021h	Programmable interrupt controller
040h-05Fh	Counter/Timer
060h-07Fh	Digital I/O
080h-09Fh	DMA controller
0A0h-0BFh	NMI reset
0C0h-0DFh	DMA controller
0E0h-0FFh	Math coprocessor
170h-178h	Hard disk (Secondary IDE drive or CD-ROM drive)
1F0h-1F8h	Hard disk (Primary IDE drive)
200h-20Fh	Game I/O adapter
210h-217h	Expansion unit
278h-27Fh	Second parallel port (LPT2:)
2F8h-2FFh	Second serial port (COM2:)
300h-31Fh	Prototype card
378h-37Fh	Primary parallel port (LPT1:)
380h-38Ch	SDLC interface
3A0h-3AFh	Primary binary synchronous port
3B0h-3BFh	Graphics adapter
3C0h-3DFh	Graphics adapter
3F0h-3F7h	Floppy disk controller
3F8h-3FFh	Primary serial port (COM1:)

Inputting a byte from an I/O port

The assembly language command to input a byte is

```
IN AL,DX
```

where `DX` is the data register which contains the address of the input port. The 8-bit value loaded from this address is put into the register `A`

For Turbo/Borland C the equivalent function is `inportb()`. Its general syntax is as follows:

```
value=inportb(PORTADDRESS);
```

where `PORTADDRESS` is the address of the input port and `value` is loaded with the 8-bit value from this address. This function is prototyped in the header file `dos.h`.

For Turbo Pascal the equivalent is accessed via the `port[]` array. Its general syntax is as follows:

```
value:=port[PORTADDRESS];
```

where `PORTADDRESS` is the address of the input port and `value` the 8-bit value at this address. To gain access to this function the statement `uses dos` requires to be placed near the top of the program.

Microsoft C++ uses the equivalent `_inp()` function (which is prototyped in `conio.h`).

Inputting a word from a port

The assembly language command to input a word is

```
IN AX,DX
```

where `DX` is the data register which contains the address of the input port. The 16-bit value loaded from this address is put into the register `AX`.

For Turbo/Borland C the equivalent function is `inport()`. Its general syntax is as follows:

```
value=inport(PORTADDRESS);
```

where `PORTADDRESS` is the address of the input port and `value` is loaded with the 16-bit value at this address. This function is prototyped in the header file `dos.h`.

For Turbo Pascal the equivalent is accessed via the `portw[]` array. Its general syntax is as follows:

```
value:=portw[PORTADDRESS];
```

where `PORTADDRESS` is the address of the input port and `value` is the 16-bit value at this address. To gain access to this function the statement `uses dos` requires to be placed near the top of the program.

Microsoft C++ uses the equivalent `_inpw()` function (which is prototyped in `conio.h`).

Outputting a byte to an I/O port

The assembly language command to output a byte is

```
OUT DX,AL
```

where `DX` is the data register which contains the address of the output port. The 8-bit value sent to this address is stored in register `AL`.

For Turbo/Borland C the equivalent function is `outportb()`. Its general syntax is as follows:

```
outportb(PORTADDRESS,value);
```

where `PORTADDRESS` is the address of the output port and `value` is the 8-bit value to be sent to this address. This function is prototyped in the header file `dos.h`.

For Turbo Pascal the equivalent is accessed via the `port[]` array. Its general syntax is as follows:

```
port[PORTADDRESS]:=value;
```

where `PORTADDRESS` is the address of the output port and `value` is the 8-bit value to be sent to that address. To gain access to this function the statement `uses dos` requires to be placed near the top of the program.

Microsoft C++ uses the equivalent `_outp()` function (which is prototyped in `conio.h`).

Outputting a word

The assembly language command to input a byte is:

```
OUT DX,AX
```

where `DX` is the data register which contains the address of the output port. The 16-bit value sent to this address is stored in register `AX`.

For Turbo/Borland C the equivalent function is `outport()`. Its general syntax is as follows:

```
outport(PORTADDRESS,value);
```

where `PORTADDRESS` is the address of the output port and `value` is the 16-bit value to be sent to that address. This function is prototyped in the header file `dos.h`.

For Turbo Pascal the equivalent is accessed via the `port[]` array. Its general syntax is as follows:

```
portw[PORTADDRESS]:=value;
```

where `PORTADDRESS` is the address of the output port and `value` is the 16-bit value to be sent to that address. To gain access to this function the statement `uses dos` requires to be placed near the top of the program.

Microsoft C++ uses the equivalent `_outp()` function (which is prototyped in `conio.h`).

In-line assembly language

Most modern C++ development systems use an inline assembler which allows assembly language code to be embedded with C++ code. This code can use any C variable or function name that is in scope. The `__asm` keyword invokes the inline assembler and can appear wherever a C statement is legal. The following code is a simple `__asm` block enclosed in braces.

```
__asm
{
      /* Initialize serial port */
      mov dx,0x01;    /* COM2:                    */
      mov al,0xD2;    /* serial port parameters   */
      mov ah,0x0;     /* initialize serial port   */
      int 14h;
      line_status=ah;
      modem_status=al;
}
```

Note these statements can also be inserted after the `__asm` keyword, such as:

```
__asm mov dx,0x01;    /* COM2:                  */
__asm mov al,0xD2;    /* serial port parameters */
__asm mov ah,0x0;     /* initialize serial port */
__asm int 14h;
__asm line_status=ah;
__asm modem_status=al;
```

2.4 PC Systems

In selecting a PC many different components must be considered, especially in the way that they connect. Figure 2.21 outlines some of the component parts and the decisions that have to be made on each component.

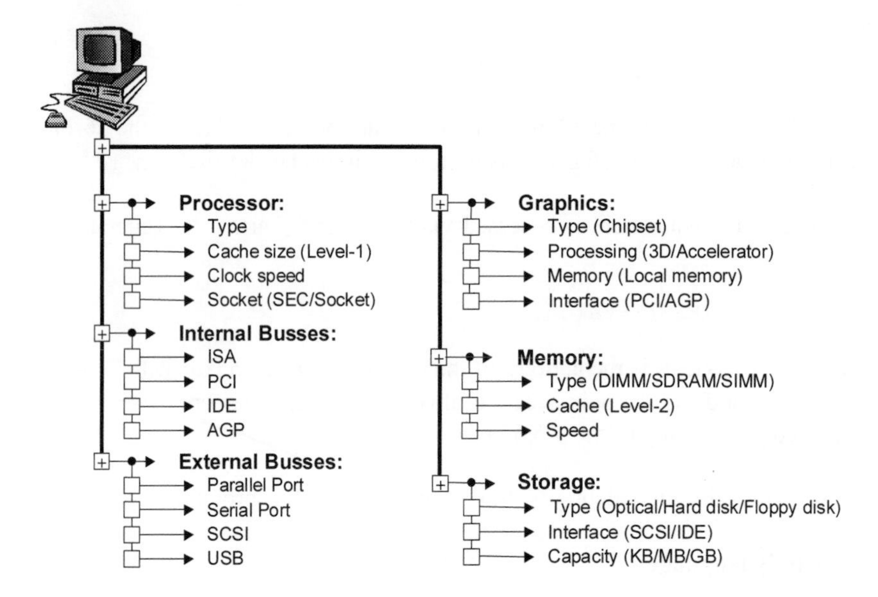

Figure 2.21 PC components

The Top 5 things that affect the *general* performance of a PC (in ranked order) are:

1. **Processor**. The type of the processor, its speed, its socket (which helps in upgrading in the future), its interface to Level-2 cache, and so on. Additionally, MMX™, (which is an Intel trademark, but many read it as MultiMedia eXtension) can speed-up multi-media applications.
2. **Local Memory**. Most operating systems can run multiple programs, each of which require their own memory space. When the system runs out of electronic memory (local memory), it uses the hard disk for an extra storage (to create a virtual memory). Hard disk accesses are much slower than electronic memory, thus the system is severely slowed down if there is a lack of local electronic memory. Most modern operating sys-

tems require a great deal of local electronic memory to operate.

3. **Graphics adaptor**. The graphics adaptor can be a major limiting factor on the performance of a system. New interfaces, such as AGP, considerably speed-up graphics performance. Another limiting factor is the amount of local memory on the graphics adaptor. The more memory, the higher the resolution that can be used, and the more colours that can be displayed. AGP is overcoming this limiting factor, as it allows the main electronic memory to be used to store graphics images.
4. **Cache capacity**. Cache memory has caused a great increase in the performance of a system. If a cache controller makes a correct guess, the processor merely has to examine the contents of the cache to get the required information. A level-1 cache is the fastest and is typically connected directly to the processor (normally inside the processor package), and the level-2 cache is on the motherboard.
5. **Hard disk capacity/interface**. The hard disk typically has an affect on the running of a program, as the program and its component parts must be loaded from the disk. The interface is thus extremely important as it defines the maximum data rate. SCSI has fast modes which give up to 40 MB/s, while IDE gives a maximum rate of 33 MB/s. The capacity of the disk also can lead to problems as the system can use unused disk capacity of a virtual memory capacity.

Obviously, applications that are more specific will be affected by other factors, such as:

- **Internet access**. Affected mainly by the network connection (especially if a modem is used).
- **CD-ROM access**. Affected by the interface to the CD-ROM.
- **Modelling software**. Affected by mathematical processing.
- **3D game playing**. Affected mainly by the graphics adaptor and graphics processing (and possibly the network connection, if playing over a network).

2.8 Practical PC system

At one time PCs were crammed full of microchips, wires and connectors. These days they tend to be based on just a few microchips, and contain very few interconnecting wires. The main reason for this is that much of the functionality of the PC has been integrated into several key devices. In the future, PCs may only require one or two devices to make them operate.

The architecture of the PC has changed over the past few years. It is now mainly based on the PCI bus. Figure 2.22 shows the architecture of a modern PC. The system controller is the real heart of the PC, as it transfers data to and from the processor to the rest of the system. Bridges are used to connect one type of bus to another. There are two main bridges: the system controller (the north bridge), and the bus bridge (the south bridge).

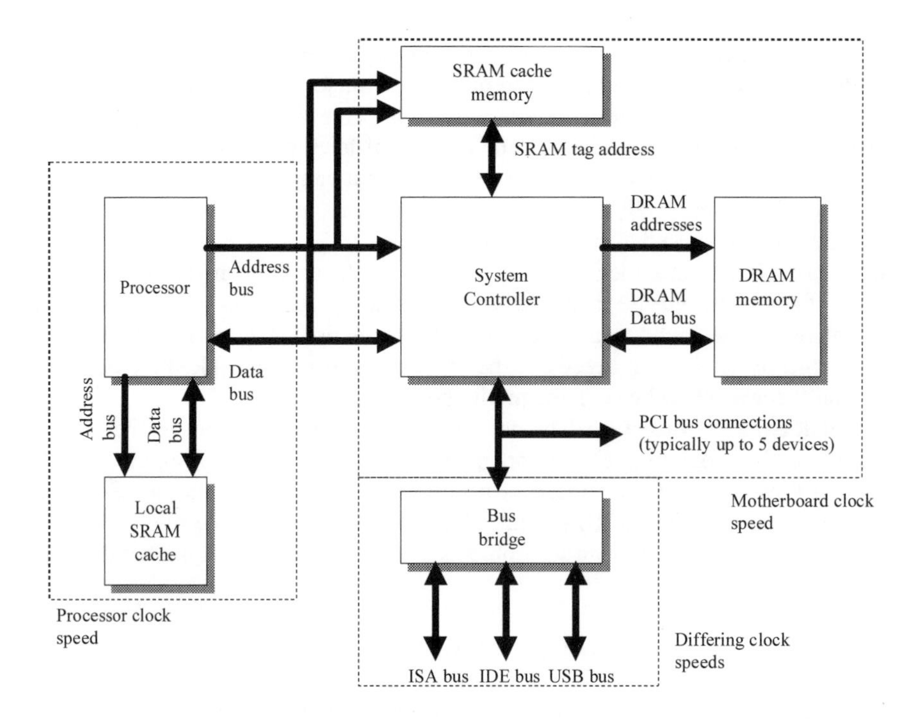

Figure 2.22 Local bus architecture

An example PC motherboard is illustrated in Figure 2.23. The main components are:

- **Processor**. The processor is typically a Pentium processor, which has a SEC (single-edge connector) or fits into a socket. The processor can run at a faster rate than the rest of the motherboard (called clock multiplication). Typically, the motherboard runs at 50MHz, and the clock rate is multiplied by a given factor, such as 500MHz (for a ×10 clock multiplier).
- **System controller**. Controls the interface between the processor, memory and the PCI bus.
- **PCI/ISA/IDE Xcelerated Controller**. Controls the interface between the PCI bus and the ISA, USB and IDE busses.
- **I/O controller**. Controls the interface between the ISA and the other busses, such as the parallel bus, serial bus, floppy disk drive, keyboard, mouse, and infrared transmission.
- **DIMM sockets**. This connects to the main memory of the computer. Typically it uses either EDO DRAM and SDRAM (Synchronous DRAM). SDRAM transfers data faster than EDO DRAM as its uses the clock rate of the processor, rather than the clock rate of the motherboard.
- **Flash memory**. Used to store the program which starts the computer up (the boot process).
- **PCI connectors**. Used to connect to PCI-based interface adaptors, such as a network card, sound card, and so on.
- **ISA connectors**. Used to connect to ISA-based interface adaptors, such as a sound cards.
- **IDE connectors**. Used to connect to hard disks or CD-ROM drives. Up to two drives

can connect to each connector (IDE0 or IDE1) as a master or a slave. Thus, the PC can support up to four disk drives on the IDE bus.

- **TV out socket**. Used to provide an output which will interface to a TV, using either PAL (for the UK) or NSTC (for the US).
- **Level-2 cache (SRAM)**. Used to store information from DRAM memory.
- **Video memory**. Used to store video information.
- **Graphics controller**. Used to control the graphics output.
- **Audio codec**. Used to process audio data.

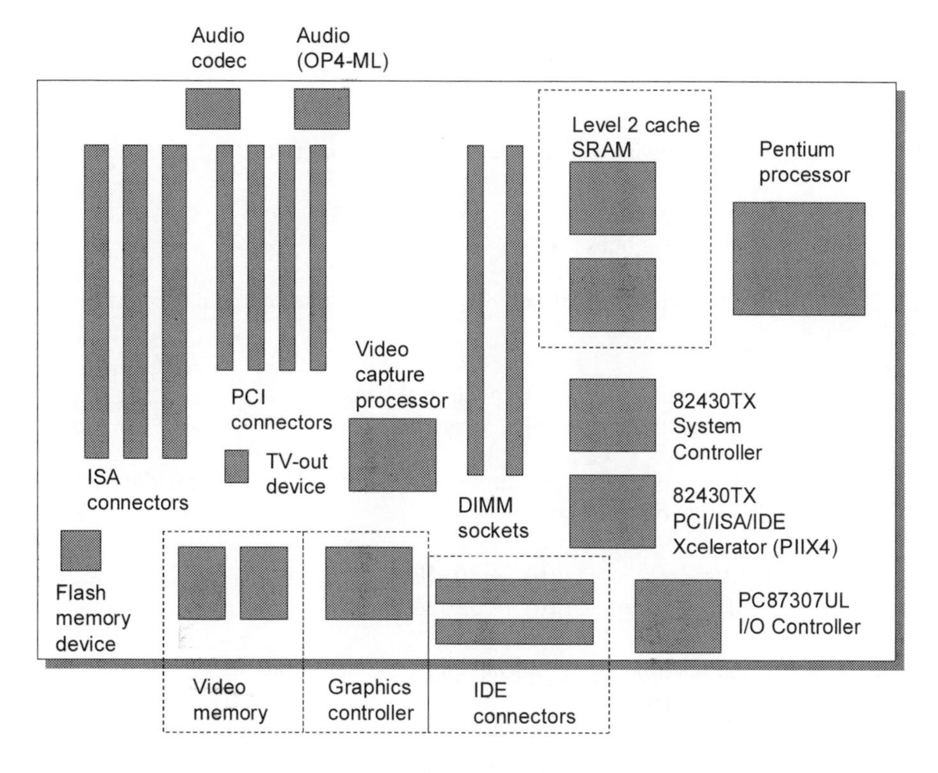

Figure 2.23 AN430TX board

2.5 Exercises

The following questions are multiple choice. Please select from a–d.

2.5.1 Which type of memory does not lose its contents when the power is withdrawn:

(a)	ROM	(b)	RAM
(c)	DRAM	(d)	SRAM

2.5.2 Which type of memory uses a single capacitor and a transistor to store a single bit of data:

(a)	EPROM	(b)	ERAM
(c)	DRAM	(d)	SRAM

2.5.3 Which type of memory requires its memory of be refreshed at regular intervals:

(a)	EPROM	(b)	ERAM
(c)	DRAM	(d)	SRAM

2.5.4 If a processor can operate on four bytes at a time, which is its classification:

(a)	8-bit	(b)	16-bit
(a)	32-bit	(b)	64-bit

2.5.5 Which of the following defines the amount of memory that can be accessed:

(a)	Address bus	(b)	Control lines
(c)	Handshaking lines	(d)	Data bus

2.5.6 Which of the following defines the number of bits that can be transmitted at a time:

(a)	Address bus	(b)	Control lines
(c)	Handshaking lines	(d)	Data bus

2.5.7 Which is the maximum data throughput for a 32-bit parallel data bus with a clocked data rate of 10 MHz:

(a)	4 MB/s	(b)	40 MB/s
(c)	32 MB/s	(d)	320 MB/s

2.5.8 Which is the maximum data throughput for a serial bus which has a bit transmission time of 69.44 μs:

(a)	6944 bps	(b)	9600 bps
(c)	1440 bps	(d)	14 400 bps

2.5.9 How much memory can be accessed with a 20-bit address bus:

(a)	20 B	(b)	20 KB
(c)	1 MB	(d)	20 MB

2.5.10 How much memory can be accessed with a 32-bit address bus:

(a)	32 B	(b)	32 KB
(c)	1 GB	(d)	32 MB

2.5.11 Which interrupt does the primary serial port of a PC (COM1:) normally use:

(a)	IRQ0	(b)	IRQ3

(c) IRQ4 (d) IRQ7

2.5.12 Which interrupt does the secondary serial port of a PC (COM2:) normally use:

(a) IRQ0 (b) IRQ3
(c) IRQ4 (d) IRQ7

2.5.13 Which interrupt does the system timer on the PC use:

(a) IRQ0 (b) IRQ3
(c) IRQ4 (d) IRQ7

2.5.14 Which interrupt was used to increase the amount of interrupts from 8 to 16:

(a) IRQ0 (b) IRQ1
(c) IRQ2 (d) IRQ15

2.5.15 Which interrupt is used by the keyboard:

(a) IRQ0 (b) IRQ1
(c) IRQ2 (d) IRQ15

2.5.16 What does ISR stand for:

(a) Interval Status Register (b) Interrupt Status Register
(c) Interrupt Service Routine (d) Interrupt Standard Routine

2.5.17 How is isolated memory differentiated from memory added I/O:

(a) Different address bus (b) Different data bus
(c) Control line differentiates between them (Memory/Isolated)
(d) There is no differentiation as they are physically the same

2.5.18 How many addresses can be accessed in the address range 0000h to FFFFh:

(a) 32 768 (32 kB) (b) 65 536 (64 kB)
(c) 262 144 (256 kB) (d) 1 048 576 (1 MB)

2.5.19 How much physical memory can a DOS-compatible program access:

(a) 32 768 (32 kB) (b) 65 536 (64 kB)
(c) 262 144 (256 kB) (d) 1 048 576 (1 MB)

2.5.20 Which address is the interrupt control port register:

(a) 0002h (b) 0020h
(c) 0200h (d) 2000h

2.5.21 Which is normally the base address for the primary parallel port:

(a) 0378h (b) 0278h
(c) 03F8h (d) 02F8h

2.5.22 Contrast the operation of polling and interrupt-driven software when interfacing to external equipment.

2.5.23 Access a PC and determine the following:

Interrupt	Device connected
IRQ1	
IRQ3	
IRQ5	
IRQ7	
IRQ9	
IRQ11	
IRQ13	
IRQ15	
I/O address	**Device connected**
0060h, 0064h	
0070h	
0090h	
00F0h	
0278h	
02F8h	
0378h	
03F8h	
DMA channel	**Device connected**
DMA0	
DMA1	
DMA2	
DMA3	

2.6 Notes from the author

This chapter has introduced some of the key concepts used in defining computer systems. So, what is it that differentiates one PC system from another? It is difficult to say, but basically its all about how well bolted together systems are, how compatible the parts are with the loaded software, how they organise the peripherals, and so on. The big problem though is compatibility, and compatibility is all about peripherals looking the same, that is, having the same IRQ, the same I/O address, and so on.

The PC is an amazing device, and has allowed computers to move from technical specialists to, well, anyone. However, they are also one of the most annoying of pieces of technology of all time, in terms of their software, their operating system, and their hardware. If we bought a car and it failed at least a few times every day, we would take it back and demand another one. When that failed, we would demand our money back. Or, sorry I could go on forever here, imagine a toaster that failed half way through making a piece of toast, and we had to turn the power off, and restart it. We just wouldn't allow it.

So why does the PC lead such a privileged life. Well it's because it's so useful and multi-talented, although it doesn't really excel at much. Contrast a simple games computer against the PC and you find many lessons in how to make a computer easy-to-use, and to configure.

One of the main reasons for many of its problems is the compatibility with previous systems both in terms of hardware compatibility and software compatibility (and dodgy software, of course). The big change on the PC was the introduction of proper 32-bit software, Windows 95/NT.

In the future systems will be configured by the operating system, and not by the user. How many people understand what an IRQ is, what I/O addresses are, and so on. Maybe if the PC faced some proper competition it would become easy to use and become totally reliable. Then when they were switched on they would configure themselves automatically, and you could connect any device you wanted and it would understand how to configure (we're nearly there, but it's still not perfect). Then we would have a tool which could be used to improve creativity and you didn't need a degree in computer engineering to use one (in your dreams!). But, anyway, it's keeping a lot of technical people in a job, so, don't tell anyone our little secret. The Apple Macintosh was a classic example of a well-designed computer that was designed as a single unit. When initially released it started up with messages like I'm glad to be out of that bag and Hello, I am Macintosh. Never trust a computer you cannot lift.

So, apart from the IBM PC, what are the all-time best computers? A list by Byte *in September 1995 stated the following:*

1.	*MITS Altair8800*	*11.*	*IBM AT*
2.	*Apple II*	*12.*	*Commodore Amiga 1000*
3.	*Commodore PET*	*13.*	*Compaq Deskpro 386*
4.	*Radio Shack TRS-80*	*14.*	*Apple Macintosh II*
5.	*Osborne 1 Portable*	*15.*	*Next Nextstation*
6.	*Xerox Star*	*16.*	*NEC UltraLite*
7.	*IBM PC*	*17.*	*Sun SparcStation 1*
8.	*Compaq Portable*	*18.*	*IBM RS/6000*
9.	*Radio Shack TRS-80 Model 100*	*19.*	*Apple Power Macintosh*
10.	*Apple Macintosh*	*20.*	*IBM ThinkPad 701C*

And the Top 10 computer people as:

1.	*Dan Bricklin (VisiCalc)*	*11.*	*Philippe Kahn (Turbo Pascal)*
2.	*Bill Gates (Microsoft)*	*12.*	*Mitch Kapor (Lotus 123)*
3.	*Steve Jobs (Apple)*	*13.*	*Donald Knuth (TEX)*
4.	*Robert Noyce (Intel)*	*14.*	*Thomas Kurtz*
5.	*Dennis Ritchie (C Programming)*	*15.*	*Drew Major (NetWare)*
6.	*Marc Andreessen (Netscape Communications)*	*16.*	*Robert Metcalfe (Ethernet)*
7.	*Bill Atkinson (Apple Mac GUI)*	*17.*	*Bjarne Strousstrup (C++)*
8.	*Tim Berners-Lee (CERN)*	*18.*	*John Warnock (Adobe)*
9.	*Doug Engelbart (Mouse/Windows/etc)*	*19.*	*Niklaus Wirth (Pascal)*
10.	*Grace Murray Hopper (COBOL)*	*20*	*Steve Wozniak (Apple)*

One of the classic comments of all time was by Ken Olson at DEC, who stated, that there is no reason anyone would want a computer in their home. This seems farcical now, but at the time, in the 1970s, there were no CD-ROMs, no microwave ovens, no automated cash dispensers, and no Internet. Few people predicted them, so, predicting the PC was also difficult. But the two best comments were:

Computers in the future may weigh no more than 1.5 tons. Popular Mechanics.
I think there is a world market for maybe five computers, Thomas Watson, chairman of IBM, 1943.

3 Interfacing Standards

3.1 Introduction

The type of interface card used greatly affects the performance of a PC system. Early models of PCs relied on expansion options to improve their specification. These expansion options were cards that plugged into an expansion bus. Eight slots were usually available and these added memory, video, fixed and floppy disk controllers, printer output, modem ports, serial communications and so on.

There are eight main types of interface busses available for the PC. The number of data bits they handle at a time determines their classification. They are:

- PC (8-bit) ISA (16-bit)
- EISA (32-bit) MCA (32-bit)
- VL-Local Bus (32-bit) PCI bus (32/64-bit)
- SCSI (16/32-bit) PCMCIA (16-bit)

3.2 PC bus

The PC bus uses the architecture of the Intel 8088 processor which has an external 8-bit data bus and 20-bit address bus. A PC bus connector has a 62-pin printed circuit card edge connector and a long narrow or half-length plug-in card. As it uses a 20-bit address bus, it can address a maximum of 1 MB of memory. The transfer rate is fixed at 4.772 727 MHz; thus, a maximum of 4 772 727 bytes can be transferred every second. Dividing a crystal oscillator frequency of 14.318 18 MHz by three derives this clock speed. Figure 3.1 shows a PC card. Figure 3.2 defines the signal connections. The direction of the signal is taken as input if a signal comes from the ISA bus controller. An output comes from the slave device and input/output identifies that the signal can originate from either the ISA controller or the slave device.

The following gives the 8-bit PC bus connections:

`SA0-SA19` Address bus (input/output). The lower 20 bits of the system address bus.

`D0-D7` Data bus (input/output). The eight data bits that allow a transfer between the busmaster and the slave.

`AEN` Address enable (output). The address enable allows for an expansion bus board to disable its local I/O address decode logic. It is active high. When active, address enable indicates that either DMA or refresh are in control of the busses.

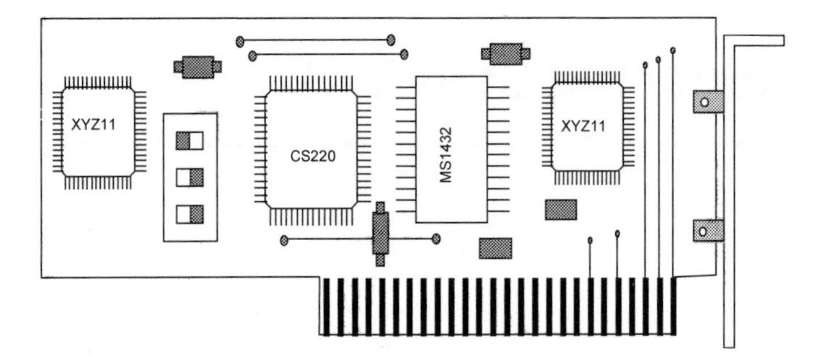

Figure 3.1 PC card

CLK — Clock (output). The bus CLK is set to 4.772 727 MHz (for PC bus and 8.33 MHz for ISA bus) and provides synchronisation of the data transmission (it is derived from the OSC clock).

ALE — Address latch (output). The bus address latch indicates to the expansion bus that the address bus and bus cycle control signals are valid. It thus indicates the beginning of a bus cycle on the expansion bus.

$\overline{\text{IOR}}$ — I/O read (input/output). I/O read command signal indicates that an I/O read cycle is in progress.

$\overline{\text{IOW}}$ — I/O write (input/output). I/O write command signal indicates that an I/O write bus cycle is in progress.

$\overline{\text{SMEMR}}$ — System memory read (output). System memory read signal indicates a memory read bus cycle for the 20-bit address bus range (0h to FFFFFh).

$\overline{\text{SMEMW}}$ — System memory write (output). System memory write signal indicates a memory read bus cycle from the 20-bit address bus range (0h to FFFFFh).

IO CH RDY — Bus ready (input). The bus ready signal allows a slave to lengthen the amount of time required for a bus cycle.

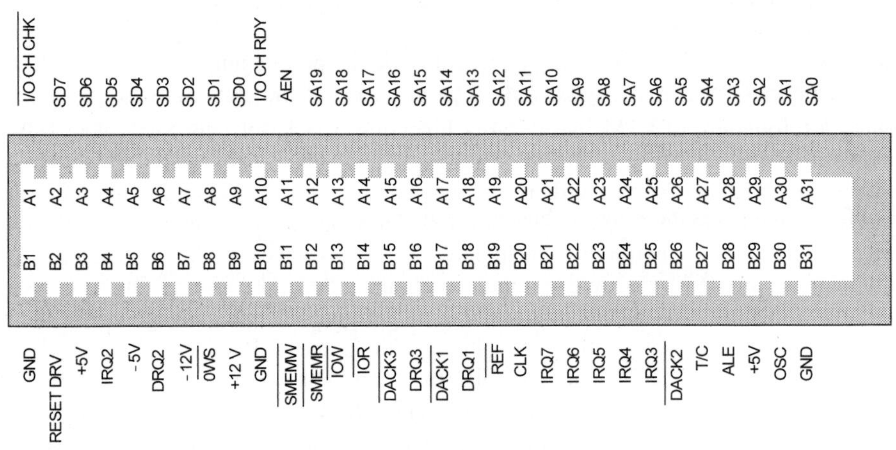

Figure 3.2 PC card connections

$\overline{\texttt{0WS}}$	Zero wait states (input). The zero wait states (or no wait state) allows a slave to shorten the amount of time required for a bus cycle.
`DRQ1-DRQ3`	DMA request (input). The DMA request indicates that a slave device is requesting a DMA transfer.
$\overline{\texttt{DACK1}}$ - $\overline{\texttt{DACK3}}$	DMA acknowledge (output). The DMA acknowledge indicates to the requesting slave that the DMA is handling its request.
`T/C`	Terminal count (input). The terminal count indicates that the DMA transfer has been successful and all the bytes have been transferred.
$\overline{\texttt{REF}}$	Refresh (output). The refresh signal is used to inform a memory board that it should perform a refresh cycle.
`IRQ2-IRQ7`	Interrupt request. The interrupt request signals indicate that the slave device is requesting service by the processor.
`RESET DRV`	Reset drive (output). The reset drive resets and plug-in boards connected to the ISA bus.
`OSC`	Crystal oscillator (output). The crystal oscillator signal is 14.318 18 MHz signal provided for use by expansion boards. This clock speed is three times the CLK speed.
$\overline{\texttt{IO CH CHK}}$	I/O check (input). The I/O check signal indicates that a memory slave has detected a parity error.
`±5V`, `±12V` and `GND`	Power (output).

3.3 ISA bus

IBM developed the ISA (Industry Standard Architecture) for their 80286-based AT (Advanced Technology) computer. It had the advantage of being able to deal with 16 bits of data at a time. An extra edge connector gives compatibility with the PC bus. This gives an extra 8 data bits and 4 address lines. Thus, the ISA bus has a 16-bit data and a 24-bit address bus. This gives a maximum of 16 MB of addressable memory and like the PC bus it uses a fixed clock rate of 8 MHz. The maximum data rate is thus 2 bytes (16 bits) per clock cycle, giving a maximum throughput of 16 MB/sec. In machines that run faster than 8 MHz the ISA bus runs slower than the rest of the computer.

A great advantage of PC bus cards is that they can be plugged into an ISA bus connector. ISA cards are very popular as they give good performance for most interface applications. The components used are extremely cheap and it is a well-proven reliable technology. Typical applications include serial and parallel communications, networking cards and sound cards. Figure 3.3 illustrates an ISA card and Figure 3.4 gives the pin connections for the bus. It can be seen that there are four main sets of connections, the A, B, C and D sections (Figure 3.4). The standard PC bus connection contains the A and B sections. The A section includes the address lines A0–A19 and 8 data lines, D0–D7. The B section contains interrupt lines, IRQ0–IRQ7, power supplies and various other control signals. The extra ISA lines are added with the C and D section; these include the address lines, A17–A23, data lines D8–D15 and interrupt lines IRQ10–IRQ14.

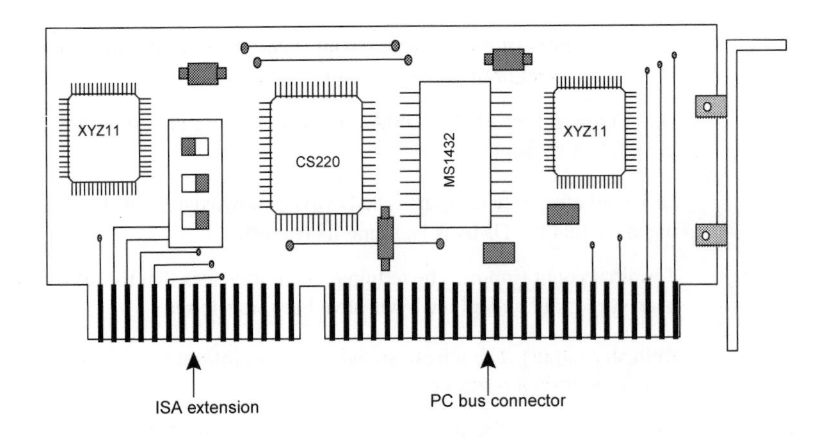

Figure 3.3 ISA card

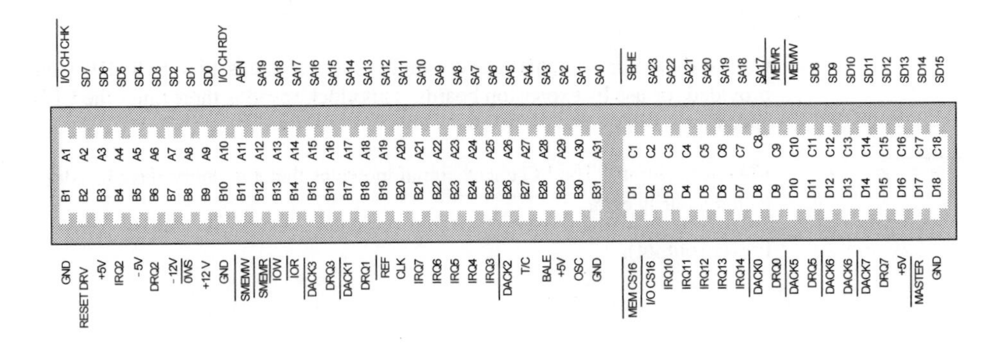

Figure 3.4 ISA bus connections

The Industry Standard Architecture (ISA) bus uses a 16-bit data bus (D0-D15) a 24-bit address bus (A0–A24) and the CLK signal is set to 8.33 MHz. The $\overline{\text{SMEMR}}$ and $\overline{\text{SMEMW}}$ lines are used to transfer data for the lowest 1 MB (`0h` to `FFFFFh`) of memory (where the S prefix can be interpreted as small memory model) and the signals $\overline{\text{MEMR}}$ and $\overline{\text{MEMW}}$ are used to transfer data between 1 MB (`FFFFFh`) and 16 MB (`FFFFFFh`). For example if reading from the address is `001000h` then the $\overline{\text{SMEMR}}$ the line is made active low, while if the address `1F0000h` then the $\overline{\text{MEMR}}$ line is made active. For a 16-bit transfer the $\overline{\text{M16}}$ and $\overline{\text{IO16}}$ lines are made active.

The extra 16-bit ISA bus connections are:

`A17-A23`	Address bus (input/output). The upper 7 bits of the address of the system address bus.
$\overline{\text{SBHE}}$	System byte high enable (output). The system byte high enable indicates that data is expected on the upper 8 bits of the data bus (D8–D15).
`D8-D15`	Data bus (input/output). The upper 8 bits of the data bus provides for the second half of the 16-bit data bus.

$\overline{\texttt{MEMR}}$	Memory read (input/output). The memory read command indicates a memory read when the memory address is in the range 100000h – FFFFFFh (16 MB of memory).
$\overline{\texttt{MEMW}}$	Memory write (input/output). The memory write command indicates a memory write when the memory address is in the range 100000h – FFFFFFh (16 MB of memory).
$\overline{\texttt{M16}}$	16-bit memory slave. Indicates that the addressed slave is a 16-bit memory slave.
$\overline{\texttt{IO16}}$	16-bit I/O slave (input/output). Indicates that the addressed slave is a 16-bit I/O slave.
`DRQ0, DRQ5-DRQ7`	DMA request lines (input). Extra DMA request lines that indicate that a slave device is requesting a DMA transfer.
$\overline{\texttt{DACK0}}$, $\overline{\texttt{DACK5}}$ - $\overline{\texttt{DACK7}}$	DMA acknowledge lines (output). Extra DMA acknowledge lines that indicate to the requesting slave that the DMA is handling its request.
$\overline{\texttt{MASTER}}$	Bus ready (input). This allows another processor to take control of the system address, data and control lines.
`IRQ9-IRQ12, IRQ14, IRQ15`	Interrupt requests (input). Additional interrupt request signals that indicate that the slave device is requesting service by the processor. Note that the `IRQ13` line is normally used by the hard disk and included in the IDE bus.

3.3.1 Handshaking lines

Figure 3.5 shows a typical connection to the ISA bus. The ALE (or sometimes known as BALE) controls the address latch and, when active low, it latches the address lines A2–A19 to the ISA bus. The address is latched when ALE goes from a high to a low.

The Pentium's data bus is 64 bits wide, whereas the ISA expansion bus is 16-bits wide. It is the bus controller's function to steer data between the processor and the slave device for either 8-bit or 16-bit communications. For this purpose the bus controller monitors $\overline{\mathrm{BE0}}$ – $\overline{\mathrm{BE3}}$, W/$\overline{\mathrm{R}}$, $\overline{\mathrm{M16}}$ and $\overline{\mathrm{IO16}}$ to determine the movement of data.

When the processor outputs a valid address it sets address lines (AD2–AD31), the byte enables ($\overline{\mathrm{BE0}}$ – $\overline{\mathrm{BE3}}$) and sets ADS active. The bus controller then picks up this address and uses it to generate the system address lines, SA0–SA19 (which are just a copy of the lines A2–A19. The bus controller then uses the byte enable lines to generate the address bits SA0 and SA1.

The EADS signal returns an active low signal to the processor if the external bus controller has sent a valid address on address pins A2–A21.

It can be seen from Figure 3.6 that the BE0 line accesses the addresses ending with 0h, 4h, 8h and Ch, the BE1 line accesses addresses ending with 1h, 5h, 9h and Dh, the BE2 line accesses addresses ending with 02, 5h, Ah and Eh, and so on.

Thus if the BE0 line is asserted and the SBHE line is high then a single byte is accessed through the D0–D7. If a word is to be accessed then SBHE is low and D0–D15 contains the data.

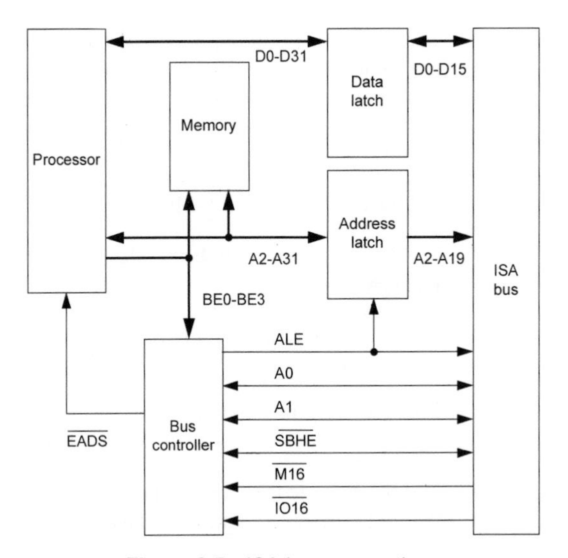

Figure 3.5 ISA bus connections

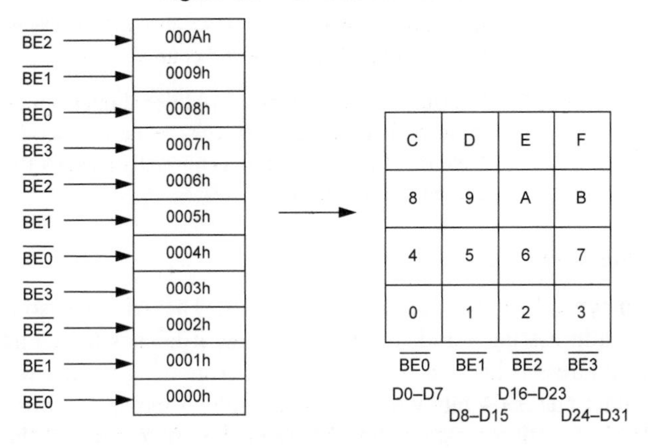

Figure 3.6 Address decoding

Table 3.1 shows three examples of handshaking lines. The first is an example of a byte transfer with an 8-bit slave at an even address. The second example gives a byte transfer for an 8-bit slave at an odd address. Finally, the table shows a 2-byte transfer with a 16-bit slave at an even address.

Table 3.1 Example handshaking lines

BE0	*BE1*	*BE2*	*BE3*	*IO16*	*M16*	*SBHE*	*SA0*	*SA1*	*Data*
0	1	1	1	1	1	1	0	0	SD0–SD7
1	0	1	1	1	1	0	1	0	SD8–SD15
0	0	1	1	0	1	0	0	0	SD0–SD15

If 32-bit data is to be accessed then BE0–BE3 will each be 0000 which makes 4 bytes active. The bus controller will then cycle through SA0, SA1 = 00 to SA0, SA1 = 11. Each time the 8-bit data is placed into a copy buffer which is then passed to the processor as 32 bits.

3.3.2 82344 IC

Much of the electronics in a PC has been integrated onto single ICs. The 82344 IC is one that interfaces directly to the ISA bus. Figure 3.7 shows its pin connections.

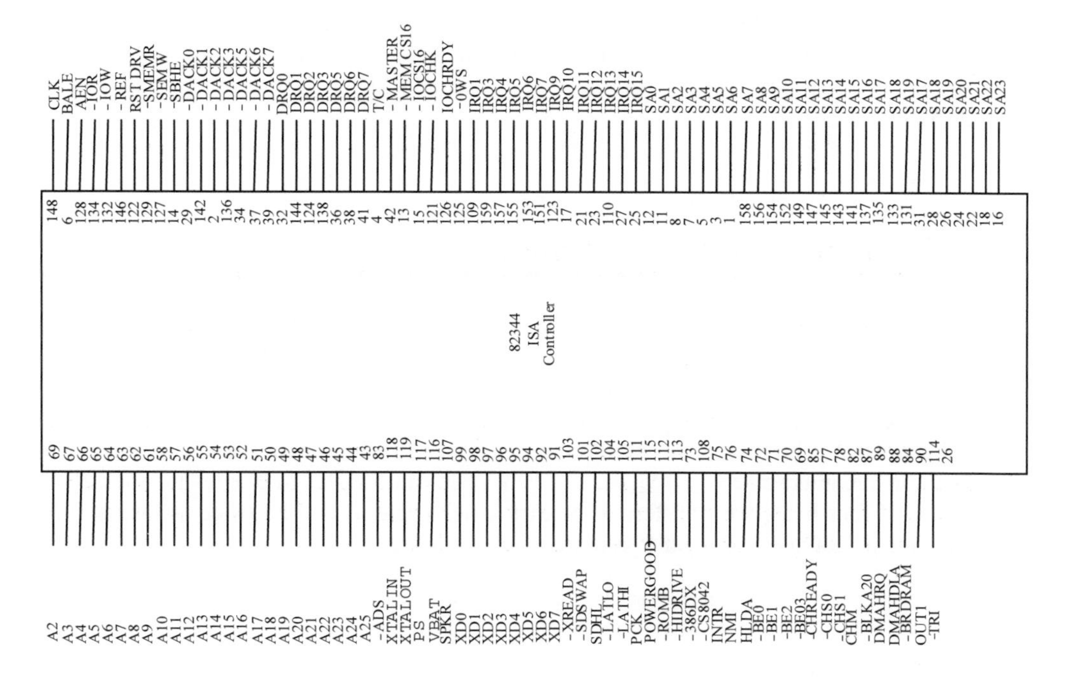

Figure 3.7 82344 IC connections

3.4 Other legacy busses

Two other busses which were used in the past are:

- **MCA**. IBM developed the Microchannel Interface Architecture (MCA) bus for their PS/2 computers. This bus is completely incompatible with ISA bus. It can operate as a 16-bit or 32-bit data bus. The main technical difference between the MCA and PC/ISA (and EISA) is that the MCA is an asynchronous bus whereas PC/ISA/EISA use a synchronous bus. An synchronous bus works at a fixed clock rate whereas an asynchronous bus data transfer is not dependent on a fixed clock. Asynchronous buses take their timings from the devices involved in the data transfer (that is, the processor or system clock). The original MCA specification resulted in a maximum transfer rate of 160 MB/sec. Very few manufacturers have adopted MCA technology and it is mainly found in IBM PS/2 computers.
- **EISA**. Several manufacturers developed the EISA (Extended Industry Standard Architecture) bus in direct competition to the MCA bus. It provides compatibility with PC/ISA but not MCA. The EISA connector looks like an ISA connector. It is possible to plug an ISA card into an EISA connector, but a special key allows the EISA card to be inserted

deeper into the EISA bus connector. It then makes connections with a 32-bit data and address bus. An EISA card has twice the number of connections over an ISA card and there are extra slots that allow it to be inserted deeper into the connector. The ISA card only connects with the upper connectors because it has only a single key slot. EISA uses a synchronous transfer at a clock speed of 8 MHz. It has a full 32-bit data and address bus and can address up to 4 GB of memory. In theory the maximum transfer rate is 4 bytes for every clock cycle. As the clock runs at 8 MHz, the maximum data rate is 32 MB/s.

3.5 Comparison of different types

Data throughput depends on the number of bytes being communicated for each transfer and the speed of the transfer. With the PC, ISA and EISA buses this transfer rate is fixed at 8 MHz, whereas the PCI and VL local buses use the system clock. For many applications the ISA bus offers the best technology as it has been around for a long time, it gives a good data throughput and it is relatively cheap and reliable. It has a 16-bit data bus and can thus transfer data at a maximum rate of 16 MB/s. The EISA bus can transfer four bytes for each clock cycle, thus if four bytes are transferred for each clock cycle, it will be twice as fast as ISA. The maximum data rates for the different interface cards are:

PC	8 MB/s	
ISA	16 MB/s	
EISA	32 MB/s	
VL-Local bus	132 MB/s	(33 MHz system clock using 32-bit transfers)
PCI	264 MB/s	(33 MHz system clock using 64-bit transfers)
MCA	20 MB/s	(160 MB/s burst)

The type of interface technology used depends on the data throughput. Table 3.2 shows some typical transfer data rates. The heaviest usage on the system are microprocessor to memory and graphics adaptor transfers. These data rates depend on the application and the operating system used. Graphical user interface (GUI) programs have much greater data throughput than programs running in text mode. Notice that a high specification sound card with recording standard quality (16-bit samples at 44.1 kHz sampling rate) only requires a transfer rate of 172 kB/s. The transfer rate for audio is:

$$
\begin{aligned}
\text{Transfer rate (hi - fi)} &= \text{Number of samples per second} \times \text{Number of bytes per sample (B/s)} \\
&= (44100 \times 2) \times 2 \text{ B/s} \\
&= 176\,400 \text{ B/s} \\
&= \frac{166\,400}{1024} = 172.26 \text{ kB/s}
\end{aligned}
$$

A standard Ethernet local area network card transfers at data rates of up 10 Mbps (about 1 MB/s), although new fast Ethernet cards can transfer at data rates of up to 100 Mbps (about 10 MB/s). These transfers thus require local bus type interfaces.

Table 3.2 Example transfer rates

Device	*Transfer rate*	*Application*
Hard disk	4 MB/s	Typical transfer
Sound card	88 KB/s	16-bit, 44.1 kHz sampling
LAN	1 MB/s	10 Mbps Ethernet
RAM	66 MB/s	Microprocessor to RAM
Serial Communications	1 KB/s	9600 bps
Super VGA	15 MB/s	1024×768 pixels with 256 colours

For a graphics adaptor with a screen resolution of 1024×640, 64k colours (16-bit colour) which is updated 20 times per second (20 Hz), the maximum transfer rate will be:

$$\begin{aligned}\text{Transfer rate (max)} &= \text{No. of pixels per screen} \times \text{No. of bits per pixel} \times \text{No. of screen updates (b/s)} \\ &= (1024\times640)\times16\times20 \text{ b/s} \\ &= 209\,715\,200 \text{ b/s} \\ &= \frac{209\,715\,200}{8} = 26\,214\,400 \text{ B/s} \\ &= \frac{26\,214\,400}{1024\times1024} = 246.2 \text{ MB/s}\end{aligned}$$

The PCI Local bus has become a standard on most new PC systems and has replaced the VL-local bus for graphics adaptors. It has the advantage over the VL-local bus in that it can transfer at much higher rates. Unfortunately, most available software packages cannot use the full power of the PCI bus because they do not use the full 64-bit data bus. PCI and VL-local bus are discussed in the next chapter.

3.6 Exercises

3.6.1 How many bits are transferred in a single clock operation with the PC bus:

(a) 1 (b) 8
(c) 16 (d) 32

3.6.2 What is the standard clock frequency which is used in ISA transfers:

(a) 4.77 MHz (b) 8 MHz
(c) 10 MHz (d) 16 MHz

3.6.3 What is the maximum transfer rate for the ISA bus:

(a) 8 MB/s (b) 16 MB/s
(c) 32 MB/s (d) 64 MB/s

3.6.4 What is the maximum transfer rate for the EISA bus:

(a) 8 MB/s (b) 16 MB/s

(c) 32 MB/s (d) 64 MB/s

3.6.5 What is the main disadvantage of PC, ISA and EISA busses:

(a) They are incompatible with each other.
(b) They use a fixed clock frequency.
(c) They are not supported in PC systems.
(d) They are expensive to implement.

3.6.6 What is the maximum transfer rate (in B/s) for a 10 Mbps Ethernet adaptor:
(a) 1 MB/s (b) 1.221 MB/s
(c) 10 MB/s (d) 100 MB/s

3.6.7 Prove, apart from the MCA bus, the transfer rates given in Section 3.5.

3.6.8 If an audio card is using 16-bit sampling at a rate of 44.1kHz. Prove that the transfer rate for stereo sound will be 176.4 KB/sec. Show, also that this is equivalent to 1.411 Mbps (note that this is approximately the standard rate for CD-ROMs). Can this rate be transferred using the ISA bus? Using this transfer rate, determine the maximum transfer speed of a ×32 CD-ROM drive.

3.6.9 Determine amount of data for a single screen that required to be transferred for the following screen resolutions:

(a) 800×600, 65 536 colours (960 000 B/s).
(b) 800×600, 16 777 216 colours (1 440 000 B/s).
(c) 1024×768, 65 536 colours (1 572 864 B/s).

Determine the maximum number of screen updates that is required for a 32-bit PCI bus for each of the above.

3.6.10 Identify the main ISA signal lines and how they are used to transfer data. Answer clearly the following:

(a) The main differences between a PC card and an ISA card.
(b) How the byte enable lines are used.
(c) How is a read or write transfer identified?
(d) How is a memory read/write or isolated memory read/write transfer identified?
(e) The interrupt lines which are available on a PC card and an ISA card.

Typically what devices could be supported by a PC card (that is, what devices use the interrupts that a PC card can support). How does this related to the original specification of the PC?

3.7 Summary of interface bus types

Devices connect to the microprocessor using a computer bus. The specification of this bus defines the transfer speed between the microprocessor and the connected device. Peripherals can connect to the computer using either an internal or an external interface. Table 3.3 and Table 3.4 give some examples of typical PC interfaces.

Table 3.3 Internal PC busses

Bus	*Description*	*Typical devices connected*
ISA	The ISA (International Standards Architecture) bus uses an interface card which has two edge connectors (as one of the connectors was used on the original PC bus). Typical ISA connections are network interface adaptors, video camera interfaces and sound cards. It can transfer up to 16 bits at a time, and uses a fixed transfer rate of 8 MHz (8 000 000 transfers every second). Max: 16 MB/s.	Network adaptor Video camera adaptor Sound card
PCI	The PCI (Peripheral Component Interconnection) bus is used connect internal devices in the PC. Typically, modern PCs have at least four PCI adaptors, which are used to connect to network interface cards, graphics adaptors and sound cards. It can transfer up to 32 bits at a time. Max: 132 MB/s (more typically, 66 MB/s).	Network adaptor Video camera adaptor Sound card
AGP	The AGP (Accelerated Graphics Port) bus is used solely to connect to video cards. It uses a special connector, but builds on the standard PCI bus. It is optimised so that it uses the main memory of the computer, and does not depend on memory on the graphics card. Max: 500 MB/s.	Graphics adaptor
IDE	The Integrated Drive Electronics (IDE) bus is used solely to connect to either hard disk drives, or CD-ROM drives. There are two IDE connections: IDE0 and IDE1. Up to two devices can connect to each IDE connector, thus up to four disk drives can connect to the IDE bus. Max: 16.6 MB/s (IDE, Mode 4).	Hard disk drive CD-ROM drive
PCMCIA	The Personal Computer Memory Card International Association (PCMCIA) interface allows small thin cards to be plugged into laptop, notebook or palmtop computers. It was originally designed for memory cards (Version 1.0) but has since been adopted for many other types of adapters (Version 2.0), such as fax/modems, sound-cards, local area network cards, CD-ROM controllers, digital I/O cards, and so on. Most PCMCIA cards comply with either PCMCIA Type II or Type III. Type I cards are 3.3 mm thick, Type II take cards up to 5 mm thick, Type III allows cards up to 10.5 mm thick. A new standard, Type IV, takes cards which are greater than 10.5 mm. Type II interfaces can accept Type I cards, Type III accept Types I and II and Type IV interfaces accept Types I, II and III. It uses a 16-bit data transfer.	Network adaptor Modem adaptor Sound card CD-ROM drive Memory upgrade Hard disk drive

Table 3.4 External PC busses

Bus	*Description*	*Typical devices connected*
SCSI	The SCSI (Small Computer System Interface) bus is used to connect to a wide range of device, and is typically used in workstations and Apple computers. It allows devices to connect using cables which connect from one to the next (a daisy chain). In its standard form, it allows for up to seven devices to be connected (SCSI-I), but new standards (SCSI-II/III) allow up to 15 devices to connect. It can also be used as an internal bus system or as an external one. In Apple Macs and workstations, SCSI is used to connect hard disk drives. Max: 5 MB/s (SCSI-I), 20 MB/s (SCSI-II), 40 MB/s (SCSI-III).	Hard disk drive CD-ROM drive Scanner Back-up device
RS-232	RS-232 is a standard interface on most computer systems. It uses serial communications, to send data one bit at a time. The speed of the transfer is set by the bit rate. Typical bit rates are 9600 bps (bits per second), 14 400 bps, 28 800 bps and 56 000 bps. It is typically used to transfer files from one computer to another, and to connect to a modem. In the past, it was also used to connect to a serial mouse, but mice typically connect using the PS/2 mouse connector. Typically PCs have one or two serial port, which are given the names: COM1: and COM2:. Max: 7 KB/s (56 000 bps)	Modem Mouse File Transfer (with Null Modem cable)
Parallel port	The parallel port transfers 8 bits of data at a time. In its standard form, it only supports a maximum rate of 150kbps, with only one connected device. It also slows down the processor, as it must involve itself with the transfer of the data. In its standard form, it uses a 25-pin D-type connector to connect to the PC. As technology has improved a new standard named ECP (Extended Capabilities Port Protocol)/EPP (Enhanced Parallel Port) was been developed to increase the data rate, and also to connect multiple devices to it (as the SCSI bus). These allow the transfer of data to be automatically controlled by the system, and not by the processor. Typically, now with ECP/EPP, several devices can connect to the port, such as a printer, external CD-ROM drive, scanner, and so on. Its main advantage is that it is standard on most PCs, but suffers from many disadvantages. Typically PCs have a single parallel port, which is given the name LPT1:. In many cases, it is being replaced by USB. Max: 150 kbps (Standard), 1.5 Mbps (ECP/EPP)	Printer CD-ROM drive Scanner File Transfer (with Parallel Port cable)
USB	USB (Universal Serial Bus) allows for the connection of medium bandwidth peripherals such as keyboards, mice, tablets, modems, telephones, CD-ROM drives, printers and other low to moderate speed external peripherals in a tiered-star topology. It is typically used to connect to printers, scanners, external CD-ROM drives, digital speakers, and so on. It is likely to replace the printer port and the serial port for connecting external devices. Max: 12 Mbps.	Digital speakers Scanner Printer Video camera Modem Joystick Monitor
PS/2 Port/ Keyboard	Initially, on PCs, the serial port was used to connect a mouse to, which reduced the number of connections to the serial port. Typically, these days, a mouse connects to the PS/2 mouse port, which has a small 5-pin DIN-like connector. This is the same connector that is used to connect to the keyboard.	Keyboard Mouse

3.8 The fall of the MCA bus

The leading computer companies of 1987 were Intel, IBM, Compaq, and Microsoft, but a special mention must go to Apple, Commodore and Sun Microsystems, who fought bravely against the growing IBM PC market. With the release of the IBM AT and now the PS/2, IBM had presented Intel with a large market for their 80286 design, but it was Compaq who increased it even more with the release of the DeskPro 386. The new Intel processors were now so successful that Intel had little to do, but try to keep up with demand, and try to stop cloners from copying their designs. They could now consolidate on their success with other support devices, such as the 80387-math coprocessor. By the end of the year, their only real mass-market competitor was Motorola, who released their excellent Motorola 68030 microprocessor, which would become the foundation of many Apple Mac computers. Within a few years, Motorola would become extremely reliant on the Apple Mac, while Intel held onto the PC market.

At IBM, things were hectic. They were phasing out their IBM PC range, and introduced their new computer range, the PS/2. IBM realised that the open architecture of the IBM PC held little long term advantages for them, as clone manufacturers could always sell computers at much less cost than themselves. IBM had large development teams, sales staff, distribution centres, training centres, back up support, and so on. They thus need to make enough profit on each computer to support all these functions. The PS/2 was their attempt at trying to close the open system, and make one that required to be licensed through themselves. It was also an attempt at trying to reduce some of the problems that were caused with the limited technology of the IBM PC. One of the main problems was the PC bus, which allowed users to easily add peripherals to the computer, by plugging them into the system with a standard card which had a standard edge connector. Initially this used an 8-bit data bus, and operated at 8 MHz, which gave a data throughput of 8 MB/s (as one byte is 8 bits). This was upgraded on the AT computer with the AT bus, which used a 16-bit data bus, giving a data throughput of 16 MB/s. The great advantage of the AT bus, was that it was still compatible with the PC bus, where PC bus cards could still be slotted into AT bus connectors (soon to be renamed the ISA, or Industry Standard Bus).

The AT bus was fine for slow devices, such as printer, modems, and so on. However, for colour graphics it was far to slow. For example, a colour monitor with a resolution of 640×480, with 256 colours (8 bits per colour), and a screen refresh of 25 Hz, requires a data throughput of 7 680 000 B/s (640×480×1×25 B/s).

IBM's concept was to use a bus, which, intentionally, had a different connector to the PC and the AT bus, which did not use a fixed clock rate, and could thus operate at the speed of the processor, which was now moving above 20 MHz. The MCA (Micro Channel Architecture) bus also used a 32-bit data bus, which allowed data throughputs of 100 MB/s.

The PS/2 was an excellent concept, and was boosted by an extensive advertising campaign that boosted performance improvement over previous systems. It was the right move, and the system looked well, with 3.5-inch disk drives, and a rugged gray plastic case. Computers had never look so professional. For many businesses, they were heaven sent. However, the fly-in-the-ointment for IBM was Compaq, who had previously released their DeskPro 386. The big problem with the PS/2 range was that the lower-end PCs were based on the 8086 and the 80286, and against the 80386-based Compaq they seem slow. The initial range was:

- Model 30, which used the, at the time, relatively slow, 8 MHz 8086. IBM also introduced the Model 25, also with an 8 MHz Intel 8086, which had no hard drive, and a reduced keyboard for $1350.

- Model 50 and 60, which used a 10 MHz 80286 with MCA.
- Model 80, which used a 20 MHz 80386 with MCA.

In 1988, the battle lines had been drawn the year before. IBM was trying to pull the market towards their architecture. The strength, of this was highlighted by John Akers, the then IBM Chairman:

We're trying to change the habits of an awful lot of people. That won't happen overnight, but it will bloody well happen.

IBM thought they would win the battle, and the older IBM PC architecture would die off. Several companies went with IBM, including Tandy (Tandy 5000MC), Dell and Olivetti. But the first signs of problems for IBM came when 61 companies developed the 32-bit version of the ISA bus, the EISA (Extended Industry Standard Architecture). This allowed 32 bits to be transferred at a time. Unfortunately, it was still based on an 8 MHz clocking rate, which gave it a data throughput of 32 MB/s. It was supported by the leading computer companies, such as Compaq Computer, AST, Epson, Hewlett-Packard, NEC Technologies, Olivetti, Tandy, Wyse, Zenith, and Microsoft. Along with this, Compaq Computer and eight other companies started developing the ISA standard to improve the AT-bus. Rod Canion, the Compaq Computer CEO, showed his company's resistance to the MCA bus:

If people are going to buy Micro Channel, they're going to buy it from IBM.

The market would eventually reject the MCA bus, mainly because of the weight of the new x86, 80386 computers on the market. It would take a company such as Intel to develop a totally new bus system: the PCI bus.

3.9 Notes from the author

Oh boy, is it confusing! There are so many different busses used in systems, especially in PCs. There are three main reasons for the number of busses: legacy, compatibility and efficiency. With legacy, busses exist because they have existed in the past, and are required to be supported for the time being. With compatibility, busses allow the segmentation of a system and provide, most of the time, a standard interface that allows an orderly flow of data and allows manufactures to interconnect their equipment. If allowed, many manufacturers would force users to use their own versions of busses, but the trend these days is to use internationally defined busses. Efficient busses have been designed to take into account the type of data that is being transmitted. So, in the end, we win.

Sometimes, though, the bus technology does always win, and manufacturers who try to develop systems on their own can often fail miserably. This has been shown with the MCA bus, which was an excellent step in bus technology, and made up for the mistakes made in the original PC design. But, IBM tried to force the market, and failed. In these days, it is international standards that are important. Products to be accepted in the market or in the industry require an international standards body to standardise them, such as the IEEE, the ISO, ANSI and so on. Without them, very few companies will accept the product. A classic case of an open standard winning against a closed system was Betamax video which was a closed standard produced by Sony, up against VHS which was an open standard. VHS was the inferior technology, but won because there were more companies willing to adopt it.

The days of having a single computer bus for internal and external connections are a long way off (if ever), as there will always be some peripherals that need to transmit data in a certain way that differs from other peripheral. Also standard technology always tends to win over newer, faster technology. Few companies can now define new standards on their own.

Before we start to look at the technology behind computer busses, here is my All-Time Best Busses (in order of their current and future usefulness).

TOP BUSSES OF ALL-TIME AWARD

1. ***PCI bus***. *An excellent internal bus that provides the backbone to most modern PCs. It has been a complete success, and provides for many modern enhancements, such as plug-and-play technology, steerable interrupts, and so on. In its acceptance speech for Best Bus of All-Time it would thank the VL-local bus for starting the trend toward local bus technology. The VL-local bus held-the-fort for a short time, and gave a short-term fix for high-speed graphics transfers, but Intel busily developed a proper bus which could support other high-speed devices. With local bus technology, low-speed devices were pushed away from the processor, and can only communicate with it over a bridge. A worthy champion that is the bedrock of modern computing. It even has a few trump cards yet to play (including increasing its transfer rate, integration with the AGP port, and increasing its data bus size).*
2. ***SCSI bus***. *The most general-purpose of the external busses and in many respects as great as the PCI bus, but it looses out to the PCI bus in that it is not used in as many computers. It provides an easy method of connecting external devices in a daisy-chain connection. New standards for the SCSI bus support fast transfer rates (over 40 MB/s), and allow up to 15 internal or external devices to be connected.*
3. ***USB***. *An external bus which shows great potential in the way that it integrates many of the low and medium bit rate devices onto a single bus system. New standards for USB are trying to also integrate high bit rate devices. It supports hot plug-and-play, which allows users to connect and disconnect peripherals from the bus, while the computer is still on.*
4. ***AGP port***. *The PCI, SCSI and USB busses are a long-way out in front to the other busses, and the forth place in the table goes to AGP port which overcomes the last great problem area of the PC: the graphics adaptor. AGP provides for fast transfer rates using the PCI bus as a foundation, and allows the PC to use local memory for graphics transfers.*
5. ***PCMCIA***. *A long way behind in fifth place comes the PCMCIA bus, which is an external bus that provides for easy upgrades on notebook technology. It highlights how small and compact interface devices can be. Typical additions are modem and network adaptors. Its future will depend on how the USB bus is used in the future.*

And let's not forget the busses which have helped us to get to this point. We may call them legacy busses, but they have allowed us to get to where we are now, and still provide a useful function. Thus, the awards for the Most Helpful Busses of the Past (in order of their previous usefulness) are given next.

MOST HELPFUL BUSSES OF ALL-TIME

1. ***ISA bus***. *The ISA bus competed head-to-head with the MCA, and although it was much slower, it triumphed, as it was compatible with the older PC bus. For many years its performance was acceptable (16MB/s), but the advent of the graphical user interface was the beginning of the end for it. It has sadly seen the number of PCI slots increasing while it has seen its own connections reduce from over five, to less than two.*
2. ***RS-232 port***. *A classic bus, which is compatible (almost) with all the other RS-232 ports on every computer in the world. It provides a standard way to talk to devices, such as instruments, other computers, and modems.*
3. ***IDE bus***. *A rather quiet and unassuming internal bus, which does its job of interfacing to disk drives well, without, these days, much troubles. It has reasonable performance (over 14MB/s) and does not really have any intentions of ever becoming anything other than a disk interfacing bus. It, like Ethernet, has overcome early retirement by increasing its transfer rate, but still keeping compatibility with previous systems. Its main competitor is SCSI, which is unlikely to ever to beat it for compatibility and cost, thus it is likely to stay around for much longer than some of its earlier PC partners. For systems which have less than four disk drives (in a combination of CD-ROM or hard disks) it is still the best choice, and is often integrated in the PC motherboard.*
4. ***VL-Local bus***. *The bus that showed the way for local bus technology, especially the PCI bus. It always knew that it was a short-term fix, but it did its job effectively and quietly. Apart from the 80486, it was one of leading factors which increased the adoption of Microsoft Windows (as it allowed the fast transfer to graphical data).*
5. ***Parallel port***. *Another classic bus which was created to purely interface to an external printer, but has now been developed to support a multi-attachment busses with reasonable transfer rates (over 150kB/s).*

Let's not forget about the great-grandfather of all the PC busses: the PC bus. It is now enjoying a well-earned retirement (but can be pulled back from retirement at any time). It is the bus that has launched a million computers.

Finally, the relegation zone for computer busses (in order of the problems that they have caused or for their lack of adoption).

RELEGATION ZONE FOR BUSSES

1. ***ISA bus***. *Like the 8088 processor and DOS, it has a Dr Jekyll and Mr Hyde appearance, and is both the top computer bus of all-time, and worst bus of all-time. It is the bus, which, in the past, has provided the foundation for upgrading the PC, and has gently handed over its mantel to the PCI bus. But, it has caused lots of problems as it quickly fossilized the connection between the processor and the peripherals. Its major problems were it fixed rate transfer rate, the way that it handled interrupts, its lack of address lines (only up to 24), its lack of data lines (only up to 16), and the way that fast, medium and slow devices all connected to the same bus (thus allowing slow devices to 'hog' the bus). It started to show its age when 32-bit processors appeared and as the motherboard speeds increased. The beginning of its end was the introduction of Microsoft Windows 3, which started to properly use a graphical user interface. The VL-local bus quickly came in as a stand-in. From there on, local bus technology*

become the standard way to transfer large amounts of data.

2. ***MCA bus****. IBM tried to pull the standard for bus technology back to a closed system with the MCA bus. It failed as it came up against the technologically inferior ISA, as the ISA bus was an open standard.*
3. ***VME bus****. Powerful, complex and very misunderstood. Avoid in the same way that you would avoid the plague. The designers decided to create a bus that had everything, all that's lacking is kitchen sink.*
4. ***RS-232 port****. Another bus, along with the ISA bus, which manages to get into the Top 5 busses of all-time, and also into the relegation zone. It is an extremely useful bus, but suffers because of a lack of speed and its incompatibility (even although everyone is working to the same standard, the level of implementation of the standard varies).*
5. ***Keyboard connection/serial port mouse connection****. Two extremely limited connectors. The keyboard connector has virtually no intelligence built into it, and provides for limited sensing of the keyboard type, or extra functionality. Its only real advantage is that it looks so different from other connectors. Another advantage is that is has allowed the integration of the new PS/2-type mouse connector. Serial port mouse connections have always caused problems, mainly because the use up one of the serial port connection.*

Well that's the bus awards. As we still have half a page to complete, here's the winners of the specialist technologies:

- ***Local Area Networking bus award****:*
  1. *Ethernet (one horse race). Although ISDN deserves a mention.*
- ***Wide Area Networking bus award****:*
  1. *ATM. A proper networking technology that integrates real-time signals (such as speech) and non-real-time data (such as computer data) into a networking technology.*
  2. *Gigabit Ethernet. Uses standard Ethernet technology to gives an extremely fast transfer rate, but still suffers from the problems of the original Ethernet specification.*
  3. *FDDI. A reliable ring-based networking technology that allows for the transmission of data over a large geographical area.*
- ***Instrumentation bus award****:*
  1. *RS-422. An excellent bus which allows multiple attachments to a single bus, and up to 1Mbps.*
  2. *IEEE-488. An easy-to-use, robust, widely used, standardised and easy-to-connect-to bus, which is rather limited, but extremely useful.*
  3. *CAN bus. An excellent multi-drop system, for closely connected devices, especially in automobile applications. It overcomes some of the problems that Ethernet has, as devices are so closely connected.*
- ***Interconnecting protocol award****.*
  1. *TCP/IP (one horse race).*

4 PCI Bus

4.1 Introduction

The PC was conceived at a time when processor clock speeds were measured in several MHz. Initially this was set at 4.77 MHz, and then increased to 8 MHz. The PC and ISA busses fossilised with these clock frequencies. In the first few years of its design, the motherboard ran at the same speed as the processor. Soon, with improvements in silicon design, the speed of the processor was increased to tens of megahertz. Soon the maximum limit of the motherboard was reached and the only way to break this limit was to double or treble the motherboard clock speed. This limit was set at 33 MHz or 50 MHz. Processor speed has since been risen to over 500 MHz. Local bus technology uses the speed of the motherboard, rather than a fixed rate. Most new PCs have a motherboard speed of 100 MHz, which is at least twice as fast as 50 MHz motherboards.

The greatest need for greater data throughput is the video adaptor. A high-resolution video screen with high screen update rate can require burst rates of over 100 MB/s. For example a screen of 1024×640 with 16.7 million colours (24-bit colour) will require the following amount of memory for a single screen:

$$\begin{aligned}\text{Memory} &= 1024\times 640\times 3 \ \ \text{B}\\ &= 1\,966\,080\ \text{B}\\ &= \frac{=1\,966\,080}{1024\times 1024} = 1.875\text{MB}\end{aligned}$$

If this screen is updated 10 times every second (10Hz) then the data throughput is
:

$$\begin{aligned}\text{Data transfer} &= 1.875\times 20 \ \ \text{MB/s}\\ &= 37.5\ \text{MB/s}\end{aligned}$$

This transfer rate is far too fast for busses such as ISA and EISA, and the only solution is a fast 32-bit bus, transferring at a rate of at least 33 MHz. The maximum transfer rates for various local bus transfers are as follows:

Data bus size	Transfer clock (MHz)	Data transfer rate (MB/s)
16	33	66
16	50	100
32	33	132
32	50	200
32	100	320
64	50	400

Intel have developed a standard interface, named the PCI (Peripheral Component Interconnection) local bus, for the Pentium processor. This technology allows fast memory, disk and video access. A standard set of interface ICs known as the 82430 PCI chipset is available to interface to the bus. Figure 4.1 shows how the PCI bus integrates into the PC. The processor runs at a multiple of the motherboard clock speed, and is closely coupled to a local SRAM cache (first-level, or primary, cache). If the processor requires data it will first look in the primary cache of its contents. If it is in this cache it will read its contents, and there is thus no need to either read from the second level cache or from DRAM memory. If the data is not in the primary cache then the processor slows downs to the motherboard clock speed, and contacts the system controller (which contains a cache controller). The controller then examines the second-level cache and if the contents are there, it passes the data onto the processor. If it does not have the contents then DRAM memory is accessed (which is a relatively slow transfer).

The system controller also interfaces to PCI bus, which is running at the motherboard clock frequency. This then bridges onto other busses, such as ISA, IDE and USB, each of which is running at different clock rates. The PCI bus thus provides a foundation bus for most of the internal and external busses.

Local bus design involves direct access to fast address and data busses. The ISA bus was a great bottleneck because it could only run at 8 MHz. This chapter discusses the VL-local bus and the PCI bus. The PCI bus is now the main interface bus used in most PCs, and is rapidly replacing the ISA bus for internal interface devices. It is a very adaptable bus and most of the external busses, such as SCSI and USB connect to the processor via the PCI bus.

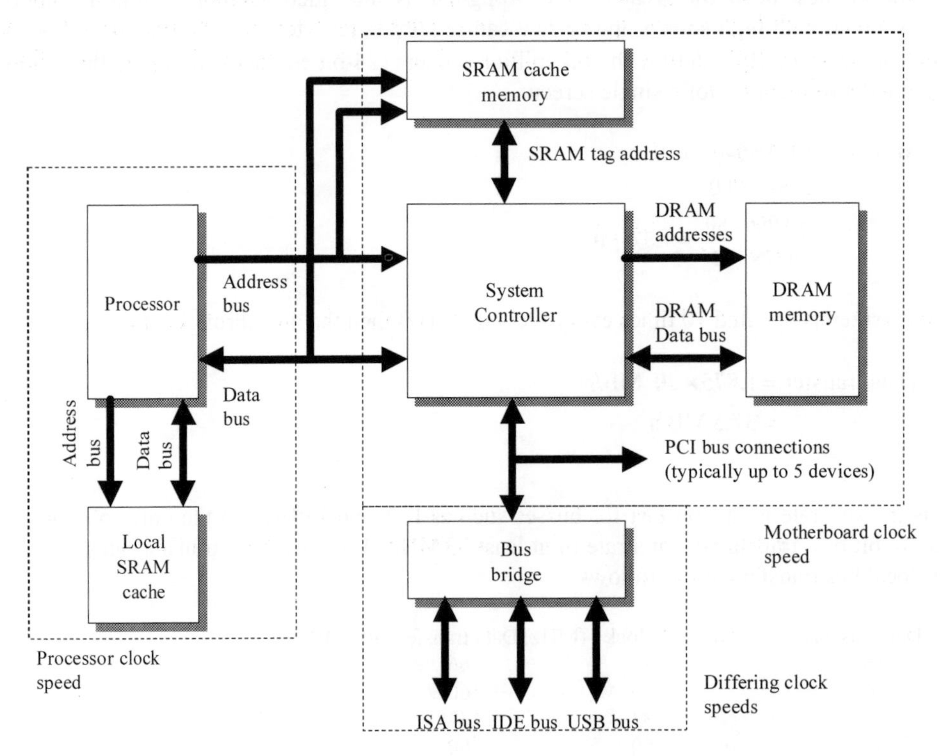

Figure 4.1 Local bus architecture

The PCI bus transfers data using the system clock, and can operate over a 32-bit or 64-bit data path. The high transfer rates used in PCI architecture machines limit the number of PCI bus interfaces to two or three (normally the graphics adapter and hard disk controller). If data is transferred at 64 bits (8 bytes) at a rate of 33 MHz then the maximum transfer rate is 264 MB/s. Figure 4.1 shows the PCI architecture. Notice that an I/O bridge gives access to ISA, IDE and USB. Unfortunately, to accommodate for the high data rates and for a reduction in the size of the interface card, the PCI connector is not compatible with PC, ISA or EISA.

The maximum data rate of the PCI bus is 264 MB/s, which can only be achieved using 64-bit software on a Pentium-based system. On a system based on the 80486 processor this maximum data rate will only be 132 MB/s (that is, using a 32-bit data bus).

The PCI local bus is a radical redesign of the PC bus technology. Table 4.1 lists the pin connections for the 32-bit PCI local bus and it shows that there are two lines of connections, the A and the B side. Each side has 64 connections giving 128 connections. A 64-bit, 2×94-pin connector version is also available. The PCI bus runs at the speed of the motherboard which for the Pentium processor is typically 33 MHz.

Table 4.1 32-bit PCI local bus connections

Pin	*Side A*	*Side B*	*Pin*	*Side A*	*Side B*
1	−12V	$\overline{\text{TRST}}$	32	AD17	AD16
2	TCK	+12V	33	$\overline{\text{C / BE2}}$	+3.3V
3	GND	TMS	34	GND	$\overline{\text{FRAME}}$
4	TDO	TDI	35	$\overline{\text{IRDY}}$	GND
5	+5V	+5V	36	+3.3V	$\overline{\text{TRDY}}$
6	+5V	$\overline{\text{INTA}}$	37	$\overline{\text{DEVSEL}}$	GND
7	$\overline{\text{INTB}}$	$\overline{\text{INTC}}$	38	GND	$\overline{\text{STOP}}$
8	$\overline{\text{INTD}}$	+5V	39	$\overline{\text{LOCK}}$	+3.3V
9	$\overline{\text{PRSNT1}}$	Reserved	40	$\overline{\text{PERR}}$	SDONE
10	Reserved	+5V(I/O)	41	+3.3V	$\overline{\text{SBO}}$
11	$\overline{\text{PRSNT2}}$	Reserved	42	$\overline{\text{SERR}}$	GND
12	GND	GND	43	+3.3V	PAR
13	GND	GND	44	$\overline{\text{C / BE1}}$	AD15
14	Reserved	Reserved	45	AD14	+3.3V
15	GND	$\overline{\text{RST}}$	46	GND	AD13
16	CLK	+5V(I/O)	47	AD12	AD11
17	GND	$\overline{\text{GNT}}$	48	AD10	GND
18	$\overline{\text{REQ}}$	GND	49	GND	AD09
19	+5V(I/O)	Reserved	50	KEY	KEY
20	AD31	AD30	51	KEY	KEY
21	AD29	+3.3V	52	AD08	$\overline{\text{C / BE0}}$
22	GND	AD28	53	AD07	+3.3V
23	AD27	AD26	54	+3.3V	AD06
24	AD25	GND	55	AD05	AD04
25	+3.3V	AD24	56	AD03	GND
26	$\overline{\text{C / BE3}}$	IDSEL	57	GND	AD02
27	AD23	+3.3V	58	AD01	AD00
28	GND	$\overline{\text{FRAME}}$	59	+5V(I/O)	+5V(I/O)
29	AD21	AD20	60	$\overline{\text{ACK64}}$	$\overline{\text{REQ64}}$
30	AD19	GND	61	+5V	+5V
31	+3.3V	$\overline{\text{TRDY}}$	62	+5V	+5V

4.2 PCI operation

The PCI bus cleverly saves lines by multiplexing the address and data lines. It has two modes (Figure 4.2):

- Multiplexed mode – the address and data lines are used alternately. First, the address is sent, followed by a data read or write. Unfortunately, this requires two or three clock cycles for a single transfer (either an address followed by a read or write cycle, or an address followed by read and write cycle). This causes a maximum data write transfer rate of 66 MB/s (address then write) and a read transfer rate of 44 MB/s (address, write then read), for a 32-bit data bus width.
- Burst mode – the multiplexed mode obviously slows down the maximum transfer rate. Additionally, it can be operated in burst mode, where a single address can be initially sent, followed by implicitly addressed data. Thus, if a large amount of sequentially addressed memory is transferred then the data rate approach the maximum transfer of 133 MB/s for a 32-bit data bus and 266 MB/s for a 64-bit data bus.

If the data from the processor is sequentially addressed data then PCI bridge buffers the incoming data and then releases it to the PCI bus in burst mode. The PCI bridge may also use burst mode when there are gaps in the addressed data and use a handshaking line to identify that no data is transferred for the implied address. For example in Figure 4.2 the burst mode could involve Address+1, Address+2 and Address+3 and Address+5, then the byte enable signal can be made inactive for the fourth data transfer cycle.

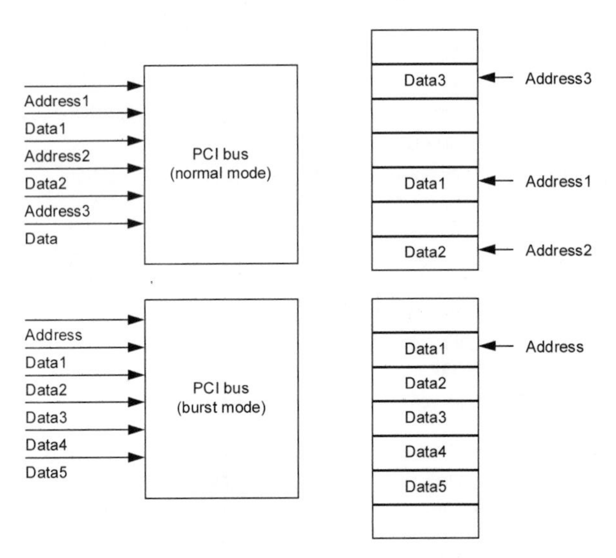

Figure 4.2 PCI bus transfer modes

To accommodate the burst mode, the PCI bridge has a prefetch and posting buffer on both the host bus and the PCI bus sides. This allows the bridge to build the data access up into burst accesses. For example, the processor typically transfers data to the graphics card with sequential accessing. The bridge can detect this and buffer the transfer. It will then transfer the data in burst mode when it has enough data. Figure 4.3 shows an example where the PCI bridge buffers the incoming data and transfers it using burst mode. The transfers between the processor and the PCI bridge, and between the PCI bridge and the PCI bus can be independent where the processor can be transferring to its local memory while the PCI bus is transferring data. This helps to decouple the PCI bus from the processor.

The primary bus in the PCI bridge connects to the processor bus and the secondary bus connects to the PCI bus. The prefetch buffer stores incoming data from the connected bus and the posting buffer holds the data ready to be sent to the connected bus.

The PCI bus also provides for a configuration memory address (along with direct memory access and isolated I/O memory access). This memory is used to access the configuration register and 256-byte configuration memory of each PCI unit.

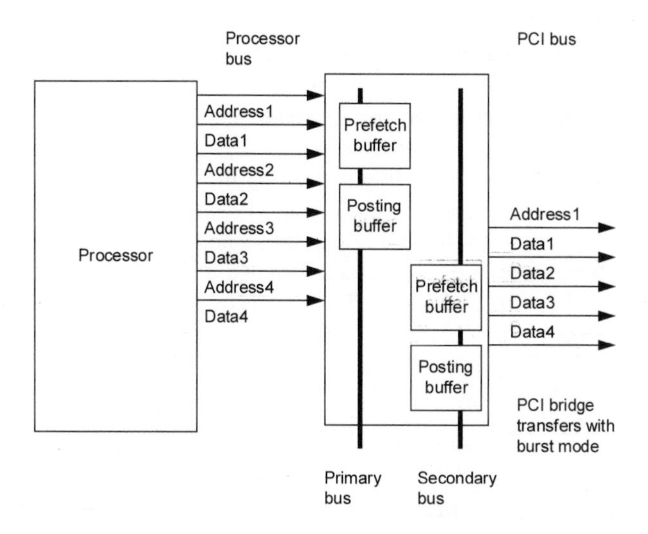

Figure 4.3 PCI bridge using buffering for burst transfer

4.2.1 PCI bus cycles

The PCI has built-in intelligence where the command/byte enable signals ($\overline{C/BE3} - \overline{C/BE0}$) are used to identify the command. They are given by:

$\overline{C/BE3}$	$\overline{C/BE2}$	$\overline{C/BE1}$	$\overline{C/BE0}$	*Description*
0	0	0	0	INTA sequence
0	0	0	1	Special cycle
0	0	1	0	I/O read access
0	0	1	1	I/O write access
0	1	1	0	Memory read access
0	1	1	1	Memory write access
1	0	1	0	Configuration read access
1	0	1	1	Configuration write access
1	1	0	0	Memory multiple read access
1	1	0	1	Dual addressing cycle
1	1	1	0	Line memory read access
1	1	1	1	Memory write access with invalidations

The PCI bus allows any device to talk to any other device, thus one device can talk to another without the processor being involved. The device that starts the conversion is known as the initiator and the addressed PCI device is known as the target. The sequence of operation for write cycles, in burst mode, is:

- Address phase – the transfer data is started by the initiator activating the $\overline{\text{FRAME}}$ signal. The command is set on the command lines ($\overline{\text{C/BE3}} - \overline{\text{C/BE0}}$) and the address/data pins (AD31–AD0) are used to transfer the address. The bus then uses the byte enable lines ($\overline{\text{C/BE3}} - \overline{\text{C/BE0}}$) to transfer a number of bytes.
- The target sets the $\overline{\text{TRDY}}$ signal (target ready) active to indicate that the data has on the AD31–AD0 (or AD62–AD0 for a 64-bit transfer) lines is valid. In addition, the initiator indicates its readiness to the PCI bridge by setting the $\overline{\text{IRDY}}$ signal (indicator ready) active. Figure 4.4 illustrates this.
- The transfer continues using the byte enable lines. The initiator can block transfers if it sets $\overline{\text{IRDY}}$ and the target with $\overline{\text{TRDY}}$.
- Transfer is ended by deactivating the $\overline{\text{FRAME}}$ signal.

The read cycle is similar but the $\overline{\text{TRDY}}$ line is used by the target to indicate that the data on the bus is valid.

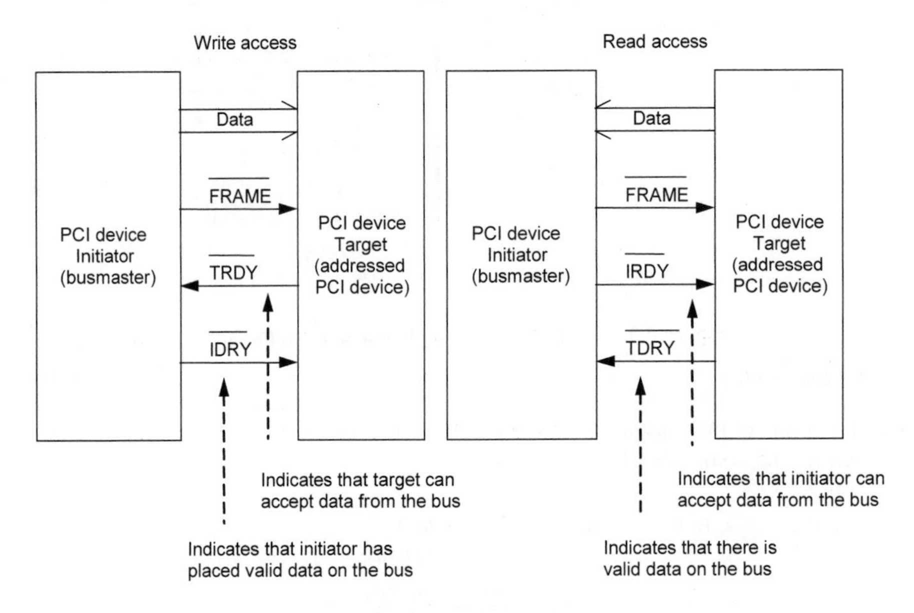

Figure 4.4 PCI handshaking

4.2.2 PCI commands

The first phase of the bus access is the command/addressing phase. Its main commands are:

- INTA sequence – addresses an interrupt controller where interrupt vectors are transferred after the command phase.
- Special cycle – used to transfer information to the PCI device about the processor's status. The lower 16 bits contain the information codes, such as 0000h for a processor

shutdown, 0001h for a processor halt, 0002h for x86specific code and 0003h to FFFFh for reserved codes. The upper 16 bits (AD31–AD16) indicate x86specific codes when the information code is set to 0002h.

- I/O read access – indicates a read operation for I/O address memory, where the AD lines indicate the I/O address. The address lines AD0 and AD1 are decoded to define whether an 8-bit or 16-bit access is being conducted.
- I/O write access – indicates a write operation to an I/O address memory, where the AD lines indicate the I/O address.
- Memory read access – indicates a direct memory read operation. The byte-enable lines ($\overline{C/BE3} - \overline{C/BE0}$) identify the size of the data access.
- Memory write access – indicates a direct memory write operation. The byte-enable lines ($\overline{C/BE3} - \overline{C/BE0}$) identify the size of the data access.
- Configuration read access – used when accessing the configuration address area of a PCI unit. The initiator sets the IDSEL line activated to select it. It then uses address bits AD7–AD2 to indicate the addresses of the double words to be read (AD1 and AD0 are set to 0). The address lines AD10–AD18 can be used for selecting the addressed unit in a multi-function unit.
- Configuration write access – as the configuration read access, but data is written from the initiator to the target.
- Memory multiple read access – used to perform multiple data read transfers (after the initial addressing phase). Data is transferred until the initiator sets the $\overline{FRAME}$ signal inactive.
- Dual addressing cycle – used to transfer a 64-bit address to the PCI device (normally only 32-bit addresses are used) in either a single or a double clock cycle. In a single clock cycle the address lines AD63–AD0 contain the 64-bit address (note that the Pentium processor only has a 32-bit address bus, but this mode has been included to support other systems). With a 32-bit address transfer the lower 32 bits are placed on the AD31–AD0 lines, followed by the upper 32 bits on the AD31–AD0 lines.
- Line memory read access – used to perform multiple data read transfers (after the initial addressing phase). Data is transferred until the initiator sets the $\overline{FRAME}$ signal inactive.
- Memory write access with invalidations – used to perform multiple data write transfers (after the initial addressing phase).

4.2.3 PCI interrupts

The PCI bus support four interrupts ($\overline{INTA} - \overline{INTD}$). The $\overline{INTA}$ signal can be used by any of the PCI units, but only a multifunction unit can use the other three interrupt lines ($\overline{INTB} - \overline{INTD}$). These interrupts can be steered, using system BIOS, to one of the IRQ*x* interrupts by the PCI bridge. For example, a 100 Mbps Ethernet PCI card can be set to interrupt with $\overline{INTA}$ and this could be steered to IRQ10.

4.3 Bus arbitration

Busmasters are devices on a bus which are allowed to take control of the bus. For this purpose, PCI uses the $\overline{REQ}$ (request) and $\overline{GNT}$ (grant) signals. There is no real standard for this arbitration, but normally the PCI busmaster activates the $\overline{REQ}$ signal to indicate a request to the PCI bus, and the arbitration logic must then activate the $\overline{GNT}$ signal so that the requesting master gains control of the bus. To prevent a bus lock-up, the busmaster is given 16 CLK cycles before a time-overrun error occurs.

4.4 Other PCI pins

The other PCI pins are:

- $\overline{RST}$ (Pin A15) – resets all PCI devices.
- $\overline{PRSNT1}$ and $\overline{PRSNT2}$ (Pins B9 and B11) – these, individually, or jointly, show that there is an installed device and what the power consumption is. A setting of 11 (that is, $\overline{PRSNT1}$ is a 1 and $\overline{PRSNT2}$ is a 1) indicates no adapter installed, 01 indicates maximum power dissipation of 25 W, 10 indicates a maximum dissipation of 15 W and 00 indicate a maximum power dissipation of 7.5 W.
- $\overline{DEVSEL}$ (Pin B37) – indicates that addressed device is the target for a bus operation.
- TMS (test mode select), TDI (test data input), TDO (test data output), $\overline{TRST}$ (test reset), and TCK (test clock) – used to interface to the JTAG boundary scan test.
- IDSEL (Pin A26) – used for device initialization select signal during the accessing of the configuration area.
- $\overline{LOCK}$ (Pin A15) – indicates that an addressed device is to be locked-out of bus transfers. All other unlocked devices can still communicate.
- PAR, $\overline{PERR}$ (Pins A43 and B40) – The parity pin (PAR) is used for even parity for AD31–AD0 and C/BE3–C/BE0, and $\overline{PERR}$ indicates that a parity error has occurred.
- SDONE, $\overline{SBO}$ (Pins A40 and A41) – used in snoop cycles. SDONE (snoop done) and $\overline{SBO}$ (snoop back off signal).
- $\overline{SERR}$ (Pin B42) – used to indicate a system error.
- $\overline{STOP}$ (Pin A38) – used by a device to stop the current operation.
- $\overline{ACK64}$, $\overline{REQ64}$ (Pins B60 and A60) – the $\overline{REQ64}$ signal is an active request for a 64-bit transfer and $\overline{ACK64}$ is the acknowledge for a 64-bit transfer.

4.5 Configuration address space

Each PCI device has 256 bytes of configuration data, which is arranged as 64 registers of 32 bits. It contains a 64-byte predefined header followed by an extra 192 bytes which contain extra configuration data. Figure 4.5 shows the arrangement of the header. The definitions of the fields are:

- Unit ID and Man. ID – a Unit ID of FFFFh defines that there is no unit installed, while any other address defines its ID. The PCI SIG, which is the governing body for the PCI specification, allocates a Man. ID. This ID is normally shown at BIOS start-up. Section 4.8 gives some example Man. IDs (and plug-and-play IDs).
- Status and command.

- Class code and Revision – the class code defines PCI device type. It splits into two 8-bit values with a further 8-bit value that defines the programming interface for the unit. The first defines the unit classification (00h for no class code, 01h for mass storage, 02h for network controllers, 03h for video controllers, 04h for multimedia units, 05h for memory controller and 06h for a bridge), followed by a subcode which defines the actual type. Typical codes are:

  - 0100h – SCSI controller
  - 0101h – IDE controller.
  - 0102h – Floppy controller
  - 0200h – Ethernet network adapter
  - 0201h – Token ring network adapter
  - 0202h – FDDI network adapter
  - 0280h – Other network adapter
  - 0300h – VGA video adapter
  - 0301h – XGA video adapter
  - 0380h – Other video adapter
  - 0400h – Video multimedia device
  - 0401h – Audio multimedia device
  - 0480h – Other multimedia device
  - 0500h – RAM memory controller
  - 0501h – Flash memory controller
  - 0580h – Other memory controller
  - 0600h – Host
  - 0601h – ISA Bridge
  - 0602h – EISA Bridge
  - 0603h – MAC Bridge
  - 0604h – PCI–PCI Bridge
  - 0680h – Other Bridge

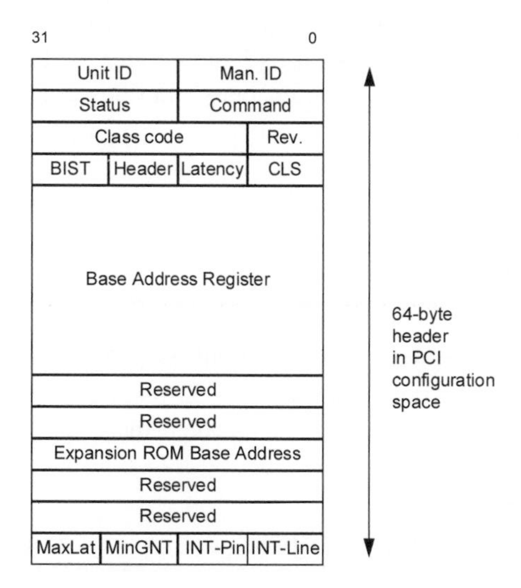

Figure 4.5 PCI configuration space

- BIST, header, latency, CLS – the BIST (built-in self test) is an 8-bit field, where the most significant bit defines if the device can carry out a BIST, the next bit defines if a BIST is to be performed (a 1 in this position indicates that it should be performed) and bits 3–0 define the status code after the BIST has been performed (a value of zero indicates no error). The header field defines the layout of the 48 bytes after the standard 16-byte header. The most significant bit of the header field defines whether the device is a multifunction device or not. A 1 defines a multi-function unit. The CLS (cache line size) field defines the size of the cache in units of 32 bytes. Latency indicates the length of time for a PCI bus operation, where the amount of time is the latency+8 PCI clock cycles.

- Base address register – this area of memory allows the device to be programmed with an I/O or memory address area. It can contain a number of 32- or 64-bit addresses. The format of a memory address is

  Bit 64–4 Base address.
  Bit 3 PRF. Prefetching, 0 identifies not possible, 1 identifies possible.
  Bit 2, 1 Type. 00 – any 32-bit address, 01 – less than 1MB, 10 – any 64-bit address and 11 – reserved.
  Bit 0 0. Always set to a 0 for a memory address.

  For an I/O address space it is defined as:

  Bit 31–2 Base address.
  Bit 1, 0 01. Always set to a 01 for an I/O address.

- Expansion ROM base address – allows a ROM expansion to be placed at any position in the 32-bit memory address area.
- MaxLat, MinGNT, INT-pin, INT-line – the MinGNT and MaxLat registers are read-only registers that define the minimum and maximum latency values. The INT-Line field is a 4-bit field that defines the interrupt line used (IRQ0–IRQ15). A value of 0 corresponds to IRQ0 and a value of 15 corresponds to IRQ15. The PCI bridge can then redirect this interrupt to the correct IRQ line. The 4-bit INT-pin defines the interrupt line that the device is using. A value of 0 defines no interrupt line, 1 defines $\overline{\text{INTA}}$, 2 defines $\overline{\text{INTB}}$, and so on.

4.6 I/O addressing

The standard PC I/O addressing ranges from 0000h to FFFFh, which gives an addressable space of 64 KB, whereas the PCI bus can support a 32-bit or 64-bit addressable memory. The PCI device can be configured using one of two mechanisms.

Configuration mechanism 1

Passing two 32-bit values to two standard addresses configures the PCI bus:

Address	*Name*	*Description*
0CF8h	Configuration address	Used to access the configuration address area.
0CFCh	Configuration data	Used to read or write a 32-bit (double word) value to the configuration memory of the PCI device.

The format of the configuration address register is

Bit 31 ECD (`Enable CONFIG_DATA`) bit. A 1 activates the `CONFIG_DATA` register, while a 0 disables it.

Bit 30–24 Reserved.
Bit 23–16 PCI bus number. Defines the number of the number of the PCI bus (to a maximum of 256).
Bit 15–11 PCI unit. Selects a PCI device (to a maximum of 32). PCI thus supports a maximum of 256 attached buses with a maximum of 32 devices on each bus.
Bit 10–8 PCI function. Selects a function within a PCI multifunction device (one of eight functions).
Bit 7–2 Register. Selects a Dword entry in a specified configuration address area (one of 64 Dwords).
Bit 1, 0 Type. 00 – decode unit, 01 – `CONFIG_ADDRESS` value copy to `AD`*x*.

Configuration mechanism 2

In this mode, each PCI device is mapped to a 4 KB I/O address range between C000h and CFFFh. This is achieved by used in the activation register CSE (configuration space enable) for the configuration area at the port address 0CF8h. The format of the CSE register is located at 0CF8h and is defined as

Bit 7–4 Key. 0000 – normal mode, 0001…1111 – configuration area activated. A value other than zero for the key activates the configuration area mapping, that is, all I/O addresses to the 4 KB range between C000h and CFFFh would be performed as normal I/O cycles.
Bit 3–1 Function. Defines the function number within the PCI device (if it represents a multifunction device).
Bit 0 SCE. 0 defines a configuration cycle, 1 defines a special cycle.

The forward register is stored at address 0CFAh and contains

Bit 7–0 PCI bus.

The I/O address is defined by:

Bit 31–12 Contains the bit value of 0000Ch.
Bit 11–8 PCI unit.
Bit 7–2 Register index.
Bit 1, 0 Contains the bit value of 00 (binary).

4.6.1 Sample test program

PCI bridge test

An example BASIC program to test the PCI bridge device is given next.

Program 4.1

```
130  Print "Host PCI bridge test"
160  Print " PCI Configuration Address &80000000": Print
170  IOWRITE &CF8,2,&80000000
180  IOREAD &CFC,2
190  IF B1<>&10000E11 THEN GOTO 410
```

```
200  Print "Test1 Passed....Component ID Test"

210  Dim TEST(4)
220  TEST(1)=&FFFFFFFF
230  TEST(2)=&AAAAAAAA
240  TEST(3)=&055555555
250  TEST(4)=&00

260  D9=&80000000
270  REG = &060

280  REPEAT
290    TST = &01
300    IOWRITE &CF8,2,D9 + REG
310    REPEAT
320            IOWRITE &CFC,2,TEST(TST)
330            IOREAD &CFC,2
340            If B1 <> TEST(TST) Then GoTo 450
350            TST = TST + &01
360    UNTIL TST=&5
370    REG = REG + &04
380  UNTIL REG=&68

390  Print "Test2 Passed...Internal Register Test"
400  GoTo 480
410  Print "Test1 Failed...."
420  Print "Component ID Test...."
430  PRINT "Expected ID &10000E11  Actual ID "~B1
440  GoTo 480
450  Print "Test2 Failed...."
460  Print "Internal register test...."
470  PRINT "Register "~REG", Expected "~TEST(TST)"  Actual "~B1
```

The code:

```
170  IOWRITE &CF8,2,&80000000
180  IOREAD &CFC,2
190  IF B1<>&10008086 THEN GOTO 410
```

writes the value 80000000h (1000 0000 ... 0000b) to the CF8h register (configuration address), where the most signification bit activates the configuration data register. Next the program reads from the CFCh register (configuration data), after this the B1 value contains the 32-bit value read from the configuration data register. In this case the value will be the first 32 bits from the configuration memory of the PCI device. The value tested in this case is 10000E11h, where 1000h identifies the unit ID and 8086h identifies the manufacturer ID (Intel).

The values written to the registers are FFFFFFFFh (1111 1111 ... 1111), AAAAAAAAh (1010 1010 ... 1010), 55555555h (0101 0101 ... 0101) and 00000000h. These values are then read back and tested to determine if they match the values that where written.

Video device test

An example BASIC program to test the video adaptor on the PCI bus is given next.

Program 4.2

```
112  Print "PCI test: Component ID and PCI Register Test"
122  IOWRITE &CF8,2,&80005000
```

```
124  IOREAD &CFC,2
126  IF B1<>&00A81013 THEN PRINT : GOTO 172
128  Print "     Passed....Component ID Test"

130  Dim TEST(4)
132  TEST(1)=&FF000000
134  TEST(2)=&AA000000
136  TEST(3)=&055000000
138  TEST(4)=&00


140  ADDR=&80005000
142  REG =&010

144  REPEAT
146     TST = &01
148     IOWRITE &CF8,2,ADDR + REG
150     REPEAT
152             IOWRITE &CFC,2,TEST(TST)
154             IOREAD &CFC,2
156             If B1 <> TEST(TST) Then Print: GoTo 180
158             TST = TST + &01
160     UNTIL TST=&5

162     REG = REG + &020
164  UNTIL REG=&50

166  Print "     Test02 Passed....PCI Register Test"
168  GoTo 188

172  Print "     FAIL: Component ID Test"
174  PRINT "     Expected ID &00A81013  Actual ID "~B1
176  Print:      GoTo 130
180  Print "     FAIL: PCI Register Test...."
182  PRINT "     Register "~REG", Expected "~TEST(TST)"  Actual "~B1
188             etc
```

The code:

```
122  IOWRITE &CF8,2,&80005000
124  IOREAD &CFC,2
126  IF B1<>&00A81013 THEN PRINT : GOTO 172
```

writes the value 80005000h (1000 0000 ... 0000b) to the CF8h register (configuration address), where the most signification bit activates the configuration data register. Next the program reads from the CFCh register (configuration data), after this the B1 value contains the 32-bit value read from the configuration data register. In this case the value will be the first 32 bits from the configuration memory of the PCI device. The value tested in this case is 00A81013h, where 00A8h identifies the Unit ID and 1013h identifies the manufacturer ID (Cirrus Logic).

The following code tests four 32-bit words from the configuration memory. The values written are:

FF000000h, AA000000h, 55000000h, 00h

These values are then read back and tested against the values actual written. It should be noted that the least significant 24 bits are read-only registers, thus they cannot be written to.

```
144  REPEAT
146    TST = 1
148    IOWRITE &CF8,2,ADDR + REG
150    REPEAT
152          IOWRITE &CFC,2,TEST(TST)
154          IOREAD &CFC,2
156          If B1 <> TEST(TST) Then Print: GoTo 180
158          TST = TST + 1
160    UNTIL TST=5

162    REG = REG + &020
164  UNTIL REG=&50
```

Note, C/C++ can only access 8 or 16 bits at a time, thus the code:

```
122  IOWRITE &CF8,2,&80005000
124  IOREAD &CFC,2
126  IF B1<>&00A81013 THEN PRINT : GOTO 172
```

would be replaced with:

```
#include <conio.h>

int main(void)
{
unsigned int         val1, val2, val3, val4;
unsigned long int    val;

     _outp(0xcf8,0x00);      /* least significant byte */
     _outp(0xcf9,0x00);
     _outp(0xcfa,0x00);
     _outp(0xcfb,0x80);      /* most significant byte */

     val1=_inp(0xcfc) & 0xff;      val2=_inp(0xcfd) & 0xff;
     val3=_inp(0xcfe) & 0xff;      val4=_inp(0xcff) & 0xff;

     val= val1 + (val2<<8) + (val3<<16)+ (val4<<24);

     if (val==0x00a81013)
     {
        printf("Success");
     }

     etc
     return(0);
}
```

4.7 Exercises

4.7.1 What is the maximum data throughput for a 33 MHz, 32-bit data PCI bus:

(a) 33 MB/s (b) 66 MB/s
(c) 132 MB/s (d) 264 MB/s

4.7.2 Which I/O register address is used to access PCI configuration address space:

(a)	1F8h	(b)	CF8h
(c)	3F8h	(d)	2F8h

4.7.3 Which I/O register address is used read and write to registers in the PCI configuration address space:

(a)	1FCh	(b)	CFCh
(c)	3FCh	(d)	1FCh

4.7.4 How many bits can be accessed, at a time, with the configuration address register:

(a)	8	(b)	16
(c)	32	(d)	64

4.7.5 Which company has the manufacture ID of 8086:

(a)	Compaq	(b)	Motorola
(c)	NCR	(d)	Intel

4.7.6 Explain how PCI architecture uses bridges.

4.7.7 Outline the operation of Program 4.1 and Program 4.2. Highlight the range of addresses used. Why does Program 4.2 write the bit pattern FF000000h and not FFFFFFFFh?

4.7.8 Explain how the 32-bit PCI bus transfers data. Prove that the maximum data rate for a 32-bit PCI in its normal mode is only 66 MB/s. Explain the mechanism that the PCI bus uses to increase the maximum data rate to 132 MB/s.

4.7.9 How does buffering in the PCI bridge aid the transfer of data to and from the processor.

4.7.10 Explain how the PCI bus uses the command phase to set up a peripheral.

4.7.11 How are interrupt lines used in the PCI bus. Explain how these interrupts can be steered to the ISA bus interrupt lines.

4.7.12 Outline the concept of bus mastering and how it occurs on the PCI bus. What signal lines are used?

4.7.13 Explain how the PCI bus uses configuration addresses.

4.8 Example manufacturer and plug-and-play IDs

Manufacturer	*Man. ID*	*PNP ID*	*Manufacturer*	*Man. ID*	*PNP ID*
NCR	1000	4096	ULSI	1003	4099
VLSI	1004	4100	ALR	1005	4101
Reply Group	1006	4102	Netframe	1007	4103
EPSON	1008	4104	Phoenix	100a	4106
National Semi	100b	4107	Tseng Labs	100c	4108
AST	100d	4109	Weitek	100e	4110
Video Logic Ltd	1010	4112	Digital	1011	4113
Micronics	1012	4114	Cirrus Logic	1013	4115
IBM	1014	4116	ICL	1016	4118
Spea Software	1017	4119	UNISYS	1018	4120
Elite	1019	4121	NCR	101a	4122
Vitesse	101b	4123	Western Digital	101c	4124
American Mega	101e	4126	PictureTel	101f	4127
Hitachi	1020	4128	Oki	1021	4129
AMD	1022	4130	Trident	1023	4131
Acer	1025	4133	Dell	1028	4136
Siemens	1029	4137	LSI	102a	4138
Matrox	102b	4139	Chips and Tech.	102c	4140
Wyse	102d	4141	Olivetti	102e	4142
Toshiba	102f	4143	Miro Computer	1031	4145
Compaq	1032	4146	NEC	1033	4147
Future Domain	1036	4150	HITACHI	1037	4151
AMP	1038	4152	Seiko Epson	103a	4154
Tatung	103b	4155	HP	103c	4156
Genoa	1047	4167	Fountain	1049	4169
SGS Thomson	104a	4170	Buslogic	104b	4171
TI	104c	4172	SONY	104d	4173
OAK	104e	4174	Hitachi	1054	4180
ICL	1056	4182	Motorola	1057	4183
Vtech	105e	4190	United Micro	1060	4192
Mitsubishi	1067	4199	Apple	106b	4203
Hyundai	106c	4204	Sequent	106d	4205
Daewood	1070	4208	Mitac	1071	4209
Yamaha	1073	4211	Nexgen	1074	4212
Cyrix	1078	4216	I-BUS	1079	4217
Networth	107a	4218	Gateway 2000	107b	4219
Goldstar	107c	4220	Leadtek	107d	4221
Interphase	107e	4222	Tulip	1085	4229
Data General	1089	4233	Elonex	108c	4236
Intergraph	1091	4241	Diamond	1092	4242
National Instruments	1093	4243	Quantum Designs	1098	4248
Samsung	1099	4249	Packard Bell	109a	4250
Gemlight	109b	4251	Megachips	109c	4252
3COM	10b7	4279	SMC	10b8	4280
Acer	10b9	4281	Mitsubishi	10ba	4282
Tsenglabs	10be	4286	Samsung	10c3	4291
Award	10c4	4292	Xerox	10c5	4293
Neomagic	10c8	4296	Fujitsu	10ca	4298
Fujitsu	10d0	4304	Newbridge	10e3	4323
Tandem	10e4	4324	Micro Industries	10e5	4325
Xilinx	10ee	4334	Creative	10f6	4342
Matsushita	10f7	4343	Altos	10f8	4344
PC Direct	10f9	4345	Truevision	10fa	4346
Creative Labs	1102	4354	Santa Cruz	1111	4369
Rockwell	1112	4370	Zilog	1121	4385
S3	5333	21299	Intel	8086	32902
Adaptec	9004	36868			

4.9 Notes from the author

There is an amusing statement that was made in 1981, in the book 30 Hour BASIC Standard, 1981:

Microcomputers are the tool of the 80's. BASIC is the language that all of them use. So the sooner you learn BASIC, the sooner you will understand the microcomputer revolution

Well, as it has proven, a good knowledge of BASIC will not really help your understanding of microcomputers, but if there is one bus that you need to understand in the PC, it is the PCI bus. This is because it is the main interface bus within the PC. Most external devices eventually connect to the PCI through bridge devices. There were several short-term fixes for local bus technology, but the PCI was the first attempt at a properly designed system bus. It allows the PC to be segmented into differing data transfer rates. PCI provides a buffer between the main system core, such as the processor and its memory, and the slower peripherals, such as the hard-disk, serial ports, and so on.

With interrupts, the PCI has the great advantage over ISA in that it allows interrupts to be properly assigned at system start-up. The BIOS or the operating system can communicate with the PCI-connected bus with the configuration address area. From this, the system can determine the type of device it is, whether it be a graphics card or a network card. The system can then properly configure the device and grant it the required resources. The days of users having to assign interrupts (using card jumpers, in some cases) and I/O addresses are reducing (thankfully!).

The great leap forward in PC systems happened with local bus technology. The demand came from graphics cards as Windows 3.0 was being adopted. The ISA bus was far too slow, as it only supported 8MHz transfers. Graphic card manufacturers got together and developed the VESA-backed VL-local bus standard. It showed how fast transfer devices could be connected to a local bus, while other slower devices had to access the processor through a bridge, which allowed a different clock speed, and a different data and address bus. Most PCs are now based around this local bus idea, and they can be split into there main areas:

- *Local processor bus. Direct connection of the processor to its local cache memory (either Level-1 or Level-2 cache.*
- *Local bus. Connection onto the PCI bus. This connects to the local processor bus via a bridge.*
- *External bus. ISA, IDE, RS-232, and so on. This connects to the local bus via a bridge.*

There is great potential in the PCI bus. At present, most systems use 32-bit data transfers, but there is scope for 64-bit data transfers. Also, the 33 MHz clock can be increased to 66MHz with double edge clocking. A new enhanced PCI-based system called the AGP (Advanced Graphics Port) has been developed which allows for data transfers of over 500 MB/s.

I'm slightly annoyed with the success of the PCI bus, as I've got an ISA-based sound card, an ISA-based Ethernet card and an ISA-based video camera, and I've only got two ISA slots. So, every so often, I have to swap the sound card for the video camera, and vice-versa. At present, I've got four empty PCI slots, and I think one of them is waiting for a PCI-based Ethernet card. Then I'll be able to have a proper video conference, with sound and video. But, never mind, I've just got myself a lovely new Dell notebook, and a USB-based video camera, and a single PCMCIA card for my modem and network connections, so I may never need my desktop computer again (here's hoping).

5 Motherboard Design

5.1 Introduction

This chapter analyses a Pentium-based motherboard. An example board is the Intel 430HX motherboard which supports most Pentium processors and has the following component parts:

- PCIset components – 82438 System Controller (TXC) and 82371SB PCI ISA Xcelerator (PIIX3).
- 82091AA (AIP) for serial and parallel ports, and floppy disk controller.
- DRAM main memory.
- L2 cache SRAM.
- Universal serial bus (USB).
- Interface slots (typically 4 PCI and 3 ISA).
- 1 Mbit flash RAM.

Figure 5.1 illustrates the main connections of the PCIset (which are the TXC and PIIX3 devices). The TCX allows for a host-to-PCI bridge, whereas the PIIX3 device supports:

- PCI-to-ISA bridge.
- Fast IDE.
- APIC interface.
- USB host/hub controller.
- Power management.

The 430HX board has 3 V and 5 V busses. PCI bus connections are 5 V and the Pentium bus is 3V.

5.1.1 Pentium-II/III processor

Figure 5.2 illustrates the main connections to the Pentium II/III processor. It can be seen that it has:

- 64-bit data bus (D0–D63) which connects to the TXC (HD0–HD63).
- 32-bit address bus (A0–A31) which connects to the TXC (HA0–HA31).
- 8-byte address lines (BE0#–BE7#) to allow the processor to access from 1 to 8 bytes (64 bits) at a time, which connects to TXC (HBE0#–HBE7#).
- Read/write line (W/R#) which connects to TXC (HW/R#).
- Memory/IO (M/IO#) which connects to TXC (HM/IO#).
- Data/control (D/C#) which connects to TXC (HD/C#).

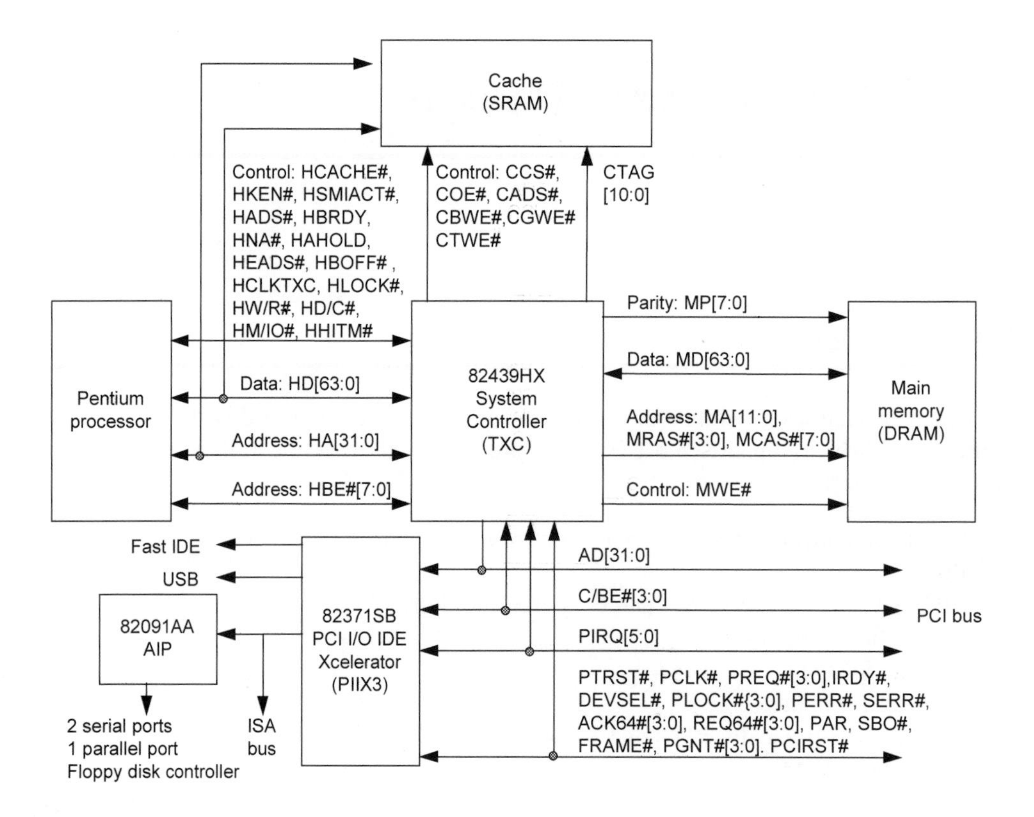

Figure 5.1 PCIset system architecture

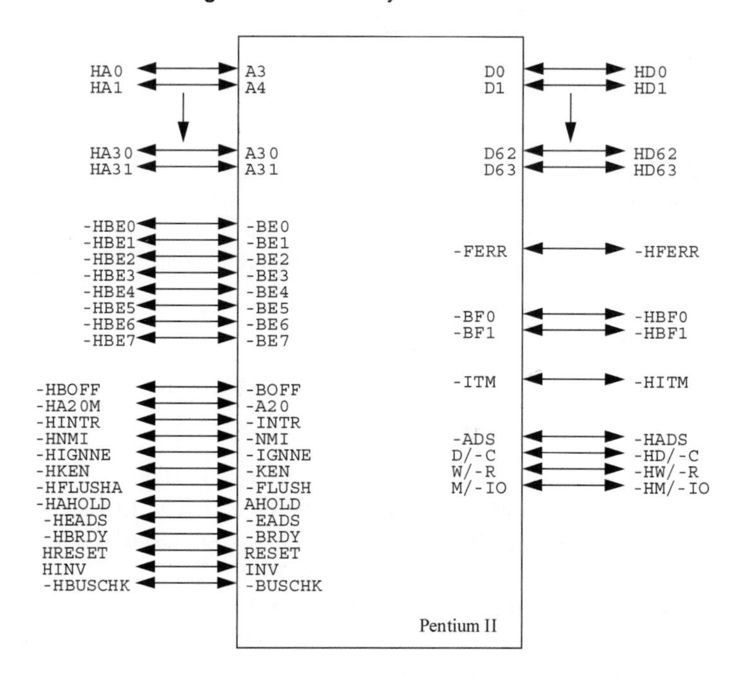

Figure 5.2 Pentium II/III connections

5.1.2 82371SB PCI ISA Xcelerator (PIIX3)

The PIIX3 is a 208-pin QFP (quad flat pack) IC that integrates much the functionality of the ISA bus interface onto a single device. Table 5.1 outlines the main connections to the PIIX3 IC.

Table 5.1 PIIX3 connections

Address lines		**IRQ Lines**		**ISA Lines**		**ISA Lines**	
Signal	*Pin*	*Signal*	*Pin*	*Signal*	*Pin*	*Signal*	*Pin*
AD0	206	IRQ1	4	BALE	64	SA8/DD0 55	
AD1	205	IRQ3	58	AEN	20	SA9/DD1 50	
AD2	204	IRQ4	56	LA17	86	SA10/DD2	49
AD3	203	IRQ5	34	LA18	84	SA11/DD3	48
AD4	202	IRQ6	33	LA19	82	SA12/DD4	47
AD5	201	IRQ7	32	LA20	80	SA13/DD5	46
AD6	200	-IRQ8	5	LA21	76	SA14/DD6	45
AD7	199	IRQ9	10	LA22	74	SA15/DD7	44
AD8	197	IRQ10	73	LA23	72	SA16/DD8	43
AD9	194	IRQ11	75	SA0	69	SA17/DD9	41
AD10	193	IRQ12/M	77	SA1	68	SA18/DD10	40
AD11	192	IRQ14	83	SA2	67	SA19/DD11	39
AD12	191	IRQ15	81	SA3	66	SA20/DD12	38
AD13	190			SA4	63	SA21/DD13	37
AD14	189			SA5	61	SA22/DD14	36
AD15	188			SA6	59	SA23/DD15	35
AD16	177			SA7	57	-OWS	15
AD17	176			DRQ0	87	-SMEMW	22
AD18	175			DRQ1	30	-SMEMR	19
AD19	174			DRQ2	12	-IOW	24
AD20	173			DRQ3	25	-IOR	23
AD21	172			DRQ5	91	-REFRESH	31
AD22	171			DRQ6	95	T/C	62
AD23	168			DRQ7	99	OSC	
AD24	166			-DACK0	85	-MEMCS16	70
AD25	165			-DACK1	29	-IOCS16 71	
AD26	164			-DACK2	60	-MASTER	
AD27	163			-DACK3	21	IOCHK	6
AD28	162			-DACK5	89	IOCHRDY 18	
AD29	161			-DACK6	93	-SBHE (DD12)	
AD30	160			-DACK7	97	-MEMR	88
AD31	159			RSTISA		-MEMW	90

USB	
Signal	Pin
USBP1-	143
USBP1+	142
USBP0-	145
USBP0+	144

PIIX3's functionality includes:

- Enhanced 7-channel DMA with two 8237 controllers. This is supported with the hand-shaking lines DRQ0–DRQ7 and DRQ0#–DRQ7#.
- ISA-PCI bridge.
- Fast IDE support for up to four disk drives (two masters and two slaves). It supports mode four timings, which gives transfer rates of up to 22MB/s.
- I/O APIC (advanced programmable interrupt controller) support.
- Implementation of PCI 2.1.

- Incorporates 82C54 timer for system timer, refresh request and speaker output tone.
- Non-maskable interrupts (NMI).
- PCI clock speed of 25/33 MHz. Motherboard configurable clock speed (normally 33 MHz).
- Plug-and-play support with one steerable interrupt line and one programmable chip select. The motherboard interrupt MIRQ0 can be steered to any one of 11 interrupts (IRQ3–IRQ7, IRQ9–IRQ12, IRQ14 and IRQ15).
- Steerable PCI interrupts for PCI device plug-and-play. The PCI interrupt lines (PIRQA-PIRQD) can be steered to one of 11 interrupt (IRQ3–IRQ7, IRQ9–IRQ12, IRQ14 and IRQ15).
- Support for PS/2-type mouse and serial port mouse. IRQ12/M can be enabled for the PS/2-type mouse or disable for a serial port mouse.
- Support for five ISA slots. Typical applications for ISA include 10M bps Ethernet adaptor cards, serial/parallel port cards, sound cards, and so on.
- System power management. Allows the system to operate in a low power state without being powered down. This can be triggered either by a software, hardware or external event.
- Math coprocessor error function. The FERR# line goes active (LOW) when a math coprocessor error occurs. The PIIX3 device automatically generates an IRQ13 interrupt and sets the INTR line to the processor. The PXII3 device then sets the IGNNE# active and INTR inactive when there is a write to address F0h.
- Two 82C59 controllers with 14 interrupts. The interrupts lines IRQ1, IRQ3–IRQ15 are available (IRQ0 is used by the system time and IRQ2 by the cascaded interrupt line).
- Universal serial bus with root hub and two USB ports. With the USB the host controller transfers data between the system memory and USB devices. This is achieved by processing data structures set up to by the host software and generated the transaction on USB.

The address lines (AD0–AD22) connect to the TXC IC and the available interrupt lines at IRQ1, IRQ2–IRQ12, IRQ14 and IRQ15 (IRQ0 is generated by the system timer and IRQ2 is the cascaded interrupt line). The PS/2-type mouse uses the IRQ12/M line.

5.1.3 82438 System Controller (TXC)

The 324-pin TXC BGA (ball grid array) provides an interface between the processor, DRAM and the external buses (such as the PCI, ISA, and so on). Table 5.2 outlines its main pin connections. The TXC's functionality includes:

- Supports 50 MHz, 60 MHz and 66 MHz host bus.
- Integrated DRAM controller. Supports four CAS lines and eight RAS lines. The memory supports symmetrical and asymmetrical addressing for 1 MB, 2 MB and 4 MB-deep SIMMs and symmetrical addressing for 16 MB-deep SIMMs.
- Integrated second level cache controller. Supports up to 512 MB of second-level cache with synchronous pipelined burst SRAM.
- Dual processor support.
- Optional parity.
- Optional error checking and correction on DRAM. The ECC mode is software configurable and allows for single bit error correction and multibit error detection on single nibbles in DRAM.

- Swapable memory bank support. This allows memory banks to be swapped out.
- PCI 2.1-compliant bus.
- Supports USB.

The TXC controls the processor cycles for:

- Second-level cache transfer – the processor directly sends data to the second level cache and the TXC controls its operation.
- All other processor cycles – the TXC directs all other processor cycles to their destination (DRAM, PCI or internal TXC configuration space).

Table 5.2 TXC connections

PCI Memory Addresses		**Cache Memory Addresses**		**Cache Memory Data**			
Signal	*Pin*	*Signal*	*Pin*	*Signal*	*Pin*	*Signal*	*Pin*
AD0	15			HD0	305	HD32	179
AD1	14			HD1	307	HD33	178
AD2	33			HD2	306	HD34	149
AD3	13	HA3	275	HD3	308	HD35	180
AD4	52	HA4	315	HD4	285	HD36	136
AD5	32	HA5	252	HD5	286	HD37	135
AD6	12	HA6	316	HD6	265	HD38	138
AD7	51	HA7	312	HD7	212	HD39	125
AD8	11	HA8	272	HD8	245	HD40	126
AD9	50	HA9	271	HD9	287	HD41	115
AD10	30	HA10	311	HD10	267	HD42	137
AD11	10	HA11	291	HD11	288	HD43	117
AD12	49	HA12	251	HD12	225	HD44	128
AD13	29	HA13	310	HD13	268	HD45	114
AD14	9	HA14	270	HD14	247	HD46	127
AD15	48	HA15	290	HD15	266	HD47	102
AD16	47	HA16	250	HD16	248	HD48	101
AD17	27	HA17	309	HD17	247	HD49	116
AD18	7	HA18	289	HD18	246	HD50	104
AD19	46	HA19	269	HD19	214	HD51	103
AD20	26	HA20	249	HD20	228	HD52	81
AD21	6	HA21	273	HD21	213	HD53	84
AD22	45	HA22	254	HD22	226	HD54	82
AD23	25	HA23	253	HD23	201	HD55	61
AD24	66	HA24	294	HD24	215	HD56	83
AD25	44	HA25	293	HD25	203	HD57	63
AD26	24	HA26	274	HD26	202	HD58	62
AD27	4	HA27	313	HD27	191	HD59	41
AD28	23	HA28	314	HD28	204	HD60	42
AD29	3	HA29	255	HD29	193	HD61	43
AD30	22	HA30	295	HD30	192	HD62	21
AD31	2	HA31	292	HD31	194	HD63	1

PCI control lines

C/BE0#	21	FRAME#	86	PREQ0#	67	PGNT0#	68
C/BE1#	31	DEVSEL#	89	PREQ1#	69	PGNT1#	70
C/BE2#	8	IRDY#	88	PREQ2#	71	PGNT2#	72
C/BE3#	5	STOP#	91	PREQ3#	73	PGNT3#	74
		LOCK#	85				
		PHOLD#	64				
		PHLDA#	65				
		PAR	92				
		SERR#	93				

Cache Memory Tag		**Parity**		**Address Lines**			
Signal	*Pin*	*Signal*	*Pin*	*Signal*	*Pin*	*Signal*	*Pin*
CTAG0	207	MP0	133	MD32	74		
CTAG1	260	MP1	123	MD33	75		
CTAG2	261	MP2	146	MA2	317	MD34	76
CTAG3	281	MP3	113	MA3	297	MD35	76
CTAG4	238	MP4	132	MA4	277	MD36	76
CTAG5	282	MP5	124	MA5	257	MD37	76
CTAG6	302	MP6	134	MA6	237	MD38	76
CTAG7	322	MP7	122	MA7	298	MD39	76
CTAG8	303			MA8	258		
CTAG9	323			MA9	319		
CTAG10	324			MA10	318		
				MA11	278		
Cache address lines							
MRASR0#	121	MCASR0#	145	MAA0	300		
MRASR1#	110	MCASR1#	159	MAA1	300		
MRASR2#	109	MCASR2#	131	MAB0	300		
MRASR3#	96	MCASR3#	173	MAB1	300		
		MCASR4#	130				
		MCASR5#	144				
		MCASR6#	120				
		MCASR7#	172				
Cache control lines							
CBWE#	321	COE#	259	CCS#	300	CADS#	299
CGWE#	320	CADV#	279				

5.1.4 82091AA (AIP)

The AIP device integrates the serial ports, parallel ports and floppy disk interfaces. Figure 5.3 shows its connections and Figure 5.4 shows the interconnection between the AIP and the PIIX3 device. The OSC frequency is set to 14.218 18 MHz. It can be seen that the range of interrupts for the serial, parallel and floppy disk drive is IRQ3, IRQ4, IRQ5, IRQ6 and IRQ7. Normally the settings are:

- IRQ3 – secondary serial port (COM2/COM4).
- IRQ4 – primary serial port (COM1/COM3).
- IRQ6 – floppy disk controller.
- IRQ7 – parallel port (LPT1).

Figure 5.4 shows the main connections between the TXC, PIIX3 and the AIP. It can be seen that the AIP uses many of the ISA connections (such as 0WS#, IOCHRDY, and so on). The interface between the TCX and the PIIX3 defines the PCI bus and the interface between the PIIX3 and AIP defines some of the ISA signals.

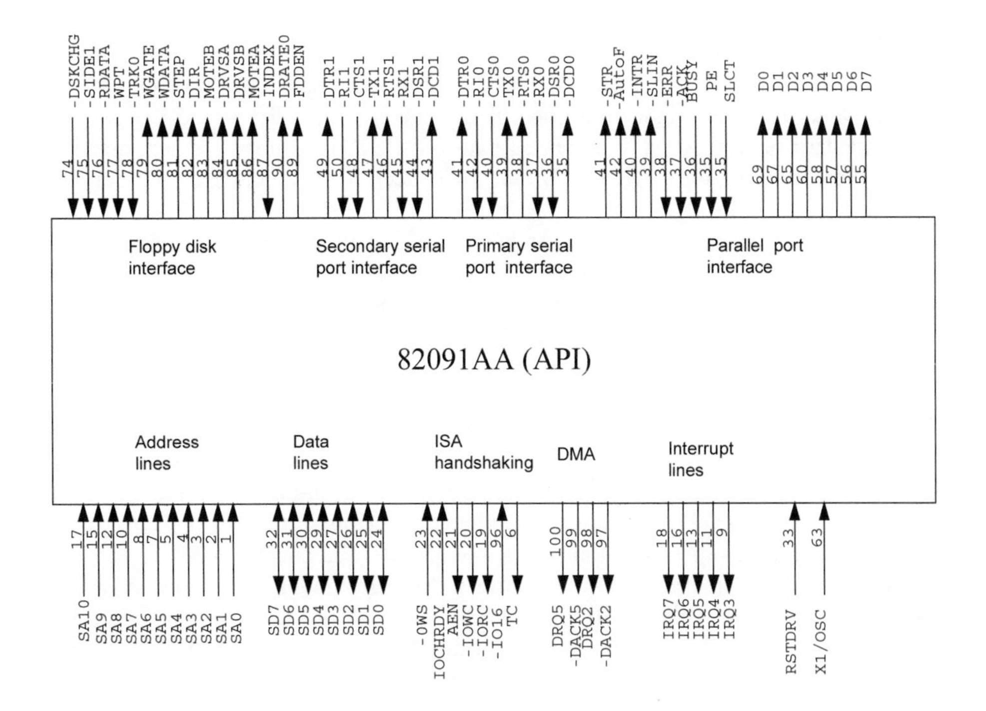

Figure 5.3 API IC

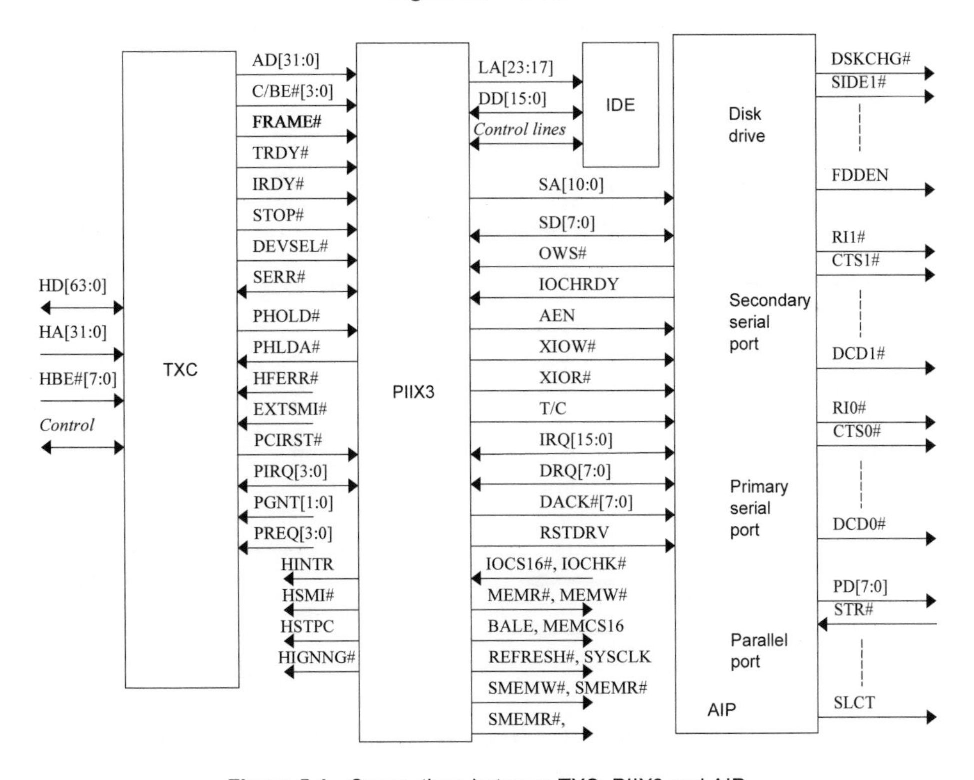

Figure 5.4 Connections between TXC, PIIX3 and AIP

5.1.5 DRAM interface

The DRAM interface supports from 4 MB to 512 MB with eight RAS lines (RAS0–RAS7) and a 64-bit data path with eight parity bits. It can use either a 3.3 V or a 5 V power supply and both standard page mode and extended data out (EDO) memory are supported with a mixture of memory sizes for 1 MB, 2 MB and 4 MB-deep SIMMs and symmetrical addressing for 16 MB-deep SIMMs.

Each SIMM (single in-line memory module) has 12 input address lines and has a 32-bit data output. They are normally available with 72 pins (named tabs) on each side. These pins can read the same signal because they are shorted together on the board. For example, tab 1 (pin 1) on side A is shorted to tab 1 on side B. Thus, the 144 tabs only gives 72 useable signal connections.

Figure 5.5 shows how the DRAM memory is organized. It shows bank 1 and 2 (and does not show banks 3 and 4). Each bank has two modules, such as modules 0 and 1 are in bank 0. The bank is selected with the MRAS lines, for example bank 1 is selected with MRAS0 and MRAS1, bank 1 by MRAS2 and MRAS3, and so on. An even-numbered module gives the lower 32 bits (MD0–MD31) and the odd number modules give the upper 32 bits (MD32–MD63). Each module also provides four parity bits (MP0–MP3 and MP4–MP7).

DIMMs (dual in-line memory modules) have independent signal lines on each side of the module and are available with 72 (36 tabs on each side), 88 (44 tabs on each side), 144 (72 tabs on each side), 168 (84 tabs on each side) or 200 tabs (100 tabs on each side). They give greater reliability and density and are used in modern high performance PC servers.

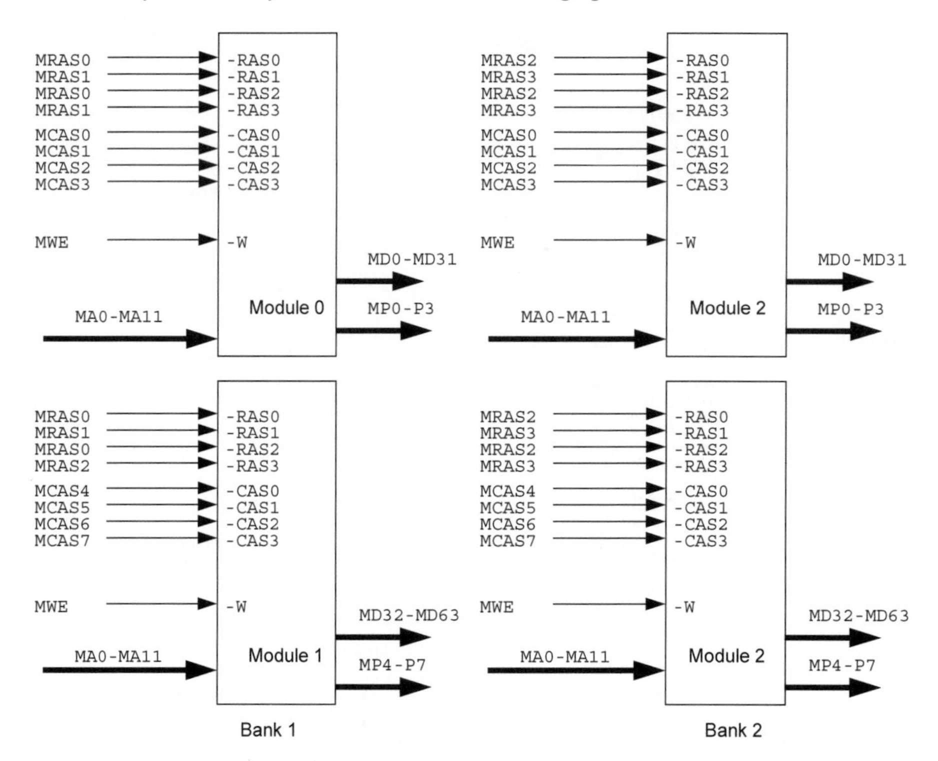

Figure 5.5 DRAM memory interface

5.1.6 Clock rates

The system board runs at several clock frequencies. These are:

- Processor speed – the processor, TXC and SRAM run at the system frequency (such as 66 MHz),
- PCI bus speed – TCX, PIIX3 and PCI slots.
- 24 or 48 MHz – USB (Universal Serial Bus).
- 12 MHz – keyboard.
- 24 MHz – floppy clock.
- 14 MHz – ISA bus OSC.
- 8 MHz – ISA bus clock.

5.1.7 ISA/IDE interface

The IDE and ISA buses share several data, address and control lines. Figure 5.6 shows the connections to the buses. The IDE interface uses the DD[12:0] and LA[23:17] lines, and the ISA uses these lines as SBHE#, SA[19:8}, CS1S, CS3S, CS1P, CS3P and DA[2:0]. A multiplexor (MUX) is used to select either the ISA or IDE interface lines.

5.1.8 DMA interface

The PIIX3 device incorporates the functionality of two 8237 DMA controllers to give seven independently programmable channel (channels 0–3 and Channels 5–7). DMA channel 4 is used to cascade the two controllers and defaults to cascade mode in the DMA channel mode (DCM) register. Figure 5.7 shows the interface connections.

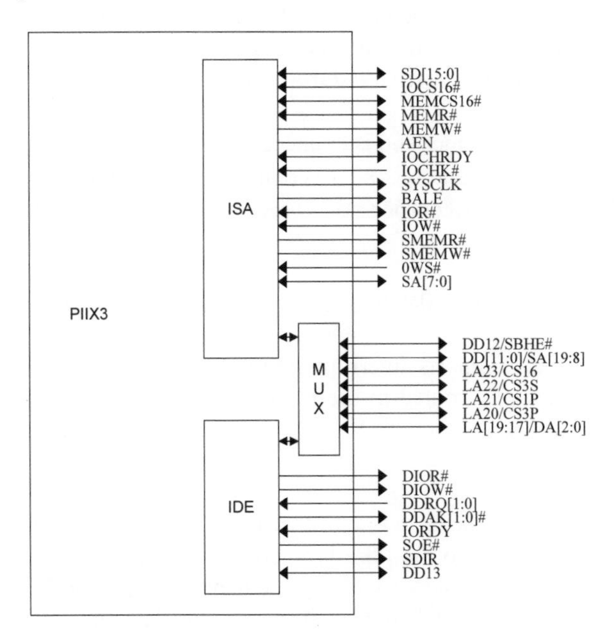

Figure 5.6 IDE/ISA interface with PIIX3

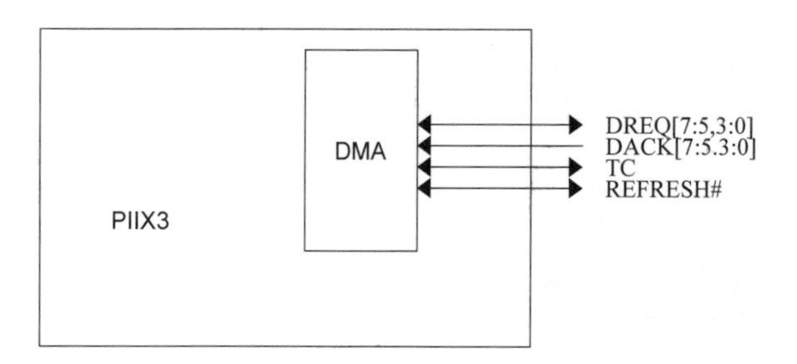

Figure 5.7 DMA interface

5.1.9 Interval timer

The PIIX3 contains three 8251-compatible counters. The three counters are contained in one PIIX3 timer unit, referred to as Timer 1. Each counter provides an essential system function. The functions of the counters are:

- Counter 0 – connects to the IRQ0 line and provides a system timer interrupt for a time-of-day, diskette time-out, and so on. The input to the counter is a 14.218 18 MHz clock (OSC). This is then used to increment a 16-bit register, which rolls over every 55 ms.
- Counter 1 – generates a refresh request signal.
- Counter 2 – generates the speaker tone.

5.1.10 Interrupt controller

The PXII3 incorporates two 8259-compatible interrupt controllers which provide an ISA-compatible interrupt controller. These are cascaded to give 13 external and three internal interrupts. The primary interrupt controller connects to IRQ0–IRQ7 and the secondary connects to IRQ8–IRQ15. The three internal interrupts are:

- IRQ0 – used by the system timer and is connected to Timer 1, Counter 0.
- IRQ2 – used by the primary and secondary controller (see Figure 2.2 in Section 2.3.2).
- IRQ13 – used by the math coprocessor, which is connected to the FERR pin on the processor.

Figure 5.8 shows that the PC uses IRQ0 as the system timer and IRQ2 by the programmable interrupt controller.

The interrupt unit also supports interrupt steering. The four PCI active low interrupts (PIRQ#[D:A]) can be internally routed in the PIIX3 to one of 11 interrupts (IRQ15, IRQ14, IRQ12–IRQ9, IRQ7–IRQ3).

5.1.11 Mouse function

The mouse normally either connects to one of the serial ports (COM1: or COM2:) or a PS/2-type connector. If they connect to the PS/2-type connector then IRQ12 is used, else a serial port connected mouse uses the serial interrupts (such as IRQ4 for COM1 and IRQ3 for COM2). Thus, a system with a serial connected mouse must have the IRQ12/M interrupt disabled. This is normally done with a motherboard jumper (to enable or disable the mouse interrupt). Figure 5.8 shows an example of a mouse using IRQ12.

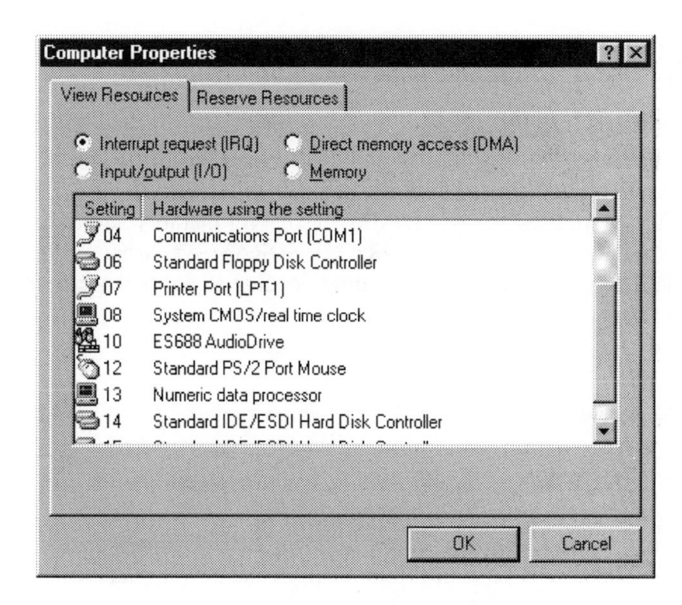

Figure 5.8 Interrupts usage shows PS/2 port mouse

5.1.12 Power management

PIIX3 has extensive power management capability permitting a system to operate in a low power state without being powered down. In a typical desktop PC there are two states – power on and power off. Leaving a system powered on when not in use wastes power. PIIX3 provides a fast on/off feature that creates a third state called fast off. When in the fast off state, the system consumes less power than the power on state.

The PIIX3's power management function is based on two modes:

- System management mode (SMM).
- Advanced power management (APM).

Software (called SMM code) controls the transitions between the power on state and the fast off state. PIIX3 invokes this software by generating an SMI (system management interrupt) to the CPU (asserting the $\overline{\text{SMI}}$ signal). The SMM code places the system in either the power on state or the fast off state.

5.1.13 Graphics subsystem

The 430HX incorporates the S3 ViRGE graphics device with 2 MB of graphics memory, which has:

- High performance 64-bit 2D/3D graphics engine.
- RAMDAC/clock synthesiser capable of pixel rates of 135 MHz.
- S3 streams processor, enabling the device to convert YUV formatted video data to RGB format.
- 3D features including flat shading and texture mapping support.
- Fast linear addressing scheme.
- VESA (Video Electronics Standards Association) capability.

5.2 TX motherboard

The Intel 430HX motherboard only supports up to 128 MB of memory and has a relatively small second level cache (256 kB). The Intel 430TX board has many enhanced devices, such as standardised USB connections and enhanced super I/O device. Figure 5.9 shows the main layout. The 430TX board uses 168-pin DIMM sockets for memory addition. It supports both EDO DRAM and SDRAM (synchronous DRAM). SDRAM synchronous data transfers using the system clock. This simplifies memory timing, leading to an increase in memory transfer. The 430TX motherboard supports a 64-bit data path to memory.
The 430TX board uses the 82430TX PCI chipset, which has:

- 82439TX Xcelerated Controller (MTXC), which replaces the TXC (82439HX) in the HX board.
- 82371AB PCI/ISA IDE Xcelerator (PIIX4), which replaces the PIIX3 (82371SB) in the HX board. This is a 324-pin BGA that integrates PCI-to-ISA bridge (two 82C37 DMA controllers, two 82C59 interrupt controllers, an 82C54 timer/counter and a real-time clock), PCI/IDE interface, USB host/hub function and power management functions.

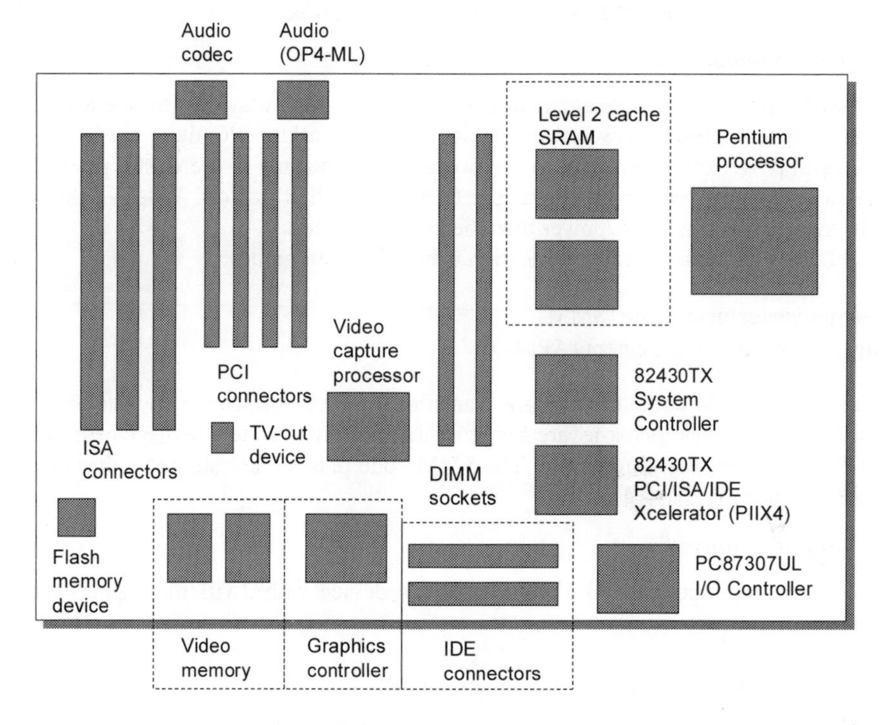

Figure 5.9 AN430TX board

5.2.1 PIIX4

The PIIX4 supports two types of PCI DMA protocol, PC/PCI DMA which uses dedicated request and grant signals to permit PCI devices to request transfers associated with specific

DMA channels, and distribute DMA which is based on monitoring CPU accesses to the 8237 controller (this was not implemented on the PIIX3).

The PIIX4 also supports Ultra DMA/33 Synchronous DMA mode transfers up to 33MB/s. Ultra DMA is a newer protocol for the IDE hard drive interface that doubles the burst data rate from 16.6 MB/s (as supported by the PIIX3). Ultra DMA widens the path to the hard drive by transferring twice as much data per clock cycle, so doubling the performance. The Ultra DMA protocol lets host computers send and receive data faster, removing bottlenecks associated with data transfers.

In addition to speed improvements, the protocol brings new data integrity capabilities to the IDE interface such as improved timing margins and data protection verification. Ultra DMA protocol also allows drives and system to retain backward compatibility with the previous ATA standard.

Real time clock

The real-time clock provides a data-and-time keeping device with alarm features and battery-backed operation. The RTC counts seconds, minutes, hours, days, day of the week, date, month and year. It counts 256 bytes of battery-backed SRAM in two banks. The RTC module requires an external oscillator source of 32.768 kHz connected between TXCX1 and RTCX2.

5.2.2 I/O controller

The PC87306B I/O controller is similar to the 82091AA (AIP) used in the 430HX board. It has:

- Floppy disk interface – provides support for several different floppy disk capacities and sizes.
- Multimode parallel port – supports for output only compatibility mode, bidirectional mode, EPP mode and ECP mode.
- Two FIFO serial ports. Giving transfers rates up to 921Kbps.
- Real-time clock – provides time-of-day, 100-year calendar and alarm features.
- Keyboard and mouse controller – keyboard and mouse interfaces (as well as power on/reset password protection).
- Infrared support – connection to infrared transmitter/receiver.

5.2.3 Graphics subsystem

The 430HX incorporates the ATI-241 Rage II+ graphics controller which has:

- Drawing coprocessor that operates concurrently with the host processor.
- Video coprocessor.
- Video scalar, colour space converter, true colour palette.
- ATI multimedia.
- Enhanced power features.
- VGA/VESA capability.

The 430HX also has an optional video capture processor for digitising analogue inputs from

VCRs, cameras, TVs, and so on. It also has an optional ATI-ImpactTC NSTC/PAL encoder which provides a TV output for the graphics accelerators.

5.2.4 DRAM interface

The DRAM interface is a 64-bit data path that supports fast page mode (FPM) and extended data out (EDO) memory. The integrated DRAM controller supports from 4 MB to 256 MB of main memory. The 12 multiplexed address lines (MA[11:0]) allow the chips to support 4-bit, 16-bit and 64-bit memory, both symmetrical and asymmetrical addressing. The MTXC has six RAS lines which enables support to up to six rows of DRAM (the TXC has eight RAS lines).

The MTXC supports SRAM. The 14 multiplexed address lines (MA[13:0]) allow the MTXC to support 16-bit and 64-bit SDRAM devices. The MTXC has six CS (chip select) lines (muxed into RAS[5:0]) which allows six rows of the faster SDRAM modules to be installed.

All these memory types FPM, EDO and SDRAM can be mixed on the 430TX (but only the FPM and EDO are supported in the 430HX board). The extra lines that have been added in the MTXC are:

- SRAS [A,B] – SRAM row address strobe.
- SCAS [A,B] – DRAM column address.

5.2.5 Second- level cache

The MTXC supports cache memory area of 64 MB using either 8 K×8 or 16 K×8 SRAM blocks to store the cache tags for either 256 KB or 512 KB SRAM cache. (8 K×8 is used for 256 KB and 16 K×8 is used for 512 KB). Each cache entry is 32 bits (4 bytes) thus the total cache memory size is 512 KB (16 K×8×4). The signals are:

$\overline{\text{CCS}}$ Cache chip select – set active upon power-up and allows access to the cache.

$\overline{\text{TWE}}$ Tag write enable – allows new state and tag addresses to be written into the cache.

$\overline{\text{COE}}$ Cache output enable – puts the cache data onto the data bus.

$\overline{\text{GWE}}$ Global write enable – causes all bytes to be written to.

$\overline{\text{CADS}}$ Cache address strobe – cache loads the address register from the address pins.

$\overline{\text{CADV}}$ Cache advance – the address is automatically increment to the next word.

TIO[7 : 0] Tag address – input lines for tag addresses.

KRQAK/$\overline{\text{CS4\_64}}$. Cache chip select – KRQAK specifies DRAM cache, else implements a 64 MB main memory cache.

$\overline{\text{BWE}}$ Byte write enable – enables up to eight bytes from the data bus.

Figure 5.10 shows the interface between the MTXC and the second-level cache. Note that four 32 K×32 devices make up the 512 KB (4×32×4) SRAM cache. Only two are shown in Figure 5.10, as the other two are connected in parallel with the two shown.

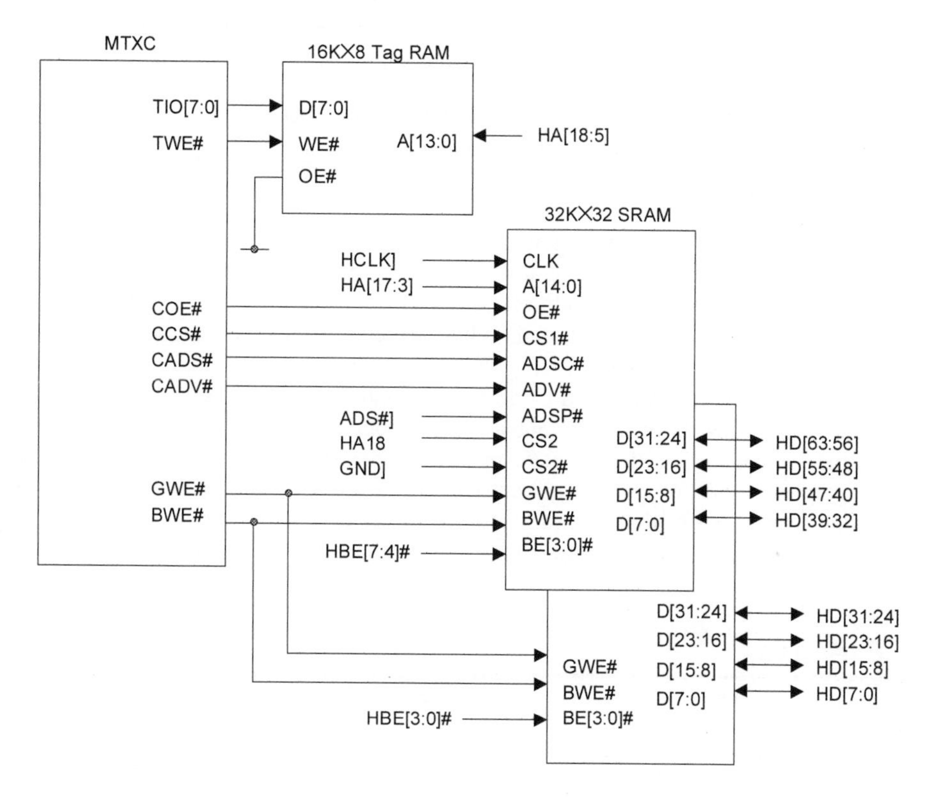

Figure 5.10 Second-level cache interface

Cache control register (CC)

This is an 8-bit register which is located at 52h in the I/O memory. It defines secondary cache operations. Its format is:

Bit	*Description*
7:6	Secondary cache size – 00 (disabled), 01 (256K), 10 (512K), 11 (reserved).
5:4	SRAM cache type – 00 (pipelined burst SRAM), 01 (reserved), 10 (reserved), 11 (two banks of pipelined burst).
3	NA disable – 1 (disable), 0 (enabled); normally enabled.
2	Reserved.
1	Secondary cache force miss or invalidated (SCFMI). When set to a 1, the Level 2 hit/miss facility is disabled, else it is enabled.
0	First-level cache enable (FLCE) – 1 (enable), 0 (disable). When it is set to a 1, the control responds to processor cycles with KEN# active. Normal mode for FLCE, SCFMI is 1, 0.

Extended cache control register (CEC)

This is an 8-bit register which is located at 53h in the I/O memory. It defines the refresh rate for DRAM Level 2. Its format is:

Bit	Description
7:6	Reserved.
5	Defines if DRAM cache is present – 1 (present), 0 (not present).
4:0	DRAM cache refresh timer value.

5.2.6 Power management

The PIIX4 has enhanced power management over the PIIX3 and can detect when a specific device is idle. The system management software is then informed, which then can place the idle device into a power managed condition (such as local standby or powered off). Accesses targeted to that device are then monitored. When detected, an $\overline{\text{SMI}}$ is generated to allow the software to restore the device to operation.

The PIIX4 supports the Advanced Configuration and Power Interface (ACPI) specification. The software consists of system management mode (SMM) BIOS for legacy control and operating system for ACPI. The basic operation consists of software setting up the desired configurations and power management mode and corresponding power saving levels. The hardware then performs the necessary actions to maintain the power mode. The I/O chip also monitors the system for events which may require changing the system power mode.

5.3 Exercises

5.3.1 The data bus signals which connect to the processor are:

(a) HD0–HD63 (b) D0–D63
(c) HA0–HA63 (d) AD0–AD63

5.3.2 Which device provides the bridge between the processor, second-level cache, DRAM and the PCI bus:

(a) PIIX3/4 (b) HXC
(c) RFC/MRFC (d) TXC/MTXC

5.3.3 Which device provides the bridge between the PCI bus and other busses, such as the IDE, ISA and USB:

(a) PIIX3/4 (b) HXC
(c) RFC (d) TXC

5.3.4 The maximum achievable data throughput for a 33 MHz, 32-bit PCI is 132 MB/s. Why is this not achievable in the normal multiplexed mode:

(a) Half of the bus is used for addresses, the other half for data
(b) The bus must slow down because of synchronisation problems
(c) The address and data line are shared (multiplexed address then data).
(d) The clock rate is halved for all transfers

5.3.5 How does a cache identify the address of the data it has in its memory:

(a) The full address is stored along with the data
(b) It is tagged (CTAG).

(c) It guesses the address
(d) It checks the address with the contents of the DRAM

5.3.6 How many data bits are transferred between the processor and the second-level cache:

(a) 16 (b) 32 (c) 64 (c) 128

5.3.7 Outline the importance of the TXC (system controller) device in the PC. Outline also the main ICs that are used in a PC.

5.3.8 Describe, in detail, the architecture of the HX PCI chip set, and how the Pentium processor communicates with: DRAM memory, Level-2 SRAM cache, the PCI bus, the ISA bus and the IDE bus.

5.3.9 Explain, with reference to the PIIX3 and Pentium processor, how interrupts on the PCI and ISA busses are dealt with.

5.3.10 Explain, with reference to the level-1 cache, the level-2 cache and DRAM, how the processor accesses memory. What advantage does level-1 have over level-2 cache, and what advantage do these have over DRAM.

5.3.11 Discuss the power management modes supported by the PXII3, and also by the PXII4.

5.3.12 Which interrupts are supported with the AIP and where are they typically used?

5.3.13 Explain how the ISA and IDE busses share the same control and data lines.

5.3.14 Contrast the HX motherboard with the LX motherboard.

5.4 Notes from the author

I hope that this chapter was not too heavy. It is important to realise that it is not just the speed of the processor that defines the performance of a system – it is the cache controllers, the bridge devices, the PCI bus, and so on. So have tried to give you an understanding of the segmentation that is used in typical PCs. The devices used will change, but the basic concept is likely to stay the same (I hope!). The days of a PC on a chip will happen, someday.

The most amazing thing about modern PC systems is that they are almost completely compatible with the original PC, the big change has happened in the integration of many of the components parts. The great strength of the PC is its availability, durability and up-gradeability of its components. I find it amazing that it can disconnect the cable to the disk drives, turn it round, and connect it and the system will not be damage, in any way (although it won't start). I can even put the processor in the wrong way, and it will not damage it.

The other amazing thing about PCs is the way that new peripherals are quickly adopted, and become standard parts of the PC. This has included CD-ROM drives, USB connectors, PS/2-type mouse connectors, PCMCIA connectors (in notebooks), VGA graphics adaptors, TV output, DVD drives, network cards, sound cards, and so on. Who would have believed that such a basic system as the original PC would support all this expansion, without ever the need to redesign it (although the PCI bus provided a new architectural design).

6 IDE and Mass Storage

6.1 Introduction

This chapter and the next chapter discuss IDE and SCSI interfaces which are used to interface to disk drives and mass storage devices. Disks are used to store data reliably in the long term. Typical disk drives either store binary information as magnetic fields on a fixed disk (as in a hard disk drive), a plastic disk (as in a floppy disk or tape drive), or as optical representation (on optical disks).

The main sources of permanent read/writeable storage are:

- Magnetic tape – where the digital bits are stored with varying magnetic fields. Typical devices are tape cartridges, DAT and 8 mm video tape.
- Magnetic disk – as with the magnetic tape the bits are stored as varying magnetic fields on a magnetic disk. This disk can either be permanent (such as a hard disk) or flexible (such as a floppy disk). Large capacity hard disks allow storage of several gigabytes of data. Normally fixed disks are designed to a much higher specification than floppy disks and can thus store much more information.
- Optical disk – where the digital bits are stored as pits on an optical disk. A laser then reads these bits. This information can either be read only (CD-ROM), write once read many (WORM) or can be reprogrammable. A standard CD-ROM stores up to 650 MB of data. Their main disadvantage in the past has been their relative slowness as compared with Winchester hard disks; this is now much less of a problem as speeds have steadily increased over the years.

6.2 Tracks and sectors

A disk must be formatted before it is used, which allows data to be stored in a logical manner. The format of the disk is defined by a series of tracks and sectors on either one or two sides. A track is a concentric circle around the disk where the outermost track is track 40 and the innermost track is 0. The next track is track 1 and so on, as shown in Figure 6.1. Each of these tracks is divided into a number of sectors. The first sector is named sector 1, the second is sector 2, and so on. Most disks also have two sides: the first side of the disk is called side 0 and the other is side 1.

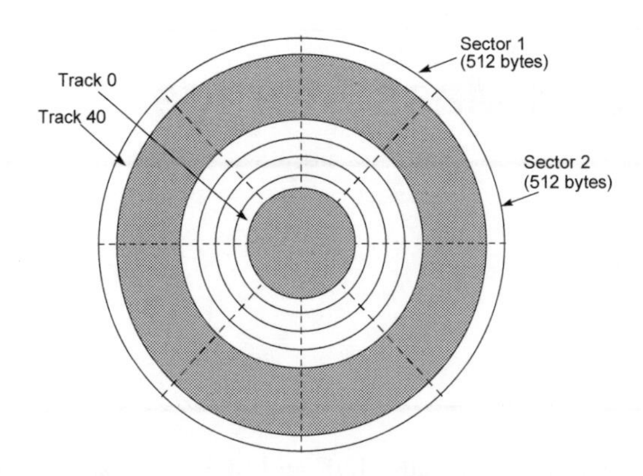

Figure 6.1 Tracks and sectors on a disk

Figure 6.1 also shows how each track is split into a number of sectors, in this case there are eight sectors per track. Typically each sector stores 512 bytes. The total disk space, in bytes, will thus be given by

$$\text{Disk space} = \text{Number of sides} \times \text{tracks} \times \text{sectors per track} \times \text{bytes per sector}$$

For example, a typical floppy disk has two sides, 80 tracks per side, 18 sectors per track and 512 bytes per sector, so

$$\begin{aligned} \text{Disk capacity} &= 2\times80\times18\times512 && = 1\,474\,560 \text{ B} \\ &= 1\,474\,560/1\,024 \text{ KB} && = 1\,440 \text{ KB} \\ &= 1440/1024 \text{ MB} && = 1.4 \text{ MB} \end{aligned}$$

6.3 Floppy disks

A 3.5-inch DD (double density) disk can be formatted with two sides, nine sectors per track and 40 tracks per side. This gives a total capacity of 720 KB. A 3.5 inch HD (high density) disk has a maximum capacity when formatted with 80 tracks per side.

A 5.25-inch DD disk can be formatted with two sides, nine sectors per disk with either 40 or 80 tracks per side. The maximum capacity of these formats is 360 KB (40 tracks) or 720 KB (80 tracks). A 5.25-inch HD disk can be formatted with 15 sectors per track which gives a total capacity of 1.2 MB. When reading data the disks rotate at 300 rpm. Table 6.1 outlines the differing formats.

Table 6.1 Capacity of different disk types

Size	*Tracks per side*	*Sectors per track*	*Capacity*
5.25-inch	40	9	360 KB
5.25-inch	80	15	1.2 MB
3.5-inch	40	9	720 KB
3.5-inch	80	18	1.44 MB

6.4 Fixed disks

Fixed disks store large amounts of data and vary in their capacity, from several MB to several GB. A fixed disk (or hard disk) consists of one or more platters which spin at around 3000 rpm (10 times faster than a floppy disk). A hard disk with four platters is shown in Figure 6.2. Data is read from the disk by a flying head which sits just above the surface of the platter. This head does not actually touch the surface as the disk is spinning so fast. The distance between the platter and the head is only about 10 μin (which is no larger than the thickness of a human hair or a smoke particle). It must thus be protected from any outer particles by sealing it in an airtight container. A floppy disk is prone to wear as the head touches the disk as it reads but a fixed disk has no wear as its heads never touch the disk.

One problem with a fixed disk is head crashes, typically caused when the power is abruptly interrupted or if the disk drive is jolted. This can cause the head to crash into the disk surface. In most modern disk drives the head is automatically parked when the power is taken away. Older disk drives that do not have automatic head parking require a program to park the heads before the drive is powered down.

There are two sides to each platter and, like floppy disks, each side divides into a number a tracks which are subdivided into sectors. A number of tracks on fixed disks are usually named cylinders. For example a 40 MB hard disk has two platters with 306 cylinders, four tracks per cylinder, 17 sectors per track and 512 bytes per sector, thus each side of a platter stores

$$
\begin{aligned}
306\times4\times17\times512\text{ B} &= 10\,653\,696\text{ B}\\
&= 10\,653\,696/\,1\,048\,576\text{ MB}\\
&= 10.2\text{ MB}
\end{aligned}
$$

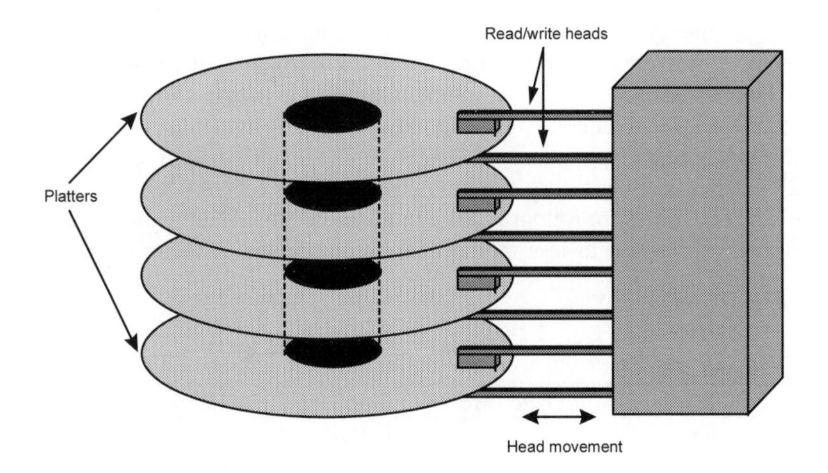

Figure 6.2 Hard disk with four platters

6.5 Drive specifications

Access time is the time taken for a disk to locate data. Typical access times for modern disk drives range from 10 to 30 ms. The average access time is the time for the head to travel half way across the platters. Once the head has located the correct sector then there may be another wait until it locates the start of the data within the sector. If it is positioned at a point after the start of the data, it requires another rotation of the disk to locate the data. This average wait, or latency time, is usually taken as half of a revolution of the disk. If the disk spins at 3600 rpm then the latency is 8.33 ms.

The main parameters which affect the drive specification are the data transfer rate and the average access time. The transfer rate is dependent upon the interface for the controller/disk drive and system/controller and the access time is dependent upon the disk design.

6.6 Hard disk and CD-ROM interfaces

There are two main interfaces involved with a hard disks (and CD-ROMs). One connects the disk controller to the system (system–controller interface) and the other connects the disk controller to the disk drive (disk–controller interface).

The controller can be interfaced by standards such as ISA, EISA, MCA, VL-Local bus or PCI bus. For the interface between the disk drive and the controller then standards such as ST-506, ESDI, SCSI or IDE can be used. ST-506 was developed by Seagate Technologies and is used in many older machines with hard disks of a capacity less than 40 MB. The enhanced small disk interface (ESDI) is capable of transferring data between itself and the processor at rates approaching 10 MB/s.

The small computer system interface (SCSI) allows up to seven different disk drives or other interfaces to be connected to the system through the same interface controller. SCSI is a common interface for large capacity disk drives and is illustrated in Figure 6.3.

The most popular type of PC disk interface is the integrated drive electronics (IDE) standard. It has the advantage of incorporating the disk controller in the disk drive, and attaches directly to the motherboard through an interface cable. This cable allows many disk drives to be connected to a system without worrying about bus or controller conflicts. The IDE interface is also capable of driving other I/O devices besides a hard disk. It also normally contains at least 32K of disk cache memory. Common access times for an IDE are often less than 16 ms, where as access times for a floppy disk are about 200 ms. With a good disk cache system the access time can reduce to less than 1 ms. A comparison of the maximum data rates is given in Table 6.2.

Table 6.2 Capacity of different disk types

Interface	*Maximum data rate (MB/s)*	*Interface*	*Maximum data rate (MB/s)*
ST-506	0.6	E-IDE	16.6
ESDI	1.25	SCSI	4.0
IDE	8.3	SCSI-II	10.0

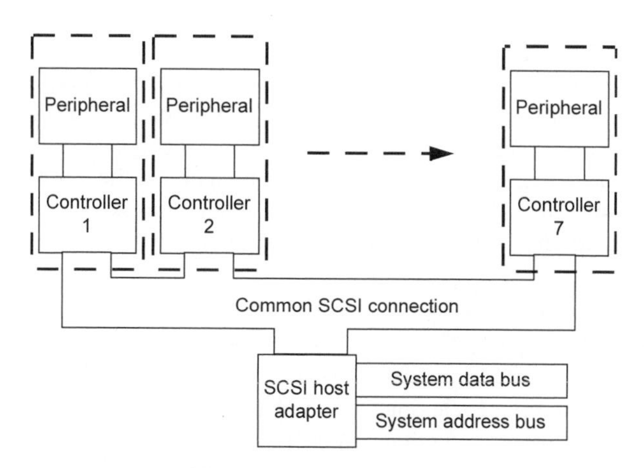

Figure 6.3 SCSI interface

A typical modern PC contains two IDE connections on the motherboard, named IDE0 and IDE1. The IDE0 connection connects to the master drive (C:) and IDE1 to the slave drive (D:). These could connect either to two hard disks or, possibility, to one hard disk and a CD-ROM drive (or even a tape backup system). Unfortunately, the IDE standard only allows disk access up to 528 MB. A new standard called Enhanced-IDE (E-IDE) allows for disk capacities of over this limit. The connector used is the same as IDE but the computers' BIOS must be able to recognise the new standard. Most computers manufactured since 1993 are able to fully access E-IDE disk drives.

The specifications for the IDE and EIDE are

IDE

- Maximum of two devices (hard disks).
- Maximum capacity for each disk of 528 MB.
- Maximum cable length of 18 inches.
- Data transfer rates of 3.3, 5.2 and 8.3 MB/s.

EIDE

- Maximum of four devices (hard disks, CD-ROM and tape).
- Uses two ports (for master and slave).
- Maximum capacity for each disk is 8.4 GB.
- Maximum cable length of 18 inches.
- Data transfer rates of 3.3, 5.2, 8.3, 11.1 and 16.6 MB/s.

6.7 IDE interface

The most popular interface for hard disk drives is the Integrated Drive Electronics (IDE) interface. Its main advantage is that the hard disk controller is built into the disk drive and the interface to the motherboard simply consists of a stripped-down version of the ISA bus. The most common standard is the ANSI-defined ATA-IDE standard. It uses a 40-way ribbon cable to connect to 40-pin header connectors. Table 6.3 lists the pin connections. It has a 16-bit data bus (D0–D15) and the only available interrupt line used is IRQ14 (the hard disk uses

IRQ14).

The standard allows for the connection of two disk drives in a daisy-chain configuration. This can cause problems because both drives have controllers within their drives. The primary drive (Drive 0) is assigned as the master and the secondary driver (Drive 1) as the slave. A drive is set as a master or a slave by setting jumpers on the disk drive. They can also be set by software using the cable select (CSEL) pin on the interface.

EIDE has various modes (ANSI modes) of operation, these are

- Mode 0 – 600 ns read/write cycle time, 3.3 MB/s burst data transfer rate.
- Mode 1 – 383 ns read/write cycle time, 5.2 MB/s burst data transfer rate.
- Mode 2 – 240 ns read/write cycle time, 8.3 MB/s burst data transfer rate.
- Mode 3 – 180 ns read/write cycle time, 11.1MB/s burst data transfer rate.
- Mode 4 – 120 ns read/write cycle time, 16.6MB/s burst data transfer rate.

Table 6.3 IDE connections

Pin	*IDE signal*	*AT signal*	*Pin*	*IDE signal*	*AT signal*
1	RESET	RESET DRV	2	GND	–
3	D7	SD7	4	D8	SD8
5	D6	SD6	6	D9	SD9
7	D5	SD5	8	D10	SD10
9	D4	SD4	10	D11	SD11
11	D3	SD3	12	D12	SD12
13	D2	SD2	14	D13	SD13
15	D1	SD1	16	D14	SD14
17	D0	SD0	18	D15	SD15
19	GND	–	20	KEY	–
21	DRQ3	DRQ3	22	GND	–
23	$\overline{IOW}$	$\overline{IOW}$	24	GND	–
25	$\overline{IOR}$	$\overline{IOR}$	26	GND	–
27	IOCHRDY	IOCHRDY	28	CSEL	–
29	$\overline{DACK3}$	$\overline{DACK3}$	30	GND	–
31	IRQ14	IRQ14	32	$\overline{IOCS16}$	$\overline{IOCS16}$
33	Address bit 1	SA1	34	$\overline{PDIAG}$	–
35	Address bit 0	SA0	36	Address bit 2	SA2
37	$\overline{CS1FX}$	–	38	$\overline{CS3FX}$	–
39	SP / $\overline{DA}$	–	40	GND	–

6.8 IDE communication

The IDE (or AT bus) is the de facto standard for most hard disks in PCs. It has the advantage over older type interfaces that the controller is integrated into the disk drive. Thus the computer only has to pass high-level commands to the unit and the actual control can be achieved with the integrated controller. Several companies developed a standard command set for an ATA (AT attachment). Commands include:

- Read sector buffer – reads contents of the controller's sector buffer.
- Write sector buffer – writes data to the controller's sector buffer.

- Check for active.
- Read multiple sectors.
- Write multiple sectors.
- Lock drive door.

The control of the disk is achieved by passing a number of high-level commands through a number of I/O port registers. Table 6.3 outlined the pin connections for the IDE connector. Typically pin 20 is missing on the connector cable so that it cannot be inserted the wrong way, although most systems buffer the signals so that the bus will not be damaged if the cable is inserted the wrong way. The five control signals which are unique to the IDE interface (and not the AT bus) are

- $\overline{CS3FX}$, $\overline{CS1FX}$ – these are used to identify either the master or the slave.
- $\overline{PDIAG}$ (passed diagnostic) – used by the slave drive to indicate that it has passed its diagnostic test.
- $SP/\overline{DA}$ (slave present/drive active) – used by the slave drive to indicate that it is present and active.

The other signals are

- IOCHRDY – This signal is optional and is used by the drive to tell the processor that it requires extra clock cycles for the current I/O transfer. A high level informs the processor that it is ready, while a low informs it that it needs more time.
- DRQ3, $\overline{DACK3}$ – These are used for DMA transfers.

6.8.1 AT task file

The processor communicates with the IDE controller through data and control registers (typically known as the AT task file). The base registers used are between 1F0h and 1F7h for the primary disk (170h and 177h for secondary), and 3F6h (376h for secondary), as shown in Figure 6.4. Their function is:

Port	*Function*	*Bits*	*Direction*
1F0h	Data register	16	R/W
1F1h	Error register	8	R
	Precompensation	8	W
1F2h	Sector count	8	R/W
1F3h	Sector number	8	R/W
1F4h	Cylinder LSB	8	R/W
1F5h	Cylinder MSB	8	R/W
1F5h	Drive/head	8	R/W
1F6h	Status register	8	R
	Command register	8	W
3F6h	Alternative status register	8	R
	Digital output register	8	W
3F7h	Drive address	8	R

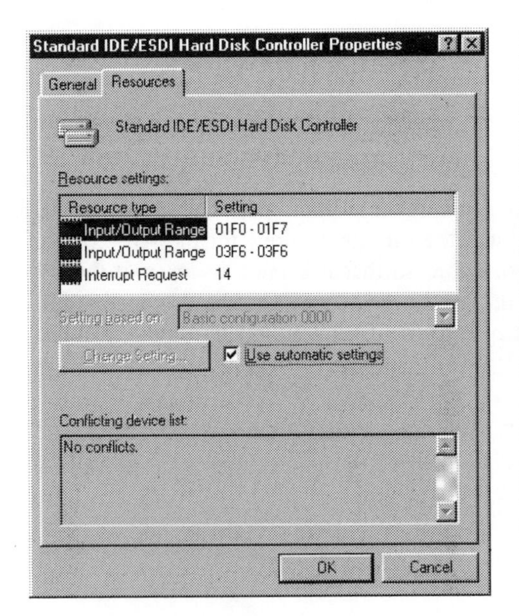

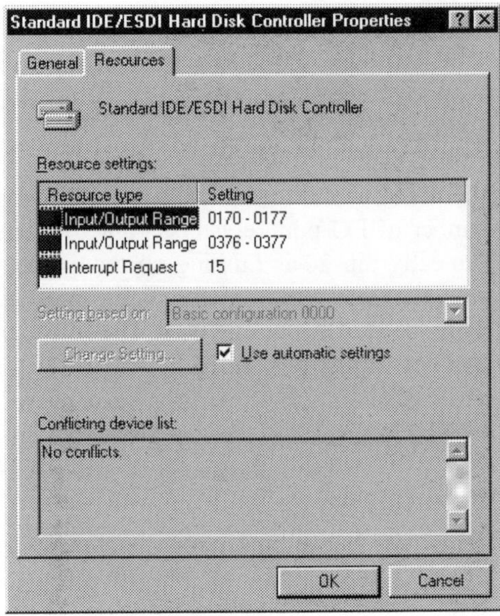

Figure 6.4 Typical hard-disk controller settings for the primary and secondary drive

Data register (1F0h)

The data register is a 16-bit register which is used to read/write data from/to the disk.

Error register (1F1h)

The error register is read-only and contains error information relating to the last command. Its definitions are

b_7	b_6	b_5	b_4	b_3	b_2	b_1	b_0
BBK	UNC	MCNID	MCR	ABT	NT0		NDM

where

- BBK – set to 1 if the sector is bad.
- UNC – set to 1 if there is an unrecoverable error.
- NID – set to 1 if mark not found.
- ABT – set to 1 if command aborted.
- NT0 – set to 1 if track 0 not found.
- MC – set to 1 identifies that the medium has changed (EIDE only). The EIDE standard support disks which can be changed while the system is running (such as CD-ROMs, tape drives, and so on).
- MCR – set to 1 identifies that the medium requires to be changed (EIDE only).

Sector count register (1F2h)

This is a read/write 8-bit register which defines the number of sectors to be read, written or verified. After each transfer to/from the disk causes the register value to be decremented by one.

Sector number register (1F3h)

This is a read/write 8-bit register which defines the start sector to be read, written or verified. After each transfer to/from the disk, the register contains the last processed sector.

Cylinder register (1F4h/1F5h)

These are read/write 8-bit registers which define the LSB (1F4h) and MSB (1F5h) of the cylinder number. The two registers are capable of containing a 16-bit value. In standard IDE the cylinder number is 10-bit and can only vary from 0 to 1023 (0 to $2^{10}-1$). For E-IDE the value can be a 16-bit value and can thus vary from 0 to 65 535 (0 to $2^{16}-1$). This is one of the main reasons that E-IDE can address much more data than IDE.

Drive/head register (1F6h)

This is a read/write 8-bit register which defines the currently used head. Its definitions are:

b_7	b_6	b_5	b_4	b_3	b_2	b_1	b_0
1	L	1	DRV	HD_3	HD_2	HD_1	HD_0

where:

- L – set to 1 if LBA (logical block addressing) mode else set to 0 if CHS (EIDE only).
- DRV – set to 1 for the slave, else it is master.
- HD_3–HD_0 – identifies the head number, where 0000 identifies head 0, 0001 identifies head 1, and so on.

Status register (1F7h)

The 1F7h register has two modes. If it is written to then it is a command register (see next section) else, if it is read from, then it is a status register. The status register is a read-only 8-bit register which contains status information from the previously issued command. Its definitions are:

b_7	b_6	b_5	b_4	b_3	b_2	b_1	b_0
BUSY	RDY	WFT	SKT	DRQ	COR	IDX	ERR

where:

- BUSY – set to 1 if the drive is busy.
- RDY – set to 1 if the drive is ready.
- WFT– set to 1 if there is a write fault.
- SKT– set to 1 if head seek positioning is complete.
- DRQ – set to 1 if data can be transferred.
- COR – set to 1 if there is a correctable data error.
- IDX – set to 1 identifies that the disk index has just passed.
- ERR – set to 1 identifies that the error register contains error information.

Command register (1F7h)

If the 1F7h register is written-to then it is a command register. The command register is a 8-bit register can contain commands, such as:

Command	b_7	b_6	b_5	b_4	b_3	b_2	b_1	b_0	*Related registers*
Calibrate drive	0	0	0	1	–	–	–	–	1F6h
Read sector	0	0	1	0	–	–	L	R	1F2h–1F6h
Write sector	0	0	1	1	–	–	L	R	1F2h–1F6h
Verify sector	0	1	0	0	–	–	–	R	1F2h–1F6h
Format track	0	1	0	1	–	–	–	–	1F3h–1F6h
Seek	0	1	1	1	–	–	–	–	1F4h–1F6h
Diagnostics	1	0	0	1	–	–	–	–	1F2h, 1F6h
Read sector buffer	1	1	1	0	0	1	0	0	1F6h
Write sector buffer	1	1	1	0	1	0	0	0	1F6h
Identify drive	1	1	1	0	1	1	–	–	1F6h

where R is set to 0 if the command is automatically retried and L identifies the long-bit.

Digital output register (3F6h)

This is a write-only 8-bit register which allows drives to be reset and also IRQ14 to be masked. Its definitions are

b_7	b_6	b_5	b_4	b_3	b_2	b_1	b_0
–	–	–	–	–	SRST	$\overline{\text{IEN}}$	–

where

- SRST– set to a 1 to reset all connected drives, else accept the command.
- $\overline{\text{IEN}}$ – controls the interrupt enable. If set to 1 then IRQ14 is always masked, else interrupted after each command.

Drive address register (3F7h)

The drive address register is a read-only register which contains information on which drive and which head is currently active. Its definitions are:

b_7	b_6	b_5	b_4	b_3	b_2	b_1	b_0
–	$\overline{\text{WTGT}}$	$\overline{\text{HS3}}$	$\overline{\text{HS2}}$	$\overline{\text{HS1}}$	$\overline{\text{HS0}}$	$\overline{\text{DS1}}$	$\overline{\text{DS0}}$

where

- $\overline{\text{WTGT}}$ – set to 1 if the write gate is closed, else the write gate is open.
- $\overline{\text{HS3}}$ – $\overline{\text{HS0}}$ –1's complement value of currently active head.
- $\overline{\text{DS1}}$ – $\overline{\text{DS0}}$ – identifies the selected drive.

6.8.2 Command phase

The IRQ14 line is used by the disk to when it wants to interrupt the processor, either when it wants to read or write data to/from memory. For example, using Microsoft C++ (for Borland replace `_outp()` and `_inp()` with `outport()` and `inport()`) to write to a disk at cylinder 150, head 0 and sector 7:

```
#include <conio.h>
int main(void)
{
int         sectors=4, sector_no=7, cylinder=150, drive=0, command=0x33, i;
unsigned    int buff[1024], *buff_pointer;
      do
      {
         /* wait until BSY signal is set to a 1 */

      } while (( _inp(0x1f7) & 0x80) != 0x80);

      _outp(0x1f2,sectors);            /* set number of sectors        */
      _outp(0x1f3,sector_no);          /* set sector number            */
      _outp(0x1f4,cylinder & 0x0ff);   /* set cylinder number LSB      */
      _outp(0x1f5,cylinder & 0xf00);   /* set cylinder number MSB      */
      _outp(0x1f6,drive);              /* set DRV=0 and head=0         */
      _outp(0x1f7,command);            /* 0011 0011 (write sector)     */

      do
      {
         /* wait until BSY signal is set to a 1 and DRQ is set to a 1 */

      } while ( ((_inp(0x1f7) & 0x80) != 0x80) &&
                                  ((_inp(0x1f7) & 0x08) !=0x08) );
      buff_pointer= buff;
      for (i=0;i<512;i++,buff_pointer++)
      {
         _outp(0x1f0,*buff_pointer); /* output 16-bits at a time */
      }
      return(0);
}
```

Note that if the L bit is set then an extra four ECC (error correcting code) bytes must be written to the sector (thus a total of 516 bytes are written to each sector). The code used is cyclic redundancy check, which, while it cannot correct errors is very powerful at detecting them.

6.8.3 E-IDE

The main differences between IDE and E-IDE are:

- E-IDE support removable media.
- E-IDE supports a 16-bit cylinder value, which gives a maximum of 65 636 cylinders.
- Higher transfer rates. In mode 4, E-IDE has a 120 ns read/write cycle time, which gives a 16.6 MB/s burst data transfer rate.
- E-IDE supports LBA (logical block addressing) which differs from CHS (cylinder head sector) in that the disk drive appears to be a continuous stream of sequential blocks. The addressing of these blocks is achieved from within the controller and the system does not have to bother about which cylinder, header and sector is being used.

IDE is limited to 1024 cylinders, 16 heads (drive/head register has only four bits for the number of heads) and 63 sectors, which gives

$$\text{Disk capacity} = 1024 \times 16 \times 63 \times 512 = 504\ \text{MB}$$

With enhanced BIOS this is increased to 1024 cylinders, 256 heads (8-bit definition for the number of heads) and 63 sectors, to give

Disk capacity = 1024×256×63×512 = 7.88 GB

With E-IDE the maximum possible is 65 536 cylinders, 256 heads and 63 sectors, to give

Disk capacity = 65536×256×63×512 = 128 GB

Normally a 3.5-inch hard disk would be limited to around two platters, with four heads. Thus, the capacity is around 8.1GB.

6.9 Optical storage

Optical storage devices can store extremely large amounts of digital data. They use a laser beam which reflects from an optical disk. If a pit exists in the disk then the laser beam does not reflect back. Figure 6.5 shows the basic mechanism for reading from optical disks. A focusing lens directs the laser light to an objective lens which focuses the light onto a small area on the disk. If a pit exists then the light does not reflect back from the disk. If the pit does not exist then it is reflected and directed through the objective lens and a quarter-wave plate to the polarised prism. The quarter-wave polarises the light by 45° thus the reflected light will have a polarisation of 90°, with respect to the original incident light in the prism. The polarised prism then directs this polarised light to the sensor.

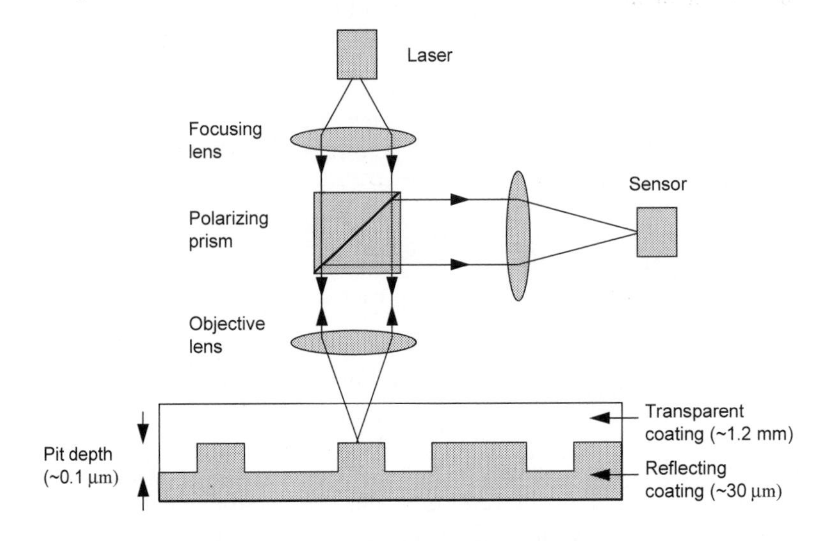

Figure 6.5 Reading from an optical disk

6.9.1 CD-ROM

In a permanent disk (also known as compact disk or CD) the pits are set up by pressing them onto the disk at production. The data on this type of disk is permanent and cannot be reprogrammed to store different data, and is known as CD-ROM (compact disk read-only memory). This type of disk is normally only cost effective in large quantities.

Standard CD-ROM disks have a diameter of 120 mm (4.7 inch) and a thickness of

1.2 mm. They can store up to 650 MB of data which gives around 74 minutes of compressed video (MPEG format with near VCR quality) or uncompressed hi-fi audio. The reflective coating (normally aluminium) on the disk is approximately 30 µm and the pits are approximately 0.1 µm long and deep. A protective transparent coating is applied on top of the reflective coating with a depth of 1.2 mm (the approximate thickness of the disk). The protective coating also helps to focus the light beam from about 0.7 mm on the surface of the coating to the 0.1 µm pit. Data is stored on the disk as a spiral starting from the inside and ending at the outside (which is opposite to hard disk. The thickness of the track is 1.6 µm, which gives a total spiral length of 5.7 km.

6.9.2 WORM drives

WORM (write once read many) disks allow data to be written to the optical disk once. The data is then permanent and thus cannot be altered. They are typically used in data logging applications and in making small volumes of CD-ROMs. A 350 mm (14 inch) WORM disk can store up to 10 GBs of data (5 GB per side). This gives around 15 hours of compressed video (MPEG format with near VCR quality).

WORM disks consist of two pieces of transparent material (normally glass) with a layer of metal (typically tellurium) sandwiched in between. Initially the metal recording surface is clear. A high intensity laser beam then writes information to the disk by burning small pits into the surface.

6.9.3 CD-R and CD-RW disks

CD-R (CD-recordable) disks are write-once disks that can store up to 650 MB of data or 74 minutes of audio. For a disk to be read by any CD-ROM drive they must comply with ISO 9660 format. A CD-R disk can also be made multisession where a new file system is written each time the disk is written to. Unfortunately, this takes up around 14 MB of header data for each session. Typical parameters for sessions are:

No. of sessions	*Header information*	*Data for each session*
1	approx. 14MB	636MB one session
5	approx. 70MB	116MB each session
10	approx. 140MB	51MB each session
30	approx. 420MB	7.7MB each session

Typically CD recorders write at two (or even four) times the standard writing/playback speed of 150 KB (75 sectors) per second.

A CD-RW (CD-rewriteable) disk allows a disk to be written-to many times, but the file format is incompatible with standard CD-ROM systems (IS0 9660). The formatting of the CD-RW disk (which can take a few hours) takes up about 157 MB of disk space, which only leaves about 493 MB for data.

New CD-R and CD-RW writing systems incorporate a smart laser system that eradicates the problem of dirt on the disk. It does this by adjusting the write power of the laser using Automatic Power Control. This allows the unit to continue to write when it encounters minor media errors such as dirt, smudges, small scratches, and so on.

6.9.4 CD-ROM disk format

The two main standards for writing a CD-ROM are ISO 9660 and UDF (universal disk format). The ISO 9660 disk unfortunately uses 14 MB for each write to the disk.

In 1980, Philips NV and Sony Corporation first announced the CD-DA (digital audio) and in 1983 released the standard for CD-ROM. Then in 1988, they released the Red Book standard for recordable CD audio disks (CD-DA)

This served as a blueprint for the Yellow Book specification for CD-ROMs (CD-ROM and CD-ROM-XA data format) and the Orange Book Parts 1 and 2 specifications for CD-Recordable (CD-R/CD-E (CD-recordable/CD-erasable)). In the Red Book standard a disk is organized into a number of segments:

- Lead in – contains the table of contents for the disk that specifies the physical location of each track.
- Program area – contains the actual disk data or audio data and is divided up into 99 tracks, with a two-second gap between each track.
- Lead out – contains a string of zeros which is a legacy of the old Red Book standard. These zeros enabled old CD players to identify the end of a CD.

The CD is laid out in a number of sectors. Each of these sectors contains 2352 bytes, made up of 2048 bytes of data and other information such as headers, sub-headers, error detection codes and so on. The data is organised into logical blocks. After each session a logical block has a logical address, which is used by the drive to find a particular logical block number (LBN).

Within the tracks the CD can contain either audio or computer data. The most common formats for computer data are ISO 9660, hierarchical file system (HFS) and the Joliet file system.

The ISO 9660 was developed at a time when disks required to be mass replicated. It thus wrote the complete file system at the time of creation, as there was no need for incremental creation. Now, with CD-R technology, it is possible to incrementally write to a disk. This is described as multisession. Unfortunately, after each session a new lead in and lead out must be written (requiring a minimum of 13 MB of disk space). This consists of:

- 13.2 MB for the lead out for the first session and 4.4 MB for each subsequent session.
- 8.8 MB for lead in for each session comprising 8.8 MB.

Thus multisession is useful for writing large amounts of data for each session, but is not efficient when writing many small updates. Most new CD-R systems now use a track-at-once technique which stores the data one track at a time and only writes the lead in and lead out data when the session is actually finished. In this technique the CD can be built up with data over a long period of time. Unfortunately the disk cannot be read by standard CD-ROM drives until the session is closed (and written with the ISO 9660 format). Another disadvantage is that the Red Book only specifies up to 99 tracks for each CD.

Unfortunately the ISO 9660 is not well-suited for packet writing and is likely to be phased out over the coming years.

6.9.5 Magneto-optical (MO) disks

As with CD-R disks, magneto-optical (MO) disks allow the data to be rewritten many times. These disks use magnetic and optical fields to store the data. Unfortunately the disk must first be totally erased before data is written (although new developments are overcoming this limitation).

6.9.6 Transfer rates

Optical disks spin at variable speeds, they spin at a lower rate on the outside of the disk than on the inside. Thus the disk increases its speed progressively as the data is read from the disk. The actual rate at which the drive reads the data is constant for the disk. The basic transfer rate for a typical CD-ROM is 150 KB/s. This has recently been increased to 300 KB/s (×2 CD drives), 600 KB/s (×4), 900 KB/s (×6), 1.5 MB/s (×10) and even 6 MB/s (×40).

6.9.7 Standards

Data disks are described in the following standards books, each of them specific to an area or type of data application. These books can be obtained by becoming a licensed CD developer with Philips. These standards apply to media, hardware, operating systems, file systems and software.

Red Book	World standard for all compact disks (CD-DA) (audio).
Yellow Book	Covers CD-ROM and CD-ROM-XA data formats.
Green Book	Covers CD-I data formats and operating systems (photo).
White Book	CD-I (video)
Orange Book	Covers CD-R/CD-E (CD-Recordable/CD-Eraseable).
Blue Book	CD-Enhanced (CD Extra, CD Plus).

6.9.8 Silver, green, blue or gold

CD-ROMs are available in a number of colours, these are:

- **Silver**. These are read-only disks which are a stamped as an original disk.
- **Gold**. These are recordable disks which use a basic phthalocyanine formulation which was patented by Mitsui Toatsu Chemicals (MTC) of Japan, and is licensed to other phthalocyanine media manufacturers. They generally work better with 2m writing speeds as some models of disk can not be written to at 1m writing speed.
- **Green**. These are recordable disks which are based on cyanine-based formulations. They are not covered by a governing patent, and are more or less unique to the individual manufacturers. An early problem was encountered with cyanine-based disk as the dye became chemically unstable in the presence of sunlight. Other problems included a wide variation in electrical performance depending on write speed and location (inner or outer portion of the disk). Eventually, in 1995, some stabilising compounds were added. The best attempt produced a metal-stabilised cyanine dye formulation that gave excellent overall performance. Gradually the performance of these disks is approaching gold disk performance.
- **Blue**. These are recordable disks which are based on an azo media. This was designed and manufactured by Mitsubishi Chemical Corporation (MCC) and marketed through its US subsidiary, Verbatim Corporation.

6.10 Magnetic tape

Magnetic tapes use a thin plastic tape with a magnetic coating (normally of ferric oxide). Most modern tapes are either reel-to-reel or cartridge type. A reel-to-reel tape normally has two interconnected reels of tape with tension arms (similar to standard compact audio cassettes). The cartridge type has a drive belt to spin the reels, this mechanism reduces the strain on the tape and allows faster access speeds.

Magnetic tapes have an extremely high capacity and are relatively cheap. Data is saved in a serial manner with one bit (or one record) at a time. This has the disadvantage that they are relatively slow when moving back and forward within the tape to find the required data. Typically, it may take many seconds (or even minutes) to search from the start to the end of a tape. In most applications, magnetic tapes are used to back up a system. This type of application requires large amounts of data to be stored reliably over time but the recall speed is not important.

The most common types of tape are:

- Reel-to-reel tapes – the tapes have two interconnected reels with an interconnecting tape which is tensioned by tension arms. They were used extensively in the past to store computer-type data but have been replaced by the following three types (8 mm, QIC and DAT tapes).
- 8 mm video cartridge tapes – this type of tape was developed to be used in video cameras and is extremely compact. As with videotapes the tape wraps round the read/write head in a helix.
- Quarter inch cartridge (QIC) tapes – a QIC is available in two main sizes: 5.25 inch and 3.5 inch. They give capacities of 40 MB to tens of GB.
- Digital audio tapes (DAT) – this type of tape was developed to be used in hi-fi applications and is extremely compact. As with the 8 mm tape, the tape wraps round the read/write head in a helix. The tape itself is 4 mm wide and can store several GBs of data with a transfer rate of several hundred kbps.

6.10.1 QIC tapes

QIC tapes are available in two sizes: 5.25-inch and 3.5-inch. The tape length ranges from 200 to 1000 feet, with a tape width of 0.25-inch. Typical capacities range from 40 MB to tens of GB. A single capstan drive is driven by the tape drive. Figure 6.6 illustrates a QIC tape.

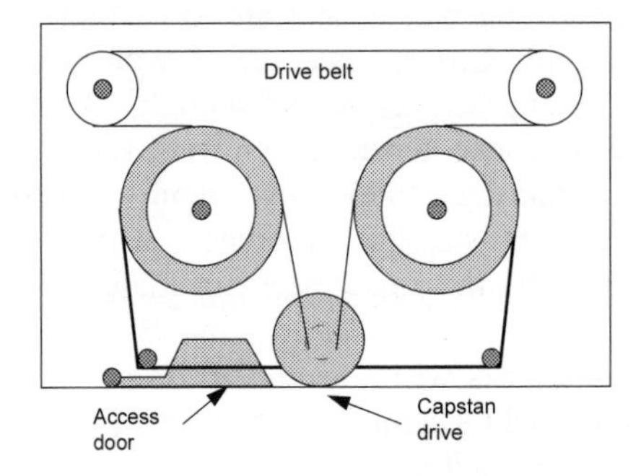

Figure 6.6 QIC tape

6.10.2 8 mm video tape

The 8 mm video tape is a high specification tape and was originally used in video cameras. These types are also known as Exabyte after the company that originally developed a back up system using 8 mm videotapes. They can be used to store several GBs of data with a transfer rate of 500 kbps. In order to achieve this high transfer rate the read/write head spins at 2000 rpm and the tape passes it at a relatively slow speed.

6.11 Exercises

6.11.1 What is a typical data capacity for a CD-ROM disk:

(a) 100 MB (b) 650 MB
(c) 800 MB (d) 1 GB

6.11.2 Which interface is most often used in PCs to connect to hard-disk drives:

(a) ST-506 (b) ISA
(c) EISA (d) IDE

6.11.3 Which bus does the IDE share many of its signals with:

(a) ST-506 (b) ISA
(c) EISA (d) PCI

6.11.4 How many devices can a single IDE bus support:

(a) 1 (b) 2
(c) 4 (d) 7

6.11.5 What is the base address for the primary IDE controller:

(a) 1F0h (b) 170h
(c) 2F0h (d) 270h

6.11.6 What is the base address for the secondary IDE controller:

(a) 1F0h (b) 170h
(c) 2F0h (d) 270h

6.11.7 What is the main advantage of E-IDE over IDE:

(a) It is compatible with more hard-disk drives
(b) It allows for larger hard-disk capacities
(c) It is easier to interface to
(d) It has a large data bus size

6.11.8 What is the main advantage, apart from increased transfer rate, that IDE has over older interface standards, such as ST-506 and ESDI?

6.11.9 Explain how IDE differs from E-IDE and how E-IDE supports larger disk capacities. How does E-IDE use modes to define the maximum transfer rate. Which mode is the fastest?

6.11.10 Show that the maximum capacity of IDE is 528 MB and that the maximum capacity (per disk) is 8.4 GB for E-IDE.

6.11.11 Which IRQ does an IDE connected disk drive normally use and what is the size of its data bus.

6.11.12 A floppy disk ribbon cable has a cable twist to differentiate between the A: drive and the B: drive. How does the ribbon cable that connects two IDE connected drives differ. In addition, how many wires does the ribbon cable have.

6.11.13 Outline how three hard disks and a CD-ROM can be connected to the IDE bus. What settings are required for the disks to connect properly? Which signal line differentiates between a master and a slave?

6.11.14 How are I/O addresses used to communicate with hard disks. How is data transferred to and from the disk? What are the standard address ranges for the primary and the secondary? If possible, check these on an available PC.

6.11.15 Which register is used to identify a hard disk error. Explain its operation.

6.11.16 Which is the IDE signal line that identifies if a slave device exists?

6.11.17 Prove that, 16-bit, 44.1 kHz sampled, stereo audio gives over 65 minutes for a 650 MB optical disk.

6.12 Notes from the author

The IDE bus. What can you say about it? Not much really. It has no future plans for glory and is looking forward to a graceful retirement. It works, it's reliable, it's standard, it's cheap, blah, blah, and relatively easy to set up. I've spent many a happy hour (not!) setting the jumpers on CD-ROM drives and secondary hard-disk drives which I want to add to a PC system. Luckily, these days, modern disk drives and BIOS cope well with adding and deleting disk drives to systems.

On its negative side, IDE is not really that fast, but it really doesn't have to be, as disk drives do not require high data rates. E-IDE improved IDE a great deal and only required a simple change in the BIOS. In conclusion, SCSI is the natural choice for disk drives and allows for much greater flexibility in configuration and also high data rates. But, it tends to be more expensive, and we'd miss IDE, wouldn't we?

In Chapter 3, I voted the IDE bus as the third most helpful bus of all-time. It merited this position as, over the years, it has quietly interfaced to disk drives, and has even supported the addition of CD-ROM drives. By the flick of a BIOS chip, it supported large capacity disk drives (EIDE). It also requires very little to set it up, as the BIOS tends is able to determine the capacity of the disk drive, and properly set it up. At present, there are no real plans to phase the IDE out, thus it is likely to stay a standard part of the motherboard.

Unix workstations and Apple computers have always used the SCSI bus, as it gives easy external disk upgrades, but, as few users of PC require to add external disk drives to their computer, there has never really been a great demand for SCSI-based disk drives for the PC. IDE drives have two interrupts lines set aside for themselves, so why not use them to interface to disk drives. The SCSI bus, though, now offers high data rates, improved connectivity, improved command and message structure, and easy-of-upgrade. So why isn't it the standard bus for PC system. Well it costs more, doesn't it, and well, it isn't PC, is it? It's an Apple thing, isn't it. When has the PC ever done anything in the right way?

7 SCSI

7.1 Introduction

SCSI has many advantages over IDE, these include:

- A single bus system for up to seven connected devices.
- It supports many different peripherals, such as hard disks, tape drives, CD-ROMs, and so on.
- It supports device priority where a higher SCSI-ID has priority over a lower SCSI-ID.
- It supports both high-quality connectors and cables, and low-quality connection and ribbon cable.
- It supports differential signals, which gives longer cable lengths.
- Extended support for commands and messaging.
- Devices do not need individual IRQ lines (as they do in IDE) as the controller communicates with the devices.
- It has great potential for faster transfer and enhanced peripheral support.

7.2 SCSI types

SCSI has an intelligent bus subsystem and can support multiple devices cooperating currently. Each device is assigned a priority. The main types of SCSI are:

- SCSI-I. Transfer rate of 5 MB/s with an 8-bit data bus and seven devices per controller.
- SCSI-II. Support for SCSI-1 and with one or more of the following:
  - Fast SCSI which uses a synchronous transfer to give 10 MB/s transfer rate. The initiator and target initially negotiate to see if they can both support synchronous transfer. If they can they then go into a synchronous transfer mode.
  - Fast/wide SCSI-2 which doubles the data bus width to 16 bits to give 20 MB/s transfer rate.
  - 15 devices per master device.
  - Tagged command queuing (TCQ) which greatly improves performance and is supported by Windows NT, NetWare and OS/2.
  - Multiple commands sent to each device.
  - Commands executed in whatever sequence will maximize device performance.
- Ultra SCSI (SCSI-III). Operates either as 8-bit or 16-bit with either 20 MB/s or 40 MB/s transfer rate.

7.2.1 SCSI-II

SCSI-II supports fast SCSI which is basically SCSI-I operating at a rate of 10 MB/s (using synchronous versus asynchronous) and Wide SCSI which uses a 64-pin connector and a 16-bit data bus. The SCSI-II controller is also more efficient and processes commands up to seven times faster than SCSI-I.

The SCSI-II drive latency is also much less than SCSI-I due mainly to tag command queuing (TCQ) which allows multiple commands to be sent to each device. Each device then holds its own commands and executes them in whatever sequence that will maximize performance (such as by minimizing the latency associated with disk rotation). Table 7.1 contrasts Fast SCSI-II and Fast/Wide SCSI-II. It can be seen that both disks have predictive failure analysis (PFA) and automatic defect reallocation (ADR).

The normal 50-core cable is typically known as A-cable, while the 68-core cable is known as B-cable.

Table 7.1 Comparison of SCSI-II disks

	Seek time (ms)	*Latency* (ms)	*Rotational speed* (rpm)	*Sustained data read* (MB/s)	*PFA*	*ADR*
1 GB SCSI-II fast	10.5	5.56	5400	4	✓	✓
4.5 GB SCSI-II fast/ wide	8.2	4.17	7200	12	✓	✓

7.2.2 Ultra SCSI

Ultra SCSI (or SCSI-III) allows for 20 MB/s burst transfers on an 8-bit data path and 40 MB/s burst transfer on a 16-bit data path. It uses the same cables as SCSI-II and the maximum cable length is 1.5 m. Ultra SCSI disks are compatible with SCSI-2 controllers; however the transfer will be at the slower speed of the SCSI controller. SCSI disks are compatible with UltraSCSI controllers; however, the transfer will be at the slower speed of the SCSI disk.

SCSI-I and Fast SCSI-II use a 50-pin 8-bit connector, whereas fast/wide SCSI-II and Ultra SCSI uses a 68-pin 16-bit connector. The 16-bit connector is physically smaller than the 8-bit connector and the 16-bit connector cannot connect directly to the 8-bit connector. The cable used is called P-cable and replaces the A/B-cable.

Note that SCSI-II, and Ultra SCSI require an active terminator on the last external device. Table 7.2 compares the main types of SCSI.

Table 7.2 SCSI types

	Data bus (bits)	*Transfer rate* (MB/s)	*Tagged command queuing*	*Parity checking*	*Maximum devices*	*Pins on cable and connector*
SCSI-I	8	5	×	×/✓ (optional)	7	50
SCSI-II Fast	8	10 (10MHz)	✓	✓	7	50
SCSI-II fast/ wide	16	20 (10MHz)	✓	✓	15	68
Ultra SCSI	16	40 (20MHz)	✓	✓	15	68

7.3 SCSI interface

In its standard form the small computer systems interface (SCSI) standard uses a 50-pin header connector and a ribbon cable to connect to up to eight devices. It overcomes the problems of the IDE, where devices have to be assigned as a master and a slave. SCSI and fast SCSI transfer one byte at a time with a parity check on each byte. SCSI-II, wide SCSI and Ultra SCSI use a 16-bit data transfer and a 68-pin connector. Table 7.3 lists the pin connections for SCSI-I (single-ended cable) and Fast SCSI (differential cable) and Table 7.4 lists the pin connections for SCSI-II, wide SCSI and ultra SCSI.

Table 7.3 SCSI-I and Fast SCSI connections

	Single-ended cable				Differential cable		
Pin	*Signal*	*Pin*	*Signal*	*Pin*	*Signal*	*Pin*	*Signal*
1	GND	2	$\overline{D0}$	1	GND	2	GND
3	GND	4	$\overline{D1}$	3	$+\overline{D0}$	4	$-\overline{D0}$
5	GND	6	$\overline{D2}$	5	$+\overline{D1}$	6	$-\overline{D1}$
7	GND	8	$\overline{D3}$	6	$+\overline{D2}$	8	$-\overline{D2}$
9	GND	10	$\overline{D4}$	8	$+\overline{D3}$	10	$-\overline{D3}$
11	GND	12	$\overline{D5}$	11	$+\overline{D4}$	12	$-\overline{D4}$
13	GND	14	$\overline{D6}$	13	$+\overline{D5}$	14	$-\overline{D5}$
15	GND	16	$\overline{D7}$	15	$+\overline{D6}$	16	$-\overline{D6}$
17	GND	18	$\overline{D(PARITY)}$	17	$+\overline{D7}$	18	$-\overline{D7}$
19	GND	20	GND	19	D(PARITY)	20	$-\overline{D(PARITY)}$
21	GND	22	GND	21	DIFFSEN	22	GND
23	RESERVED	24	RESERVED	23	RESERVED	24	RESERVED
25	Open	26	TERMPWR	25	TERMPWR	26	TEMPWR
27	RESERVED	28	RESERVED	27	RESERVED	28	RESERVED
29	GND	30	GND	29	$+\overline{ATN}$	30	$-\overline{ATN}$
31	GND	32	$\overline{ATN}$	31	GND	32	GND
33	GND	34	GND	33	$+\overline{RST}$	34	$-\overline{RST}$
35	GND	36	$\overline{BSY}$	35	$+\overline{ACK}$	36	$-\overline{ACK}$
37	GND	38	$\overline{ACK}$	37	$+\overline{RST}$	38	$-\overline{RST}$
39	GND	40	$\overline{RST}$	39	$+\overline{MSG}$	40	$-\overline{MSG}$
41	GND	42	$\overline{MSG}$	41	$+\overline{SEL}$	42	$-\overline{SEL}$
43	GND	44	$\overline{SEL}$	43	$+\overline{C}/D$	44	$-\overline{C}/D$
45	GND	46	$\overline{C}/D$	45	$+\overline{REQ}$	46	$-\overline{REQ}$
47	GND	48	$\overline{REQ}$	47	$+\overline{I}/O$	48	$-\overline{I}/O$
49	GND	50	$\overline{I}/O$	49	GND	50	GND

7.3.1 Signals

A SCSI bus is made up of a SCSI host adapter connected to a number of SCSI units via a SCSI bus. As all units connect to a common bus, only two units can transfer data at a time, either from one SCSI unit to another or from one SCSI unit to the SCSI host. The great advantage of this transfer is that is does not involve the processor.

Table 7.4 SCSI-II, wide SCSI and ultra SCSI

Pin	*Signal*	*Pin*	*Signal*
1	GND	35	GND
2	GND	36	$\overline{\text{D8}}$
3	GND	37	$\overline{\text{D9}}$
4	GND	38	$\overline{\text{D10}}$
5	GND	39	$\overline{\text{D11}}$
6	GND	40	$\overline{\text{D12}}$
7	GND	41	$\overline{\text{D13}}$
8	GND	42	$\overline{\text{D14}}$
9	GND	43	$\overline{\text{D15}}$
10	GND	44	$\overline{\text{D(PARITY1)}}$
11	GND	45	$\overline{\text{ACKB}}$
12	GND	46	GND
13	GND	47	$\overline{\text{REQB}}$
14	GND	48	$\overline{\text{D16}}$
15	GND	49	$\overline{\text{D17}}$
16	GND	50	$\overline{\text{D18}}$
17	TERMPWR	51	TERMPWR
18	TERMPWR	52	TERMPWR
19	GND	53	$\overline{\text{D19}}$
20	GND	54	$\overline{\text{D20}}$
21	GND	55	$\overline{\text{D21}}$
22	GND	56	$\overline{\text{D22}}$
23	GND	57	$\overline{\text{D23}}$
24	GND	58	$\overline{\text{D(PARITY2)}}$
25	GND	59	$\overline{\text{D24}}$
26	GND	60	$\overline{\text{D25}}$
27	GND	61	$\overline{\text{D26}}$
28	GND	62	$\overline{\text{D27}}$
29	GND	63	$\overline{\text{D28}}$
30	GND	64	$\overline{\text{D29}}$
31	GND	65	$\overline{\text{D30}}$
32	GND	66	$\overline{\text{D31}}$
33	GND	67	$\overline{\text{D(PARITY3)}}$
34	GND	68	GND

Each unit on a SCSI is assigned a SCSI ID address. In the case of SCSI-I this ranges from 0 to 7 (where 7 is normally reserved for a tape drive). The host adapter takes one of the addresses thus a maximum of seven units can connect to the bus. Most systems allow the units to take on any SCSI ID address, but older systems required boot drives to be connected to a specific SCSI address. When the system is initially booted, the host adapter sends out a Start Unit command to each SCSI unit. This allows each of the units to start in an orderly manner (and not overloading the local power supply). The host will start with the highest priority address (ID=7) and finishes with the lowest address (ID=0). Typically, the ID is set with a rotating switch selector or by three jumpers.

SCSI defines an initiator control and a target control. The initiator requests a function from a target, which then executes the function, as illustrated in Figure 7.1. The initiator effectively takes over the bus for the time to send a command and the target executes the command and then contacts the initiator and transfers any data. The bus will then be free for other transfers.

The main signals are:

- $\overline{\text{BSY}}$ – indicates that the bus is busy, or not (an OR-tied signal).
- $\overline{\text{ACK}}$ – activated by the initiator to indicate an acknowledgement for a $\overline{\text{REQ}}$ information transfer handshake.
- $\overline{\text{RST}}$ – when active (low) resets all the SCSI devices (an OR-tied signal).
- $\overline{\text{ATN}}$ – activated by the initiator to indicate the attention state.
- $\overline{\text{MSG}}$ – activated by the target to indicate the message phase.
- $\overline{\text{SEL}}$ – activated by the initiator and is used to select a particular target device (an OR-tied signal).
- $\overline{\text{C}}/\text{D}$ (control/data) – activated by the target to identify if there is data or control on the SCSI bus.
- $\overline{\text{REQ}}$ – activated by the target to acknowledge to indicate a request for an $\overline{\text{ACK}}$ information transfer handshake.
- $\overline{\text{I}}/\text{O}$ (input/output) – activated by the target to show the direction of the data on the data bus. Input defines that data is an input to the initiator, else it is an output.

Each of the control signals can be true or false. They can be:

- OR-tied driven, where the driver does not drive the signal to the false state. In this case the bias circuitry of the bus terminators pulls the signal false whenever it is released by the drivers at every SCSI device. If any driver is asserted, then the signal is true. The $\overline{\text{BSY}}$, $\overline{\text{SEL}}$, and $\overline{\text{RST}}$ signals are OR-tied. In the ordinary operation of the bus, the $\overline{\text{BSY}}$ and $\overline{\text{RST}}$ signals may be simultaneously driven true by several drivers.
- Non-OR-tied driven, where the signal may be actively driven false. No signals other than $\overline{\text{BSY}}$, $\overline{\text{RST}}$ and $\overline{\text{D(PARITY)}}$ are simultaneously driven by two or more drivers.

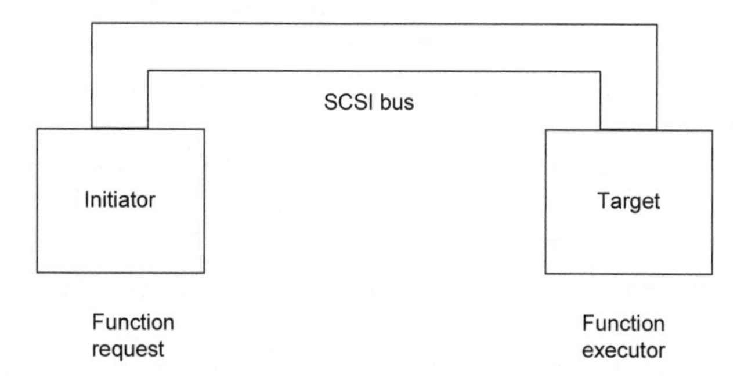

Figure 7.1 Initiator and target in SCSI

7.4 SCSI operation

The SCSI bus allows any unit to talk to any other unit, or the host to talk to any unit. Thus there must be some way of arbitration where units capture the bus. The main phases that the bus goes through are:

- **Free-bus**. In this state there are no units which either transfer data or have control of the bus. It is identified by disactive $\overline{SEL}$ and $\overline{BSY}$ (both will be high). Thus, any unit can capture the bus.
- **Arbitration.** In this state a unit can take control of the bus and become an initiator. To do this it activates the $\overline{BSY}$ signal and puts its own ID address on the data bus. Next, after a delay, it tests the data bus to determine if a high-priority unit has put its own address on the bus. If it has then it will allow the other unit access to the bus. If its address is still on the bus then it asserts the $\overline{SEL}$ line. After a delay, it then has control of the bus.
- **Selection**. In this state the initiator selects a target unit and gets the target to carry out a given function, such as reading or writing data. The initiator outputs the OR value of its SCSI-ID and the SCSI-ID of the target onto the data bus (for example, if the initiator is 2 and the target is 5 then the OR-ed ID on the bus will be 00100100.). The target then determines that its ID is on the data bus and set the $\overline{BSY}$ line active. If this does not happen within a given time then the initiator deactivates the $\overline{SEL}$ signal, and the bus will be free. The target determines that it is selected when the $\overline{SEL}$ signal and its SCSI ID bit are active and the $\overline{BSY}$ and $\overline{I}/O$ signals are false. It then asserts the $\overline{BSY}$ signal within a selection abort time.
- **Reselection**. When the arbitration phase is complete, the winning SCSI device asserts the $\overline{BSY}$ and $\overline{SEL}$ signals and has delayed at least a bus clear delay plus a bus settle delay. The winning SCSI device sets the DATA BUS to a value that is the logical OR of its SCSI ID bit and the initiator's SCSI ID bit. Sometimes the target takes some time to reply to the initiators request. The initiator determines that it is reselected when the $\overline{SEL}$ and $\overline{I}/O$ signals and its SCSI ID bit are true and the $\overline{BSY}$ signal is false. The reselected initiator then asserts the $\overline{BSY}$ signal within a selection abort time of its most recent detection of being reselected. An initiator does not respond to a reselection phase if other than two SCSI ID bits are on the data bus. After the target detects that the $\overline{BSY}$ signal is true, it also asserts the $\overline{BSY}$ signal and waits a given time delay and then releases the $\overline{SEL}$ signal. The target may then change the $\overline{I}/O$ signal and the data bus. After the reselected initiator detects the $\overline{SEL}$ signal is false, it releases the $\overline{BSY}$ signal. The target continues to assert the $\overline{BSY}$ signal until it gives up the SCSI bus.
- **Command**. The command phase is used by the target to request command information from the initiator. The target asserts the $\overline{C}/D$ signal and negates the $\overline{I}/O$ and $\overline{MSG}$ signals during the $\overline{REQ}/\overline{ACK}$ handshake(s) of this phase.
- **Data**. The data phase covers both the data in and data out phases. In the data in phase the target requests that data be sent to the initiator from the target. For this purpose the target asserts the $\overline{I}/O$ signal and negates the $\overline{C}/D$ and $\overline{MSG}$ signals during the $\overline{REQ}/\overline{ACK}$ handshake(s) of this phase. In the data out phase, the target requests that data be sent from the initiator to the target. The target negates the $\overline{C}/D$, $\overline{I}/O$ and $\overline{MSG}$ signals during the $\overline{REQ}/\overline{ACK}$ handshake(s) of this phase.

- **Message**. The message phase covers both the message out and message in phase. The first byte transferred in either of these phases can be either a single-byte message or the first byte of a multiple-byte message. Multiple-byte messages are completely contained within a single message phase.
- **Status**. The status phase allows the target to request that status information be sent from the target to the initiator. The target shall assert the $\overline{\text{C}}/\text{D}$ and $\overline{\text{I}}/\text{O}$ signals and negate the $\overline{\text{MSG}}$ signal during the $\overline{\text{REQ}}/\overline{\text{ACK}}$ handshake of this phase.

Typical times are:

- Arbitration delay, 2–4 μs. This is the minimum time that a SCSI device waits from asserting $\overline{\text{BSY}}$ for arbitration until the data bus can be examined to see if arbitration has been won.
- Power-on to selection time, 10 s. This is the maximum time from power start-up until a SCSI target is able to respond with appropriate status and sense data.
- Selection abort time, 200 μs. This is the maximum time that a target (or initiator) takes from its most recent detection of being selected (or reselected) until asserting a $\overline{\text{BSY}}$ response. This is required to ensure that a target (or initiator) does not assert $\overline{\text{BSY}}$ after a select (or reselection) phase has been aborted.
- Selection time-out delay, 250ms. The minimum time that a SCSI device should wait for a $\overline{\text{BSY}}$ response during the selection or reselection phase before starting the time-out procedure.
- Disconnection delay, 200 μs. The minimum time that a target shall wait after releasing $\overline{\text{BSY}}$ before participating in an arbitration phase when honouring a disconnect message from the initiator.
- Reset hold time, 25 μs. The minimum time for which $\overline{\text{RST}}$ is asserted.

The signals $\overline{\text{C}}/\text{D}$, $\overline{\text{I}}/\text{O}$, and $\overline{\text{MSG}}$ distinguish between the different information transfer phases, as summarised in Table 7.5 (where a 1 identifies an active signal and a 0 identifies a false signal). The target drives these three signals and therefore controls all changes from one phase to another. The initiator can request a message out phase by asserting the $\overline{\text{ATN}}$ signal, while the target can cause the bus free phase by releasing the $\overline{\text{MSG}}$, $\overline{\text{C}}/\text{D}$, $\overline{\text{I}}/\text{O}$, and $\overline{\text{BSY}}$ signals.

Table 7.5 Information transfer phases

$\overline{\text{MSG}}$	$\overline{\text{C}}/\text{D}$	$\overline{\text{I}}/\text{O}$	*Phase*	*Direction*
0	0	0	Data out	Initiator→target
0	0	1	Data in	Initiator←target
0	1	0	Command	Initiator→target
0	1	1	Status	Initiator←target
1	0	0	–	–
1	0	1	–	–
1	1	0	Message out	Initiator→target
1	1	1	Message in	Initiator←target

The information transfer phases use one or more $\overline{\mathrm{REQ}}/\overline{\mathrm{ACK}}$ handshakes to control the information transfer. Each $\overline{\mathrm{REQ}}/\overline{\mathrm{ACK}}$ handshake allows the transfer of one byte of information. During the information transfer phases the $\overline{\mathrm{BSY}}$ signal shall remain true and the $\overline{\mathrm{SEL}}$ signal shall remain false. Additionally, during the information transfer phases, the target shall continuously envelope the $\overline{\mathrm{REQ}}/\overline{\mathrm{ACK}}$ handshake(s) with the $\overline{\mathrm{C}}/\mathrm{D}$, $\overline{\mathrm{I}}/\mathrm{O}$ and $\overline{\mathrm{MSG}}$ signals in such a manner that these control signals are valid for a bus settle delay before the assertion of the $\overline{\mathrm{REQ}}$ signal of the first handshake and remain valid until after the negation of the $\overline{\mathrm{ACK}}$ signal at the end of the handshake of the last transfer of the phase.

The $\overline{\mathrm{I}}/\mathrm{O}$ signal allows the target to control the direction of information, when its $\overline{\mathrm{I}}/\mathrm{O}$ signal is true then the information is transferred from the target to the initiator and when false, the transfer is from the initiator to the target.

The handshaking operation for a transfer to the initiator is as follows:

- The $\overline{\mathrm{I}}/\mathrm{O}$ signal is asserted as a true.
- The target sets the data bus lines.
- The target asserts the $\overline{\mathrm{REQ}}$ signal.
- The initiator reads the data bus.
- The initiator then indicates its acceptance of the data by asserting the $\overline{\mathrm{ACK}}$ signal.
- The target may change or release the data bus.
- The target negates the $\overline{\mathrm{REQ}}$ signal.
- The initiator shall then negate the $\overline{\mathrm{ACK}}$ signal.
- The target may continue the transfer by driving the data bus and asserting the $\overline{\mathrm{REQ}}$ signal, and so on.

The handshaking operation for a transfer from the initiator is as follows:

- The $\overline{\mathrm{I}}/\mathrm{O}$ signal is asserted as a false.
- The target asserts the $\overline{\mathrm{REQ}}$ signal (requesting information).
- The initiator sets the data bus lines.
- The initiator asserts the $\overline{\mathrm{ACK}}$ signal.
- The target then reads the data bus.
- The target negates the $\overline{\mathrm{REQ}}$ signal (acknowledging transfer).
- The initiator may then set the data bus, and so on.

7.5 SCSI pointers

SCSI provides for three pointers for each I/O process (called saved pointers), for command, data and status. When an I/O process becomes active, its three saved pointers are copied into the initiator's set of three current pointers. These current pointers point to the next command, data or status byte to be transferred between the initiator's memory and the target.

7.6 Message system description

The message system allows the initiator and the target to communicate over the interface connection. Each message can be one, two, or multiple bytes in length. In a single message phase, one or more messages can be transmitted, (but a message cannot be split between multiple message phases). Table 7.6 lists the message format, where the first byte of the message determines the format. The initiator ends the message out phase (by negating $\overline{\text{ATN}}$) when it sends certain messages identified in Table 7.7.

Single-byte messages consist of a single byte transferred during a message phase. Table 7.7 defines the message type.

Table 7.6 Message format

Value	*Message format*
00h	One byte message (command complete)
01h	Extended messages
02h–1Fh	One-byte messages
20h–2Fh	Two-byte messages
30h–7Fh	Reserved
80h–FFh	One-byte message (identify)

Table 7.7 Message codes

Code	*Message*	*Direction*	*Description*
00h	Command complete	In	Sent from a target to an initiator to indicate that the execution of an I/O process has completed and that valid status has been sent to the initiator. After successfully sending this message, the target shall go to the bus free phase by releasing the $\overline{\text{BSY}}$ signal. The target considers the message transmission to be successful when it detects the negation of $\overline{\text{ACK}}$ for the command complete message with the $\overline{\text{ATN}}$ signal false.
03h	Restore pointers	In	
04h	Disconnect	In/Out	Sent from a target to inform an initiator that the present connection is going to be broken (the target plans to disconnect by releasing the $\overline{\text{BSY}}$ signal), but that a later reconnect will be required in order to complete the current I/O process. This message shall not cause the initiator to save the data pointer. After successfully sending this message, the target shall go to the bus free phase by releasing the $\overline{\text{BSY}}$ signal. The target shall consider the message transmission to be successful when it detects the negation of the $\overline{\text{ACK}}$ signal for the disconnect message with the $\overline{\text{ATN}}$ signal false.

05h	Initiator-detected error	Out	
06h	Abort	Out	Sent from the initiator to the target to clear any I/O process. The target goes to the bus-free phase following successful receipt of this message.
07h	Message reject	Out	Sent from either the initiator or target to indicate that the last message or message byte it received was inappropriate or has not been implemented.
08h	No operation	Out	Sent from an initiator in response to a target's request for a message when the initiator does not currently have any other valid message to send.
09h	Message parity error	Out	
0Ah	Linked command complete	In	
0Bh	Linked command complete (with flag)	In	
0Ch	Bus device reset	Out	Sent from an initiator to direct a target to clear all I/O processes on that SCSI device. This message forces a hard reset condition to the selected SCSI device.
0Dh	Abort tag	Out	
0Eh	Clear queue	Out	
0Fh	Initiate recovery	In/Out	
10h	Release recovery	Out	
11h	Terminate I/O process	Out	
12h–1Fh	Reserved		
23h	Ignore wide residue (2 bytes)		
24h–2Fh	Reserved for two-byte messages		
30h–7Fh	Reserved		
80h–FFh	Identify	In/Out	

7.7 SCSI commands

A command is sent from the initiator to the target. The first byte of all SCSI commands contains an operation code, followed by a command descriptor block and finally the control byte.

The format of the command descriptor block for 6-byte commands is:

- Byte 0 – operation code.
- Byte 1 – logical unit number (MSB, if required).
- Byte 2 – logical bock address.
- Byte 3 – logical bock address (LSB, if required).
- Byte 4 – transfer length (if required) / Parameter list length (if required) / allocation length (if required).
- Byte 5 – control.

7.7.1 Operation code

Figure 7.2 shows the operation code of the command descriptor block. It has a group code field and a command code field. The 3-bit group code field provides for eight groups of command codes and the 5-bit command code field provides for 32 command codes in each group.

The group code specifies one of the following groups:

- Group 0 – 6-byte commands.
- Group 1/2 – 10-byte commands.
- Group ¾ – reserved.
- Group 5 – 12-byte commands.
- Group 6/7 – vendor-specific.

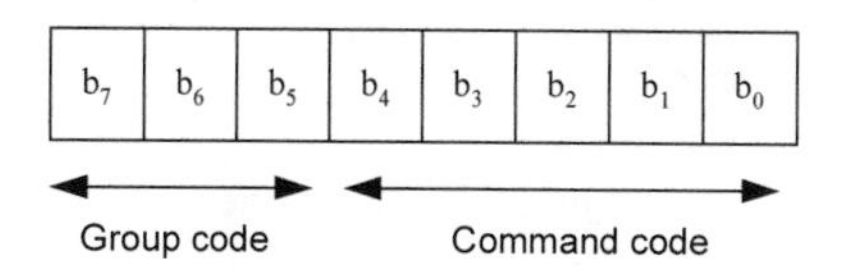

Figure 7.2 Operation code

7.7.2 Logical unit number

The logical unit number (LUN) is defined in the identify message. The target ignores the LUN specified within the command descriptor block if an identify message was received (normally the logical unit number in the command descriptor block to be set to zero).

7.7.3 Logical block address

The logical block address (LBA) on logical units or within a partition on device volumes begins with block zero and is contiguous up to the last logical block on that logical unit or within that partition.

A 10-byte and a 12-byte command descriptor blocks contain 32-bit logical block addresses, whereas a 6-byte command descriptor block contains a 21-bit logical block address.

7.7.4 Transfer length

The transfer length field specifies the amount of data to be transferred (normally the number of blocks). For several commands the transfer length indicates the requested number of bytes to be sent as defined in the command description. A command that uses 1 byte for the transfer length will thus allow up to 256 blocks of data for one command (a value of 0 identifies a transfer bock of 256 blocks).

7.7.5 Parameter list length

The parameter list length specifies the number of bytes to be sent during the data-out phase. It is typically used in command descriptor blocks for parameters that are sent to a target (such as, mode parameters, diagnostic parameters, log parameters, and so on).

7.7.6 Allocation length

The allocation length field specifies the maximum number of bytes that an initiator has allocated for returned data. The target terminates the data in phase when allocation length bytes have been transferred or when all available data have been transferred to the initiator, whichever is less. The allocation length is used to limit the maximum amount of data (for example, sense data, mode data, log data, diagnostic data, and so on) returned to an initiator.

7.7.7 Control field

The control field is the last byte of every command descriptor block. Its format is shown in Figure 7.3. The flag bit specifies which message the target returns to the initiator if the link bit is a 1 and the command completes without error. If the link bit is 0 then the flag bit should be a 0, else the target returns check condition status.

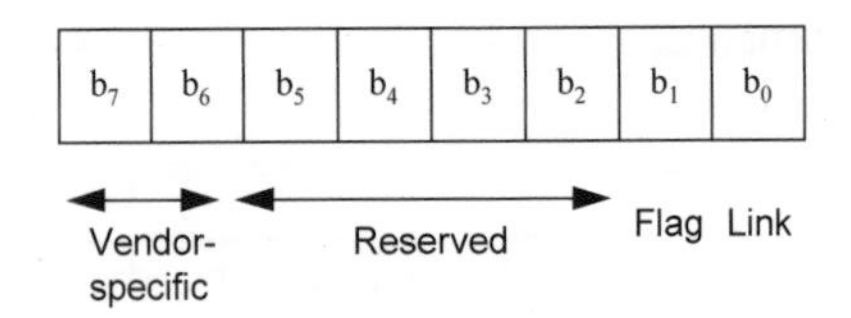

Figure 7.3 Control field

7.7.8 Command code

Commands for all device types are (bold type identifies the mandatory commands and the operation code is given in brackets):

- Change definition (40h). This command modifies the operating definition of the selected logical unit or target with respect to commands from the selecting initiator or for all initiators.
- Compare (39h). This command allows for a compare operation of data on one logical unit with another or the same logical unit in a manner similar to the copy command.
- Copy (18h). This command allows the copying of data from one logical unit to another or the same logical unit. The logical unit that receives and performs the copy command is the copy manager. It is responsible for copying data from the source device to the destination device.
- Copy and compare (3Ah). This command performs the same function as the COPY

command, except that a verification of the data written to the destination logical unit is performed after the data is written.

- Inquiry (12h). This command requests that information regarding parameters of the target and its attached peripheral device(s) be sent to the initiator.
- Log select (4Ch). This command provides a means for the initiator to manage statistical information maintained by the device about the device or its logical units. Targets that implement the log select command shall also implement the log sense command. Structures in the form of log parameters within log pages are defined as a way to manage the log data. The log select command provides for sending zero or more log pages during a data out phase.
- Log sense (4Dh). This command allows the initiator to retrieve statistical information maintained by the device about the device or its logical units. It is a complementary command to the log select command.
- Mode select (15h). This command provides a means for the initiator to specify medium, logical unit, or peripheral device parameters to the target. Targets that implement the mode select command shall also implement the mode sense command.
- Mode sense (1Ah). This command allows a target to report parameters to the initiator and is a complementary command to the mode select command.
- Read buffer (3Ch). This command is used in conjunction with the write buffer command as a diagnostic function for testing target memory and the SCSI bus integrity.
- Receive diagnostic results (1Ch). This command requests analysis data be sent to the initiator after completion of a send diagnostic.
- Send diagnostic (1Dh). This command requests the target to perform diagnostic operations on itself, on the logical unit, or on both.
- Test unit ready (00h). This command provides a means to check if the logical unit is ready. This is not a request for a self-test. If the logical unit would accept an appropriate medium-access command without returning check condition status, this command shall return a good status.
- Write buffer (3Bh). This command is used in conjunction with the read buffer command as a diagnostic for testing target memory and the SCSI bus integrity.

7.8 Status

The status phase normally occurs at the end of a command (although in some cases may occur before transferring the command descriptor block). Figure 7.4 shows the format of the status byte and Table 7.8 defines the status byte codes. This status byte is sent from the target to the initiator during the status phase at the completion of each command unless the command is terminated by one of the following events:

- Abort message.
- Abort tag message.
- Bus device reset message.
- Clear queue message.
- Hard reset condition.
- Unexpected disconnect.

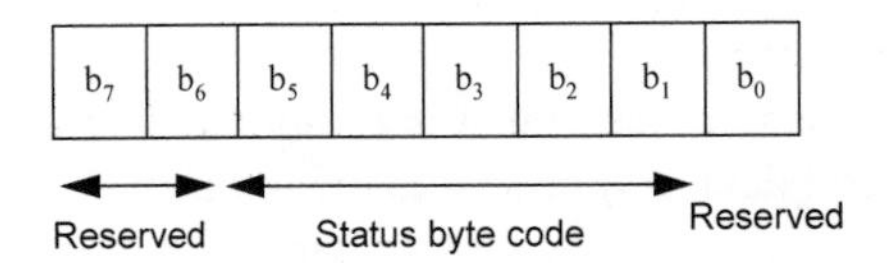

Figure 7.4 Status field

Table 7.8 Status byte codes

Bit values of status byte								*Status*	*Description*
7	6	5	4	3	2	1	0		
R	R	0	0	0	0	0	R	Good	Indicates that the target has successfully completed the command.
R	R	0	0	0	0	1	R	Check condition	Indicates that a contingent allegiance condition has occurred.
R	R	0	0	0	1	0	R	Condition met	This status or INTERMEDIATE-CONDITION MET is returned whenever the requested operation is satisfied.
R	R	0	0	1	0	0	R	Busy	Indicates that the target is busy. This status shall be returned whenever a target is unable to accept a command from an otherwise acceptable initiator (that is, no reservation conflicts).
R	R	0	1	0	0	0	R	Immediate	This status or INTERMEDIATE-CONDITION MET shall be returned for every successfully completed command in a series of linked commands (except the last command).
R	R	0	1	0	1	0	R	Immediate-condition met	This status is the combination of the CONDITION MET and INTERMEDIATE status's.
R	R	0	1	1	0	0	R	Reservation conflict	This status occurs whenever an initiator attempts to access a logical unit that is reserved with a conflicting reservation type for another SCSI device.
R	R	1	0	0	0	1	R	Command conflict	This status occurs whenever the target terminates the current I/O process after receiving a TERMINATE I/O PROCESS message.
R	R	1	0	1	0	0	R	Queue full	This status shall be implemented if tagged queuing is implemented.
R	R	R	R	R	R	R	R	Reserved	

7.9 Exercises

7.9.1 What is the maximum number of devices that can connect to a standard SCSI bus:

(a) 1 (b) 4
(c) 7 (d) 8

7.9.2 How many data bits does the SCSI-I bus use:

(a) 8 (b) 16
(c) 32 (d) 64

7.9.3 How many data bits does the SCSI-II fast/wide bus use:

(a) 8 (b) 16
(c) 32 (d) 64

7.9.4 How is device priority implemented on the SCSI bus:

(a) by active polling (b) by interrupt priority
(c) by brute force (d) by unit IDs

7.9.5 What method does the SCSI bus use to prevent devices from hogging the bus:

(a) Time-outs (b) Interrupts
(c) Active polling (d) Memory mapping

7.9.6 The transfer clock for a SCSI bus is 20 MHz. Which is the transfer rate for a 16-bit data bus:

(a) 10 MB/s (b) 20 MB/s
(c) 40 MB/s (d) 80 MB/s

7.9.7 Explain the main differences between SCSI-I, SCSI-II and ultra SCSI. Outline their maximum data throughput, the connectors used and the size of their data busses. Also, outline some of the advantages of SCSI over busses such as the ISA bus.

7.9.8 State the SCSI lines that are used for simple error detection. Why is it not possible to detect which bits are in error?

7.9.9 Discuss the main system lines that are used in the SCSI bus and the operation of OR-tied driven signals.

7.9.10 Outline the main phases that the initiator and target go through in setting up a connection. Also, outline the importance of device time-outs for the different SCSI phases.

7.9.11 Discuss how the $\overline{MSG}$, $\overline{C}/D$ and $\overline{I}/O$ signals are used to set up different transfer phases.

7.9.12 Explain how SCSI uses the SCSI-ID address to set up a device priority system.

7.9.13 Discuss the usage of the message phase in SCSI and cite typical examples of its usage.

7.9.14 Discuss the usage of the command phase in SCSI and cite typical examples of its usage.

7.9.15 Discuss the usage of the status phase in SCSI and cite typical examples.

7.10 Notes from the author

Well I did it. I covered SCSI in a single chapter. It wasn't easy, but its here. SCSI is a massive area, and one which could fill this book three or four times over. So, as I do not have enough space for the full specification, I've tried to give a flavour of the bus.

SCSI's full grown-up name is the small computer systems interface. It is difficult to define exactly what a small computer system is[1]*, but SCSI has outgrown its original application of interfacing to 'small' systems and to external disk drives. It now has the potential of being able to interface virtually any external peripheral to a system. It can also be used to connect devices internally within a system. Typically, it takes a bit longer to initially boot the system, but once it has, it should be as reliable as any non-SCSI device.*

An important concept in SCSI is the prioritisation of devices using SCSI IDs. Few busses allow the system to prioritise peripherals. Thus, in a properly configured system, fast devices which require to be quickly serviced will always get access onto the bus before slow devices which do not require fast servicing. Unfortunately, the method SCSI uses limits the number of devices to one less than the number of bits on the data bus (seven for an 8-bit data bus and 15 for a 16-bit data bus). In most cases, this is not a major problem. For example, two hard disks, two CD-ROM drives, a tape backup system, a zip drive and a midi keyboard could all be attached to a standard SCSI-I bus.

In most PCs the IDE drive is still used in the majority of systems, as it is relatively easy to set up and its cheap. It is also dedicated to interfacing to the disk drives; thus, no other peripheral can hog the disk drive bus. However, for most general-purpose applications, SCSI is best. New standards for SCSI give a 16-bit data bus, at a transfer rate of 20 MHz, giving a maximum data throughput of 40 MB/s, which is much faster than IDE. It is also much easier to configure a SCSI system than it is connecting peripherals internally in a PC. A SCSI system only requires a single interrupt line, for all the devices that are connected.

Ask someone who has set up a Unix network, or who has configured an Apple computer, and they will tell you that there is little to beat a well set up SCSI bus. It's reliable, and it is easy-to-upgrade.

---

[1] Probably, 'small computer' means 'not a mainframe computer' or 'a less powerful computer'. One must remember that SCSI was developed at a time when mainframe computers were kings and PCs were seen as glorified typewriters.

8 PCMCIA

8.1 Introduction

The Personal Computer Memory Card International Association (PCMCIA) interface allows small thin cards to be plugged into laptop, notebook or palmtop computers. It was originally designed for memory cards (Version 1.0) but has since been adopted for many other types of adapters (Version 2.0), such as fax/modems, sound cards, local area network cards, CD-ROM controllers, digital I/O cards, and so on. Most PCMCIA cards comply with either PCMCIA Type II or Type III. Type I cards are 3.3 mm thick, Type II take cards up to 5 mm thick, Type III allows cards up to 10.5 mm thick. A new standard, Type IV, takes cards which are greater than 10.5 mm. Type II interfaces can accept Type I cards, Type III accept Types I and II and Type IV interfaces accept Types I, II and III.

The PCMCIA standard uses a 16-bit data bus (D0–D15) and a 26-bit address bus (A0–A25), which gives an addressable memory of 2^{26} bytes (64 MB). The memory is arranged as:

- Common memory and attribute memory, which gives a total addressable memory of 128 MB.
- I/O addressable space of 65 536 (64 k) 8-bit ports.

The PCMCIA interface allows the PCMCIA device to map into the main memory or into the I/O address space. For example, a modem PCMCIA device would map its registers into the standard COM port addresses (such as 3F8h–3FFh for COM1 or 2F8h–2FF for COM2). Any accesses to the mapped memory area will be redirected to the PCMCIA rather that the main memory or I/O address space. These mapped areas are called windows. A window is defined with a START address and a LAST address. The PCMCIA control register contains these addresses.

8.2 PCMCIA signals

Table 8.1 shows the pin connections. The main PCMCIA signals are:

- A25–A0, D15–D0 – data bus (D15–D0) and a 26-bit memory address (A25–A0) or 16-bit I/O memory address (A15–A0).
- $\overline{\text{CARD DETECT 1}}$, $\overline{\text{CARD DETECT 2}}$ – used to detect if a card is present in a socket. When a card is inserted one of these lines is pulled to a low level.

Table 8.1 PCMCIA connections

Pin	*Signal*	*Pin*	*Signal*	*Pin*	*Signal*	*Pin*	*Signal*
1	GND	18	Vpp1	35	GND	52	Vpp2
2	D3	19	A16	36	*See below*	53	A22
3	D4	20	A15	37	D11	54	A23
4	D5	21	A12	38	D12	55	A24
5	D6	22	A7	39	D13	56	A25
6	D7	23	A6	40	D14	57	RFU
7	$\overline{\text{CARD ENABLE 1}}$	24	A5	41	D15	58	RESET
8	A10	25	A4	42	*See below*	59	$\overline{\text{WAIT}}$
9	$\overline{\text{OUTPUT ENABLE}}$	26	A3	43	REFRESH	60	$\overline{\text{INPACK}}$
10	A11	27	A2	44	$\overline{\text{IOR}}$	61	$\overline{\text{REGISTER SELECT}}$
11	A9	28	A1	45	$\overline{\text{IOW}}$	62	$\overline{\text{SPKR}}$
12	A8	29	A0	46	A17	63	$\overline{\text{STSCHG}}$
13	A13	30	D0	47	A18	64	D8
14	A14	31	D1	48	A19	65	D9
15	*See below*	32	D2	49	A20	66	D10
16	READY / $\overline{\text{BUSY}}$	33	$\overline{\text{IOIS16}}$	50	A21	67	$\overline{\text{CARD DETECT 2}}$
17	+5V	34	GND	51	+5V	68	GND

Pin 15 $\overline{\text{WRITE ENABLE / PROGRAM}}$ Pin 33 $\overline{\text{IOIS16}}$ (Write Protect)
Pin 36 $\overline{\text{CARD DETECT 1}}$ Pin 42 $\overline{\text{CARD ENABLE 2}}$

- $\overline{\text{CARD ENABLE 1}}$, $\overline{\text{CARD ENABLE 2}}$ – used to enable the upper 8-bits of the data bus ($\overline{\text{CARD ENABLE 1}}$) and/or the lower 8 bits of the data bus ($\overline{\text{CARD ENABLE 2}}$).
- $\overline{\text{OUTPUT ENABLE}}$ – set low by the computer when reading data from the PCMCIA unit.
- $\overline{\text{REGISTER SELECT}}$ – set high when accessing common memory or a low when accessing attribute memory.
- RESET – used to reset the PCMCIA card.
- REFRESH – used to refresh PCMCIA memory.
- $\overline{\text{WAIT}}$ – used by the PCMCIA device when it cannot transfer data fast enough and requests a wait cycle.
- $\overline{\text{WRITE ENABLE / PROGRAM}}$ – used to program the PCMCIA device.
- Vpp1, Vpp2 – programming voltages for flash memories.
- READY / $\overline{\text{BUSY}}$ – used by the PCMCIA card when it is ready to process more data (when a high) or is still occupied by a previous access (when it is a low).
- $\overline{\text{IOIS16}}$ – used to indicate the state of the write-protect switch on the PCMCIA card. A high level indicates that the write-protect switch has been set.
- $\overline{\text{INPACK}}$ – used by the PCMCIA card to acknowledge the transfer of a signal.
- $\overline{\text{IOR}}$ – used to issue an I/O read access from the PCMCIA card (must be used with an active $\overline{\text{REGISTER SELECT}}$ signal).
- $\overline{\text{IOW}}$ – used to issue an I/O write access to the PCMCIA card (must be used with an active $\overline{\text{REGISTER SELECT}}$ signal).
- $\overline{\text{SPKR}}$ – used by PCMCIA card to send audio data to the system speaker.
- $\overline{\text{STSCHG}}$ – used to identify that the card has changed its status.

8.3 PCMCIA registers

A typical PCMCIA interface controller (PCIC) is the 82365SL. Figure 8.1 shows the main registers for the first socket. The second socket index values are simply offset by 40h. Figure 8.2 shows that the base address of the PCIC is, in Windows, set to 3E0h, by default. Figure 8.3 shows an example of a FIRST and LAST memory address. The PCIC is accessed using two addresses: 3E0h and 3E1h. The I/O windows 0/1 are accessed through:

- 08h/0Ch for the low byte of the FIRST I/O address.
- 09h/0Dh for the high byte of the FIRST I/O address.
- 0Ah/0Eh for the high byte of the LAST I/O address.
- 0Bh/0Fh for the high byte of the LAST I/O address.

The registers are accessed by loading the register index into 3E0h and then the indexed register is accessed through the 3E1h. The memory windows 0/1/2/3/4 are accessed through:

- 10h/18h/20h/28h/30h for the low byte of the FIRST memory address.
- 11h/19h/21h/29h/31h for the high byte of the FIRST memory address.
- 12h/1Ah/22h/2Ah/32h for the low byte of the LAST memory address.
- 13h/1Bh/23h/2Bh/33h for the high byte of the LAST memory address.
- 14h/1Ch/24h/2Ch/34h for the low byte of the card offset.
- 15h/1Dh/25h/2Dh/35h for the high byte of the card offset.

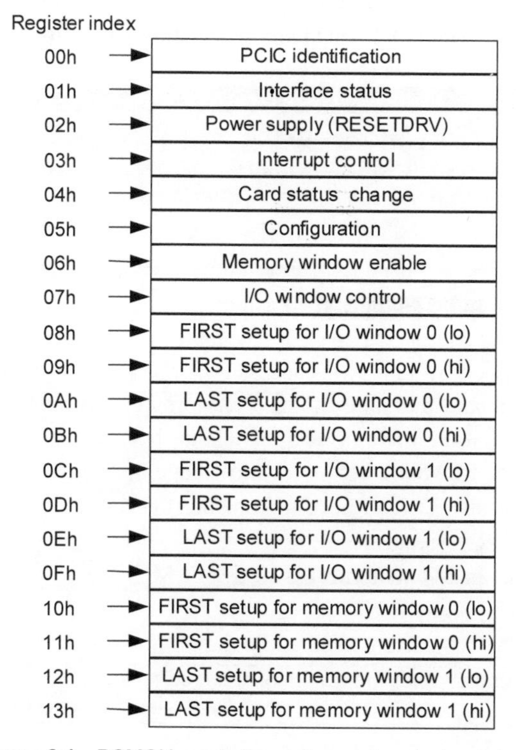

Figure 8.1 PCMCIA controller status and control registers

For example, to load a value of 22h into the Card status change register, the following would be used:

```
_outp(0x3E0,5h);  /* point to Card status change register   */
_outp(0x3E1,22h);    /* load 22h into Card status change register */
```

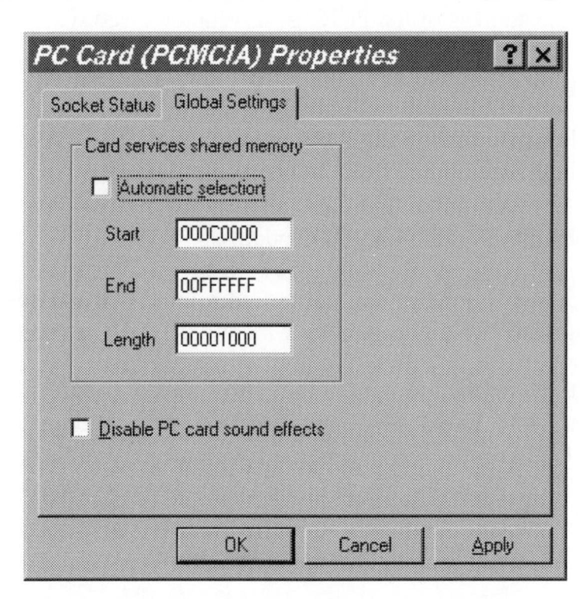

Figure 8.2 Start and end of shared memory

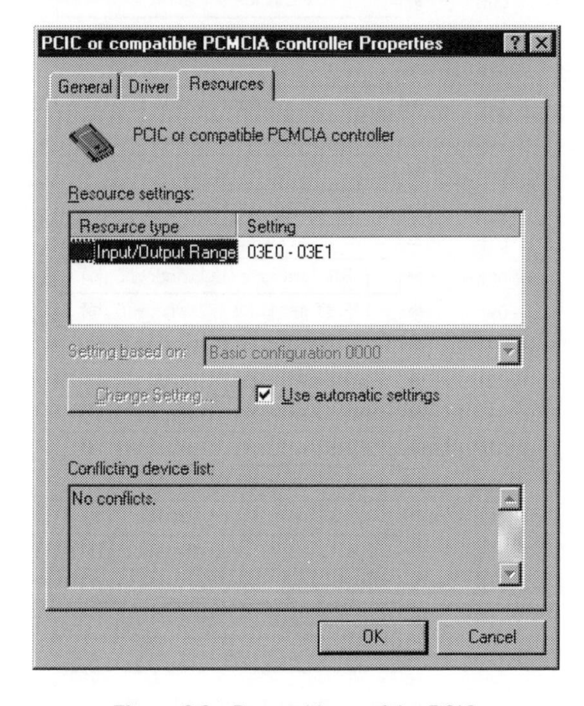

Figure 8.3 Base address of the PCIC

8.3.1 Window enable register

The window enable register has a register index of 06h (and 46h for the second socket). The definition of the register is

Bit 7	IOW1	I/O window 1 enable (1)/ disable (0).
Bit 6	IOW0	I/O window 0 enable (1)/ disable (0).
Bit 5	DEC	If active (1) $\overline{\text{MEMCS16}}$ generated from A23–A12, else from A23–A17.
Bit 4	MW4	Memory window 4 enable (1)/ disable (0).
Bit 3	MW3	Memory window 3 enable (1)/ disable (0).
Bit 2	MW2	Memory window 2 enable (1)/ disable (0).
Bit 1	MW1	Memory window 1 enable (1)/ disable (0).
Bit 0	MW0	Memory window 0 enable (1)/ disable (0).

8.3.2 FIRST set up for memory window

The FIRST window memory address is made up of a low byte and a high byte. The format of the high-byte register is

Bit 7	DS	Data bus size: 16-bit (1)/ 8-bit (0).
Bit 6	0WS	Zero wait states: no wait states (1)/ additional wait states (0).
Bit 5	SCR1	Scratch bit (not used).
Bit 4	SCR0	Scratch bit (not used).
Bit 3–0		Window start address A23–A20.

The format of the low-byte register is

Bit 7–0 A19–A12. Window start address A19–A12.

8.3.3 LAST set up for memory window

The LAST window memory address is made up of a low byte and a high byte. The format of the high-byte register is

Bit 7, 6	WS1, WS0	Wait state.
Bit 5, 4		Reserved.
Bit 3–0	A23–A20	Window start address A23–A20.

The format of the low-byte register is

Bit 7–0 Window start address A19–A12.

8.3.4 Card offset set up for memory window

The card offset memory address is made up of a low byte and a high byte. The format of the high-byte register is

Bit 7	WP	Write protection: protected (1)/ unprotected (0).
Bit 6	REG	$\overline{\text{REGISTER SELECT}}$ enabled. If set to a 1 then access to attribute memory, else common memory.
Bit 5–0		Window start address A25–A20.

The format of the low-byte register is

Bit 7–0	Window start address A19–A12

8.3.5 FIRST set up for I/O window

The FIRST window I/O address is made up of a low byte and a high byte. The format of the high-byte register is

Bit 7–0	A15–A8

The format of the low-byte register is

Bit 7–0	A7–A8

8.3.6 LAST setup for I/O window

The LAST window I/O address is made up of a low byte and a high byte. The format of the high-byte register is

Bit 7–0	A15–A8.

The format of the low-byte register is

Bit 7–0	A7–A8

8.3.7 Control register for I/O address window

The control register for the I/O address window is made up from a single byte. Its format is

Bit 7, 3	WS1, WS0	Wait states for window 1 and 0.
Bit 6, 2	0WS1, 0WS0	Zero wait states for window 1 and 0.
Bit 5, 1	CS1, CS0	$\overline{\text{IOIS16}}$ source. Select $\overline{\text{IOIS16}}$ from PC (1) or select data size from DS1 and DS0 (0).
Bit 4, 0	DS1, DS0.	Data size: 16-bit (1)/ 8-bit (0).

8.3.8 Examples

A typical application of the PCMCIA socket is to use it for a modem. This is an example of a program to set up a modem on the COM2 port. For this purpose, the socket must be set up to map into the I/O registers from 02F8h to 02FFh. The following code will achieve this:

```
/* load 02f8 into FIRST and 02FFh into LAST registers    */
_outp(0x3E0,08h);    /* point to FIRST low byte          */
_outp(0x3E1,f8h);    /* load f8h into FIRST low byte     */
_outp(0x3E0,09h);    /* point to FIRST high byte         */
_outp(0x3E1,02h);    /* load 02h into FIRST high byte    */

_outp(0x3E0,0Ah);    /* point to LAST low byte           */
_outp(0x3E1,ffh);    /* load ffh into LAST low byte      */

_outp(0x3E0,0Bh);    /* point to LAST high byte          */
_outp(0x3E1,02h);    /* load 02h into LAST high byte        */

/*setup control register: no wait states, 8-bit data access   */
```

```
_outp(0x3E0,07h);     /* point to I/O Control register          */
_outp(0x3E1,00h);     /* load 00h into register                 */

/* enable window 0 */
_outp(0x3E0,06h);     /* point to memory enable window          */
_outp(0x3E1,04h);     /* load 0100 0000b to enable I/O window 0 */
```

8.4 Exercises

8.4.1 How many data bits does the PCMCIA bus have:

(a) 8 (b) 16 (c) 24 (d) 32

8.4.2 How are devices typically added to the system:

(a) They are mapped into the I/O memory address
(b) They directly into the physical address of the system
(c) They use polled interrupts
(d) They interface to a main controller

8.4.3 What is the base address of the registers that are used to program the PCMCIA device:

(a) 1E0h (b) 2E0h
(c) 3E0h (d) 4E0h

8.4.4 Prove that the maximum address memory with PCMCIA is 64 MB.

8.4.5 Explain how I/O registers are used to program the PCMCIA device.

8.4.6 Show the lines of C code that would be required to mount a primary serial port (3F8h–3FFh) and an ECP printer port (378h–37Ah).

8.4.7 Show the lines of C code that would be required to mount a primary (1F0h–1F7h) and a secondary hard disk (170h–177h).

8.4.8 How would the programming for extra memory differ from an isolated I/O device.

8.5 Notes from the author

PCMCIA devices – To save paper, I've got seven lines to tell you about them. Well, in summary, they're really good, but tend to be relatively expensive. Their principle use is to add a network adapter or a modem to a notebook computer. They are typically not used to add to the memory of the notebook or to increase its hard disk space (an internal upgrade is much better for these). Personally, I find them a little too thin, and I do not believe they can get all the required electronics into them (but I remember when simple logic ICs, like AND and OR gates, were as big as your thumb and they could heat it if you required).

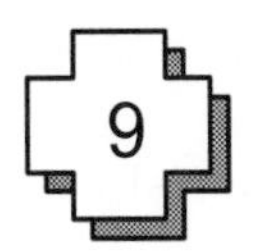

USB and Firewire

9.1 Introduction

The PC is now evolving into a powerful system through:

- Microprocessor developments.
- Improved graphics systems, such as AGP.
- The PCI bus architecture, especially the PCI bridge.
- Improved plug-and-play technology and automated set-up. The USB port aids in its ease of connection.

USB (Universal Serial Bus) allows for the connection of medium bandwidth peripherals such as keyboards, mice, tablets, modems, telephones, CD-ROM drives, printers and other low to moderate speed external peripherals in a tiered-star topology. Its basic specification is:

- Isochronous ('continuous') transfers which supports audio and video. With isochronous data transfers, devices transmit and receive data in a guaranteed and predictable fashion. USB also supports non-isochronous devices (the highest priority), and both isochronous and non-isochronous can exist at the same time.
- Standardised industry-wide plug-and-play specification, cables and connections.
- Multiple-tiered hubs with almost unlimited expansion (with up to 127 physical devices), and concurrent operations.
- 12 Mbps transfer rate and different packet sizes. It supports many device bandwidth requirements from a few kbps to 12 Mbps.
- Wide range of device data rates by accommodating packet buffer size and latencies.
- A hot-plug capability which allows peripherals to be connected without powering down the computer. Dynamically attachable and reconfigurable peripherals.
- Enhanced power management with system hibernation and sleep modes.
- Self-identifying peripherals, automatic mapping of function to driver and configuration.
- Support for compound devices which have multiple functions.
- Flow control for buffer handling built into protocol.
- Error handling/fault recovery mechanism.
- Support for identification of faulty devices.
- Simple protocol to implement and integrate.

USB is a balanced bus architecture which hides the complexity of the operation from the devices connected to the bus. The USB host controller controls system bandwidth. Each device is assigned a default address when the USB device is first powered or reset. Hubs and functions are assigned a unique device address by USB software.

Typical examples of USB connected devices are:

- Digital speakers/ microphones.
- Joysticks.
- Scanners/ modems/ printers/ monitors.
- Game controllers/ graphics tablets.
- Video conferencing cameras.
- Musical interfaces, such as MIDI.

9.2 USB

9.2.1 Physical USB connection

USB uses a four-wire cable to connect to devices. One pair of the twisted-pair lines gives the differential data lines (D+ and D–), while the other two gives a 5 V and a GND supply rail, as given in Table 9.1.

Data transfer rate is up to 12 Mbps, with a 1.5 Mbps subchannel for low-data-rate devices (such as a mouse). A single unit can connect directly to the PC, but a hub is required when more than one device is connected. Each peripheral can extend up to 5 m from each hub connection, with a maximum of 127 different devices to a single PC.

Table 9.1 USB connections

Pin	*Name*	*Description*
1	V_{CC}	+5 V_{DC}
2	D–	Data–
3	D+	Data+
4	GND	Ground

9.2.2 Bus protocol

Each bus transaction involves the transmission of up to three packets. These are

- Token packet transmission – on a scheduled basis, the host controller sends a USB packet which describes the type and direction of a transaction, the USB device address and endpoint number. The addressed USB device selects itself by decoding the appropriate address fields.
- Data packet transmission – the source of the transaction then sends a data packet, or indicates it has no data to transfer.
- Handshake packet transmission – destination device responds with a handshake packet to indicate whether the transfer was successful.

USB supports two types of transfers: stream and message. A stream has no defined structure, whereas a message does. At start-up one message pipe, Control Pipe 0, always exists as it provides access to the device's configuration, status and control information.

The USB protocol supports hardware or software error handling. In hardware error handling the host controller retries three times before informing the client software of the error.

Each packet includes a CRC field which detects all single and double bit errors, as well as many multibit errors. Typically error conditions are short term.

A major advantage of USB is the hot attachment and detachment of devices. USB does this by sensing when a device is attached or detached. When this happens, the host system is notified, and system software interrogates the device. It then determines its capabilities, and automatically configures the device. All the required drivers are then loaded and applications can immediately make use of the connected device.

9.2.3 Data transfers types

USB optimizes large data transfers and real-time data transfers. When a pipe is established for an endpoint, most of the pipe's transfer characteristics are determined and remain fixed for the lifetime of the pipe. Transfer characteristics that can be modified are described for each transfer type.

USB defines four transfer types:

- Control transfers – bursty, non-periodic, host software initiated request/response communication typically used for command/status operations.
- Isochronous transfers – periodic, continuous communication between host and device typically used for time relevant information. This transfer type also preserves the concept of time encapsulated in the data. This does not imply, however, that the delivery needs of such data are always time critical.
- Interrupt transfers – small data, non-periodic, low frequency, bounded latency, device initiated communication typically used to notify the host of device service needs.
- Bulk transfers – non-periodic, large bursty communication typically used for data that can use any available bandwidth and also is delayed until bandwidth is available.

9.2.4 USB implementation

There are two main ways to implement USB. These are:

- **OHCI (open host controller interface)**. This method defines the register level interface that enables the USB controller to communicate with the host computer and the operating system. OHCI is an industry-standard hardware interface for operating systems, device drivers, and the basic input output system (BIOS) to manage the USB. It optimises performance of the USB bus while minimising central processing unit (CPU) overhead to control the USB. Its main features are:

  - Scatter/gather bus master hardware support reduces CPU overhead to handle multiple data transfers across the USB.
  - Efficient isochronous data transfers allow for high USB bandwidth without slowing down the host CPU.
  - Assurance of full compatibility with all USB devices.

- **UHCI (universal host controller interface)**. This method defines how the USB controller talks to the host computer and its operating system. It is optimised to minimise host computer design complexity and uses the host CPU to control the USB bus. Its main features are:
  - Simple design reduces the transistor count required to implement the USB interface on the host computer, thus reducing system cost.
  - Assurance of full compatibility with all USB devices.

The PCI bridge device (PIIX3/PIIX4) contains a USB host controller (HC) with a root hub with two USB ports. This allows two USB peripheral devices to directly communicate with the PCI bridge without an external hub. When more than two USB devices require to be connected then an external hub can be added. The USB's PCI configuration registers are located in the PCI configuration space.

The host controller uses the UHCI standard and thus uses UHCI standard software drivers. It basically consists of two parts:

- **Host controller driver (HCD)**. This is the software that manages the host controller operation and is responsible for scheduling the traffic on USB by posting and maintaining transactions in system memory. It interprets requests from the USBD and builds frame list, transfer descriptor, queue head, and data buffer data structures for the host controller. These data structures are built in system memory and contain all necessary information to provide end-to-end communication between client software in the host and devices on the USB. The host controller moves data between system memory and devices on the USB by processing these data structures and generating the transaction on USB. The host controller executes the schedule lists generated by HCD and reports the status of transactions on the USB to HCD. Command execution includes generating serial bus token and data packets based on the command and initiating transmission on USB. For commands that require the Host Controller to receive data from the USB device, the host controller receives the data and then transfers it to the system memory pointed to by the command. The UHCI's HCD provides sufficient commands and data to keep ahead of the host controller execution and analyses the results as the commands are completed.
- **Host controller (HC)**. The host controller interfaces to the USB system software in the host via the HCD.

Attachment of USB devices

All USB devices attach to the USB via a port on specialised USB devices known as hubs. Hubs indicate the attachment or removal of a USB device in its per port status. The host queries the hub to determine the reason for the notification. The hub responds by identifying the port used to attach the USB device. The host enables the port and addresses the USB device with a control pipe using the USB Default Address. All USB devices are addressed using the USB Default Address when initially connected or after they have been reset. The host determines if the newly attached USB device is a hub or a function and assigns a unique USB address to the USB device. The host establishes a control pipe for the USB device using the assigned USB address and endpoint number zero. If the attached USB device is a hub and USB devices are attached to its ports, then the above procedure is followed for each of the attached USB devices. If the attached USB device is a function, then attachment notifications will be dispatched by USB software to interested host software.

Removal of USB devices

When a USB device has been removed from one of its ports, the hub automatically disables the port and provides an indication of device removal to the host. Then the host removes knowledge of the USB device. If the removed USB device is a hub, the removal process must be performed for all of the USB devices that were previously attached to the hub. If the removed USB device is a function, removal notifications are sent to interested host software.

USB host: hardware and software

The USB host interacts with USB devices through the host controller. The host is responsible for the following:

- Detecting the attachment and removal of USB devices
- Managing control flow between the host and USB devices
- Managing data flow between the host and USB devices
- Collecting status and activity statistics
- Providing a limited amount of power to attached USB devices

USB system software on the host manages interactions between USB devices and host-based device software. There are five areas of interactions between USB system software and device software, they are:

- Device enumeration and configuration.
- Isochronous data transfers.
- Asynchronous data transfers.
- Power management.
- Device and bus management information.

Whenever possible, USB software uses existing host system interfaces to manage the above interactions. For example, if a host system uses Advanced Power Management (APM) for power management, USB system software connects to the APM message broadcast facility to intercept suspend and resume notifications.

9.2.5 USB host controller registers

VID (vendor identification register)

Address offset	00–01h
Default value	8086h
Attribute	Read only

The VID register contains the vendor identification number. This register, along with the device identification register, uniquely identifies any PCI device. Writes to this register have no effect. Bit description 15:0 vendor identification number. This is a 16-bit value assigned to Intel.

DID (device identification register)

Address offset	02–03h
Default value	7112h
Attribute	Read only

The DID register contains the device identification number. This register, along with the VID register, defines the USB host controller.

9.3 Firewire

The main competitor to USB is the Firewire standard (IEEE 1394-1995) which is a high-speed serial bus for desktop peripheral devices, typically for video transfers. It supports rates of approximately 100, 200 and 400 Mbps, known as S100, S200 and S400 respectively. Future standards promise higher data rates, and ultimately it is envisaged that rates of 3.2Gbps will be achieved when optical fibre is introduced into the system. It is generally more expensive than USB to implement for both the host computer and peripherals.

Its main features are:

- 100/200/400 Mbps transfer rate.
- Point-to-point interconnect with a tree topology; 1000 buses with 64 nodes gives 64 000 nodes.
- Automatic configuration and hot plugging.
- Isochronous data transfer, where a fixed bandwidth is dedicated to a particular peripheral.
- Maximum cable length of 4.5 m.

Firewire also complements USB in that it supports high-speed peripherals, whereas USB supports low-to-medium speed peripherals. It is an attractive alternative to technologies such as SCSI and it may provide a universal connection to replace many of the older connectors normally found at the back of a standard PC. This should subsequently reduce the costs of production of computer interfaces and peripheral connectors, as well as simplifying the requirements placed on users when setting up their devices. This is made possible by the following features of the IEEE-1394 bus:

- Hot pluggable – devices can be added or removed while the bus is still active.
- Easy to use – there are no terminators, device addressing or elaborate configuration often associated with technologies like SCSI.
- Flexible topology – devices can be connected together in many configurations, thus the user need not consider logical locations on the network.
- Fast – suitable for high bandwidth applications.
- Rate mixing – a single cable medium can carry a mix of different speed capabilities at the same time
- Inexpensive – targeted at consumer devices.

9.3.1 Topology

There are two bus categories:

- **Cable.** This is a bus which connects external devices via a cable, The cable environment is a non-cyclic network with finite branches consisting of bus bridges and nodes (cable devices). Non-cyclic networks contain no loops and results in a tree topology, with devices daisy-chained and branched (where more than one device branch is connected to a device). Figure 9.1 shows an example of an IEEE-1394 Splitter which has three branches and the telephone is daisy-chained from the digital camera.

  The finite branches restriction imposes a limit of 16 cable hops between nodes. Therefore branching should be used to take advantage of the maximum number of nodes on a bus. 6-bit node addressing allows up to 63 nodes on each bus, while 10-bit bus addressing allows up to 1023 buses, interconnected using IEEE-1394 bridges. Devices on the bus are identified by node IDs. Configuration of the node IDs is performed by the self ID and tree ID processes after every bus reset. This happens every time a device is added to or removed from the bus, and is invisible to the user.

  A final restriction is that, using standard cables, the length between nodes is limited to 4.5 m. This can be increased by adding repeaters between nodes, but lengths are expected to improve as work on the standard ensues. Although a PC is shown in Figure 9.1, a principal advantage of IEEE-1394 is that, unlike USB, no PC is actually required to form a bus, and devices can talk to each other without intervention from a computer.
- **Backplane**. This type is an internal bus. An internal IEEE-1394 device can be used alone, or incorporated into another backplane bus. For example, two pins are reserved for a serial bus by various ANSI and IEEE bus standards. Implementation of the backplane specification lags the development of the cable environment, but one could image internal IEEE-1394 hard disks in one computer being directly accessed by another IEEE-1394 connected computer.

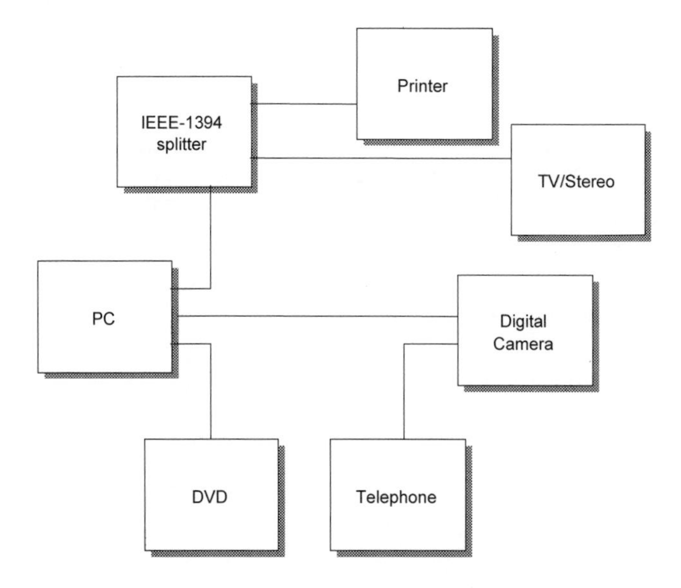

Figure 9.1 IEEE-1394 topology example

9.3.2 Asynchronous and isochronous transfer

One of the key capabilities of IEEE-1394 is isochronous data transfer. Both asynchronous and isochronous are supported, and are useful for different applications. Isochronous transmission transmits data like real-time speech and video, both of which must be delivered uninterrupted, and at the rate expected, whereas asynchronous transmission is used to transfer data that is not tied to a specific transfer time. With IEEE-1394, asynchronous is the conventional transfer method of sending data to an explicit address, and receiving confirmation when it is received. Isochronous, however, is an unacknowledged guaranteed-bandwidth transmission method, useful for just-in-time delivery of multimedia-type data.

An isochronous 'talker', requests an amount of bandwidth and a channel number. Once the bandwidth has been allocated, it can transmit data preceded by a channel ID. The isochronous *listeners* can then listen for the specified channel ID and accept the data following. If the data is not intended for a node, it will not be set to listen on the specific channel ID. Up to 64 isochronous channels are available, and these must be allocated, along with their respective bandwidths, by an isochronous resource manager on the bus.

Figure 9.2 shows an example situation where two isochronous channels are allocated. These have a guaranteed bandwidth, and any remaining bandwidth is used by pending asynchronous transfers. Thus isochronous traffic takes some priority over asynchronous traffic.

By comparison, asynchronous transfers are sent to explicit addresses on the 1394 bus (Figure 9.3). When data is to be sent, it is preceded by a destination address, which each node checks to identify packets for itself. If a node finds a packet addressed to itself, it copies it into its receive buffer. Each node is identified by a 16-bit ID, containing the 10-bit bus ID and 6-bit node or physical ID. The actual packet addressing, however, is 64 bits wide, providing a further 48 bits for addressing a specific offset within a node's memory. This addressing conforms to the control and status register (CSR) bus architecture standard. The ISO/IEC 13213:1994 minimises the amount of circuitry required by 1394 ICs to interconnect with standard parallel buses. The 48-bit offset allows for the addressing of 256 terabytes of memory and registers on each node.

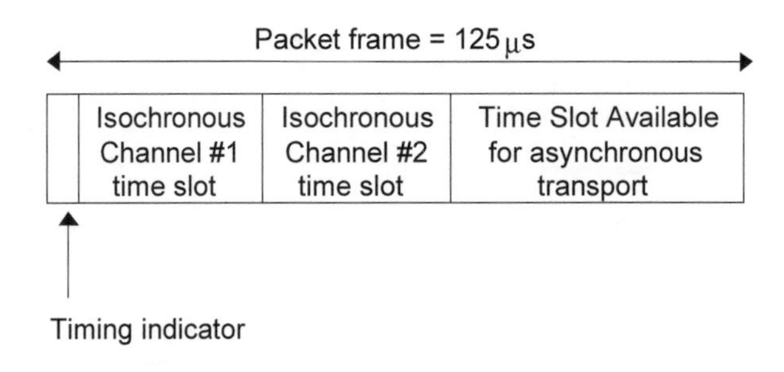

Figure 9.2 Bandwidth allocation on the IEEE-1394 bus

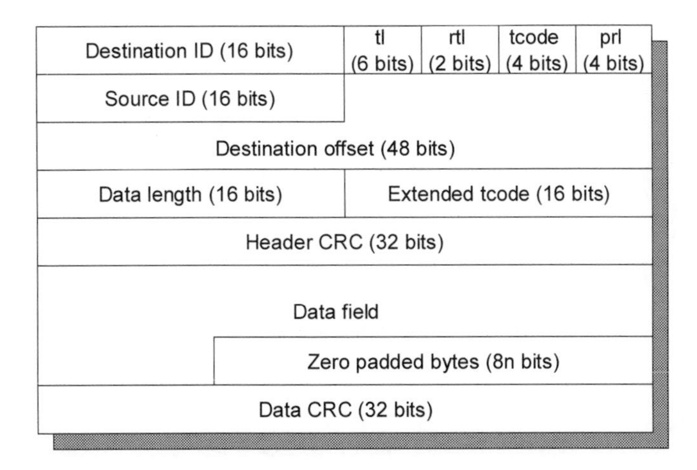

Figure 9.3 Asynchronous write data block payload

9.3.3 IEEE-1394 packet formats

There are a number of different packet formats specified in 1394-1995, however only the asynchronous block write will be presented here, as it is the main transaction type used within this project.

The asynchronous block write is described in the 1394-1995 specification as a packet type that requests a data block be written to the specified destination address. It is the packet type used on asynchronous transmits, for a variable length of data.

The destination_ID field should contain the 16-bit destination node ID, while the destination_offset field contains the remaining 48 bits required for CSR addressing. The data is sent in the data field, which can be any quadlet-aligned length up to a maximum given by the transmission speed. At 200 Mbps, for example, the data field may hold anything from 0 to 1024 bytes, in stages of four bytes. The header information is followed by a CRC (cyclic redundancy check) for error checking, as is the block of data.

9.3.4 Bus management

Two bus management entities are available in the cable environment: the isochronous resource manager and the bus manager. They provide services such as maintaining topology maps, or acting as a central resource from which bandwidth and channel allocations can be made. Further information on bus management can be found in the 1394-1995 specification.

9.3.5 Cable

Figure 9.4 shows that the 1394 cable consists of three individually shielded cable pairs. There are two power lines and two (screened) twisted pairs for data and strobe transmission.

9.3.6 Transmission rates

As already discussed, the cable rate definitions for 1394-1995 are termed S100, S200 and S400, give actually data rates of 98.304 Mbps, 196.608 Mbps and 393.216 Mbps, respectively. The high data rates are achieved by using differential non-return to zero (nrz), signalling on each shielded twisted pair.

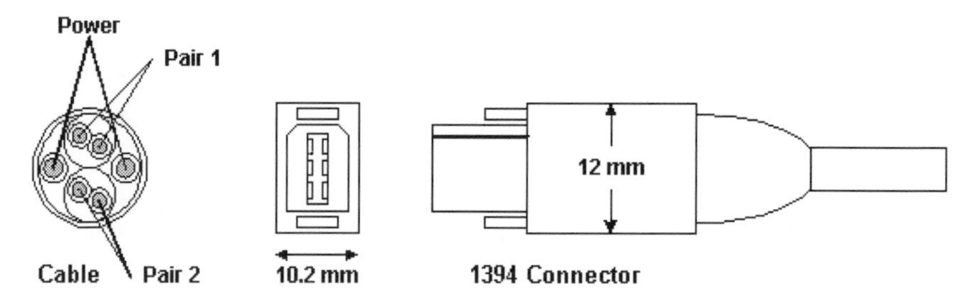

Figure 9.4 IEEE-1394 cable and connectors

9.4 Exercises

9.4.1 How many USB ports are available from the host controller on a PC (PIIX3/4):

(a) 1 (b) 2 (c) 4 (d) 8

9.4.2 Discuss the advantages of USB connected devices over:

(i) ISA devices (ii) PCI devices
(iii) Serial/parallel port connected devices

9.4.3 Outline the main difference between isochronous and asynchronous data traffic. In which applications is it isochronous.

9.4.4 Outline the main types of data transfer on the USB port.

9.4.5 By searching the Internet or a computer catalogue, locate some USB connected devices.

9.5 Notes from the author

Congratulations go to the USB port. It was the first truly generic, easy-to-use, connection bus for the PC that has mechanisms for non-real-time (such as printer data) and real-time data (such as video, audio and speech). It allows for the easy addition and removal of devices from the system, and it also supports hot plugging (adding or removing a device while the computer is on). Microsoft first supported USB in Windows 95 OSR2, and it has since become one of the most used ports, for devices such as video cameras, CD-ROM drives, printers, digital speakers, monitors, and so on. The only problem with USB is that it only gives a data throughput of 12 Mbps, and thus cannot be used for high-speed devices. Possibly, over time, this rate may be increased, or other faster busses, such as Firewire could be used for high-speed applications, such as Fast Ethernet, multimedia communications, hard disk interfaces, and so on.

The IEEE-1394 specification (or i.LINK) is now being used on some systems (especially in notebooks). Its adoption as a standard interface device will depend on whether new standard for the USB specification increase the transfer bit rate to support, at least, 100Mbps. When the USB port can do this, there will be a lesser need for IEEE-1394.

10 Games Port, Keyboard and Mouse

10.1 Introduction

PCs have traditionally been difficult to connect to and set up, for reasons such as:

- Different connectors – there are so many different types of connectors for many different types of devices that connect to the PC. For example, the keyboard uses a 5-pin DIN plug, the parallel port uses a 25-pin D-type connector, the primary serial port uses a 9-pin D-type connector, the video adaptor uses a 15-pin D-type connector, and so on. The future is likely to bring a standardisation of these connectors, possibly with the USB port.
- Different configurations – typically different peripherals required assigned interrupts and I/O addresses. For example, the keyboard uses IRQ1 and I/O ports at 60h and 64h. This is now being overcome by busses such as SCSI and USB, which only require a single interrupt and a limited range of addresses. They also cope better with hot plug-and-play devices and operating system configurable devices.
- Different data traffic rates – relatively low speed interfaces, such as the ISA bus, have often reduced the rate of other faster busses, such as the PCI bus. This is now being overcome by the use of bridges and the USB bus.

The games port, the keyboard and the mouse are also relatively slow devices which, in their standard form, all have different connectors. In the future PCs may standardise these low- and medium-speed devices on the USB port. The keyboard port and mouse port are now standard items on a PC, and most PCs now have a games port, which supports up to two joysticks.

Most PCs support either a PS/2-style mouse or one connected to the serial port (COM1: or COM2:). The operating system automatically scans all the mouse and keyboard ports to determine where the mouse is connected to, and whether there is a keyboard connected.

Typically, these days, a mouse connects to the PS/2 port, which is basically an extension of the keyboard port. The keyboard connects to either a 5-pin DIN plug, or more typically on modern PCs to a smaller 5-pin plug. With the smaller connector, the PS/2 mouse and the keyboard can share the same port (this is typical in new PCs and also for notebooks).

10.2 Games port

The PC was never really designed to provide extensive games support, but as it is so general purpose, it is now used to run arcade style games. A mouse is well designed for precise movements and to select objects, but is not a good device to play games with; thus, a joystick is typically used. The games port adapter supports up to two joysticks connected to the same port. It has 15 pins, which are outlined in Table 10.1 and connects to the system via:

- Lower eight bits of the data bus.
- Lower 10 bits of the address bus.
- $\overline{\text{IOR}}$ and $\overline{\text{IOW}}$.

Table 10.1 Game adapter connections

Pin	*Description*	*Pin*	*Description*
1	+5V	9	+5V
2	1st button for joystick A (BA1)	10	1st button for joystick B (BB1)
3	X-potentiometer of joystick A (AX)	11	X-potentiometer of joystick B (BX)
4	GND	12	GND
5	GND	13	Y-potentiometer of joystick B (BY)
6	Y-potentiometer of joystick A (AY)	14	2nd button for joystick B (BB1)
7	2nd button for joystick A (BA1)	15	+5V
8	+5V		

Each joystick has two buttons, which are normally open circuit, and two potentiometers which give a variable resistance from 0 Ω to 100 kΩ, to indicate the x- and y-position of the joystick handle. Figure 10.1 shows its connections. An unpressed button corresponds to a high level and a button press to a low level.

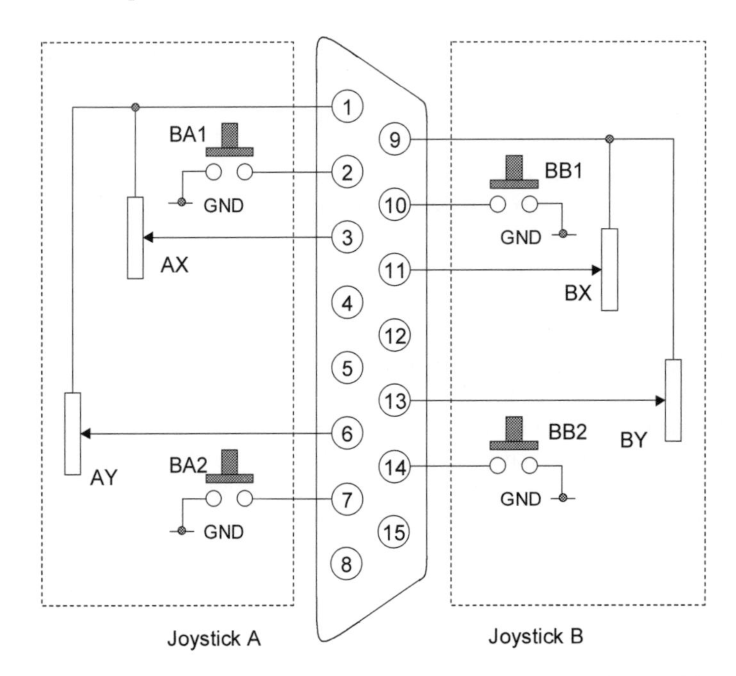

Figure 10.1 Joystick interface

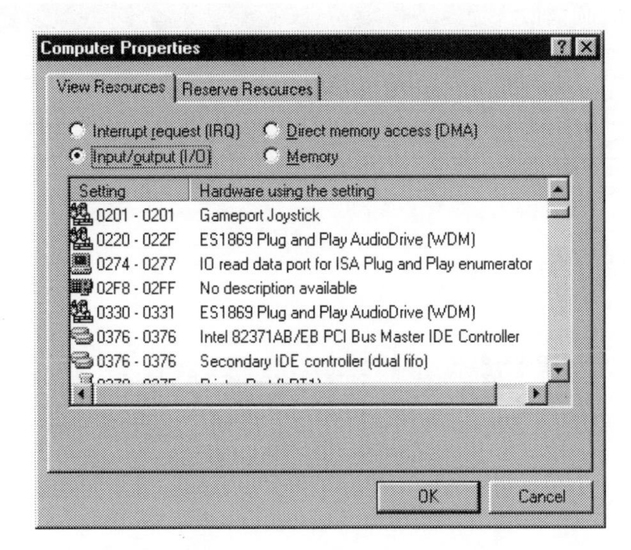

Figure 10.2 Memory map showing Gameport I/O address

The status of the button can be determined by reading the 201h address (see Figure 10.2), its format is given in Figure 10.3. Thus to test for a button press the upper four bits of the register are tested to determine if they are a zero. Figure 10.4 shows a simple C program to test the status of the buttons.

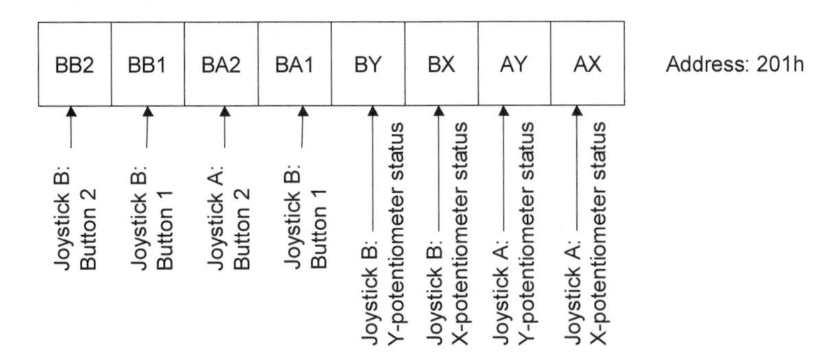

Figure 10.3 Joystick status register format

The reading of the position of the joystick is a little more difficult. For this an event is triggered by writing to the status register. This triggers a one-shot multivibrator, the status of which is given on the lower four bits of the status register, which change from a zero to a one when it has completed the single-shot. The resistance is given by

$$\text{Resistance} = \frac{\text{Time interval}\,(\mu\text{s}) - 24.2}{0.011}\,\Omega$$

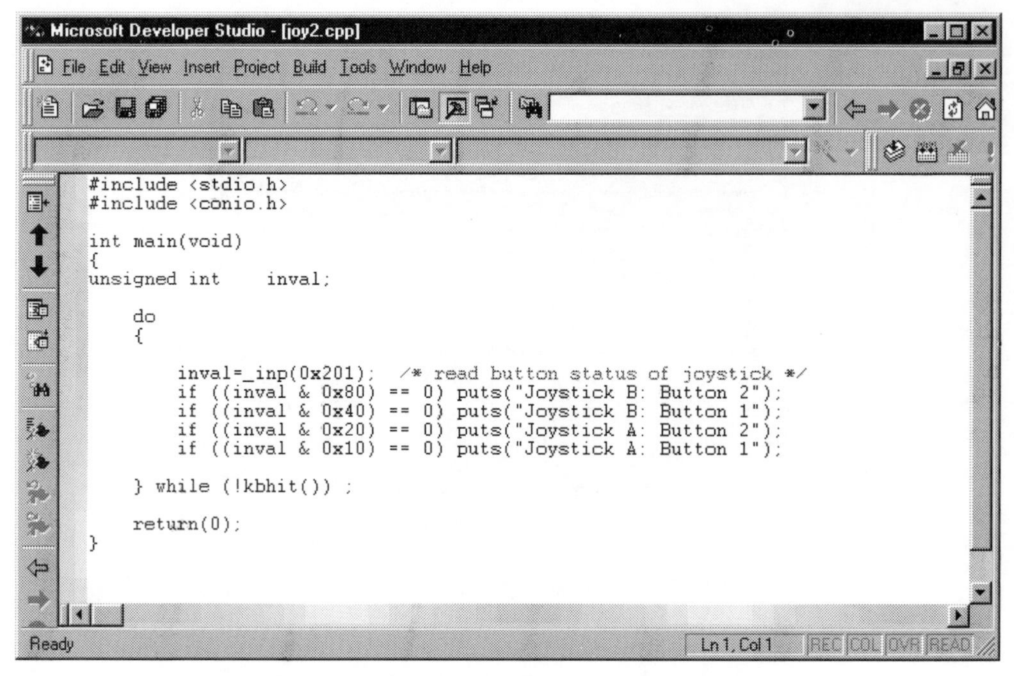

Figure 10.4 Simple C program to test joystick button status

Thus, the timing values will change from 24.2 μs (for 0 kΩ) to 1.124 ms (for 100 kΩ). A simple program that determines the time it takes for AX to be set is given next:

Program 10.1

```
#include <stdio.h>
#include <conio.h>

int main(void)
{
unsigned int   inval, start1, start2, start, end1, end2, end;

      do
      {
         _outp(0x43,0);          /* Specify Counter 0       */

         start1=_inp(0x40);      /* get LSB of Counter 0    */
         start2=_inp(0x40);      /* get MSB of Counter 0    */

         _outp(0x201,0);         /* start one-shot          */

         do
         {
            inval=_inp(0x201);   /* read button status of joystick */
         } while ((inval & 1)==1); /* wait till set to a 0 */

         _outp(0x43,0); /* Specify Counter 0 */

         end1=_inp(0x40);  /* get LSB of Counter 0 */
         end2=_inp(0x40);  /* get MSB of Counter 0 */
```

```
        start=(start1 &0xff)+((start2 &0xff)<<8);
        end=(end1 & 0xff)+((end2 & 0xff)<<8);

        if (start>end)
           printf("Value = %u\n",start-end);
        else  /* roll-over has occurred */
           printf("Value = %u\n",start+(0xffff-end));
     } while (!kbhit()) ;

     return(0);
}
```

Program 10.1 uses Counter 0 which is loaded from address 40h. It has a 16-bit counter register and has a 1.2 MHz clock as its input. It thus rolls-over every 55 ms.

In a sample test run of the above program the output value varied from 62 to 2740, with a static value of 1400. The joystick could be easily calibrated with these values, which are the extremes for either x or y. Note that AY is tested with:

```
do
{
   inval=_inp(0x201);        /* read button status of joystick */
} while ((inval & 2)==1);   /* wait till set to a 0 */
```

and BX is tested with:

```
do
{
   inval=_inp(0x201);        /* read button status of joystick */
} while ((inval & 4)==1);  /* wait till set to a 0 */
```

10.3 Keyboard

Figure 10.5 shows the main connections in the keyboard interface. It uses a 5-pin DIN socket for the connection. The data is sent from the keyboard to the PC in an 11-bit SDU (serial data unit) format over the KBD Data line. When a key has been pressed the IRQ1 interrupt line is activated. The keyboard interface IC scans the keys on the keyboard by activating the X-decoder lines and then sensing the Y-decoder lines to see if there has been a keypress. It then decodes this to sense if a key has changed its state. It then converts the keypress or release to a code which it sends to the keyboard controller on the PC. The format of the code is in the form of an RS-232 interface with eight data bits, one parity bit, one start bit and one stop bit. Unlike RS-232, it uses a synchronous transfer where the clock speed is defined by the KDB clock line.

It is very unlikely that a programmer would ever need to interface directly with the keyboard, as there are a whole host of standard functions that are well tested and interface well with the operating system. It is always advisable to use the standard input keyboard functions, over direct interfacing. Typically the operating system takes over control of all input key presses and sends these to the required process, thus it is not a good idea to interrupt the flow.

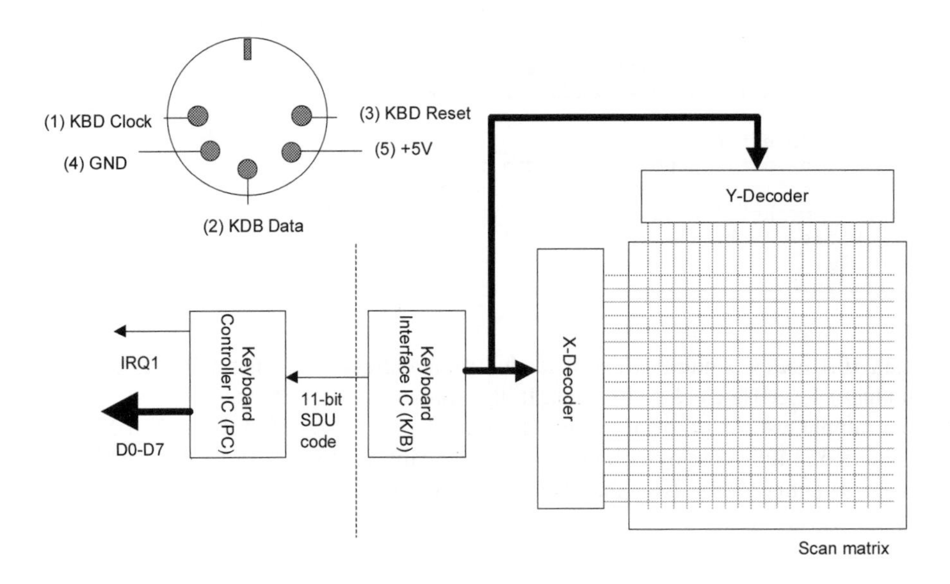

Figure 10.5 Keyboard interface

The keyboard uses two I/O addresses. These are shown in Figure 10.6, and are:

- Input/output buffer (address: 60h) – used to read the code from the keyboard.
- Control/status register (address: 64h) – used either to determine the status of the keyboard (when a value is read from the register) or to set up the keyboard (when a value is written to the register). The commands used are listed in Table 10.2. On a read operation, it acts as a status register. Figure 10.7 shows the bit definitions, these are:

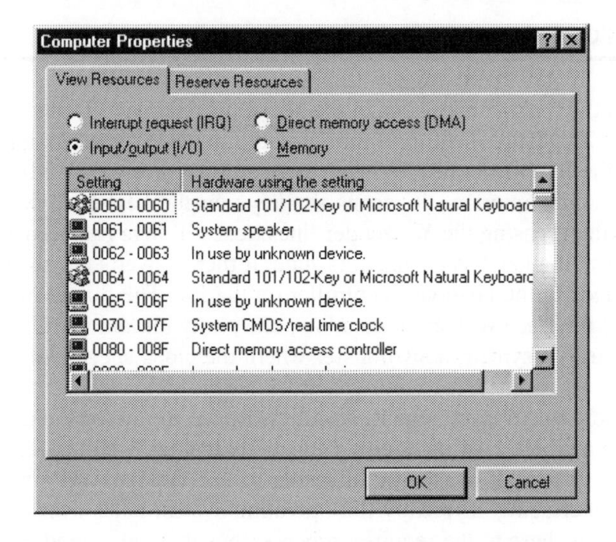

Figure 10.6 Keyboard I/O addresses (60h and 64h)

PARE	Parity bit – 1 = last byte has a parity error, 0 = no error.
TIM	General time-out – 1 = error, 0 = no error.
AUXB	Output buffer for auxiliary device – 1 = holds data for auxiliary device, 0 = holds keyboard data.
KEYL	Keyboard lock status – 1 = keyboard unlocked, 0 = keyboard locked.
C/D	Command/data – 1 command byte written via port 64h, data byte written via port 60h.
SYSF.	System flag – 1 = self-test successful, 0 = power-on reset.
INPB.	Input buffer status – 0 = Data in input buffer, 0 = no data.
OUTB.	Output buffer status – 0 = Controller data in output buffer, 0 = buffer empty.

The auxiliary device is typically a PS/2 style mouse. Program 10.2 shows an example program which reads from the keyboard buffer. It disables the IRQ1 interrupt. (Note that this may cause some systems to not respond to the keyboard if the program does not terminate properly.)

PARE	TIM	AUXB	KEYL	C/D	SYSF	INPB	OUTB

Address: 64h
Status register

Figure 10.7 Status register bits

Table 10.2 Control register commands

Code	*Command*	*Return value (in output buffer)*
a7h	Disable auxiliary device	
a8h	Enable auxiliary device	
a9h	Check interface to auxiliary device	00h = no error, 01h = clock line low, 02h = clock line high, 03h = data line low, 04h = data line high and ffh = no auxiliary device.
aah	Self-test	55h, on success
abh	Check keyboard interface	00h = no error, 01h = clock line low, 02h = clock line high, 03h = data line low, 04h = data line high and ffh = no auxiliary device
adh	Disable keyboard	
aeh	Enable keyboard	
c0h	Read input port	
c1h	Read input port (low)	
c2h	Read input port (high)	
d0h	Read output port	
d1h	Write output port	
d2h	Write keyboard output buffer	
d3h	Write output buffer of auxiliary device	
d4h	Write auxiliary device	
e0h	Read test input port	

Program 10.2

```
/* This program may not work in Windows 95/98/NT/2000     */
/* as it tries to take direct control of the keyboard     */
#include <stdio.h>
#include <conio.h>
int main(void)
{
unsigned int   inval, hit=0;
char           ch;
      _outp(0x21,0x02); /* disable IRQ1 */
      do
      {
         do
         {
            inval=_inp(0x64); /* read status register */
            if ((inval & 0x01)==0x01) /* set for output buffer */
            {
               puts("Key pressed");
               ch=_inp(0x60); /* read key from buffer */
               printf("%c",ch);
               hit=1;
            }
         } while (hit==0);
         hit=0;
         if (ch==0x1) break;  /* wait for ESC key */
      } while (1);
      _outp(0x21,0); /* enable IRQ1 */
      return(0);
}
```

10.4 Mouse and keyboard interface

Modern PCs typically use the 8242 device to provide for a PS/2 mouse and keyboard function, as illustrated in Figure 10.8. It can be seen that the two interrupts which are available are IRQ1 (the keyboard interrupt) and IRQ12 (PS/2 style mouse). If the mouse connects to the serial port then the IRQ12 line does not cause an interrupt. All clock frequencies are derived from the keyboard clock frequency (see Figure 10.8). Notice that the interface for the PS/2-style mouse is identical to the keyboard connection. They are interfaced through the same registers (60h and 64h).

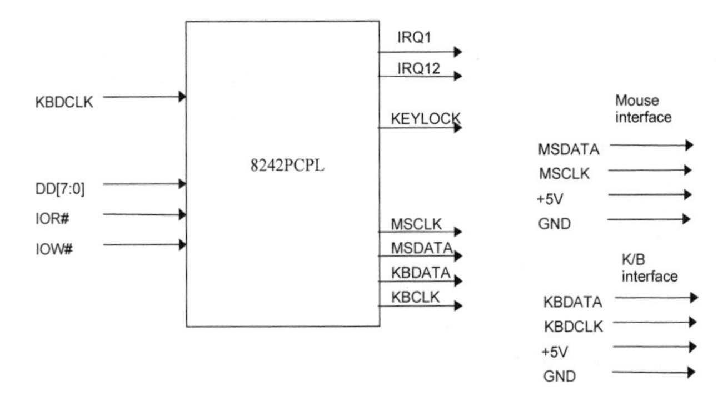

Figure 10.8 Mouse and keyboard interface

10.5 Mouse

Typically on modern PCs the PS/2-style mouse is preferred over serial port mice. PS/2-style mice free up the serial port for other uses, such as for data transfers, modem connections, and so on. Table 10.3 outlines the commands that can be used to program the mouse.

Table 10.3 Control register commands

Code	*Command*	*Description*
e6h	Reset scaling	
e7h	Set scaling	
e8h	Set resolution	Sets the resolution: 00h = 1 count/mm, 01h = 2 counts/mm, 02h = 4 counts/mm and 03 = 8 counts/mm.
e9h	Determine status	**3 status bytes** Byte 1: Bit 0: Right mouse button pressed (if 1). Bit 2: Left mouse button pressed (if 1). Bit 4: Scaling (0=1:1, 1=1:2). Bit 5: Mouse (0=enabled, 1=disabled). Bit 6: Mode (0=stream, 1=remote). Byte 2: Resolution. Byte 3: Sample rate.
eah	Set stream mode	
ebh	Read data	Reads an 8-byte data packet from the mouse.
ech	Resets mouse to normal mode	
eeh	Sets mouse to wrap mode	In wrap mode all the commands or data sent to the mouse.
f0h	Set remote mode	
f2h	Identify unit	00h = mouse
f3h	Set sampling rate	Sampling rate is then set by the value put into output buffer: 0ah = 10 samples/s, 14h = 20 samples/s, 28h = 40 samples/s, 3c = 60 samples/s, 50h = 80 samples/s, 64h = 100 samples/s and c8h = 200 samples/s.
f4h	Enable mouse	
f5h	Disable mouse	
f6h	Set standard mouse to standard values	
feh	Resend	
ffh	Reset	

The PS/2 mouse is programmed by:

- Sending the write auxiliary device (d4h) command to 64h (Control register).
- The next byte is a command code which is sent to port 60h, and then transferred to the mouse port (valid codes are given in Table 10.3). This command transfer only occurs for a single transfer.

The mouse can either be set into a stream mode or a remote mode, and writes movement data into the keyboard buffer. In stream mode, the mouse transmits movement data when it is moved by a given amount (set by the sample rate). In remote mode the mouse only transfers movement data when there is a specific read data command.

When the read data command is sent, the 8-byte data packet is read from the addresses as specified in Table 10.4. An example of programming the mouse is given next:

```
_outp(0x64,0xd4);                    /* Write aux. device              */

do
{
   inval=_inp(0x64);
} while ((inval & 0x02)==0x02);      /* wait until input buffer empty  */
_outp(0x60,0xe7);                    /* set scaling                    */
```

Table 10.4 Control register commands

Offset	*Description*	
00h	Bit 7: YOV (Y-data overflow), Bit 5: YNG (Y-value negative), Bit 1: RIG (right button pressed),	Bit 6: XOV (X-data overflow), Bit 4: XNG (X-value negative), Bit 0: LEF (left button pressed).
02h	X-data movement since last access	
04h	Y-data movement since last access	

10.6 Exercises

10.6.1 What is the base address of the joystick port:

(a) 101h (b) 201h
(c) 301h (d) 401h

10.6.2 Which I/O port addresses are used for the keyboard:

(a) 60h, 64h (b) 160h, 164h
(c) 260h, 264h (d) 360h, 364h

10.6.3 How is the *x* position and *y* position determined:

(a) The time for a single-shot (b) A voltage level
(c) An electrical current (d) A value in a register

10.6.4 What interrupt does the keyboard use:

(a)	IRQ1	(b)	IRQ3
(c)	IRQ4	(d)	IRQ12

10.6.5 What interrupt does the PS/2 style mouse use:

(a)	IRQ1	(b)	IRQ3
(c)	IRQ4	(d)	IRQ12

10.6.6 Run the program in Figure 10.4 and show that the joystick buttons are working. Modify the program so that it only displays a change of status in a button press (rather that scrolling down the screen). For example:

```
if ((inval & 0x80) == 1) && (button==0)) { button=1; puts("B:Button 2 Press");}
if ((inval & 0x80) == 0) && (button==1)) { button=0; puts("B:Button 2 Reset");}
```

10.6.7 Run Program 10.1 and test the movement detection. Modify it so that it detects the *y* movement.

10.6.8 Run Program 10.1 so that the user can calibrate the joystick. The user should be asked to move the joystick to its maximum *x* directions, and also the maximum *y* directions. From this write a program which displays the joystick movement as a value from –1 to +1.

10.7 Notes from the author

Phew. I'm glad I got these three interfaces out of the way, in a single chapter. All three are based on a legacy type system. Over time, the USB port should replace each interface type, but as they work well at the present they may be around for a while longer.

The method that the games port uses to determine position is rather cumbersome, where it uses a single-shot monostable timer to determine the x *and* y *positions. An improved method is to pass the data using a serial interface, just as the mouse does. But, it's a standard, and that's the most important thing.*

The keyboard and PS/2-style mouse connections have proved popular, as they are both now small 5-pin DIN-style connectors, and as the software automatically scans the port for devices, they can be plugged into either socket. This allows for an extra keyboard or a second mouse to be used with a notebook.

As I've got a few extra lines at the end of this chapter, I would like to review the material that has been covered up to this point. The key to understanding internal busses is contained in the Motherboard chapter, where the processor interfaces with the TXC device, which directs any requests to the second-level cache, the DRAM memory or the PCI bus. The PCI bridge device is also important as it isolates the other busses, such as ISA/IDE, USB, serial/parallel port from the PCI bus, and thus the rest of the system. The keyboard, games port and mouse interfaces are accessed via the PCI bridge.

11 AGP

11.1 Introduction

The AGP (accelerated graphics port) is a major advancement in the connection of 3D graphics applications, and is based on an enhancement of the PCI bus. One of the major motivating factors is to improve the speed of transfer between the main system memory and the local graphics card. This reduces the need for large areas of memory on the graphics card, as illustrated in Figure 11.1.

The main gain in moving graphics memory from the display buffer (on the graphics card) to the main memory is the display of text information as:

- It is generally read-only, and does not have to be displayed in any special order.
- Shifting text does not require a great deal of data transfer and can be easily cached in memory, thus reducing data transfer. A shift in text can be loaded from the cached memory.
- It is dependent on the graphics quality of the application, rather that the resolution of the display. There is thus great scope in the future for improvement in the quality of graphics images, rather than their resolution.
- It is not persistent, as it resides in memory only for the duration that it is required. When it has completed the main memory it can be assigned to another application. A display buffer, on the other hand, is permanent.

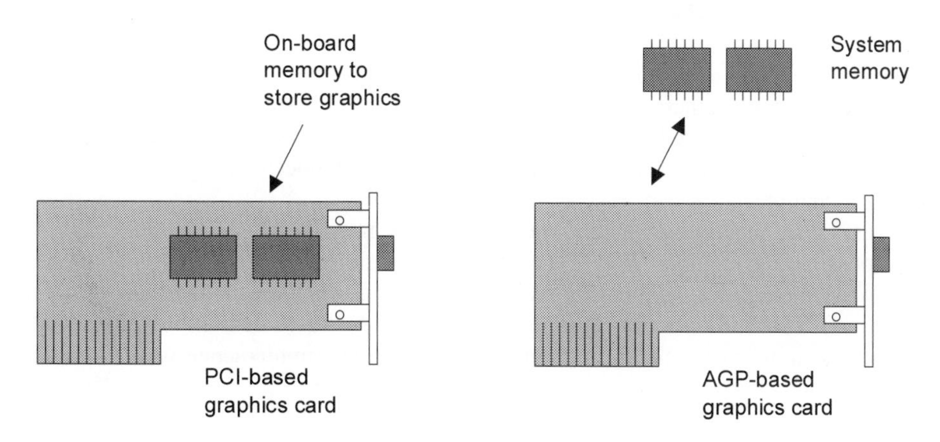

Figure 11.1 AGP card using main system memory

The 440LX is the first AGPset product designed to support the AGP interface. The HOST BRIDGE AGP implementation is compatible with the accelerated graphics port Specification 1.0. HOST BRIDGE supports only a synchronous AGP interface, coupling to the host bus frequency. The AGP interface can reach a theoretical ~532 Mbytes/sec transfer rate. The actual bandwidth will be limited by the capability of the HOST BRIDGE memory subsystem.

11.1.1 PCI interface

The HOST BRIDGE PCI interface is 33-MHz Revision 2.1 compliant and supports up to five external PCI bus masters in addition to the I/O bridge (PIIX4). HOST BRIDGE supports only synchronous PCI coupling to the host bus frequency.

HOST BRIDGE defines a sophisticated data buffering scheme to support the required level of concurrent operations and provide adequate sustained bandwidth between the DRAM subsystem and all other system interfaces (CPU, AGP and PCI).

11.2 PCI and AGP

AGP defines the master as the graphics controller and the corelogic as the graphics card. The AGP interface is based on the 66 MHz PCI standard, but has four additional extensions/enhancements. These extensions are:

- Deeply pipelined memory read and write operations, which fully hide memory access latency.
- Address bus and data bus demultiplexing, allowing for nearly 100% bus efficiency.
- Extension to the PCI timing cycle which allows for one or two data transfers per 66 MHz clock cycle. This provides a maximum data rate of 500 MB/s.
- Extension to the PCI timing cycle which allows for four data transfers per 66 MHz clock cycle. This provides for a maximum data rate of 1 GB/s.

All these enhancements are implemented using extra signal lines (sideband signals), and it is not intended as a replacement to the PCI bus. The AGP is physically, logically and electrically independent of the PCI bus, and has its own connector which is reserved solely for graphics devices (and is not interchangeable with the AGP connector). Figure 11.2 shows the main AGP signal lines.

AGP uses deep pipelining which allows the total memory READ throughput equal to that which is possible for memory WRITE (in PCI the memory read throughput is about half of memory write throughput, as memory read access time is visible as wait states on this unpipelined bus). This and optional higher transfer rates and address demultiplexing, allows for a large increase in memory read throughput over standard PCI implementations.

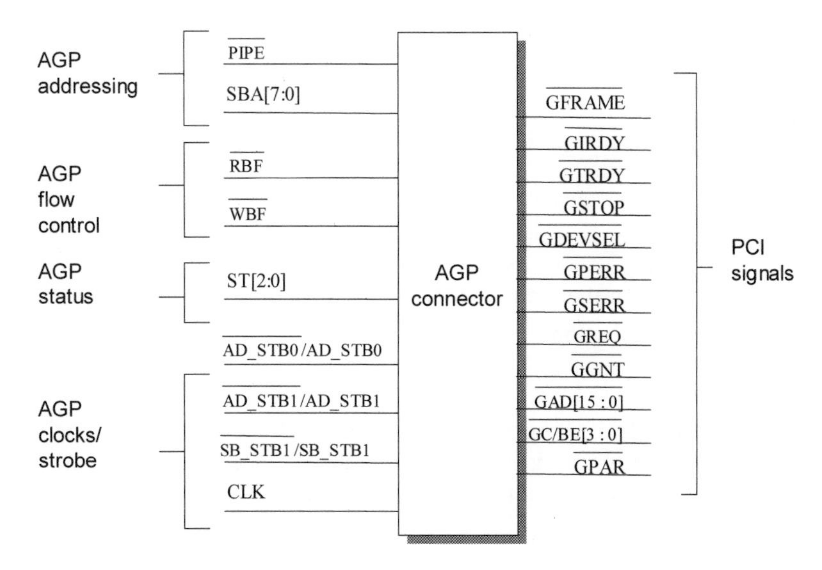

Figure 11.2 The main AGP signal lines

11.3 Bus transactions

AGP uses two types of bus operation. These are:

- **Queuing requests**. This can be done over the SBA port, or the AD bus, and is set up using Bit 9 for the status register (only one type at a time can be used). With the SBA port, the AD bus cannot be used, and vice versa. The sideband signals (SBA[7:0]) are used exclusively to transmit AGP access requests (all PCI transactions use the AD pins for both data and address), and are sent from the master to the core logic (the AGP requests are the same when sent over the AD bus or the SBA bus). A master that uses the SBA port does not require the $\overline{\text{PIPE}}$ signal which is used only to frame requests on the AD pins.
- **Address demultiplexing option**. This allows the complete AGP access request to be transmitted over the 8-bit SBA port. For this the request is broken into three parts: low-order address bits and length (type 1), mid-order address bits and command (type 2), and high-order address bits (type 3).

11.4 Pin description

AGP adds an extra 21 signal lines to the PCI specification. The basic implementation of AGP should support ×1 and ×2 transfer rates, and may optionally support ×4 data transfer rates. All devices should support low priority (LP) data writes, but optionally support fast write (FW) data transfers.

The signal lines split into four main groups:

- AGP requests.
- AGP flow control.
- AGP status.
- AGP clocking.

Also, the PCI lines are identified with a preceding G, such as GAD[31:0] for the PCI AD bus, $\overline{\text{GSTOP}}$ for $\overline{\text{STOP}}$, and so on.

11.4.1 Requests

AGP supports two methods of queuing requests by an AGP master. A master selects the required method during start-up and is not allowed to change when set up. The methods either use the $\overline{\text{PIPE}}$ signal line or they use the SBA port. These signals cannot be used at the same time. These lines are defined as:

$\overline{\text{PIPE}}$	On the master (the graphics controller), $\overline{\text{PIPE}}$ is a sustained tristate signal and is an input to the target (the core logic). When assessed by the current master it indicates a pipelined request, so that the full width request is to be queued by the target. The master queues one request each rising edge of CLK while $\overline{\text{PIPE}}$ is asserted.
SBA[7:0]	These signals are outputs from the master and are inputs to the target, and they indicate the sideband address (SBA) port which gives an additional bus to pass requests (address and command) to the target from the master.

11.4.2 Flow control

Apart from the normal PCI flow control lines, the following have been added to AGP:

$\overline{\text{RBF}}$	The read buffer full (RBF) signal indicates that the master is ready to accept previously requested low priority (LP) read data or not. When it is active (LOW), the arbiter is not allowed to initiate the return of low priority read data to the master. It is made inactive by either the AGP target or motherboard.
$\overline{\text{WBF}}$	The write buffer full (WBF) signal indicates that the master is ready to accept fast write (FW) data. When it is active (LOW), the core logic arbiter is not allowed to initiate a transaction to provide FW data. It is made inactive by either the AGP target or motherboard.

11.4.3 Status signals

The AGP status signals indicate how the AD bus is used in future transactions, such as using it to queue new requests, return previously requested read data, or send previously queued write data. These lines are always an output from the corelogic and an input to the master, and are:

ST[2:0]	These provide information from the arbiter to the master on the mode of

operation, and they only have a meaning when $\overline{\text{GNT}}$ is asserted (else they are ignored). Their settings are:

000 Previously requested low priority read or flush data is being returned to the master.
001 Previously requested high priority read data is being returned to the master.
010 Master is to provide low priority write data for a previous queued write command.
011 Master is to provide high priority write data for a previous queued write command.
100 Reserved.
101 Reserved.
110 Reserved.
111 Master has been given permission to start a bus transaction.

11.4.4 Clocks

The CLK signal provides the basic clock signal for all control signals and is based on the ×1 transfer mode. Two other strobes are used to transfer data on the AD bus or the SBA port. As the AD bus has 32 bits then two copies of the AD_STB are required. In ×4 mode is used, the compliments of the strobes are also required.

CLK Basic clock information for both AGP and PCI control signals.
AD_STB0 This strobe provides for timing in a ×2 data transfer mode on GAD[15:0] and is provided by the agent that is providing data.
$\overline{\text{AD\_STB0}}$ This strobe provides for timing in a ×4 data transfer mode on GAD[15:0] and is provided by the agent that is providing data.
AD_STB1 This strobe provides for timing in a ×2 data transfer mode on GAD[31:16] and is provided by the agent that is providing data.
$\overline{\text{AD\_STB1}}$ This strobe provides for timing in a ×4 data transfer mode on GAD[31:16] and is provided by the agent that is providing data.
SB_STB This strobe provides the strobe for the SBA[7:0] (when required). It is driven by the AGP master.
$\overline{\text{SB\_STB}}$ This strobe provides the strobe for the SBA[7:0] (when required) at ×4 data transfer mode. It is driven by the AGP master.

11.4.5 USB signals

USB+ Used to send USB data and control packets to an externally connected USB capable video monitor.
USB- Inverse of USB+.
$\overline{\text{OVRCNT}}$ The USB overcurrent indicator is set low when there is too much current being taken from the 5 V supply.

11.4.6 Other signals

$\overline{\text{PME}}$ Power management event. Not used by the AGP bus, but used by the PCI bus.
$\overline{\text{TYPEDET}}$ The type detect signal identifies whether the interface is 1.5 V or 3.3 V.

11.4.7 PCI signals and AGP

AGP supports most of the PCI signals. IDSEL, $\overline{\mathrm{LOCK}}$, $\overline{\mathrm{INTC}}$ and $\overline{\mathrm{INTD}}$ are not supported on the AGP connector, whereas, $\overline{\mathrm{FRAME}}$, IDSEL, $\overline{\mathrm{STOP}}$ and $\overline{\mathrm{DEVSEL}}$ are used in FW transactions, but not in AGP pipelined operations.

$\overline{\mathrm{FRAME}}$	Used for FW transactions, but not for AGP pipelined transaction.
$\overline{\mathrm{IRDY}}$	Used by the AGP master to indicate that it is ready to provide all write data for the current transaction. When the master asserts it, then, it cannot insert any wait states either when reading or writing blocks of data (but it can in-between blocks). In FW transactions, the core logic sets the line to indicate that there is write data on the bus. The core logic cannot insert wait states with data blocks.
$\overline{\mathrm{TRDY}}$	Used by an AGP target to indicate that it is ready to provide read data for the entire transaction or is ready to transfer a block of data when the transfer/transaction requires more than four clocks to complete the operation. In FW transactions, the AGP master uses it to indicate when it is willing to transfer a subsequent block.
$\overline{\mathrm{STOP}}$	Used in FW transactions to signal a device disconnection.
$\overline{\mathrm{DEVSEL}}$	Used in FW transactions to signal that a transaction cannot complete during the block.
IDSEL	Not used in the AGP connector, and generated internally in the graphics device.
$\overline{\mathrm{PERR}}$	Not used in the AGP transaction.
$\overline{\mathrm{SERR}}$	As PCI bus.
$\overline{\mathrm{REQ}}$	Used to request access to the bus to initiate an AGP request.
$\overline{\mathrm{GNT}}$	Same meaning as PCI (but extra information is added by ST[2:0]).
$\overline{\mathrm{RST}}$	As PCI bus.
AD[31:00]	As PCI bus.
$\overline{\mathrm{C/BE[3:0]}}$	AGP command information (see Section 11.5).
PAR	Not valid during an AGP transaction.
$\overline{\mathrm{LOCK}}$	Not supported on the AGP interface.
$\overline{\mathrm{INTA}}$, $\overline{\mathrm{INTB}}$	As PCI bus.
$\overline{\mathrm{INTC}}$, $\overline{\mathrm{INTD}}$	Not supported on the AGP connector.

11.5 AGP master configuration

The AGP master is configured in the same way as a device on the PCI bus, which requires that it responds to a PCI configuration transaction. This occurs when:

- A configuration command is decoded.
- AD01 and AD00 are '00'.

- The IDSEL signal is asserted. As the AGP connector does not support IDSEL then it is connected to AD16. This is done by connecting it internally for AGP operation, but externally for PCI operation.

Initially the AGP device asserts $\overline{\text{DEVSEL}}$ when the bus command is configuration (read or write). AD16 is set to a '1' and AD1 and AD0 are '00'. These cause the device's configuration space to be accessed. The system software then scans all configuration spaces by asserting different AD signals between AD16 and AD31, and using PCI configuration read or write commands.

11.6 Bus commands

The AGP bus uses the command lines ($\overline{\text{C/BE[3:0]}}$) to indicate the type of pipelined transaction on the AD bus or SBA port. These are:

0000	Read – Starting at the specified address, read *n* sequential Qwords, where *n* = (length_field + 1). The length_field is provided by the lower three bits on the AD bus (A2–A0).
0001	Read (hi-priority). As 'Read', but the request is queued in the high priority queue. The reply data may be returned out of order with respect to other requests.
0010	Reserved.
0011	Reserved.
0100	Write – Starting at the specified address, write *n* sequential Qwords, as enabled by the $\overline{\text{C/BE[3:0]}}$, where *n* = (length_field + 1).
0101	Write (hi-priority) – As 'Write', but indicates that the write data must be transferred from the master within the maximum latency window established for high priority accesses.
0110	Reserved.
0111	Reserved.
1000	Long read – As 'Read', except for access size, in this case, *n* = 4× (length_field + 1) allowing up to 256 byte transfers.
1001	Long read (hi-priority) – As 'Read (hi-priority)' except for access size which is the same as for 'Long Read'.
1010	Flush – Similar to 'Read'. Forces all low-priority write accesses ahead of it to the point that all the results are fully visible to all other system agents.
1011	Reserved.
1100	Fence – Creates a boundary in a single master's low-priority access stream around which writes may not pass reads.
1101	Dual address cycle (DAC) – used by the master to transfer a 64-bit address to the core logic when using the AD bus.
1110	Reserved.
1111	Reserved.

The master uses two clock periods to transfer the entire address using AD[31:0] and $\overline{C/BE[3:0]}$. Within the first clock period, the master provides the lower address bits (A31–A03) and the length encoding on (A2–A0), as with a 32-bit request, but uses the 1101 command (DAC) encoding on $\overline{C/BE[3:0]}$ instead of the actual command. The second clock of the request contains the upper address bits (A63–A32) on AD[31:0] and the actual command on $\overline{C/BE[3:0]}$.

11.7 Addressing modes and bus operations

AGP transactions differ from PCI transactions in several ways:

- In AGP, pipelined read/write transactions are disconnected from their associated access request, where the request and associated data may be separated by other AGP operations. Conversely, a PCI data phase is connected to its associated address phase, with no interventions allowed. This helps to maintain the pipe depth and allows the core logic to ensure a sufficiently large buffer for receiving the write data, before locking up the bus on a data transfer that could be blocked awaiting buffer space. The rules for the order of accesses on the AGP bus are not based on the order of the data transfer, but on the arrival order of access requests.
- AGP has different bus commands which allow access only to the main system memory. PCI allows access to multiple address spaces: memory, I/O and configuration.
- In AGP, memory addresses are always aligned in 8-byte references, whereas PCI uses 4-byte, or lower, references (the number of bytes addressed is defined with the $\overline{C/BE[3:0]}$). The reason for the increased AGP addressing granularity (from four in the PCI bus to eight in AGP) is because modern processors use a 64-bit data bus and can manipulate 64 bits at a time. The memory systems are also 64 bits wide.
- In AGP, pipelined access requests have an explicitly defined access length or size. In PCI transfer lengths are defined by the duration of $\overline{FRAME}$.

11.8 Register description

The PCI bridge supports AGP through two sets of registers, which are accessed via I/O addresses. These are:

- Configuration address (CONFADD) – Enables/disables the configuration space and determines what portion of configuration space is visible through the configuration data window.
- Configuration data (CONFDATA) – 32-bit/16-bit/8-bit read/write window into configuration space.

Configuration address register

I/O address	0CF8h accessed as a DWord (32-bit)
Default value	00000000h
Access	Read/write

CONFADD is accessed with an 8-bit or a 16-bit value, then it will 'pass through' this register and go onto the PCI bus as an I/O cycle. The register contains the bus number, device number, function number, and register number for which a subsequent configuration access is intended. Its format is:

Bit	*Description*
31	Configuration enable (CFGE) 1=enable, 0=disable.
30:24	Reserved.
23:16	Bus number (BUSNUM) – If it has a value of 00h then the target of the configuration cycle is either the HOST BRIDGE or the PCI bus that is directly connected to the HOST BRIDGE.
15:11	Device number (DEVNUM) – Selects one agent on the PCI bus selected by the bus number. In the configuration cycles this field is mapped to AD[15:11].
10:8	Function number (FUNCNUM) – This field is mapped to AD[10:8] during PCI configuration cycles. It allows for the configuration of a multifunction device.
7:2	Register number (REGNUM) – This field selects one register within a particular bus, device, and function as specified by the other fields in the configuration address register. This field is mapped to AD[7:2] during PCI configuration cycles.
1:0	Reserved.

Configuration data register

I/O address	0CFCh
Default value	00000000h
Access	Read/Write

CONFDATA is a 32-bit/16-bit/8-bit read/write window into configuration space. The portion of configuration space that is referenced by CONFDATA is determined by the settings in the CONFADD register.

11.8.2 Configuration access

The routing of configuration accesses to PCI or AGP is controlled by PCI-to-PCI bridge standard mechanism using the following:

- Primary bus number register.
- Secondary bus number register.
- Subordinate bus number register.

The PCI bus 0 is frequently known as the primary PCI.

PCI bus configuration mechanism

The PCI bus has a slot based configuration space which allows each device to contain up to eight functions, with each function containing up to 256, 8-bit configuration registers.

PCI configuration is achieved with two bus cycles: configuration read and configuration write. A device can be configured using the CONFADD and CONFDATA registers. First a DWord value is placed into the CONFADD register that enables the configuration (CONFADD[31]=1), specifies the PCI bus (CONFADD[23:16]), the device on that bus (CONFADD[15:11]), the function within the device (CONFADD[10:8]). CONFDATA then becomes a window for which four bytes of configuration space are specified by the contents of CONFADD. Any read or write to CONFDATA results in the host bridge translating CONFADD into a PCI configuration cycle.

If the bus number is 0 then a Type 0 configuration cycle is performed on primary PCI bus, where:

- CONFADD[10:2] (FUNCNUM and REGNUM) are mapped directly to AD[10:2].
- CONFADD[15:11] (DEVNUM) is decoded onto AD[31:16].

The host bridge entity within HOST BRIDGE is accessed as a Device 0 on the primary PCI bus segment and a *virtual* PCI-to-PCI bridge entity is accessed as a Device 1 on the primary PCI bus.

11.8.3 PCI configuration space

HOST BRIDGE is implemented as a dual PCI device residing within a single physical component, where:

- Device 0 is the host-to-PCI bridge, and includes PCI bus number 0 interface, main memory controller, graphics aperture control and HOST BRIDGE's specific AGP control registers.
- Device 1 is the virtual PCI-to-PCI bridge, and includes mapping of AGP space and standard PCI interface control functions of the PCI-to-PCI bridge.

Table 11.1 shows the configuration space for Device 0. Corresponding configuration registers for both devices are mapped as devices residing at the primary PCI bus (bus #0). The configuration registers layout and functionality for Device 0 is implemented with a high level of compatibility with a previous generation of PCIsets (i.e., 440FX). Configuration registers of HOST BRIDGE Device 1 are based on the standard configuration space template of a PCI-to-PCI bridge.

Table 11.1 PCI configuration space (Device 0)

Address	*Reference*	*Register name*
00–01h	VID	Vendor identification
02–03h	DID	Device identification
04–05h	PCICMD	PCI command register
06–07h	PCISTS	PCI status register
08h	RID	Revision identification
0Ah	SUBC	Subclass code
0Bh	BCC	Base class code
0Dh	MLT	Master latency timer
0Eh	HDR	Header type
10–13h	APBASE	Aperture base address
34h	CAPPTR	Capabilities pointer
50–51h	HOST BRIDGECFG	Host bridge configuration
53h	DBC	Data buffering control
55–56h	DRT	DRAM row type
57h	DRAMC	DRAM control
58h	DRAMT	DRAM timing
59–5Fh	PAM[6:0]	Programmable attribute map (7 registers)
60–67h	DRB[7:0]	DRAM row boundary (8 registers)
68h	FDHC	Fixed DRAM hole control
6A-6Bh	DRAMXC	DRAM extended mode select
6C-6Fh	MBSC	Memory buffer strength control register
70h	MTT	Multitransaction Timer
71h	CLT	CPU latency timer register
72h	SMRAM	System management RAM control
90h	ERRCMD	Error command register
91h	ERRSTS0	Error status register 0
92h	ERRSTS1	Error status register 1
93h	RSTCTRL	Reset control register
A0–A3h	ACAPID	AGP capability identifier
A4–A7h	AGPSTAT	AGP status register
A8–ABh	AGP	Command register
B0–B3h	AGPCTRL	AGP control register
B4h	APSIZE	Aperture size control register
B8–BBh	ATTBASE	Aperture translation table base register
BCh	AMTT	AGP MTT control register
BDh	LPTT	AGP low-priority transaction timer register

AGPCMD register

The AGPCMD register reports AGP device capability/status. Its main parameters are:

Address offset	A8–ABh
Default value	00000000h
Access	Read/write

Bit	*Description*
31:10	Reserved.
9	AGP side band enable – 1=enable. 0=disable (Default).
8	AGP enable – 1=enable. 0=disable (Default). When this bit is set to a 0, the HOST BRIDGE ignores all AGP operations. Any AGP operations received (queued) while this bit is 1, will be serviced even if this bit is subsequently reset to 0. If it is 1, the HOST BRIDGE responds to AGP operations delivered via $\overline{\text{PIPE}}$ (or responds to the SBA, if the AGP side band enable bit is set to 1).
7:2	Reserved.
1:0	AGP data transfer rate – One bit in this field must be set to indicate the desired data transfer rate. Bit 0 identifies $\times$1, and bit 1 identifies $\times$2.

11.8.4 AGP memory address ranges

The HOST BRIDGE can be programmed for direct memory accesses of the AGP bus interface when addresses are within the appropriate range. This uses two subranges:

- AMBASE/AMLIMIT – this method is controlled with the memory base register (AMBASE) and the memory limit register (AMLIMIT).
- APMBASE/APMLIMIT – this method is controlled with the prefetchable memory base register (APMBASE) and AGP prefetchable memory limit Register (APMLIMIT).

The decoding of these addresses is based on the top 12 bits of the memory base and memory limit registers which correspond to address bits A[31:20] of a memory address. When address decoding, the HOST BRIDGE assumes that address bits A[19:0] of the memory address are zero and that address bits A[19:0] of the memory limit address are FFFFFh. This forces the memory address range to be aligned to 1 MB boundaries and to have a size granularity of 1 MB. The base and limit addresses define the minimum and maximum range of the addresses.

11.8.5 Graphics aperture

AGP supports a graphic aperture which uses memory-mapped graphics data structures. Its starting address is defined by APBASE configuration register of HOST BRIDGE and its range is defined by the APSIZE register, such as 4 MB (default), 8 MB, 16 MB, 32 MB, 64 MB, 128 MB and 256 MB.

11.8.6 AGP address mapping

HOST BRIDGE directs I/O accesses to the AGP port in the address range defined by AGP I/O address range. This range is defined by the AGP I/O base register (AIOBASE) and AGP I/O limit register (AIOLIMIT). These are decoded, where the top four bits of the I/O base and I/O limit registers correspond to address bits A[15:12] of an I/O address. For address decoding, the HOST BRIDGE assumes that the lower 12 address bits A[11:0] of the I/O base are zero and that address bits A[11:0] of the I/O limit address are FFFh. This forces the I/O address range to be aligned to 4 KB boundary and to have a size granularity of 4 KB.

11.9 Exercises

11.9.1 Which bus is the AGP bus based on:

(a) PCI
(b) IDE
(c) ISA
(d) USB

11.9.2 How does AGP increase the data rate by x2 (and even x4):

(a) Extra clock signals
(b) Increased data bus size
(c) Direct memory accesses
(d) Increased address bus size

11.9.3 Which of the following is **not** an advantage of using the AGP bus:

(a) Faster transfers between memory and the graphics devices
(b) Increased usage of main memory (with reduced need for localized memory)
(c) Reduced requirement for interrupts
(d) Increase throughput compared with the standard PCI bus

11.9.4 Which of the following identifies the address/data lines on the AGP bus:

(a) HAD[31:0]
(b) GAD[31:0]
(c) AAD[31:0]
(d) AD[31:0]

11.9.5 Explain the main objectives of the AGP bus and outline the advantages of moving textural information into main memory.

11.9.6 Contrast the PCI and AGP busses and how AGP increases the data throughput. Also discuss the extra signal lines used with AGP, and how they are used.

11.10 Notes from the author

So, what's the biggest weakness of the PC. In the past, it has probably been the graphics facilities. This is mainly because the bus systems within the PC did not support large data throughput (ISA/EISA is way too slow). The design of the graphics system also required that the video card required to store all the data which was to be displayed on the screen. Thus no matter the amount of memory on the system, it was still limited by the amount of memory on the graphics card. AGP overcomes this by allowing graphical images to be stored in the main memory and then transferred to the video displayed over a fast bus.

The data demand for graphical displays is almost unlimited, as the faster they can be driven, the greater their application. The AGP bus is an excellent extension to the PCI bus, and gives data throughput of over 500 MB/s, whereas standard PCI devices can typically only be run at less than 100 MB/s. AGP is now a standard part of most PC motherboards, and it is still to be seen if many systems will start to use this port.

11.9 Exercises

11.10 Notes from the author

12 Fibre Channel

12.1 Introduction

The increase in demand for bandwidth requires faster server-to-storage and server-to-server networking. This is mainly due to the increase in client/server applications. Fibre Channel is one solution to this, as it is a highly reliable technology which operates at gigabit speeds. It interconnects well with other technologies, especially SCSI and TCP/IP. The main applications have been in switches, hubs, storage systems, storage devices and adapters. The term fibre is a generic term which can indicate either optical or a copper cable

Its development started in 1988 and ANSI standard approval in 1994. It has the following advantages:

- Cost-effective channel – it is a cost-effective for storage and networks.
- Reliable – it is reliable with assured information delivery.
- Gigabit bitrate – bit rate of 1.06 Gbps, but scalable to 2.12 Gbps and 4.24 Gbps.
- Multiple topologies – it has dedicated point-to-point, shared loops, and scaled switched topologies meet application requirements.
- Multiple protocols – it supports SCSI, TCP/IP, video, or raw data, and is especially suited to real-time video/audio.
- Scalable – it supports single point-to-point gigabit links to integrated enterprises with hundreds of servers.
- Congestion free – data can be sent as fast as the destination buffer can receive it.
- High efficiency – fibre channel has very little transmission overhead.

12.2 Comparison

Fibre channel is designed to support scalable gigabit technology, and provides flow control, self-management, and ultrareliability. It does not suffer from the problems associated with traditional networking technologies. Table 12.1 compares Fibre channel with gigabit ethernet and ATM.

In real-time applications, the data must have a guaranteed quality of service. gigabit ethernet does not provide for an assured quality of service, whereas ATM does. Fibre channel improves on ATM as it gives guaranteed delivery, along with gigabit bandwidth, as well as a given quality of service.

Table 12.1 Comparison between Fibre Channel and other networking technologies

Parameter	*Fibre channel*	*Gigabit ethernet*	*ATM*
Technology application	Storage, network, video, clusters	Network	Network, video
Topologies	Point-to-point loop hub, switched	Point-to-point hub, switched	Switched
Baud rate	1.06 Gbps	1.25 Gbps	622 Mbps
Scalability	2.12 Gbps, 4.24 Gbps	Not defined	1.24 Gbps
Guaranteed delivery	Yes	No	No
Congestion data loss	None	Yes	Yes
Frame size	Variable, 0–2 KB	Variable, 0–1.5 KB	Fixed, 53 B
Flow control	Credit based	Rate based	Rate based
Physical media	Copper/fibre	Copper/fibre	Copper/fibre
Protocols supported	Network, SCSI, video	Network	Network, video

12.3 Fibre channel standards

The ANSI T11 committee developed the X.3230-1994-Fibre channel physical and signaling standard (FC-PH). Its objectives where:

- Performance from 266 Mbps to over 4 Gbps.
- Support for distances up to 10 km.
- Small connectors.
- High-bandwidth utilisation with distance insensitivity.
- Greater connectivity than existing multidrop channels.
- Broad availability using standard components.
- Support for many different system types, from small computers to mainframes.
- Supports multiple existing interface command sets, such as internet protocol (IP), SCSI, IPI, HIPPI-FP and audio/video.

Fibre channel is a channel/network standard which contains networking features to provide for the required connectivity, distance and protocol multiplexing. It also supports traditional channel features for simplicity, repeatable performance, and guaranteed delivery. Fibre channel also works as a generic transport mechanism.

Fibre channel architecture is based on channel/network integration with an active, intelligent interconnection among devices. A port in Fibre channel simply has to manage a simple point-to-point connection. The transmission is isolated from the control protocol, so point-to-point links, arbitrated loops, and switched topologies are used to meet the specific needs of an application. The fabric is self-managing. Nodes do not need station management, which greatly simplifies implementation.

12.4 Cables, hubs, adapters and connectors

Fibre channel uses either fibre optic cables (either multimode or single mode) or four types of copper cables. Normally the copper cables use twin axial with DB-9 or HSSD connectors. Typically, low-cost copper cables are used for short and medium length runs, and fibre optic cable is used for longer lengths. Thus, most hubs and adapters have a standard copper interface. For fibre optic cable one of the following is used:

- **Multimode cable**. This type is used for short distances of up to 22 km. It has a 62.5 μm or 50 μm inner core diameter and allows light to propagate in multiple modes. These modes tend to disperse the signal and thus limits the distance of the cable. Typical bandwidth ratings for 62.5 μm cable are 200 MHz/km, which gives a range of 200 m at 1 Gbps.
- **Single mode cable**. This type is used for long cable runs. Its only limitation is the transmitter power and receiver sensitivity. The inner core is 7 μm or 9 μm, which only allows a single ray of light to propagate along the cable. There will thus be no dispersion of the signal.

The three main types of connectors used are:

- **SC connector**. The SC connector is the standard connector for most fibre optic cables and is also used for Fibre Channel. It is basically a push-pull connector and is preferred over the ST screw-on connector.
- **Galaxy connector**. This is a new type of connector and reduces the size of the connector by 50%, allowing increased connector densities.

There are also various connector/adapter modules, these include:

- **Gigabit interface converters (GBIC)**. These convert a copper cable connector to an optical interface. They use an HSSD connector for the copper interface and media interface converters use the DB-9 copper interface.
- **Gigabit link modules (GLM)**. These are pluggable modules which provide either a copper or fibre optic interface, and allow users to easily change the media interface from copper to fibre optics. GLMs include a serialiser-deserialiser (SERDES) and have a media independent parallel interface to the host bus adapter.
- **Extenders**. These provide for extended cable runs. They typically use multimode cable, and they convert the multimode interface to single-mode, as well as boosting the laser

power. For example, an extender can provide a single mode cable distance of 30 km.

- **Host bus adapters (HBAs)** These are similar to SCSI host bus adapters and network interface cards (NICs) They typically connect directly to the host computer using a standard bus, such as SBus, PCI, MCA, EISA, GIO, HIO, PMC and compact PCI.
- **SNA gateway** These provide gateways between from Fibre channel to SNA.
- **Switch WAN extender** These allow the interconnection to WANs using ATM or STM services.

Fibre channel systems can be integrated into virtually any network topology. It can use point-to-point dedicated connection, loop connection (with a shared bandwidth) or switched scaled bandwidth. Fibre channel devices typically either connect to one of the following:

- **Hubs**. Fibre channel hubs typically connect to a hub using copper cables. These hubs are similar to token ring/FDDI hubs with 'ring in' and 'ring out', and each port of the hub has an automatic port bypass circuit to automatically open and close the loop. Hubs thus support hot insertion and removal from the loop. If an attached node is not connected or powered on then the hub detects this and bypasses the node. Typically, a hub has 7 to 10 ports and can be stacked to the maximum loop size of 127 ports.
- **Switches** These allow the simultaneous communication or one or more connections at the same time.

Figure 12.1 shows an example network topology.

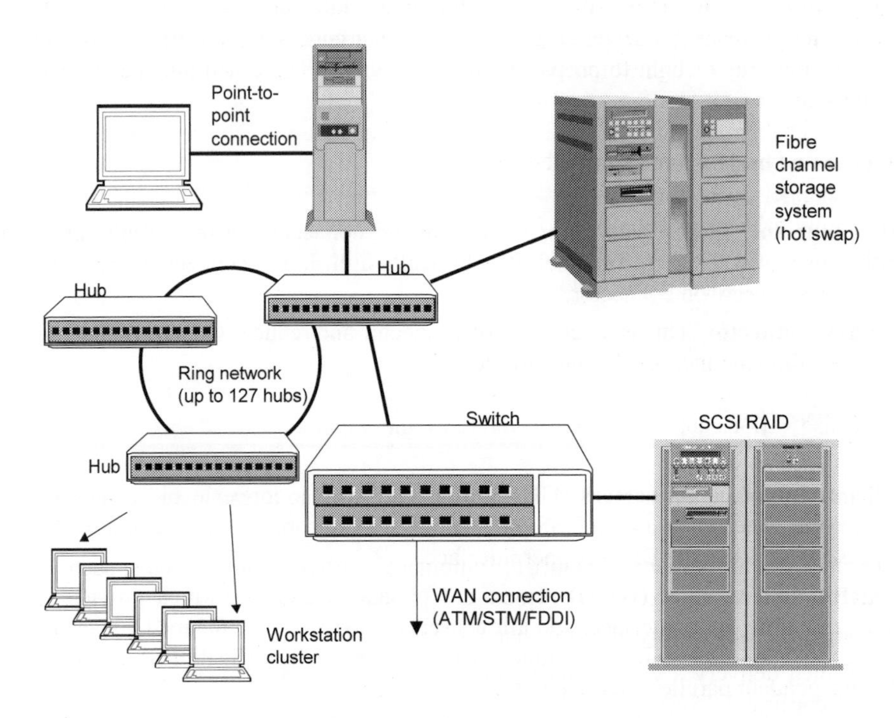

Figure 12.1 Fibre Channel connections

The networking options include:

- Point-to-point connections – these are point-to-point connections, and are not connected as a ring.
- Switch connections.
- Hub connections – these connect onto a ring, with the 'ring out' of each switch connected to the 'ring in' of the next, and so on. This makes a ring with devices connecting to the hub (and not to the ring).
- Fibre channel storage system – this contains hot swap disks with a 100 MB/s data transfer.
- SCSI RAID connection.
- Cluster connection – workstations plug directly to a hub. The hub detects if a workstation is connected and automatically connects it to the ring.
- WAN connection – connection to WAN, such as ATM, STM or FDDI.

12.5 Storage Devices and storage area networks

One of the largest uses of fibre channel is likely to be in storage interfaces. It has many advantages over SCSI, including distance, bandwidth, scalability and reliability. Many manufacturers now provide RAID-based systems with Fibre channel.

A good area of application for Fibre channel is in storage area networks. These typically contain one or more servers, which also connect to one or more storage systems. These storage systems could be RAID, tape back-up, CD-ROMs or disk drives.

In a fibre channel network, SCSI devices are interfaced to the Fibre channel using a Fibre channel to SCSI bridge, and IP is used for server to server and client/server communications.

Fibre channel disks have a back plane with a built-in fibre channel loop. At each disk location in the back-plane loop is a port bypass circuit which permits hot swapping of disks. If a disk is not present, the circuit automatically closes the loop. When a disk is inserted, the loop is opened to connect to the disk.

12.6 Networks

Fibre channel has many advantages over traditional networking technologies, these include:

- Automatic configuration – support for automatic configuration protocols, such as ARP, RARP and other self-discovery protocols.
- Automatic self-discovery of Fibre channel topology.
- Confirmed delivery – this enhances reliability, and does not rely on higher-level protocols to confirm delivery.
- Efficient, high-bandwidth low-latency transfers – this uses a variable length frame (up to 2 kB). It uses an efficient protocol which only has an overhead of up to 100 bytes.

- Fast connection time – instant circuit setup time using hardware enhanced Fibre channel protocol.
- Guaranteed Quality of Service (QoS) True connection service or fractional bandwidth, connection-oriented virtual circuits to guarantee QoS for critical back-ups or other operations.
- Hybrid topology – Supports many different topologies, such as dedicated point-to-point circuits, shared bandwidth ring networks or scalable bandwidth switched circuits.
- Low latency – extremely low latency connection and connectionless service.
- Real or virtual circuits.
- Synchronous support – this is used with video connections, and uses fractional bandwidth virtual circuits.

12.7 Exercises

12.7.1 Which of the following offers the highest potential data rate:

(a) Single-mode fibre optic (b) Multimode fibre optic
(c) Cat-3, shielded twisted pair (d) Cat-5, shielded twisted pair

12.7.2 What is the topology of a large Fibre Channel network:

(a) Back plane (b) Bus network
(c) Star network (d) Ring network

12.7.3 What devices do workstations normally connect to with a Fibre channel network:

(a) Router (b) Hub
(c) Bridge (d) Repeater

12.7.4 Which of the following is not an advantage of Fibre channel:

(a) High data rate
(b) Hot swappable local devices
(c) Can be used with many networking protocols
(d) Increased amount of ring connections

12.8 Notes from the author

Well, ask any Managing Director of a large commercial organisation about what their key resources is, and, at least, if they are honest, 9 out of 10 of them will say their IT infrastructure. Thus, if a company were to loose electronic mail or their Intranet connections, it would be very costly in lack of efficiency. What is required, then, is a fast and robust network backbone. This is what fibre channel does best. It's not cheap, but it's as good as it gets.

13 RS-232

13.1 Introduction

RS-232 is one of the most widely used techniques used to interface external equipment to computers. It uses serial communications where one bit is sent along a line, at a time. This differs from parallel communications which send one or more bytes, at a time. The main advantage that serial communications has over parallel communications is that a single wire is needed to transmit and another to receive. RS-232 is a de facto standard that most computer and instrumentation companies comply with. It was standardised in 1962 by the Electronics Industries Association (EIA). Unfortunately this standard only allows short cable runs with low bit rates. The standard RS-232 only allows a bit rate of 19600 bps for a maximum distance of 20 m. New serial communications standards, such as RS-422 and RS-449, allow very long cable runs and high bit rates. For example, RS-422 allows a bit rate of up to 10 Mbps over distances up to 1 mile, using twisted-pair, coaxial cable or optical fibres. The new standards can also be used to create computer networks. This chapter introduces the RS-232 standard and gives simple programs which can be used to transmit and receive using RS-232.

13.2 Electrical characteristics

13.2.1 Line voltages

The electrical characteristics of RS-232 defines the minimum and maximum voltages of a logic '1' and '0'. A logic '1' ranges from –3 V to –25 V, but will typically be around –12 V. A logical '0' ranges from 3 V to 25 V, but will typically be around +12 V. Any voltage between –3 V and +3 V has an indeterminate logical state. If no pulses are present on the line the voltage level is equivalent to a high level, that is –12 V. A voltage level of 0 V at the receiver is interpreted as a line break or a short circuit. Figure 13.1 shows an example transmission.

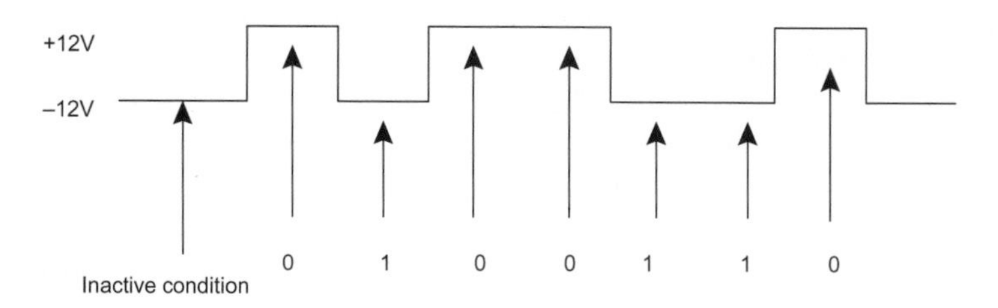

Figure 13.1 RS-232 voltage levels

13.2.2 DB25S connector

The DB25S connector is a 25-pin D-type connector and gives full RS-232 functionality. Figure 13.2 shows the pin number assignment. A DCE (the terminating cable) connector has a male outer casing with female connection pins. The DTE (the computer) has a female outer casing with male connecting pins. There are three main signal types: control, data and ground. Table 13.1 lists the main connections. Control lines are active HIGH, that is they are HIGH when the signal is active and LOW when inactive.

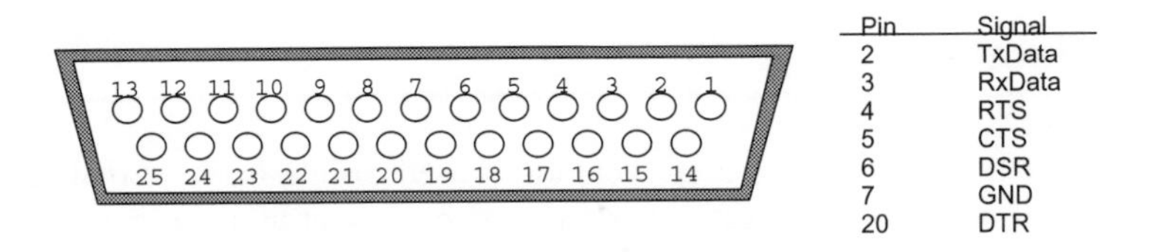

Figure 13.2 RS-232 DB25S connector

13.2.3 DB9S Connector

The 25-pin connector is the standard for RS-232 connections but as electronic equipment becomes smaller, there is a need for smaller connectors. For this purpose most PCs now use a reduced function 9-pin D-type connector rather than the full function 25-way D-type. As with the 25-pin connector the DCE (the terminating cable) connector has a male outer casing with female connection pins. The DTE (the computer) has a female outer casing with male connecting pins. Figure 13.3 shows the main connections.

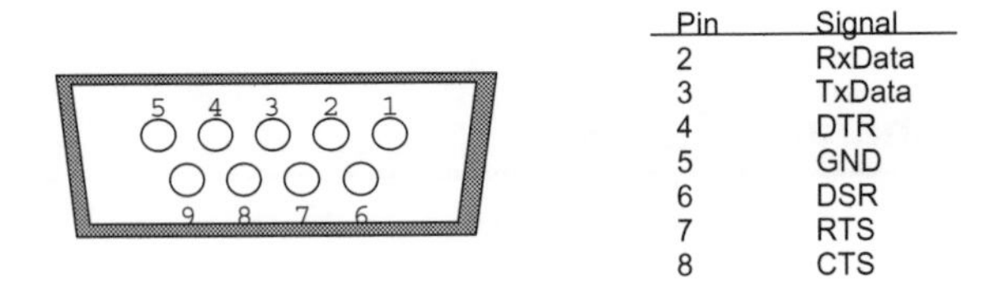

Figure 13.3 RS-232 DB9S interface

13.2.4 PC connectors

All PCs have at least one serial communications port. The primary port is named `COM1:` and the secondary is `COM2:`. There are two types of connectors used in RS-232 communications, these are the 25- and 9-way D-type. Most modern PCs use either a 9-pin connector for the primary (`COM1:`) serial port and a 25-pin for a secondary serial port (`COM2:`), or they use two 9-pin connectors for serial ports. The serial port can be differentiated from the parallel port in that the 25-pin parallel port (`LPT1:`) is a 25-pin female connector on the PC and a male connector on the cable. The 25-pin serial connector is a male on the PC and a female on the cable. The different connector types can cause problems in connecting devices. Thus a 25-to-9 pin adapter is a useful attachment, especially to connect a serial mouse to a 25-pin connector.

Table 13.1 Main pin connections used in 25-pin connector

Pin	*Name*	*Abbreviation*	*Functionality*
1	Frame ground	FG	This ground normally connects the outer sheath of the cable and to earth ground.
2	Transmit data	TD	Data is sent from the DTE (computer or terminal) to a DCE via TD.
3	Receive data	RD	Data is sent from the DCE to a DTE (computer or terminal) via RD.
4	Request to send	RTS	DTE sets this active when it is ready to transmit data.
5	Clear to send	CTS	DCE sets this active to inform the DTE that it is ready to receive data.
6	Data set ready	DSR	Similar functionality to CTS but activated by the DTE when it is ready to receive data.
7	Signal ground	SG	All signals are referenced to the signal ground (GND).
20	Data terminal ready	DTR	Similar functionality to RTS but activated by the DCE when it wishes to transmit data.

13.2.5 Frame format

RS-232 uses asynchronous communication which has a start/stop data format (Figure 13.4). Each character is transmitted one at a time with a delay between them. This delay is called the inactive time and is set at a logic level high (–12 V) as shown in Figure 13.5. The transmitter sends a start bit to inform the receiver that a character is to be sent in the following bit transmission. This start bit is always a '0'. Next, 5, 6 or 7 data bits are sent as a 7-bit ASCII character, followed by a parity bit and finally either 1, 1.5 or 2 stop bits. Figure 13.5 shows a frame format and an example transmission of the character 'A', using odd parity. The timing of a single bit sets the rate of transmission. Both the transmitter and receiver need to be set to the same bit-time interval. An internal clock on both sets this interval. These only have to be roughly synchronised at approximately the same rate as data is transmitted in relatively short bursts.

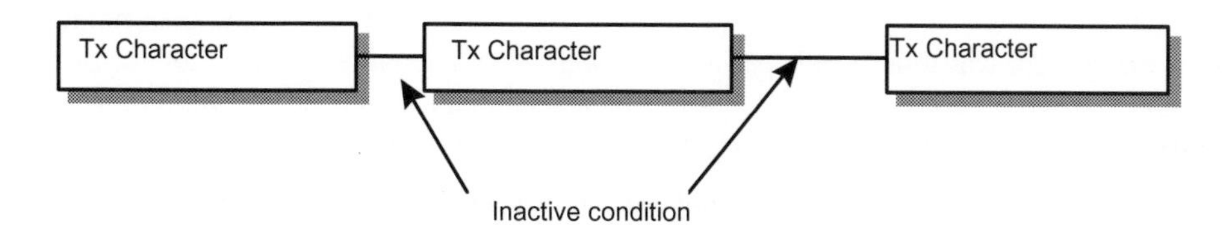

Figure 13.4 Asynchronous communications

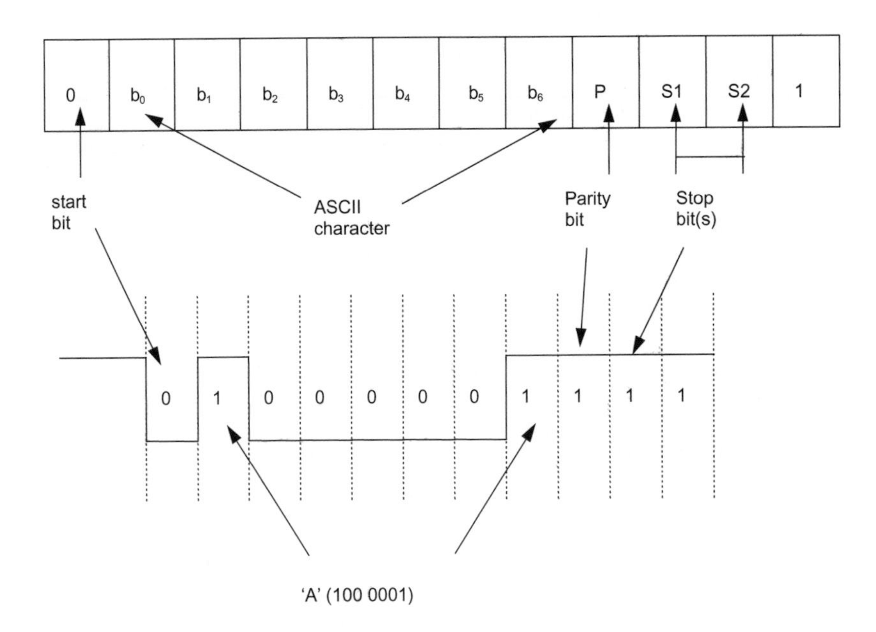

Figure 13.5 RS-232 frame format

Example

An RS-232 serial data link uses 1 start bit, 7 data bits, 1 parity bit, 2 stop bits, ASCII coding and even parity. Determine the message sent from the following bit stream.

First bit sent
⇓
`1111101000001011000001111111111111100000111111100011001111010`
`10 0111111111111`

Answer

The format of the data string sent is given next:

{idle} `11111` {start bit} `0` {'A'} `1000001` {parity bit} `0` {stop bits } `11` {start bit} `0` {'p'} `0000111` {parity bit} `1` {stop bits} `11` {idle} `11111111` {start bit} `0` {'p'} `0000111` {parity bit} `1` {stop bits} `11` {idle} `11` {start bit} `0` {'L'} `0011001` {parity bit} `1` {stop bits} `11`

The message sent was thus 'AppL'.

Parity

Error control is data added to transmitted data in order to detect or correct an error in transmission. RS-232 uses a simple technique known as parity to provide a degree of error detection.

A parity bit is added to transmitted data to make the number of 1s sent either even (even parity) or odd (odd parity). It is a simple method of error coding and only requires exclusive-OR (XOR) gates to generate the parity bit. The parity bit is added to the transmitted data by inserting it into the shift register at the correct bit position.

A single parity bit can only detect an odd number of errors, that is, 1, 3, 5, and so on. If there is an even number of bits in error then the parity bit will be correct and no error will be detected. This type of error coding is not normally used on its own where there is the possibility of several bits being in error.

Baud rate

One of the main parameters, which specify RS-232 communications, is the rate of transmission at which data is transmitted and received. It is important that the transmitter and receiver operate at, roughly, the same speed.

For asynchronous transmission the start and stop bits are added in addition to the 7 ASCII character bits and the parity. Thus a total of 10 bits are required to transmit a single character. With 2 stop bits, a total of 11 bits are required. If 10 characters are sent every second and if 11 bits are used for each character, then the transmission rate is 110 bits per second (bps). Table 13.2 lists how the bit rate relates to the characters sent per second (assuming 10 transmitted bits per character). The bit rate is measured in bits per second (bps).

	Bits
ASCII character	7
Start bit	1
Stop bit	2
Total	10

Table 13.2 Bits per second related to characters sent per second

Speed (bps)	*Characters per second*
300	30
1200	120
2400	240

In addition to the bit rate, another term used to describe the transmission speed is the baud rate. The bit rate refers to the actual rate at which bits are transmitted, whereas the baud rate relates to the rate at which signalling elements, used to represent bits, are transmitted. As one signalling element encodes one bit, the two rates are then identical. Only in modems does the bit rate differ from the baud rate.

Bit stream timings

Asynchronous communications is a stop/start mode of communication and both the transmitter and receiver must be set up with the same bit timings. A start bit identifies the start of transmission and is always a low logic level. Next, the least significant bit is sent followed by the rest of the bits in the character. After this, the parity bit is sent followed by the stop bit(s). The actual timing of each bit relates to the baud rate and can be found using the following formula:

$$\text{Time period of each bit} = \frac{1}{\text{baud rate}}\ \text{second}$$

For example, if the baud rate is 9600 baud (or bps) then the time period for each bit sent is 1/9600 s or 104 µs. Table 13.3 shows some bit timings as related to baud rate. An example of the voltage levels and timings for the ASCII character 'V' is given in Figure 13.6.

Table 13.3 Bit timings related to baud rate

Baud rate	Time for each bit (µs)
1200	833
2400	417
9600	104
19200	52

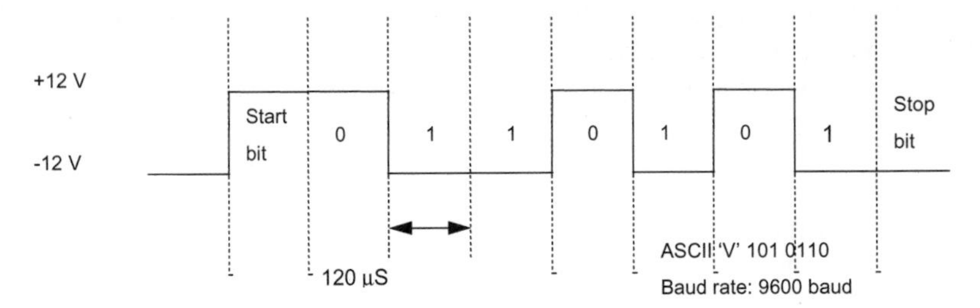

Figure 13.6 ASCII 'V' at RS-232 voltage levels

13.3 Communications between two nodes

RS-232 is intended to be a standard but not all manufacturers abide by it. Some implement the full specification while others implement just a partial specification. This is mainly because not every device requires the full functionality of RS-232, for example a modem requires many more control lines than a serial mouse.

The rate at which data is transmitted and the speed at which the transmitter and receiver can transmit/receive the data dictates whether data handshaking is required.

13.3.1 Handshaking

In the transmission of data, there can either be no handshaking, hardware handshaking or software handshaking. If no handshaking is used then the receiver must be able to read the received characters before the transmitter sends another. The receiver may buffer the received character and store it in a special memory location before it is read. This memory location is named the receiver buffer. Typically, it may only hold a single character. If it is not emptied before another character is received then any character previously in the buffer will be overwritten. An example of this is illustrated in Figure 13.6. In this case, the receiver has read the first two characters successfully from the receiver buffer, but it did not read the third character as the fourth transmitted character has overwritten it in the receiver buffer. If this condition occurs then some form of handshaking must be used to stop the transmitter sending characters before the receiver has had time to service the received characters.

Hardware handshaking involves the transmitter asking the receiver if it is ready to receive data. If the receiver buffer is empty it will inform the transmitter that it is ready to receive data. Once the data is transmitted and loaded into the receiver buffer the transmitter is informed not to transmit any more characters until the character in the receiver buffer has been read. The main hardware handshaking lines used for this purpose are:

- `CTS` – Clear to send.
- `RTS` – Ready to send.
- `DTR` – Data terminal ready.
- `DSR` – Data set ready.

Software handshaking involves sending special control characters. These include the `DC1` (Xon)-`DC4` (Xoff) control characters.

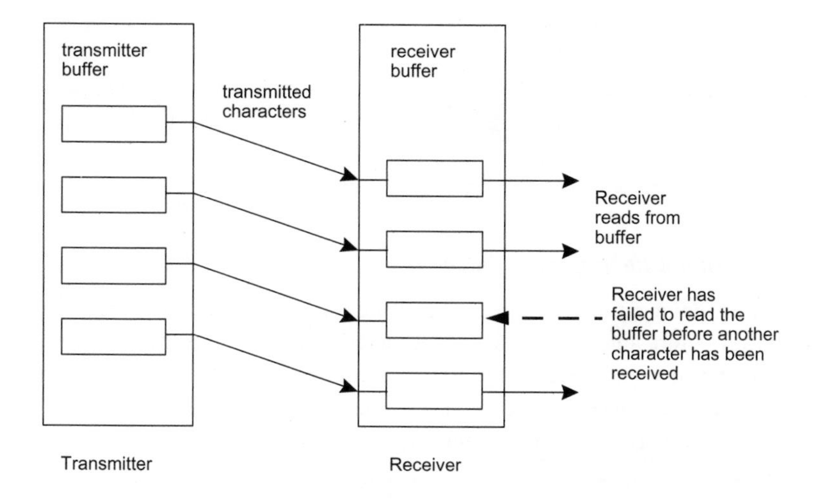

Figure 13.7 Transmission and reception of characters

13.3.2 RS-232 set-up

Microsoft Windows allows the serial port setting to be set by selecting control panel → system → device manager → ports (COM and LPT) → port settings. The settings of the communications port (the IRQ and the port address) can be changed by selecting control panel → system → device manager → ports (COM and LPT) → resources for IRQ and addresses. Figure 13.8 shows example parameters and settings. The selectable baud rates are typically 110, 300, 600, 1200, 2400, 4800, 9600 and 19 200 baud for an 8250-based device. A 16650 UART also gives enhanced speeds of 38 400, 57 600, 115 200, 230 400, 460 800 and 921 600 baud. Notice that the flow control can either be set to software handshaking (Xon/Xoff), hardware handshaking or none. The parity bit can either be set to none, odd, even, mark or space. A mark in the parity option sets the parity bit to a '1' and a space sets it to a '0'.

In this case COM1: is set at 9600 baud, 8 data bits, no parity, 1 stop bit and no parity checking.

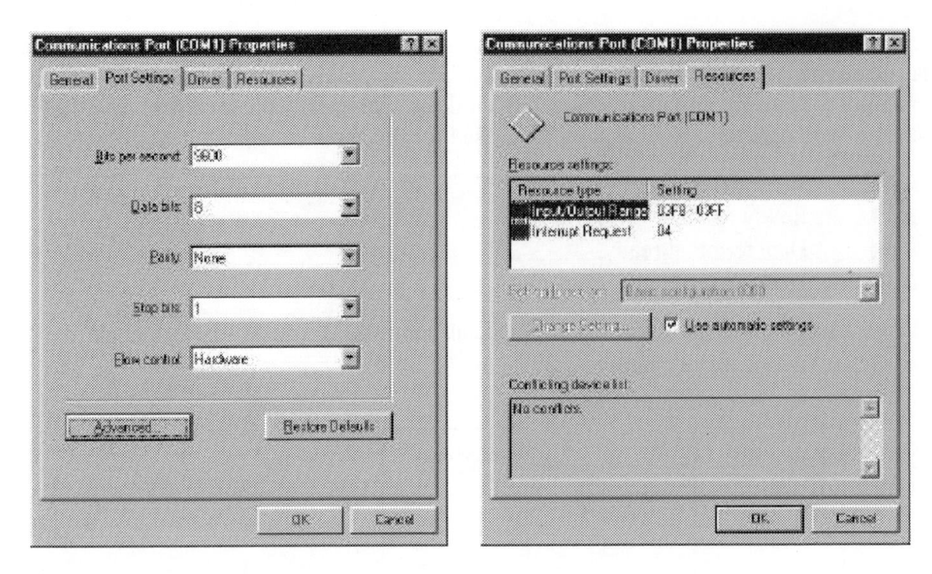

Figure 13.8 Changing port setting and parameters

13.3.3 Simple no-handshaking communications

In this form of communication it is assumed that the receiver can read the received data from the receive buffer before another character is received. Data is sent from a TD pin connection of the transmitter and is received in the RD pin connection at the receiver. When a DTE (such as a computer) connects to another DTE, then the transmit line (TD) on one is connected to the receive (RD) of the other and vice versa. Figure 13.9 shows the connections between the nodes.

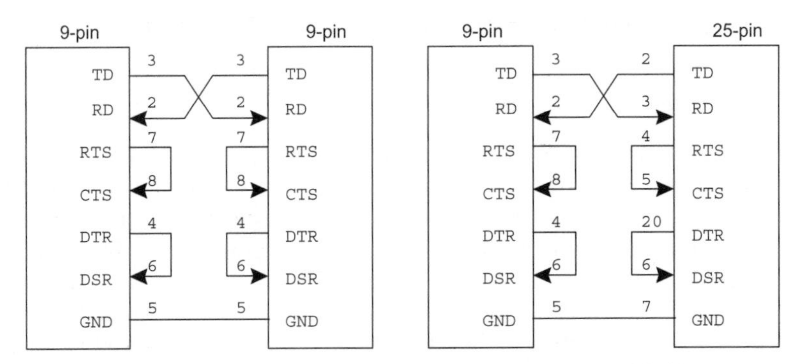

Figure 13.9 RS-232 connections with no hardware handshaking

13.3.4 Software handshaking

There are two ASCII characters that start and stop communications. These are X-ON (^S , Cntrl-S or ASCII 11) and X-OFF (^Q, Cntrl-Q or ASCII 13). When the transmitter receives an X-OFF character it ceases communications until an X-ON character is sent. This type of handshaking is normally used when the transmitter and receiver can process data relatively quickly. Normally, the receiver will also have a large buffer for the incoming characters. When this buffer is full, it transmits an X-OFF. After it has read from the buffer the X-ON is transmitted, see Figure 13.10.

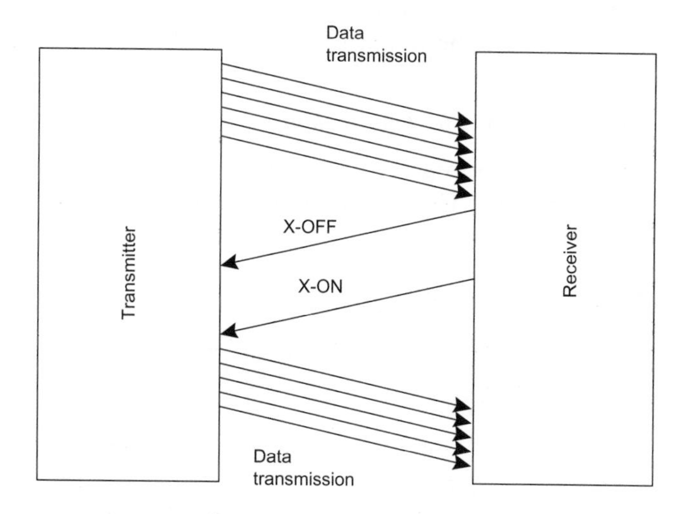

Figure 13.10 Software handshaking using X-ON and X-OFF

13.3.5 Hardware handshaking

Hardware handshaking stops characters in the receiver buffer from being overwritten. The control lines used are all active HIGH. Figure 13.11 shows how the nodes communicate. When a node wishes to transmit data it asserts the `RTS` line active (that is, HIGH). It then monitors the `CTS` line until it goes active (that is, HIGH). If the `CTS` line at the transmitter stays inactive then the receiver is busy and cannot receive data, at the present. When the receiver reads from its buffer the `RTS` line will automatically go active indicating to the transmitter that it is now ready to receive a character.

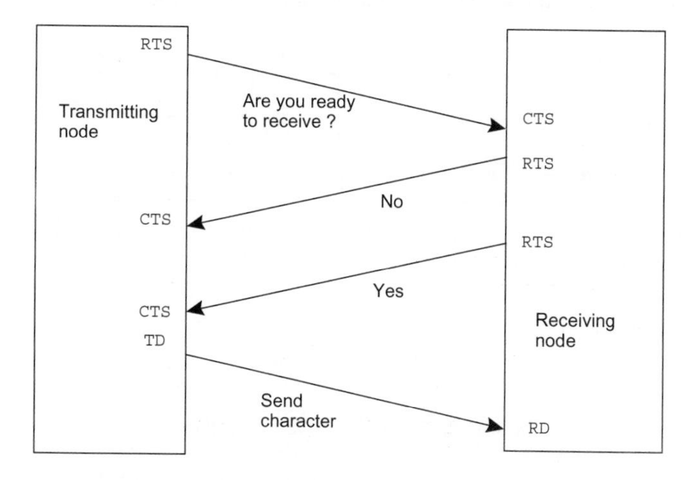

Figure 13.11 Handshaking lines used in transmitting data

Receiving data is similar to the transmission of data, but the lines DSR and DTR are used instead of RTS and CTS. When the DCE wishes to transmit to the DTE the DSR input to the receiver will become active. If the receiver cannot receive the character, it will set the DTR line inactive. When it is clear to receive it sets the DTR line active and the remote node then transmits the character. The DTR line will be set inactive until the character has been processed.

13.3.6 Two-way communications with handshaking

For full handshaking of the data between two nodes the RTS and CTS lines are crossed over (as are the DTR and DSR lines). This allows for full remote node feedback (see Figure 13.12).

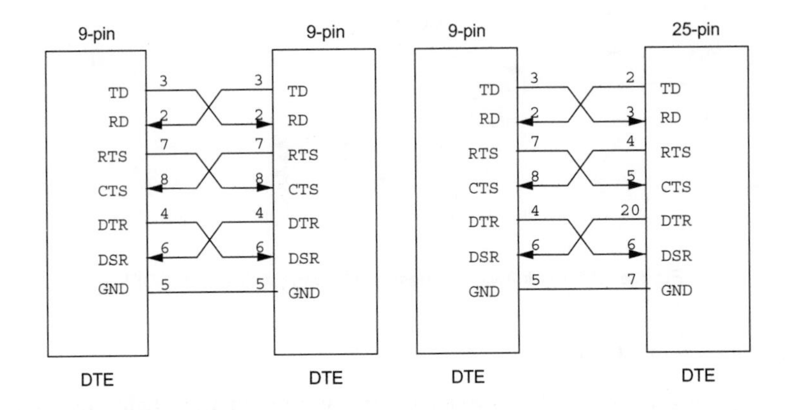

Figure 13.12 RS-232 communications with handshaking

13.3.7 DTE-DCE connections (PC to modem)

A further problem occurs in connecting two nodes. A DTE/DTE connection requires cross-overs on their signal lines, whereas DTE/DCE connections require straight-through lines. An example computer to modem connection is shown in Figure 13.13.

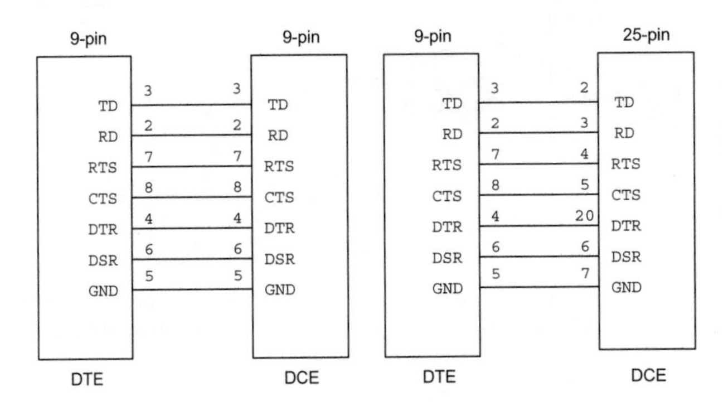

Figure 13.13 DTE to DCE connections

13.4 Programming RS-232

Normally, serial transmission is achieved via the RS-232 standard. Although 25 lines are defined usually only a few are used. Data is sent along the `TD` line and received by the `RD` line with a common ground return. The other lines, used for handshaking, are `RTS` (ready to send) which is an output signal to indicate that data is ready to be transmitted and `CTS` (clear to send), which is an input indicating that the remote equipment is ready to receive data.

The 8250/NS16650 IC is commonly used in serial communications. It is typically integrated in the PC chip set, or can be mounted on an I/O card. This section discusses how it is programmed.

Programming the serial device

The main registers used in RS-232 communications are the line control register (LCR), the line status register (LSR) and the transmit and receive buffers (see Figure 13.14). The transmit and receive buffers share the same addresses.

The base address of the primary port (`COM1:`) is normally set at 3F8h and the secondary port (`COM2:`) at 2F8h. A standard PC can support up to four COM ports. These addresses are set in the BIOS memory and the address of each of the ports is stored at address locations `0040:0000` (`COM1:`), `0040:0002` (`COM2:`), `0040:0004` (`COM3:`) and `0040:0008` (`COM4:`). Program 13.1 can be used to identify these addresses. The statement:

```
ptr=(int far *)0x0400000;
```

initializes a far pointer to the start of the BIOS communications port addresses. Each address is 16 bits thus the pointer points to an integer value. A far pointer is used as this can access the full 1 MB of memory, a near pointer can only access a maximum of 64 kB.

Program 13.1

```
#include <stdio.h>
#include <conio.h>
int   main(void)
{
int   far *ptr; /* 20-bit pointer */
      ptr=(int far *)0x0400000; /* 0040:0000 */
      clrscr();
      printf("COM1: %04x\n",*ptr);
      printf("COM2: %04x\n",*(ptr+1));
      printf("COM3: %04x\n",*(ptr+2));
      printf("COM4: %04x\n",*(ptr+3));
      return(0);
}
```

Test run 3.1 shows a sample run. In this case, there are four COM ports installed on the PC. If any of the addresses is zero then that COM port is not installed on the system.

Test run 3.1

```
COM1: 03f8
COM2: 02f8
COM3: 03e8
COM4: 02e8
```

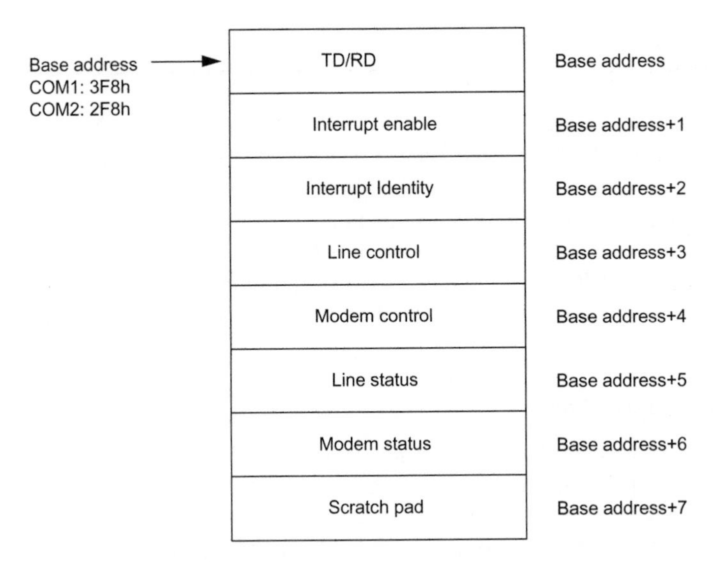

Figure 13.14 Serial communication registers

Line Status Register (LSR)

The LSR determines the status of the transmitter and receiver buffers. It can only be read from, and all the bits are automatically set by hardware. The bit definitions are given in Figure 13.15. When an error occurs in the transmission of a character one (or several) of the error bit is (are) set to a '1'.

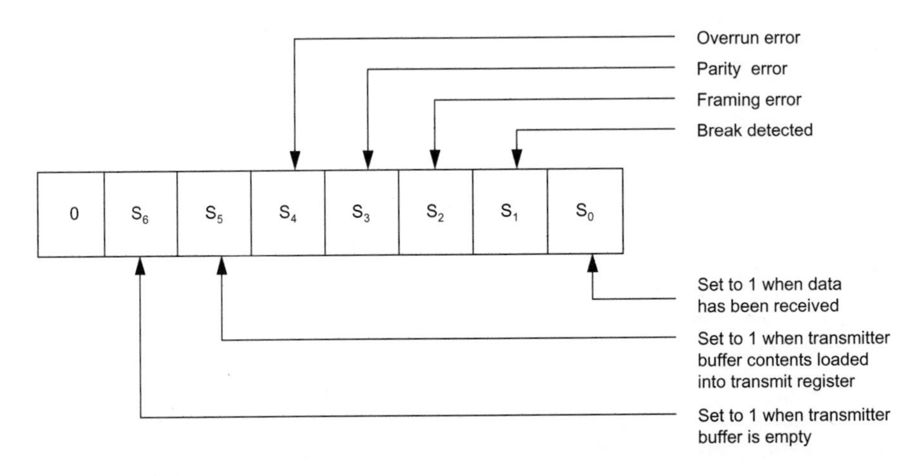

Figure 13.15 Line status register

One danger when transmitting data is that a new character can be written to the transmitter buffer before the previous character has been sent. This overwrites the contents of the character being transmitted. To avoid this the status bit S_6 is tested to determine if there is still a

character in the buffer. If there is then it is set to a '1', else the transmitter buffer is empty.
To send a character

Test bit 6 until set;
Send character;

A typical Pascal routine is

```
repeat
   status := port[LSR] and $40;
until (status=$40);
```

When receiving data the S_0 bit is tested to determine if there is a bit in the receiver buffer. To receive a character

Test bit 0 until set;
Read character;

A typical Pascal routine is

```
repeat
      status := port[LSR] and $01;
until (status=$01);
```

Figure 13.16 shows how the LSR is tested for the transmission and reception of characters.

Line control register (LCR)

The LCR sets up the communications parameters. These include the number of bits per character, the parity and the number of stop bits. It can be written to or read from and has a similar function to that of the control registers used in the PPI (programmable parallel interface) and PTC (programmable timer/counter). The bit definitions are given in Figure 13.17.

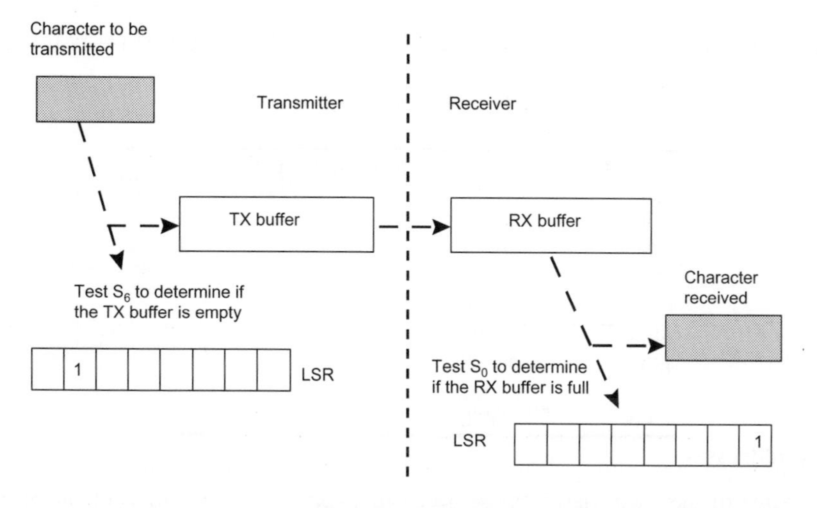

Figure 13.16 Testing of the LSR for the transmission and reception of characters

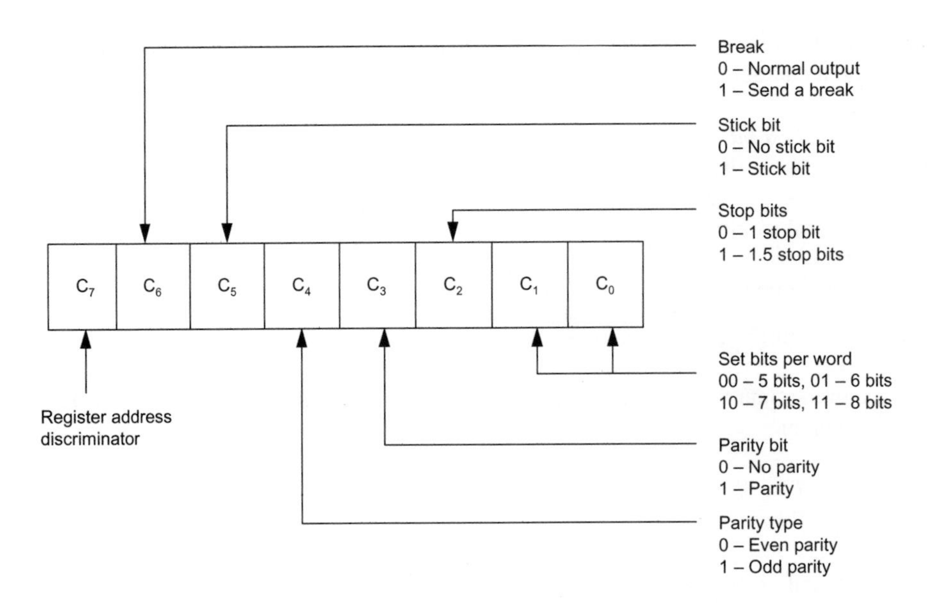

Figure 13.17 Line control register

The MSB, C_7, must to be set to a '0' in order to access the transmitter and receiver buffers, else if it is set to a '1' the baud rate divider is set up. The baud rate is set by loading an appropriate 16-bit divisor into the addresses of transmitter/receiver buffer address and the next address. The value loaded depends on the crystal frequency connected to the IC. Table 13.4 shows divisors for a crystal frequency of 1.8432 MHz. In general, the divisor, N, is related to the baud rate by:

$$\text{Baud rate} = \frac{\text{Clock frequency}}{16 \times N}$$

For example, for 1.8432 MHz and 9600 baud $N = 1.8432 \times 10^6/(9600 \times 16) = 12$ (000Ch).

Table 13.4 Baud rate divisors

Baud rate	*Divisor (value loaded into Tx/Rx buffer)*
110	0417h
300	0180h
600	00C0h
1200	0060h
1800	0040h
2400	0030h
4800	0018h
9600	000Ch
19200	0006h

Register addresses

The addresses of the main registers are given in Table 13.5. To load the baud rate divisor, first the LCR bit 7 is set to a '1', then the LSB is loaded into divisor LSB and the MSB into the divisor MSB register. Finally, bit 7 is set back to a '0'. For example, for 9600 baud, COM1 and 1.8432 MHz clock then 0Ch is loaded in 3F8h and 00h into 3F9h.

When bit 7 is set at a '0' then a read from the base address reads from the RD buffer and a write operation writes to the TD buffer. An example of this is shown in Figure 13.9.

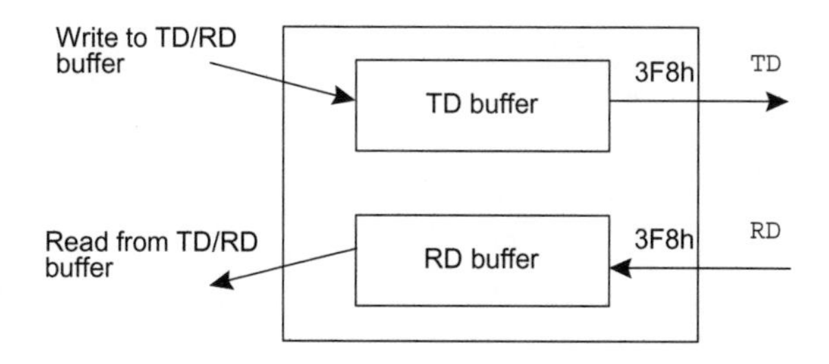

Figure 13.18 Read and write from TD/RD buffer

Table 13.5 Serial communications addresses

Primary	Secondary	Register	Bit 7 of LCR
3F8h	2F8h	TD buffer	'0'
3F8h	2F8h	RD buffer	'0'
3F8h	2F8h	Divisor LSB	'1'
3F9h	2F9h	Divisor MSB	'1'
3FBh	2FBh	Line Control Register	
3FDh	2FDh	Line Status Register	

13.5 RS-232 programs

Figure 13.19 shows the main RS-232 connections for 9 and 25-pin connections without hardware handshaking. The loopback connections are used to test the RS-232 hardware and the software, while the null modem connections are used to transmit characters between two computers. Program 13.2 uses a loop back on the TD/RD lines so that a character sent by the computer will automatically be received into the receiver buffer. This set-up is useful in testing the transmit and receive routines. The character to be sent is entered via the keyboard. A CNTRL-D (^D) keystroke exits the program.

Program 13.3 can be used as a sender program (send.c) and Program 13.4 can be used as a receiver program (receive.c). With these programs, the null modem connections shown in Figure 13.19 are used.

Note that programs 13.2 to 13.4 are written for Microsoft Visual C++. For early versions of Borland C/C++ program change _inp for inportb and _outp for outportb.

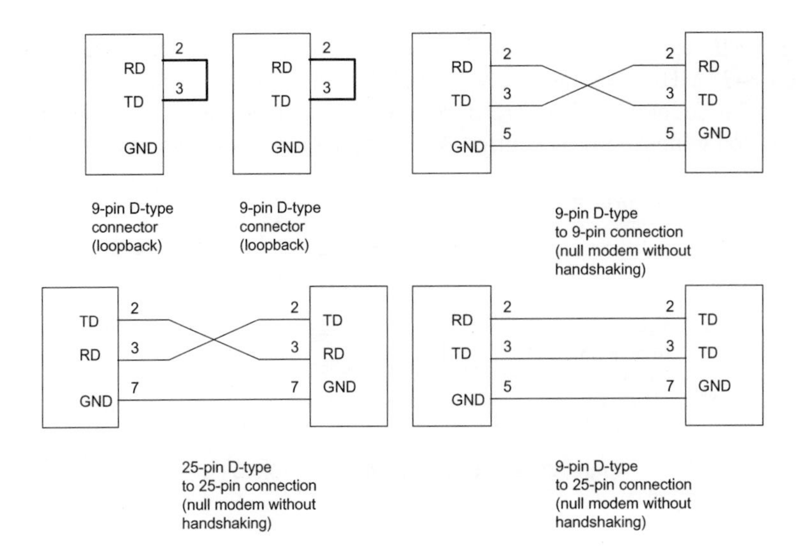

Figure 13.19 System connections

Program 13.2

```
/*    This program transmits a character from COM1: and receives        */
/*    it via this port. The TD is connected to RD.              */

#define  COM1BASE    0x3F8
#define  COM2BASE    0x2F8
#define  TXDATA      COM1BASE
#define  LCR         (COM1BASE+3) /*   0x3FB line control               */
#define  LSR         (COM1BASE+5) /*   0x3FD line status                */

#include <conio.h>   /* required for getch()                            */
#include <dos.h>     /*     */
#include <stdio.h>

/* Some ANSI C prototype definitions  */
void  setup_serial(void);
void  send_character(int ch);
int   get_character(void);

int      main(void)
{
int      inchar,outchar;

      setup_serial();
      do
      {
         puts("Enter char to be transmitted (Cntrl-D to end)");
         outchar=getch();
         send_character(outchar);
         inchar=get_character();
         printf("Character received was %c\n",inchar);
      } while (outchar!=4);
      return(0);
}

void     setup_serial(void)
{
```

```
      _outp( LCR, 0x80);
      /* set up bit 7 to a 1  to set Register address bit   */

      _outp(TXDATA,0x0C);
      _outp(TXDATA+1,0x00);
      /* load TxRegister with 12, crystal frequency is 1.8432MHz       */

      _outp(LCR, 0x0A);
      /* Bit pattern loaded is 00001010b, from msb to lsb these are:   */
      /* 0 - access TD/RD buffer ,  0 - normal output                  */
      /* 0 - no stick bit  , 0 - even parity                           */
      /* 1 - parity on,  0 - 1 stop bit                                */
      /* 10 - 7 data bits                                              */
}

void  send_character(int ch)
{
char  status;
      do
      {
         status = _inp(LSR) & 0x40;
      } while (status!=0x40);
      /*repeat until Tx buffer empty ie bit 6 set*/

      _outp(TXDATA,(char) ch);
}

int   get_character(void)
{
int   status;
      do
      {
         status = _inp(LSR) & 0x01;
      } while (status!=0x01);
      /* Repeat until bit 1 in LSR is set */

      return( (int)_inp(TXDATA));
}
```

Program 13.3

```
/*       send.c                                          */
#define  TXDATA   0x3F8
#define  LSR      0x3FD
#define  LCR      0x3FB

#include    <stdio.h>
#include    <conio.h>   /* included for getch               */
#include    <dos.h>

void     setup_serial(void);
void     send_character(int ch);

int      main(void)
{
int      ch;
      puts("Transmitter program. Please enter text (Cntl-D to end)");
      setup_serial();
      do
      {
         ch=getche();
         send_character(ch);
```

```
      } while (ch!=4);
      return(0);
}

void  setup_serial(void)
{
      _outp( LCR, 0x80);
      /* set up bit 7 to a 1  to set Register address bit       */
      _outp(TXDATA,0x0C);
      _outp(TXDATA+1,0x00);
      /* load TxRegister with 12, crystal frequency is 1.8432MHz          */
      _outp(LCR, 0x0A);
      /* Bit pattern loaded is 00001010b, from msb to lsb these are:    */
      /* Access TD/RD buffer, normal output, no stick bit      */
      /* even parity, parity on, 1 stop bit, 7 data bits       */
}
void  send_character(int ch)
{
char  status;
      do
      {
         status = _inp(LSR) & 0x40;
      } while (status!=0x40);
      /*repeat until Tx buffer empty ie bit 6 set*/
      _outp(TXDATA,(char) ch);
}
```

Program 13.4

```
/*       receive.c                                                    */
#define  TXDATA   0x3F8
#define  LSR      0x3FD
#define  LCR      0x3FB
#include    <stdio.h>
#include    <conio.h>   /* included for getch                     */
#include    <dos.h>

void     setup_serial(void);
int      get_character(void);
int      main(void)
{
int      inchar;
      setup_serial();
      do
      {
         inchar=get_character();
         putchar(inchar);
      } while (inchar!=4);
      return(0);
}
void setup_serial(void)
{
      _outp( LCR, 0x80);
      /* set up bit 7 to a 1  to set Register address bit              */
      _outp(TXDATA,0x0C);
      _outp(TXDATA+1,0x00);
      /* load TxRegister with 12, crystal frequency is 1.8432MHz        */
      _outp(LCR, 0x0A);
      /* Bit pattern loaded is 00001010b, from msb to lsb these are:  */
      /* Access TD/RD buffer, normal output, no stick bit              */
      /* even parity, parity on, 1 stop bit, 7 data bits       */
}
```

```
int    get_character(void)
{
int    status;
       do
       {
          status = _inp(LSR) & 0x01;
       } while (status!=0x01);    /* Repeat until bit 1 in LSR is set */
       return( (int)_inp(TXDATA));
}
```

13.6 Exercises

13.6.1 Which is the maximum cable length for a standard RS-232 connection:

(a) 2 m (b) 20m
(c) 200 m (d) 2 km

13.6.2 Which enhancement to RS-232 allows for 1Mbps bit rates and increased cable lengths:

(a) RS-232x (b) RS-422
(c) RS-444 (d) RS-233

13.6.3 Which of the following is not a standard RS-232 bit rate:

(a) 110 bps (b) 4800 bps
(c) 9600 bps (d) 12 200 bps

13.6.4 Which voltage range is used for a '0' bit value:

(a) –3 V to –25 V (b) 0 V to –3 V
(c) +3 V to +25 V (d) 0 V to +3 V

13.6.5 In RS-232, how is the inactive period identified:

(a) A high voltage level (b) A low voltage level
(c) Zero voltage level (d) Open circuit

13.6.6 How is a null modem cable identified:

(a) Direct connection between all the signal lines
(b) No connections to TD and RD lines
(c) Cross-over between the TD and RD, and handshaking lines
(d) No hardware handshaking lines

13.6.7 How is a modem cable identified:

(a) Direct connection between all the signal lines

(b) No connections to TD and RD lines
(c) Cross-over between the TD and RD, and handshaking lines
(d) No hardware handshaking lines

13.6.8 The main connections used to transmit data over a null modem cable with no hardware handshaking are:

(a) TD, RD, GND (b) RTS, CTS, GND
(c) DSR, DTR, GND (d) TD, RD, RTS, CTS

13.6.9 If a device transmits at 9600 bps, approximately how many characters are transmitted every minute:

(a) 5760 (b) 57600
(c) 576000 (d) 5760000

13.6.10 If a device transmits at 4800 bps, approximately what is the period of a single bit:

(a) 2.08 μs (b) 20.8 μs
(c) 208 μs (d) 2.08 ms

13.6.11 Which handshaking line is used by a transmitter to identify that it is read to send data:

(a) RTS (b) CTS
(c) DTR (d) DTE

13.6.12 Which handshaking line is used by a receiver to identify that it is ready to receive data:

(a) RTS (b) CTS
(c) DTR (d) DTE

13.6.13 Which characters are used to start and stop data transfer in software handshaking:

(a) X-ON, X-OFF (b) OFF, ON
(c) IN, OUT (d) LF, CR

13.6.14 Which is the standard I/O port address for COM1:

(a) 1F8h (b) 2F8h
(c) 3F8h (d) 4F8h

13.6.15 Which is the standard I/O port address for COM2:

(a) 1F8h (b) 2F8h
(c) 3F8h (d) 4F8h

13.6.16 The standard IC used in RS-232 communications is:

(a)	8232	(b)	8086
(c)	8088	(d)	8250

13.6.17 Which register is used to determine the status of the RS-232 connection:

(a)	LSR	(b)	LCR
(c)	STATUS	(d)	TD/RD buffer

13.6.18 Which register is used to configure the RS-232 connection:

(a)	LSR	(b)	LCR
(c)	STATUS	(d)	TD/RD buffer

13.6.19 Write a program that continuously sends the character 'A' to the serial line. Observe the output on an oscilloscope and identify the bit pattern and the baud rate.

13.6.20 Write a program that continuously sends the characters from 'A' to 'Z' to the serial line. Observe the output on an oscilloscope.

13.6.21 Modify Program 13.2 so that the program prompts the user for the baud rate when the program is started. A sample run is shown in Sample run 13.1.

Sample run 13.1

```
Enter baud rate required:
1  110          2  150       3  300          4  600
5  1200         6  2400      7  4800         8  9600
>> 8
RS232 transmission set to 9600 baud
```

13.6.22 Complete Table 13.6 to give the actual time to send 1000 characters for the given baud rates. Compare these values with estimated values.

Note that approximately 10 bits are used for each character thus 960 characters per second will be transmitted at 9600 baud.

Table 13.6 Baud rate divisors

Baud rate	*Time to send 1000 characters (s)*
110	
300	
600	
1200	
2400	
4800	
9600	
19200	

13.6.23 Modify the `setup_serial()` routine so that the RS232 parameters can be passed to it. These parameters should include the comport (either `COM1:` or `COM2:`), the baud rate, the number of data bits and the type of parity. An outline of the modified function is given in Program 13.5.

Program 13.5

```
#define     COM1BASE 0x3F8
#define     COM2BASE 0x2F8
#define     COM1        0
#define     COM2        1
enum     baud_rates  {BAUD110,BAUD300,BAUD600,BAUD1200,
                        BAUD2400,BAUD4800,BAUD9600};
enum     parity      {NO_PARITY,EVEN_PARITY,ODD_PARITY};
enum     databits    {DATABITS7,DATABITS8};
#include <conio.h>
#include <dos.h>
#include <stdio.h>
void  setup_serial(int comport, int baudrate, int parity,
                     int databits);
void  send_character(int ch);
int   get_character(void);
int   main(void)
{
int   inchar,outchar;

   setup_serial(COM1,BAUD2400,EVEN_PARITY,DATABITS7);
   ::::::::::etc.
}

void  setup_serial(int comport, int baudrate,
                     int parity, int databits)
{
int   tdreg,lcr;
   if (comport==COM1)
   {
      tdreg=COM1BASE;   lcr=COM1BASE+3;
   }
   else
   {
      tdreg=COM2BASE;   lcr=COM2BASE+3;
   }
   _outp( lcr, 0x80);
   /* set up bit 7 to a 1  to set Register address bit        */
   switch(baudrate)
   {
   case BAUD110: _outp(tdreg,0x17);_outp(tdreg+1,0x04); break;
   case BAUD300: _outp(tdreg,0x80);_outp(tdreg+1,0x01); break;
   case BAUD600: _outp(tdreg,0x00);_outp(tdreg+1,0xC0); break;
   case BAUD1200: _outp(tdreg,0x00);_outp(tdreg+1,0x40);break;
   case BAUD2400: _outp(tdreg,0x00);_outp(tdreg+1,0x30);break;
   case BAUD4800: _outp(tdreg,0x00);_outp(tdreg+1,0x18);break;
   case BAUD9600: _outp(tdreg,0x00);_outp(tdreg+1,0x0C);break;
   }
      ::::::::: etc.
}
```

13.6.24 One problem with Programs 13.2 and 13.3 is that when the return key is pressed only one character is sent. The received character will be a carriage return which returns the cursor back to the start of a line and not to the next line. Modify the receiver program so that a line feed will be generated automatically when a carriage return is received. Note a carriage return is an ASCII 13 and line feed is a 10.

13.6.25 Modify the `get_character()` routine so that it returns an error flag if it detects an error or if there is a time-out. Table 13.7 lists the error flags and the returned error value. An outline of the C code is given in Program 13.6. If a character is not received within 10 s an error message should be displayed.

Table 13.7 Error returns from get_character().

Error condition	*Error flag return*	*Notes*
Parity error	–1	
Overrun error	–2	
Framing error	–3	
Break detected	–4	
Time out	–5	`get_character()` should time out if no characters are received with 10 seconds.

Test the routine by connecting two PCs together and set the transmitter with differing RS-232 parameters.

Program 13.6

```
#include <stdio.h>
#include <dos.h>
#define  TXDATA   0x3F8
#define  LSR      0x3FD
#define  LCR      0x3FB
void     show_error(int ch);
int      get_character(void);
enum     RS232_errors   {PARITY_ERROR=-1, OVERRUN_ERROR=-2,
         FRAMING_ERROR=-3, BREAK_DETECTED=-4, TIME_OUT=-5};
int      main(void)
{
int      inchar;
   do
   {
      inchar=get_character();
      if (inchar<0) show_error(inchar);
      else printf("%c",inchar);
   } while (inchar!=4);
   return(0);
}
void     show_error(int ch)
{
   switch(ch)
   {
   case PARITY_ERROR: printf("Error: Parity error/n"); break;
   case OVERRUN_ERROR: printf("Error: Overrun error/n"); break;
   case FRAMING_ERROR: printf("Error: Framing error/n"); break;
   case BREAK_DETECTED: printf("Error: Break detected/n");break;
   case TIME_OUT: printf("Error: Time out/n"); break;
   }
}
int   get_character(void)
{
int   instatus;
   do
   {
      instatus = _inp(LSR) & 0x01;
```

```
        if (instatus & 0x02) return(BREAK_DETECTED);
                                    :::: etc
      } while (instatus!=0x01 );
      return( (int) _inp(TXDATA) );
}
```

13.7 Notes from the author

Good old RS-232. My bank manager would certainly agree with this, as I have made more consultancy income with it than any other piece of computer equipment. I have also run more RS-232 training courses than all the trendy subjects areas (such as Java and C++) put together (well, anyway, it doesn't take much to run a C++ course!). The reason for this is because it is one of the least understood connections on computer equipment. I've interfaced PCs to gas chromatographs (using an 8-port RS-232 card, heavy!), a PC to a VAX, a Sun workstation to a PC, a PC to another PC, a Honeywell TDC to a PC, a PC to a PLC, and so on. For most applications, a serial port to serial port connection is still the easiest method to transfer data from one computer to another.

RS-232 is one of the most widely used 'standards' in the world. It is virtually standard on every computer and, while it is relatively slow, it is a standard device. This over-rules its slowness, its non-standardness, its lack of powerful error checking, its lack of address facilities, and, well, need I go on. It shares its gold stars with solid performers, such as Ethernet and the parallel port. Neither of these are star performers and are far from perfect, but they are good, old robust performers who will outlast many of their more modern contenders. When their position is challenged by a young contender, the standards agency simply invest a lot of experience and brainpower to increase their performance. Who would believe that the data rate, over copper wires, could be increased to 1 Gbps for Ethernet to 1MBps for RS-422. One trusted piece of equipment I could have never done without is an RS-232 transmitter/receiver. For this, I used an old 80386-based laptop computer (which weights as much as a modern desktop computers) which ran either a simple DOS-based transmitter/receiver program (see previous chapter), or the excellent Windows 3.1 Terminal program. These I could use just as an electronic engineer would use a multimeter to test the voltages and currents in a circuit. A telltale sign that I was transmitting at the wrong bit rate or using an incorrect number of data bits was the incorrectly received characters (but at least it was receiving characters, which was an important test).

On technical questions, I get more requests on RS-232 than any other technical area. I've done a quick search of my emails, and here are my 'most requested' list:

1. *C++ student assignment problems (I seem to get more than my fair share of C++ problems from students, even although I don't actually teach the subject anymore, or use it on any of my assignments). I must admit, I really didn't enjoy teaching programming, as it allowed little scope for discussing interesting things.*
2. *Coursework questions.*
3. *Questions from students who are having problems with the PC they have at home.*
4. *Examination questions (requests for past papers, problems with previous exam questions, and so on).*
5. *Work-based problems (obviously sometimes universities provide better on-line services than my company support services). These are typically related to problems with networking.*

14 RS-422, RS-423 and RS-485

14.1 Introduction

The main standards organisations for data communications are the ITU (International Telecommunications Union), the EIA (Electronic Industry Association) and the ISO (International Standards Organisation). The ITU standards related to serial communications are defined in the V-series specifications and EIA standards as the RS-series. The EIA has defined many standards for serial communications. RS-232 has many limitations, such as:

- One transmitter and one receiver.
- Maximum connection length of 20 m.
- Maximum baud rate of 20 kbps.

The RS-422 and RS-423 standards replace the RS-232 standards and support higher data rates and greater immunity to electrical interference. The main standards are:

- RS-422A – Supports multipoint connections. It defines the electrical characteristics of balanced load voltage digital interface circuits.
- RS-423A – Supports only point-to-point connections. It defines electrical characteristics of unbalanced voltage digital interface circuits.
- RS-449 – Defines the basic interface standards and refers to the RS-422/3 standards. It defines a general-purpose 37-position and 9-position interface for DTE and DCE employing serial binary data interchange.
- RS-485 – Similar to RS-422 but can support more nodes per line because it uses lower-impedance drivers and receivers.

14.2 RS-485 (ISO 8482)

RS-485 is an upgraded version of RS-422 and extends the number of peripherals that can be interfaced. It allows for bidirectional multipoint party line communications. This can be used in networking applications. RS-422 and RS-232 facilitate simplex communication, whereas RS-485 allows for multiple receivers on a single line, facilitating half-duplex communications. The maximum data rate is unlimited and is set by the rise time of the pulses, but it is usually limited to 10 Mbps. A network using the RS-485 standard can have up to 32 transmitters/receivers with a maximum cable length of 1.2 km, as shown in Figure 14.1. The maximum cable length is 1200 m.

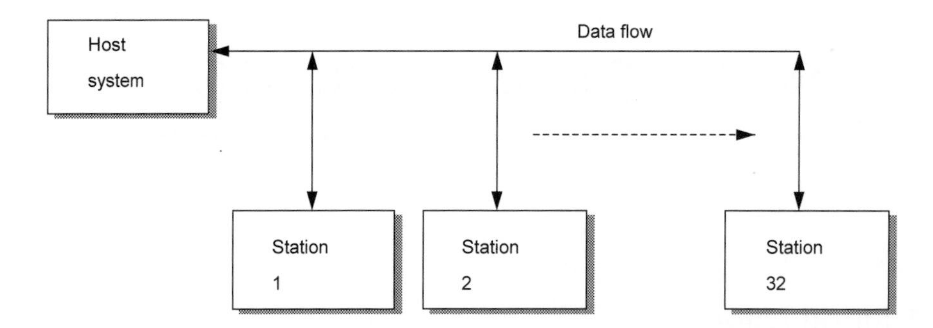

Figure 14.1 RS-485 connecting to multiple nodes

RS-485 operates in one of two modes:

- Two-wire, multidrop, party line – in this mode, a balanced transmission line is used to connect to all of the stations, which share a common communications channel. Up to 32 driver/receiver pairs can share the common channel.
- Four-wire – in this mode, each station connects to a four-wire bus, as illustrated in Figure 14.2. It is necessary in this mode that one node acts a master station and all others as slaves. The master then communicates with each of the slaves. All slave nodes communicate only with the master node. A master–slave network is useful when mixed protocols are used.

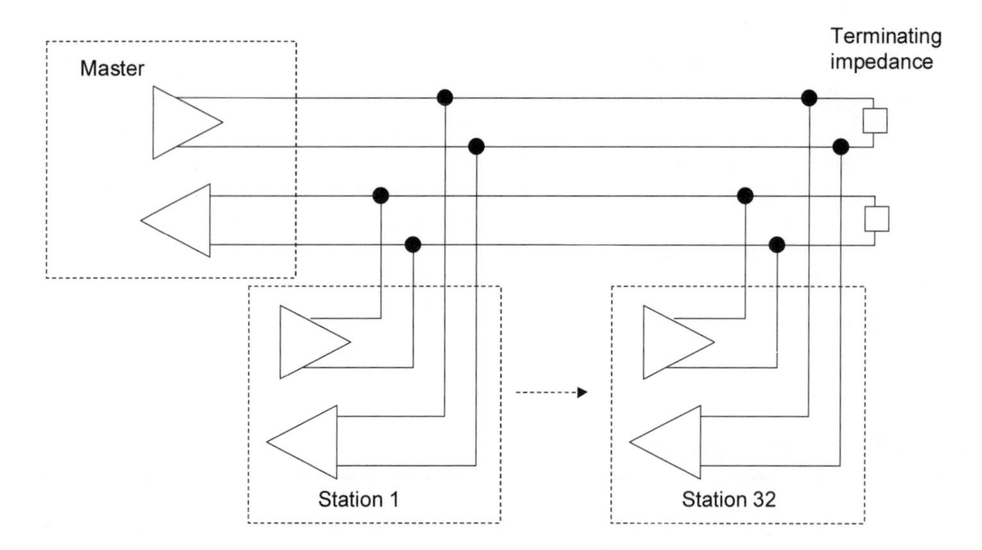

Figure 14.2 RS-485 connecting to multiple nodes

The RS-485 four-wire connection involves a half-duplex transmission mode, that is, only one device can transmit data at a time. This thus involves a polling procedure, with a master and up to 32 slaves. The slaves must wait in a high-impedance state (a stand-by mode). The control of the driver on the slave is either with:

- An active RTS line.
- Bit changes on the transmit data line.
- Sends of the X-ON/X-OFF flow control characters.

Slaves cannot send data unless they are selected.

In a four-wire operation RS-485 devices connect with two twisted pair cables with characteristic impedance of 120 Ω and general shielding. Each link requires a terminating load on the ends of the cable.

14.3 Line drivers

Transmission lines have effects on digital pulses in the following ways:

- Attenuation – The transmission line contains series resistance that causes a reduction in the pulse amplitude.
- Pulse distortion – The transmission line insulation produces a shunt capacitance on the signal path and a series resistance and inductance of the conductors. This causes the transmission line to distort the shape of the pulse. The two main effects are the block of high frequencies in the pulse and phase distortion.
- Noise – Noise is any unwanted electrical signals added to a signal. A digital system is less prone to noise as it has only two levels and it takes a relatively large change in voltage to cause an error.

Table 14.1 shows the electrical characteristics of the different serial communication standards. The two main standards agencies are the EIA and the ITU.

Balanced lines use two lines for each signal line, whereas unbalanced lines use one wire for each signal and a common return circuit (see Figure 14.3). RS-422 is a balanced interface and uses two conductors to carry the signal (see Figure 14.4). The electrical currents in each of the conductors are 180° out-of-phase with each other. Balanced lines are generally less prone to noise as any noise induced into the conductors will be of equal magnitude. At the receiver the noise will tend to cancel out.

The voltage levels for RS-232 range from ±3 to ±25 V, whereas, for RS422/ RS423 the voltage ranges are ±0.2 to ±6 V. For very high bit rates the cable is normally terminated with the characteristic impedance of the line; for example, a 50 Ω cable is terminated with a 50Ω termination.

RS-422 interface circuits can have up to 10 receivers. They have no ground connection and are thus useful in isolating two nodes. For two-way communications four connections are required, the `TX+` and `TX-` on one node connects to the `RX+` and `RX-` on the other.

Nodes may have a direct RS-422 connection or can be fitted with a special interface adaptor to convert from RS-232 to RS-422 (although the maximum data rate is likely to be limited to the maximum RS-232 rate).

It should also be noted that the maximum connection distance relates to the maximum data rate. If a lower data rate is used then the maximum distance can be increased. For example, in some situations with a good quality cable and in a low noise environment, it is possible to have cable runs of 1 km using RS-232 at 1200 bps.

Table 14.1 Main serial standards

EIA	*RS-232-C*	*RS-423-A*	*RS-422-A*	*RS-485*
ITU	*V.28*	*V.10/X.26*	*V.11/X.27*	
Data rate	20 kbps	300 kbps	10 Mbps	10 Mbps
Max distance	15 m	1200 m	1200 m	1200 m
Type	Unbalanced	Unbalanced differential	Balanced differential	Balanced differential
Number of drivers and receivers	1 driver 1 receiver	1 driver 10 receivers	1 driver 10 receivers	32 drivers 32 receivers
Driver voltages	±15 V	± 6 V	±5 V	±5 V
Number of conductors per signal	1	2	2	2

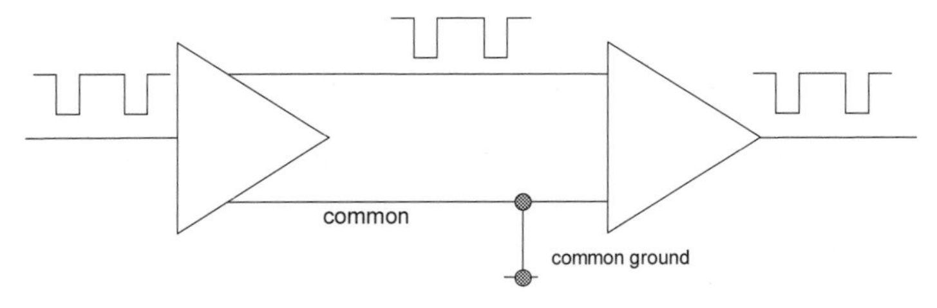

Figure 14.3 Unbalanced digital interface circuit (RS-423)

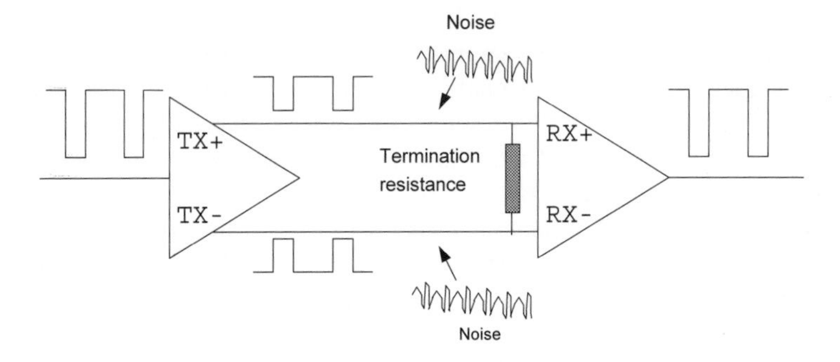

Figure 14.4 Balanced digital interface circuit (RS-422)

14.4 RS-232/485 converter

RS-232 is a standard port on many systems, including PCs and many instruments. A common requirement is to convert from RS-232 to RS-485, as this allows for long transmission lengths. If a computer connects to external equipment, it is important to isolate the grounds.

This is typically achieved with an opto-isolator which converts between the RS-232 interface and the RS-485 interface, as shown in Figure 14.5. Data is transmitted from the RS-232 port to the RS-485 line only if the RS-485 driver is in active mode. This active mode is controlled by the RTS signal from the RS-232 port (or by detecting transitions on the transmit data line, TD).

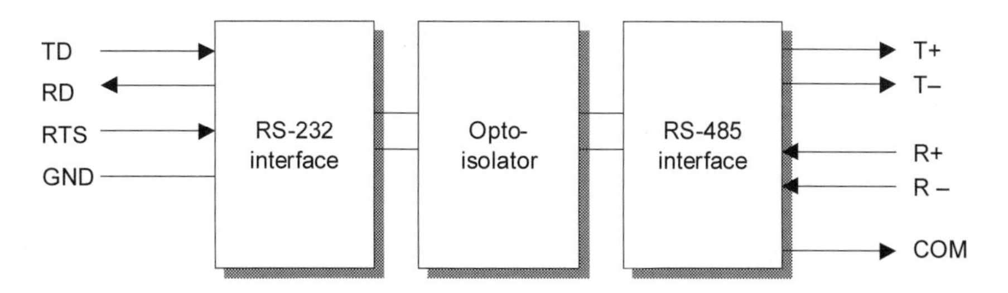

Figure 14.5 Isolation between RS-232 and RS-485

14.5 Exercises

14.5.1 Which device provides isolation between RS-232 and RS-485:

(a) Capacitor (b) Opto-isolator
(c) Integrator (d) Resistor

14.5.2 What is the maximum range for RS-422:

(a) 500 m (b) 1.2 km
(c) 5 km (d) 10 km

14.5.3 Outline the advantages of balanced lines as opposed to unbalanced lines.

14.6 Note from the author

The RS-422/RS-485 standard really does enhance the basic RS-232 standard and allows for standard RS-232 connections to be increased to over 1.2 km (although at low bit rates and unnoisy environments allows for even greater distances). A surprising thought in these days of cheap electronics, and PC motherboards that cost less than $100, is that RS-232 to RS-485 convertors are relatively expensive. In fact, it is possible to get a complete PC motherboard for the same price as an RS-232/RS485 convertor (which contains just a few internal components), but as the convertor saves lots of time and energy, they are probably worth the high costs.

14.5 Exercises

15 Modems

15.1 Introduction

Modems (MOdulator/DEModulator) connect digital equipment to a telephone line. It connects digital equipment to a speech bandwidth-limited communications channel. Typically, modems are used on telephone lines, which have a bandwidth of between 400 Hz and 3.4 kHz. If digital pulses were applied directly to these lines, they would end up severely distorted.

Modem speeds range from 300 bps to 56 kbps. A modem normally transmits about 10 bits per character (each character has 8 bits). Thus, the maximum rate of characters for a high-speed modem is 2880 characters per second. This chapter contains approximately 15 000 characters and thus to transmit the text in this chapter would take approximately 5 seconds. Text, itself, is relatively fast transfer; unfortunately, even compressed graphics can take some time to be transmitted. A compressed image of 20 KB (equivalent to 20 000 characters) will take nearly 6 seconds to load on the fastest modem.

The document that was used to store this chapter occupies, in an uncompressed form, 360 KB. Thus to download this document over a modem, on the fastest modem, would take

$$\text{Time taken} = \frac{\text{Total file size}}{\text{Characters per second}} = \frac{360\,000}{2\,800} = 125\text{ s}$$

A 14.4 kbps modem would take 250 seconds. Typically home users connect to the Internet and WWW through a modem (although increasingly ISDN is being used). The example above shows the need to compress files when transferring them over a modem. On the WWW, documents and large files are normally compressed into a ZIP file and images and video compressed in GIF and JPG.

Most modems are able to do the following:

- Automatically dial (known as auto-dial) another modem using either touch-tone or pulse dialing.
- Automatically answer (known as auto-answer) calls and make a connection with another modem.
- Disconnect a telephone connection when data transfer has completed or if an error occurs.
- Automatic speed negotiation between the two modems.
- Convert bits into a form suitable for the line (modulator).
- Convert received signals back into bits (demodulator).
- Transfer data reliably with the correct type of handshaking.

Figure 15.1 shows how two computers connect to each other using RS-232 converters and modems. The RS-232 converter is normally an integral part of the computer, while the modem can either be external or internal to the computer. If it is externally connected then it is normally connected by a cable with a 25-pin male D-type connector on either end.

Modems are either synchronous or asynchronous. A synchronous modem recovers the clock at the receiver. There is no need for start and stop bits in a synchronous modem. Asynchronous modems are, by far, the most popular types. Synchronous modems have a typical speed of 56 Kbps whereas for asynchronous modems it is 33 Kbps. A measure of the speed of the modem is the baud rate or bps (bits per second).

There are two types of circuits available from the public telephone network: either direct dial or a permanent connection. The direct dial type is a dial-up network where the link is established in the same manner as normal voice calls with a standard telephone or some kind of an automatic dial/answer machine. They can use either touch-tones or pulses to make the connection. With private line circuits, the subscriber has a permanent dedicated communication link.

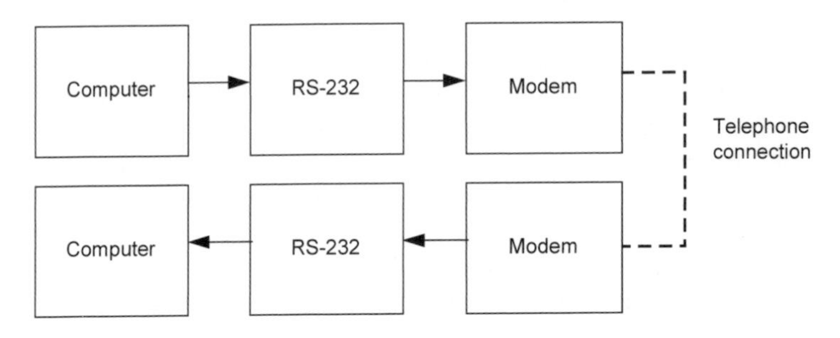

Figure 15.1 Data transfer using modems

15.2 RS-232 communications

The communication between the modem and the computer is via RS-232. RS-232 uses asynchronous communication which has a start–stop data format. Each character is transmitted one at a time with a delay between characters. This delay is called the inactive time and is set at a logic level high as shown in Figure 15.2. The transmitter sends a start bit to inform the receiver that a character is to be sent in the following bit transmission. This start bit is always a '0'. The following data bits are sent as a 7-bit ASCII character, followed by a parity bit and finally either 1, 1.5 or 2 stop bits. The rate of transmission is set by the timing of a single bit. Both the transmitter and receiver need to be set to the same bit-time interval. An internal clock on both of them sets this interval. They only have to be roughly synchronised and approximately at the same rate as data is transmitted in relatively short bursts.

15.2.1 Bit rate and the baud rate

One of the main parameters for specifying RS-232 communications is the rate at which data is transmitted and received. It is important that the transmitter and receiver operate at roughly the same speed.

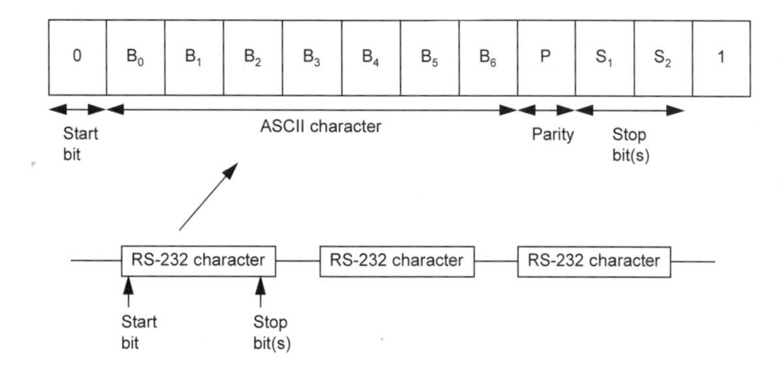

Figure 15.2 RS-232 frame format

For asynchronous transmission the start and stop bits are added in addition to the seven ASCII character bits and the parity. Thus a total of 10 bits are required to transmit a single character. With 2 stop bits, a total of 11 bits are required. If 10 characters are sent every second and if 11 bits are used for each character, then the transmission rate is 110 bits per second (bps). The fastest modem thus has a character transmission rate of 2880 characters per second.

In addition to the bit rate, another term used to describe the transmission speed is the baud rate. The bit rate refers to the actual rate at which bits are transmitted, whereas the baud rate is the rate at which signalling elements, used to represent bits, are transmitted. As one signalling element encodes 1 bit, the two rates are then identical. Only in modems does the bit rate differ from the baud rate.

15.3 Modem standards

The CCITT (now known as the ITU) has defined standards which relate to RS-232 and modem communications. Each uses a V number to define their type. Modems tend to state all the standards they comply with. An example FAX/modem has the following compatibility:

- V.32bis (14.4 Kbps). V.32 (9.6 Kbps).
- V.22bis (2.4 Kbps). V.22 (1.2 Kbps).
- Bell 212A (1.2 Kbps). Bell 103 (300 bps).
- V.17 (14.4 bps FAX). V.29 (9.6 Kbps FAX).
- V.27ter (4.8 Kbps FAX). V.21 (300 bps FAX – secondary channel).
- V.42bis (data compression). V.42 (error correction).
- MNP5 (data compression). MNP2–4 (error correction).

A 28.8 Kbps modem also supports the V.34 standard.

15.4 Modem commands

Most modems are Hayes compatible. Hayes was the company that pioneered modems and defined the standard method of programming the mode of the modem, which is the AT command language. A computer gets the attention of the modem by sending an 'AT' command. For example, 'ATDT' is the touch-tone dial command. Initially, a modem is in the command mode and accepts commands from the computer. These commands are sent at either 300 bps or 1200 bps (the modem automatically detects which of the speeds is being used).

Most commands are sent with the AT prefix. Each command is followed by a carriage return character (ASCII character 13 decimal); a command without a carriage return character is ignored (after a given time delay). More than one command can be placed on a single line and, if necessary, spaces can be entered to improve readability. Commands can be sent in either upper or lower case. Table 15.1 lists some AT commands. The complete set is defined in Appendix C.

Table 15.1 Example AT modem commands

Command	*Description*
ATDT54321	Automatically phones number 54321 using touch-tone dialing. Within the number definition, a comma (,) represents a pause and a W waits for a second dial tone and an @ waits for a 5 second silence.
ATPT12345	Automatically phones number 12345 using pulse dialing.
AT S0=2	Automatically answers a call. The S0 register contains the number of rings the modem uses before it answers the call. In this case there will be two rings before it is answered. If S0 is zero, the modem will not answer a call.
ATH	Hang up telephone line connection.
+++	Disconnect line and return to on-line command mode.
AT A	Manually answer call.
AT E0	Commands are not echoed (AT E1 causes commands to be echoed). See Table 15.2.
AT L0	Low speaker volume (AT L1 gives medium volume and AT L2 gives high speaker volume).
AT MO	Internal speaker off (ATM1 gives internal speaker on until carrier detected, ATM2 gives the speaker always on, AT M3 gives speaker on until carrier detect and while dialing).
AT QO	Modem sends responses (AT Q1 does not send responses). See Table 15.2.
AT V0	Modem sends numeric responses (AT V1 sends word responses). See Table 15.2.

The modem can enter one of two states: the normal state and the command state. In the normal state the modem transmits and/or receives characters from the computer. In the command state, characters sent to the modem are interpreted as commands. Once a command is interpreted, the modem goes into the normal mode. Any characters sent to the modem are then sent along the line. To interrupt the modem so that it goes back into command mode, three consecutive '+' characters are sent, i.e. '+++'.

After the modem has received an AT command it responds with a return code. Some return codes are given in Table 15.2 (a complete set is defined in Appendix C). For example, if a modem calls another which is busy then the return code is 7. A modem dialing another modem returns the codes for OK (when the ATDT command is received), CONNECT (when it connects to the remote modem) and CONNECT 1200 (when it detects the speed of the remote modem). Note that the return code from the modem can be suppressed by sending the AT command 'ATQ1'. The AT code for it to return the code is 'ATQ0'; normally this is the default condition

Table 15.2 Example return codes

Message	*Digit*	*Description*
OK	0	Command executed without errors
CONNECT	1	A connection has been made
RING	2	An incoming call has been detected
NO CARRIER	3	No carrier detected
ERROR	4	Invalid command
CONNECT 1200	5	Connected to a 1200 bps modem
NO DIALTONE	6	Dial-tone not detected
BUSY	7	Remote line is busy
NO ANSWER	8	No answer from remote line
CONNECT 600	9	Connected to a 600 bps modem
CONNECT 2400	10	Connected to a 2400 bps modem
CONNECT 4800	11	Connected to a 4800 bps modem
CONNECT 9600	13	Connected to a 9600 bps modem
CONNECT 14400	15	Connected to a 14 400 bps modem
CONNECT 19200	61	Connected to a 19 200 bps modem
CONNECT 28800	65	Connected to a 28 800 bps modem
CONNECT 1200/75	48	Connected to a 1200/75 bps modem

Figure 15.3 shows an example session when connecting one modem to another. Initially the modem is set up to receive commands from the computer. When the computer is ready to make a connection it sends the command 'ATDH 54321' which makes a connection with telephone number 54321 using tone dialing. The modem then replies with an OK response (a 0 value) and the modem tries to make a connection with the remote modem. If it cannot make the connection it returns back a response of NO CARRIER (3), BUSY (7), NO DIALTONE (6) or NO ANSWER (8). If it does connect to the remote modem then it returns a connect response, such as CONNECT 9600 (13). The data can then be transmitted between the modem at the assigned rate (in this case 9600 bps). When the modem wants to end the connection it gets the modem's attention by sending it three '+' characters ('+++'). The modem will then wait for a command from the host computer. In this case the command is hang-up the connection (ATH). The modem will then return an OK response when it has successfully cleared the connection.

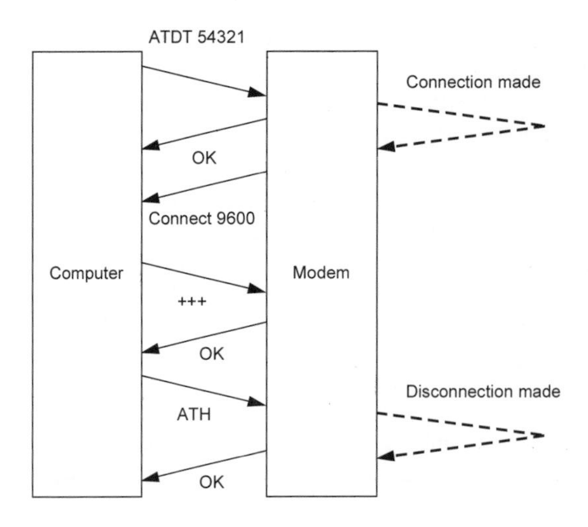

Figure 15.3 Commands and responses when making a connection

The modem contains various status registers called the S-registers which store modem settings. Table 15.3 lists some of these registers (Appendix C gives a complete listing). The S0 register sets the number of rings that must occur before the modem answers an incoming call. If it is set to zero (0) then the modem will not answer incoming calls. The S1 register stores the number of incoming rings when the modem is rung. S2 stores the escape character, normally this is set to the '+' character and the S3 register stores the character which defines the end of a command, normally the CR character (13 decimal).

Table 15.3 Modem registers

Register	*Function*	*Range (typical default)*
S0	Rings to auto-answer	0–255 rings (0 rings)
S1	Ring counter	0–255 rings (0 rings)
S2	Escape character	(43)
S3	Carriage return character	(13)
S6	Wait time for dial tone	2–255 s (2 s)
S7	Wait time for carrier	1–255 s (50 s)
S8	Pause time for automatic dialling	0–255 (2 s)

15.5 Modem set-ups

Figure 15.4 shows a sample window from the Microsoft Windows Terminal program (in both Microsoft Windows 3.*x* and Windows 95/98). It shows the modem commands window. In this case, it can be seen that when the modem dials a number the prefix to the number dialled is 'ATDT'. The hang-up command sequence is '+++ ATH'. A sample dialling window is shown in Figure 15.5. In this case, the number dialled is 9,123456789. A ',' character represents a delay. The actual delay is determined by the value in the S8 register (see Table 15.3). Typically, this value is about 2 seconds.

On many private switched telephone exchanges in the UK a '9' must prefix the number if an outside line is required (in Australia it is a '0', by contrast). A delay is normally required after the `9` prefix before dialing the actual number. To modify the delay to 5 seconds, dial the number `9 0112432` and wait 30 seconds for the carrier, then the following command line can be used:

```
ATDT 9,0112432 S8=5 S7=30
```

It can be seen in Figure 15.4 that a prefix and a suffix are sent to the modem. This is to ensure there is a time delay between the transmission prefix and the suffix string. For example, when the modem is to hang-up the connection, the '`+++`' is sent followed by a delay then the '`ATH`'.

In Figure 15.4 there is an option called Originate. This string is sent initially to the modem to set it up. In this case the string is '`ATQ0V1E1S0=0`'. The Q0 part informs the modem to return a send status code. The V1 part informs the modem that the return code message is to be displayed rather than just the value of the return code; for example, it displays CONNECT 1200 rather than the code 5 (V0 displays the status code). The E1 part enables the command message echo (E0 disables it).

Figure 15.6 shows the modem set-up windows for CompuServe access. The string in this case is:

```
ATS0=0 Q0 V1 &C1&D2^M
```

as previously seen, `S0` stops the modem from auto-answering. `V1` causes the modem to respond with word responses. `&C1` and `&D2` set up the hardware signals for the modem. Finally `^M` represents Cntrl-M which defines the carriage return character.

The modem reset command in this case is `AT &F`. This resets the modem and restores the factor default settings.

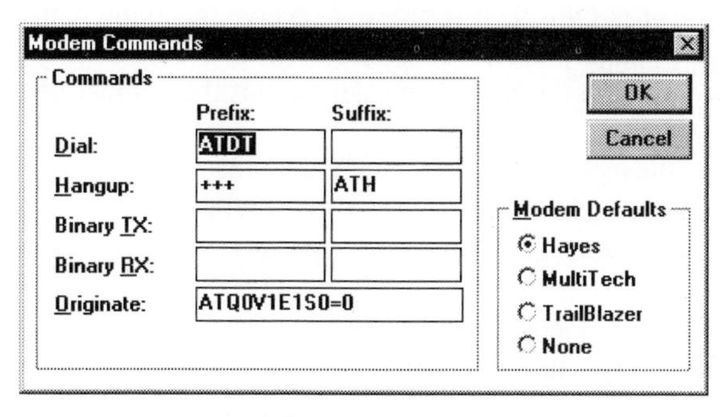

Figure 15.4 Modem commands

Phone Number
Dial: 9,123456789
Timeout If Not Connected In 30 Seconds
OK
Cancel
Redial After Timing Out
Signal When Connected

Figure 15.5 Dialling a remote modem

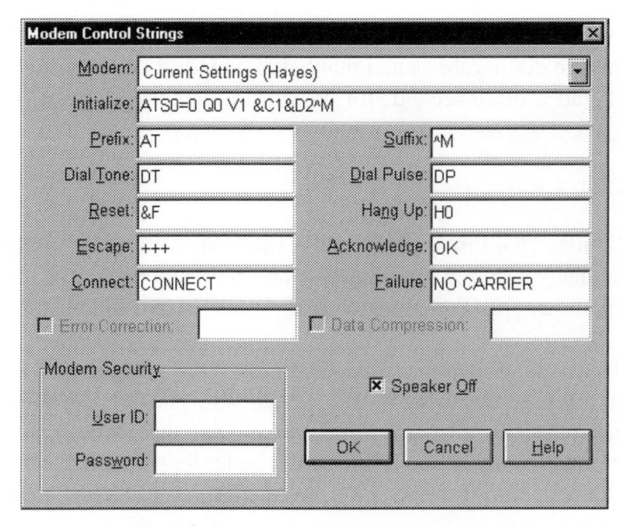

Figure 15.6 Example modem settings

15.6 Modem indicator

Most external modems have status indicators to inform the user of the current status of a connection. Typically, the indicator lights are:

- AA – is ON when the modem is ready to receive calls automatically. It flashes when a call is incoming. If it is OFF then it will not receive incoming calls. Note that if the S0 register is loaded with any other value than 0 then the modem goes into auto-answer mode. The value stored in the S0 register determines the number of rings before the modem answers.
- CD – is ON when the modem detects the remote modem's carrier, else it is OFF.
- OH – is ON when the modem is on-hook, else it is OFF.
- RD – flashes when the modem is receiving data or is getting a command from the computer.
- SD – flashes when the modem is sending data.
- TR – shows that the DTR line is active (i.e. the computer is ready to transmit or receive data).
- MR – shows that the modem is powered up.

15.7 Profile viewing

The settings of the modem can be determined by using the AT command with &V. An example is shown next (which uses a program from Chapter 13). In this it can be seen that the settings include: B0 (CCITT 300 or 1200 bps for call establishment), E1 (enable command echo), L2 (medium volume), M1 (speaker is off when receiving), Q1 (prohibits modem from sending result codes to the DTE) T (set tone dial) and V1 (display result codes in a verbose form). It can be seen that the S0 register is set to 3 which means that the modem waits for three rings before it will automatically answer the call.

```
+++
AT &V
ACTIVE PROFILE:
B0 E1 L2 M1 Q1 T V1 X4 Y0 &C1 &D0 &E0 &G2 &L0 &M0 &O0 &P1 &R0 &S0 &X0 &Y1
%A000 %C1 %D1 %E1 %P0 %S0 \A3 \C0 \E0 \G0 \J0 \K5 \N6 \Q0 \T000 \V1 \X0
S00:003 S01:000 S06:004 S07:045 S08:002 S09:006 S10:014 S11:085 S12:050
S16:1FH S18:000 S21:20H S22:F6H S23:B2H S25:005 S26:001 S27:60H S28:00H
STORED PROFILE 0:
B0 E1 L2 M1 Q0 T V1 X4 Y0 &C1 &D2 &E0 &G2 &L0 &M0 &O0 &P1 &R0 &S0 &X0
%A000 %C1 %D1 %E1 %P0 %S0 \A3 \C0 \E0 \G0 \J0 \K5 \N6 \Q3 \T000 \V1 \X0
S00:000 S16:1FH S21:30H S22:F6H S23:89H S25:005 S26:001 S27:000 S28:000
STORED PROFILE 1:
B0 E0 L2 M1 Q1 T V1 X4 Y0 &C1 &D0 &E0 &G2 &L0 &M0 &O0 &P1 &R0 &S0 &X0
%A000 %C1 %D1 %E1 %P0 %S0 \A3 \C0 \E0 \G0 \J0 \K5 \N6 \Q0 \T000 \V1 \X0
S00:003 S16:1FH S21:20H S22:F6H S23:95H S25:005 S26:001 S27:096 S28:000
TELEPHONE NUMBERS:
&Z0=
&Z1=
&Z2=
&Z3=
```

15.8 Test modes

There are several modes associated with the modems.

15.8.1 Local analogue loopback (&T1)

In the analogue loopback test the modem connects the transmit and receive lines on its output, as illustrated in Figure 15.7. This causes all transmitted characters to be received. It is initiated with the `&T1` mode. For example:

```
AT &Q0        <Enter>
AT S18=0 &T1 <Enter>
CONNECT 9600
Help the bridge is on fire <Enter>
+++
OK
AT &T0
OK
```

The initial command `AT &Q0` sets the modem into an asynchronous mode (stop–start). Next the `AT S18=0 &T1` command sets the timer test time to zero (which disables any limit to the time of the test) and `&T1` sets an analogue test. The modem responds with the message `CONNECT 9600`. Then the user enters the text `Help .... on fire` followed by an <Enter>. Next the user enters three + characters which puts the modem back into command mode. Finally, the user enters `AT &T0` which disables the current test.

If a time-limited test is required then the S18 register is loaded with the number of seconds that the test should last. For example, a test that last 2 minutes will be set up with:

```
AT S18=120 &T1
```

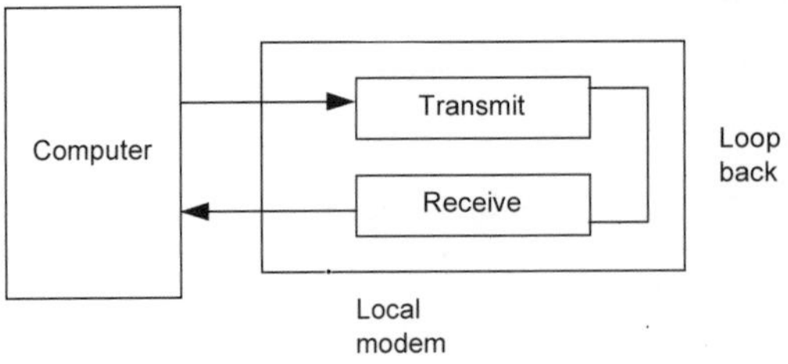

Figure 15.7 Analogue loopback with self-test

15.8.2 Local analog loopback with self-test (&T8)

In the analog loopback test with self-test the modem connects the transmit and receive lines on its output and then automatically sends a test message which is then automatically received, as illustrated in Figure 15.8. The local error checker then counts the number of errors and displays a value when the test is complete. For example, the following test has found two errors:

```
AT &Q0      <Enter>
AT S18=0 &T8 <Enter>
+++
AT &T0
002
OK
```

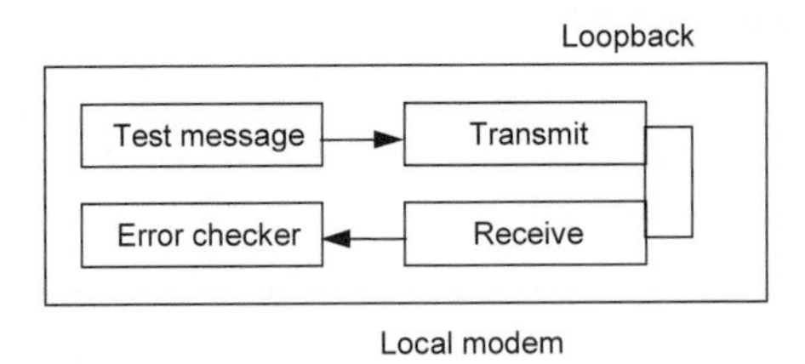

Figure 15.8 Analogue loopback with self-test

15.8.3 Remote digital loopback (&T6)

The remote digital loopback checks the local computer to modem connection, the local modem, the telephone line and the remote modem. The remote modem performs a loopback at the connection from the remote modem to its attached computer. Figure 15.9 illustrates the test set-up. An example session is:

```
AT &Q0        <Enter>
AT S18=0 &T6 <Enter>
CONNECT 9600
Help the bridge is on fire <Enter>
+++
OK
AT &T0
OK
```

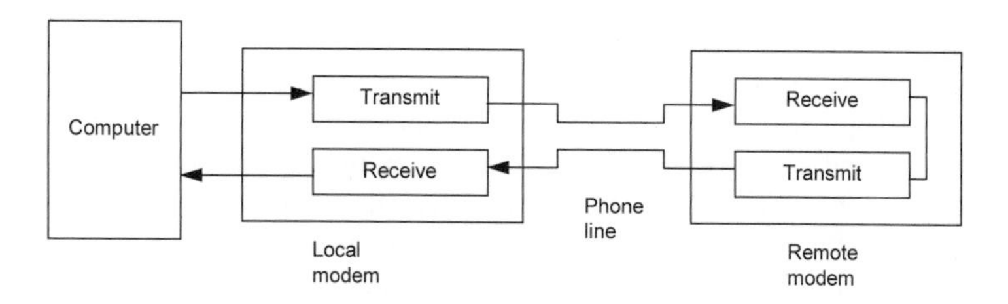

Figure 15.9 Remote digital loopback test

15.8.4 Remote digital loopback with self-test (&T7)

The remote digital loopback with self-test checks the local computer to modem connection, the local modem, the telephone line and the remote modem. The remote modem performs a loopback at the connection from the remote modem to its attached computer. The local modem sends a test message and checks the received messages for errors. On completion of the test, the local modem transmits the number of errors. Figure 15.10 illustrates the test setup. An example session is:

```
AT &Q0       <Enter>
AT S18=0 &T7 <Enter>
+++
AT &T0
004
OK
```

or with a test of 60 seconds then the user does not have to send the break sequence:

```
AT &Q0        <Enter>
AT S18=60 &T7 <Enter>
004
OK
```

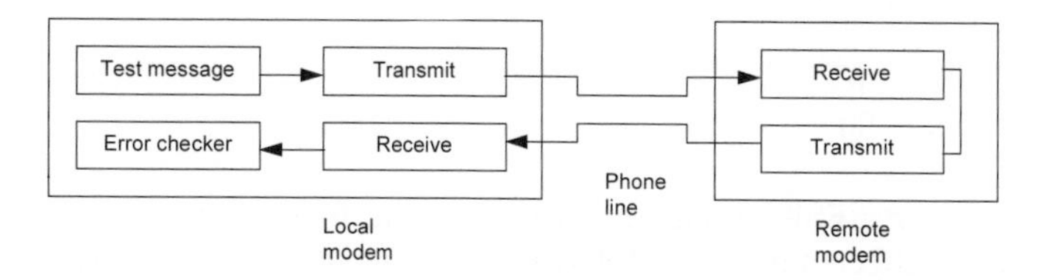

Figure 15.10 Remote digital loopback test with self-test

15.9 Digital modulation

Digital modulation changes the characteristic of a carrier according to binary information. With a sine wave carrier the amplitude, frequency or phase can be varied. Figure 15.11 illustrates the three basic types: amplitude-shift keying (ASK), frequency-shift keying (FSK) and phase-shift keying (PSK).

15.9.1 Frequency-shift keying (FSK)

FSK, in the most basic case, represents a 1 (a mark) by one frequency and a 0 (a space) by another. These frequencies lie within the bandwidth of the transmission channel.

On a V.21, 300 bps, full-duplex modem the originator modem uses the frequency 980 Hz to represent a mark and 1180 Hz a space. The answering modem transmits with 1650 Hz for a mark and 1850 Hz for a space. The four frequencies allow the caller originator and the answering modem to communicate at the same time; that is, full-duplex communication.

FSK modems are inefficient in their use of bandwidth, with the result that the maximum data rate over normal telephone lines is 1800 bps. Typically, for rates over 1200 bps, other modulation schemes are used.

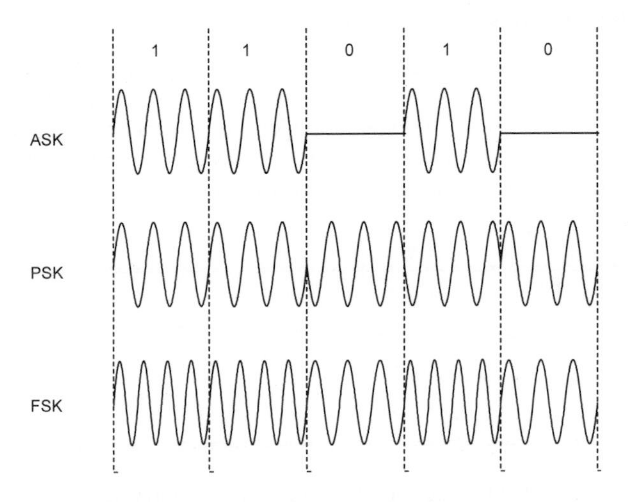

Figure 15.11 Waveforms for ASK, PSK and FSK

15.9.2 Phase-shift keying (PSK)

In coherent PSK a carrier gets no phase shift for a 0 and a 180° phase shift for a 1, as given next:

$$0 \Rightarrow 0^\circ$$
$$1 \Rightarrow 180^\circ$$

Its main advantage over FSK is that as it uses a single frequency it uses much less bandwidth. It is thus less affected by noise. It has an advantage over ASK because its information is not contained in the amplitude of the carrier, thus again it is less affected by noise.

15.9.3 M*-ary modulation*

With *M*-ary modulation a change in amplitude, phase or frequency represents one of *M* possible signals. It is possible to have *M*-ary FSK, *M*-ary PSK and *M*-ary ASK modulation schemes. This is where the baud rate differs from the bit rate. The bit rate is the true measure of the rate of the line, whereas the baud rate only indicates the signalling element rate, which might be a half or a quarter of the bit rate.

For four-phase differential phase-shift keying (DPSK) the bits are grouped into two and each group is assigned a certain phase shift. For two bits there are four combinations: a 00 is coded as 0°, 01 coded as 90°, and so on:

$$00 \Rightarrow 0^\circ \qquad 01 \Rightarrow 90^\circ$$
$$11 \Rightarrow 180^\circ \qquad 10 \Rightarrow 270^\circ$$

It is also possible to change a mixture of amplitude, phase or frequency. *M*-ary amplitude-phase keying (APK) varies both the amplitude and phase of a carrier to represent *M* possible bit patterns.

M-ary quadrature amplitude modulation (QAM) changes the amplitude and phase of the carrier. 16-QAM uses four amplitudes and four phase shifts, allowing it to code four bits at a time. In this case, the baud rate will be a quarter of the bit rate.

Typical technologies for modems are:

FSK – used up to 1200 bps
Four-phase DPSK – used at 2400 bps
Eight-phase DPSK – used at 4800 bps
16-QAM – used at 9600 bps

15.10 Typical modems

Most modern modems operate with V.22bis (2400 bps), V.32 (9600 bps) or V.32bis (14 400 bps); some standards are outlined in Table 15.4. The V.32 and V.32bis modems can be enhanced with echo cancellation. They also typically have built-in compression using either the V.42bis standard or MNP level 5.

Table 15.4 Example AT modem commands

ITU recommendation	*Bit rate (bps)*	*Modulation*
V.21	300	FSK
V.22	1 200	PSK
V.22bis	2 400	ASK/PSK
V.27ter	4 800	PSK
V.29	9 600	PSK
V.32	9 600	ASK/PSK
V.32bis	14 400	ASK/PSK
V.34	28 800	ASK/PSK

15.10.1 V.42bis and MNP compression

There are two main standards used in modems for compression. The V.42bis standard is defined by the ITU and the MNP (Microcom networking protocol) has been developed by a company named Microcom. Most modems will try to compress using V.42bis but if this fails they try MNP level 5. V.42bis uses the Lempel-Ziv algorithm, which builds dictionaries of code words for recurring characters in the data stream. These code words normally take up fewer bits than the uncoded bits. V.42bis is associated with the V.42 standard which covers error correction.

15.10.2 V.22bis modems

V.22bis modems allow transmission at up to 2400 bps. It uses four amplitudes and four phases. Figure 15.12 shows the 16 combinations of phase and amplitude for a V.22bis modem. It can be seen that there are 12 different phase shifts and four different amplitudes. Each transmission is known as a symbol, thus each transmitted symbol contains four bits. The transmission rate for a symbol is 600 symbols per second (or 600 b aud), thus the bit rate will be 2400bps.

Trellis coding tries to ensure that consecutive symbols differ as much as possible.

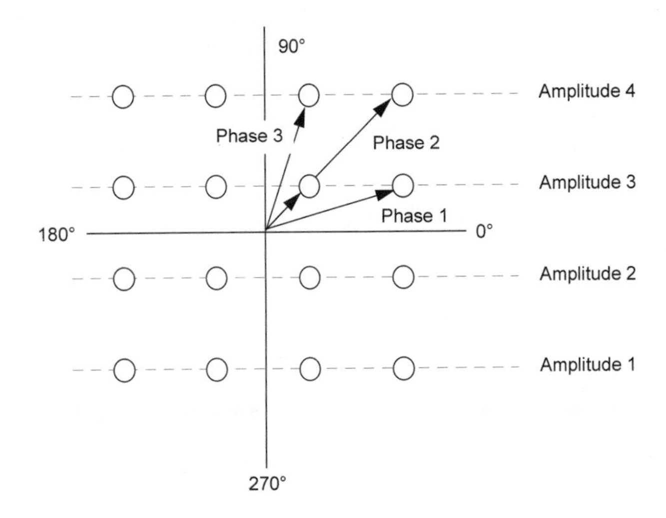

Figure 15.12 Phase and amplitude coding for V.32

15.10.3 V.32 modems

V.32 modems include echo cancellation which allows signals to be transmitted in both directions at the same time. Previous modems used different frequencies to transmit on different channels. Echo cancellation uses DSP (digital signal processing) to subtract the sending signal from the received signal.

V.32 modems use trellis encoding to enhance error detection and correction. They encode 32 signalling combinations of amplitude and phase. Each of the symbols contains four data bits and a single trellis bit. The basic symbol rate is 2400 bps; thus the actual data rate will be 9600 bps. A V.32bis modem uses seven bits per symbol; thus the data rate will be 14 400 bps (2400×6).

15.11 Fax transmission

Facsimile (fax) transmission involves the transmission of images over a telephone line using a modem. A stand-alone fax consists of:

- An image scanner.
- A graphics printer.
- A transmission/reception modem.

The fax scans an A4 image with 1142 scan lines (3.85 lines per millimetre) and 1728 pixels per line. The EIA and ITU originally produced the RS-328 standard for the transmission of analogue voltage levels to represent different brightness. The ITU recommendations are known as Group I and Group II standards. The Group III standard defines the transmission of faxes using digital transmission with 1142×1728 pixels of black or white. Group IV is an extension to Group III but allows different gray scales and also colour (unfortunately it requires a high bit rate.)

An A4 scan would consist of 1 976 832 (1142×1728) scanned elements. If each element is scanned for black and white, then, at 9600 bps, it would take over 205 s to transmit. Using RLE (run length encoding) coding can drastically reduced this transmission time.

15.11.1 Modified Huffman coding

Group III compression uses modified Huffman code to compress the transmitted bit stream. It uses a table of codes in which the most frequent run lengths are coded with a short code. Typically, documents contain long runs of white or black. A compression ratio of over 10:1 is easily achievable (thus a single-page document can be sent in under 20 s, for a 9600 bps transmission rate). Table 15.5 shows some code runs of white and Table 15.6 shows some codes for runs of black. The transmitted code always starts on white code. The codes range from 0 to 63. Values from 64 to 2560 use two codes. The first gives the multiple of 64 followed by the normally coded remainder.

Table 15.5 White run length coding

Run length	*Coding*	*Run length*	*Coding*	*Run length*	*Coding*
0	00110101	1	000111	2	0111
3	1000	4	1011	5	1100
6	1110	7	1111	8	10011
9	10100	10	00111	11	01000
12	001000	13	000011	14	110100
15	110101	16	101010	17	101011
18	0100111	19	0001100	61	00110010
62	00110011	63	00110100	EOL	00000000001

For example, if the data to be encoded is:

16 white, 4 black, 16 white, 2 black, 63 white, 10 black, 63 white

it would be coded as:

```
101010  011 101010  11 00110100  0000100  00110100
```

This would take 40 bits to transmit the coding, whereas it would take 304 bits without coding (i.e. 16+ 4+16+2+128+10+128). This results in a compression ratio of 7.6:1.

Table 15.6 Black run-length coding

Run length	*Coding*	*Run length*	*Coding*	*Run length*	*Coding*
0	0000110111	1	010	2	11
3	10	4	011	5	0011
6	0010	7	00011	8	000101
9	000100	10	0000100	11	0000101
12	0000111	13	00000100	14	00000111
15	000011000	16	0000010111	17	0000011000
18	0000001000	19	00001100111	61	000001011010
62	0000001100110	63	000001100111	EOL	00000000001

15.12 Exercises

15.12.1 What is the bandwidth of a telephone line:

(a) Almost infinite (b) 400 Hz to 3.4 kHz
(c) 400 Hz to 20 kHz (d) 400 Hz to 100 kHz

15.12.2 How does a modem transmit at 9600bps, when the symbol rate is 4800 symbols/sec (baud):

(a) It sends two bits for every symbol sent
(b) Its impossible as the bit rate is always the same as the symbol rate
(c) It hides data
(d) It uses more than one data line

15.12.3 How long does it take to transmit a 1 MB file over a 9600 bps modem connection:

(a) 13.89 min (b) 1.74 min
(c) 833.33 min (d) 104.2 min

15.12.4 What modem command is used to tone dial the number 123-456-789:

(a) AT 123456789 (b) AT 987654321
(c) ATDT 123456789 (d) DIAL 123456789

15.12.5 What character sequence is used to put the modem in the command mode:

(a) AT (b) +++
(c) HELLO? (d) +

15.12.6 Which character must appear at the end of a command string:

(a) Full stop ('.') (b) Null (ASCII, 0)
(c) Line feed (ASCII, 10) (d) Carriage return (ASCII, 13)

15.12.7 Which modem indicators would be ON when a modem has made a connection and is receiving data? Which indicators would be flashing?

15.12.8 Which modem indicators would be ON when a modem has made a connection and is sending data? Which indicators will be flashing?

15.12.9 Investigate the complete set of AT commands by referring to a modem manual or reference book.

15.12.10 Investigate the complete set of S-registers by referring to a modem manual or reference book.

15.12.11 Determine the location of modems on a network or in a works building. If possible, determine the type of data being transferred and its speed.

15.12.12 Connect a modem to a computer and dial a remote modem. If possible connect two modems together and, using a program such as `Terminal`, transfer text from one computer to the another.

15.13 Notes from the author

What a strange device a modem is. It has such as thankless task – converting information from lovely, pure digital signals into something that passes over a speech-limited voice channel. It does its job well and with compression can achieve reasonable results. But, really, it's a short-term fix, before we all use high-speed connections with proper data cables, whether they be shield twisted-pair cables or fibre optic cables. So, modems allow us to migrate our telephone connection to a full-blown network connection. The motivation for the increased bandwidth is the Internet and especially the integration of fully digital multimedia connections.

The AT command code allows for a standardisation in the modem operation, but as many have seen, modems are not quite as compatible as they may seem. Like the great RS-232 that it is based upon, it is infuriating who non-standardised modems are. I think the big problem here is that the true standard is held with a few major manufacturers, such as Hayes, and software drivers are made compatible with these modems rather than with actual standards. Sometimes industry-led standard are adopted into the market quicker than ones developed by standards organisations.

Why are modems is expensive? Why can you buy five network cards for the price of a modem, or even a whole PC motherboard? Is it because they are so useful, maybe, but it's probably because, at present, they have a virtual monopoly in the home, as their only real general-purpose competitor, ISDN, is still too expensive for its installation, running costs and costs of the equipment. So for just now, the annoying little devices that screech and whine will be around for a little longer yet. But, the people of the future will laugh when they see these archaic devices, in just the same way that we laugh at dish-washer style computers, and home computers with cassette storage and 1KB memory.

16 Parallel Port

16.1 Introduction

This chapter discusses parallel communications. The Centronics printer interface transmits eight bits of data at a time to an external device, normally a printer. A 25-pin D-type connector is used to connect to the PC and a 36-pin Centronics interface connector normally connects to the printer. This interface is not normally used for other types of interfacing as the standard interface only transmits data over the data lines in one direction, that is, from the PC to the external device. Some interface devices overcome this problem by using four of the input handshaking lines to input data and then multiplexing using an output handshaking line to multiplex them to produce eight output bits.

As technology has improved there is a great need for a bidirectional parallel port to connect to devices such as tape backup drives, CD-ROMs, and so on. The Centronics interface unfortunately lacks speed (150 kbps), has limited length of lines (2 m) and very few computer manufacturers comply with an electrical standard.

Thus, in 1991, several manufacturers (including IBM and Texas Instruments) formed a group called NPA (National Printing Alliance). Their original objective was to develop a standard for controlling printers over a network. To achieve this a bi-directional standard was developed which was compatible with existing software. This standard was submitted to the IEEE so that they could standardise it. The committee that the IEEE set up was known as the IEEE 1284 committee and the standard they produced is known as the IEEE 1284-1994 Standard (as it was released in 1994).

With this standard all parallel ports use a bidirectional link in either a compatible, nibble or byte mode. These modes are relatively slow as the software must monitor the handshaking lines (up to 100 kbps). To allow high-speed the EPP (enhanced parallel port) and ECP (extended capabilities port protocol) modes which allows high-speed data transfer using automatic hardware handshaking. In addition to the previous three modes, EPP and ECP are being implemented on the latest I/O controllers by most of the Super I/O chip manufacturers. These modes use hardware to assist in the data transfer. For example, in EPP mode, a byte of data can be transferred to the peripheral by a simple OUT instruction. The I/O controller handles all the handshaking and data transfer to the peripheral.

16.2 PC connections

Figure 16.1 shows the pin connections on the PC connector. The data lines (D0–D7) output data from the PC and each of the data lines has an associated ground line (GND).

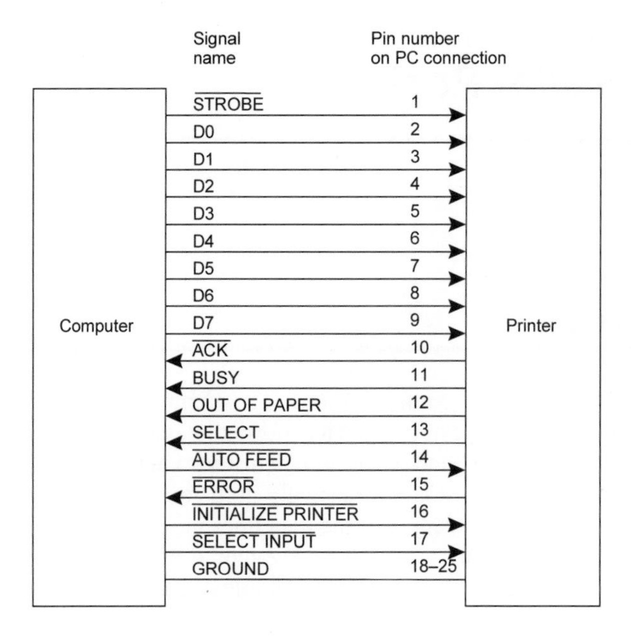

Figure 16.1 Centronics parallel interface showing pin numbers on PC connector

16.3 Data handshaking

The main handshaking lines are $\overline{\text{ACK}}$, BUSY and $\overline{\text{STROBE}}$. Initially the computer places the data on the data bus, then it sets the $\overline{\text{STROBE}}$ line low to inform the external device that the data on the data bus is valid. When the external device has read the data, it sets the $\overline{\text{ACK}}$ lines low to acknowledge that it has read the data. The PC then waits for the printer to set the BUSY line inactive, that is, low. Figure 16.2 shows a typical handshaking operation and Table 16.1 outlines the definitions of the pins.

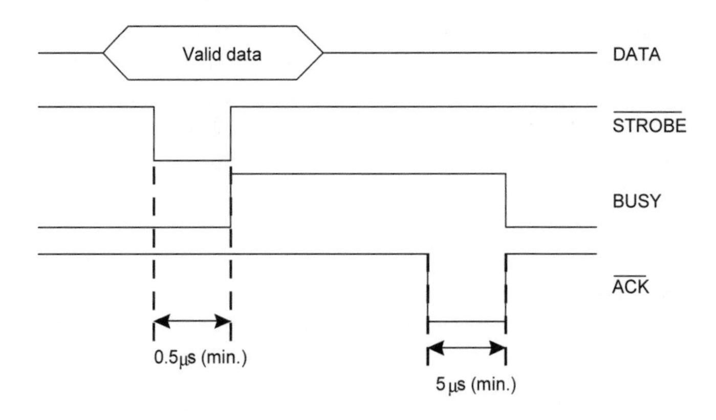

Figure 16.2 Data handshaking with the Centronics parallel printer interface

The parallel interface can be accessed either by direct reads to and writes from the I/O memory addresses or from a program which uses the BIOS printer interrupt. This interrupt allows a program either to get the status of the printer or to write a character to it. Table 16.2 outlines the interrupt calls.

Table 16.1 Signal definitions

Signal	*In/out*	*Description*
$\overline{\text{STROBE}}$	Out	Indicates that valid data is on the data lines (active low)
$\overline{\text{AUTO FEED}}$	Out	Instructs the printer to insert a line feed for every carriage return (active low)
$\overline{\text{SELECT INPUT}}$	Out	Indicates to the printer that it is selected (active low)
INIT	Out	Resets the printer
$\overline{\text{ACK}}$	In	Indicate that the last character was received (active low)
BUSY	In	Indicates that the printer is busy and thus cannot accept data
OUT OF PAPER	In	Out of paper
SELECT	In	Indicates that the printer is on line and connected
$\overline{\text{ERROR}}$	In	Indicates that an error exists (active low)

Table 16.2 BIOS printer interrupt

Description	*Input registers*	*Output registers*
Initialise printer port	AH = 01h DX = printer number (00h–02h)	AH = printer status bit 7: not busy bit 6: acknowledge bit 5: out of paper bit 4: selected bit 3: I/O error bit 2: unused bit 1: unused bit 0: timeout
Write character to printer	AH = 00h AL = character to write DX = printer number (00h–02h)	AH = printer status
Get printer status	AH = 02h DX = printer number (00h–02h)	AH = printer status

16.3.1 BIOS printer

Program 16.1 uses the BIOS printer interrupt to test the status of the printer and output characters to the printer.

Program 16.1

```
#include <dos.h>
#include <stdio.h>
#include <conio.h>

#define  PRINTERR -1

void  print_character(int ch);
int   init_printer(void);

int   main(void)
{
int   status,ch;

      status=init_printer();
      if (status==PRINTERR) return(1);

      do
      {
         printf("Enter character to output to printer");
         ch=getch();
         print_character(ch);
      } while (ch!=4);
      return(0);
}

int   init_printer(void)
{
union REGS inregs,outregs;

      inregs.h.ah=0x01; /* initialize printer */
      inregs.x.dx=0; /* LPT1: */
      int86(0x17,&inregs,&outregs);
      if (inregs.h.ah & 0x20)
      { puts("Out of paper"); return(PRINTERR); }
      else if (inregs.h.ah & 0x08)
      { puts("I/O error"); return(PRINTERR); }
      else if (inregs.h.ah & 0x01)
      { puts("Printer timeout"); return(PRINTERR); }

      return(0);
}

void  print_character(int ch)
{
union REGS inregs,outregs;

      inregs.h.ah=0x00; /* print character */
      inregs.x.dx=0; /* LPT1: */
      inregs.h.al=ch;

      int86(0x17,&inregs,&outregs);
}
```

16.4 I/O addressing

16.4.1 Addresses

The printer port has three I/O addresses assigned for the data, status and control ports. These addresses are normally assigned to:

Printer	Data register	Status register	Control register
LPT1	378h	379h	37ah
LPT2	278h	279h	27ah

The DOS debug program can be used to display the base addresses for the serial and parallel ports by displaying the 32 memory location starting at `0040:0008`. For example:

```
-d 40:00
0040:0000  F8 03 F8 02 00 00 00 00-78 03 00 00 00 00 29 02
```

The first four 16-bit addresses give the serial communications ports. In this case, there are two COM ports at address 03F8h (COM1) and 02F8h (for COM2). The next four 16-bit addresses gives the parallel port addressees. In this case there is two parallel ports. One at 0378h (LPT1) and one at 0229h (LPT4).

16.4.2 Output lines

Figure 16.3 shows the bit definitions of the registers. The data port register links to the output lines. Writing a 1 to the bit position in the port sets the output high, while a 0 sets the corresponding output line to a low. Thus to output the binary value 1010 1010b (AAh) to the parallel port data then using Visual C++:

```
_outp(0x378,0xAA);   /* in Visual C this is _outp(0x378,0xAA); */
```

The output data lines are each capable of sourcing 2.6 mA and sinking 24 mA; it is thus essential that the external device does not try to pull these lines to ground.

The control port also contains five output lines, of which the lower four bits are $\overline{\text{STROBE}}$, $\overline{\text{AUTO FEED}}$, INIT and $\overline{\text{SELECT INPUT}}$, as illustrated in Figure 16.3. These lines can be used as either control lines or as data outputs. With the data line, a 1 in the register gives an output high, while the lines in the control port have inverted logic. Thus a 1 to a bit in the register causes an output low.

Program 16.2 outputs the binary pattern 0101 0101b (55h) to the data lines and sets $\overline{\text{SELECT INPUT}}$=0, INIT=1, $\overline{\text{AUTO FEED}}$=1, and $\overline{\text{STROBE}}$=0, the value of the data port will be 55h and the value written to the control port will be XXXX 1101 (where X represents don't care). The value for the control output lines must be invert, so that the $\overline{\text{STROBE}}$ line will be set to a 1 so that it will be output as a LOW.

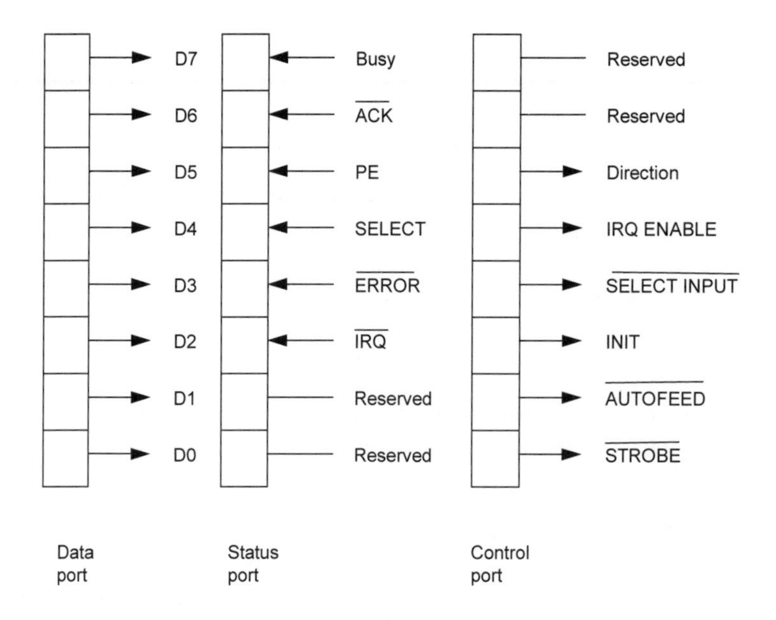

Figure 16.3 Port assignments

Program 16.2

```
#define DATA           0x378
#define STATUS         DATA+1
#define CONTROL        DATA+2

int    main(void)
{
int out1,out2;
      out1 = 0x55;                       /* 0101 0101 */
      _outp(DATA, out1);
      out2 = 0x0D;                       /* 0000 1101 */
      _outp(CONTROL, out2);              /* STROBE=LOW, AUTOFEED=HIGH, etc */
      return(0);
}
```

The setting of the output value (in this case, `out2`) looks slightly confusing as the output is the inverse of the logical setting (that is, a 1 sets the output low). An alternative method is to exclusive-OR (EX-OR) the output value with $B which will invert the 1st, 2nd and 4th least significant bits ($\overline{\text{SELECT INPUT}}$=0, $\overline{\text{AUTO FEED}}$=1, and $\overline{\text{STROBE}}$=0), while leaving the 3rd least significant bit (INIT) untouched. Thus the following will achieve the same as the previous program:

```
      out2 = 0x06;                          /* 0000 0110 */
      _outp(CONTROL, out2 ^ 0xb);   /* STROBE=LOW, AUTOFEED=HIGH, etc */
```

If the 5th bit on the control register (IRQ enable) is written as 1 then the output on this line will go from a high to a low which will cause the processor to be interrupted.

The control lines are driven by open collector drivers pulled to +5 Vdc through 4.7 kΩ resistors. Each can sink approximately 7 mA and maintain 0.8 V down-level.

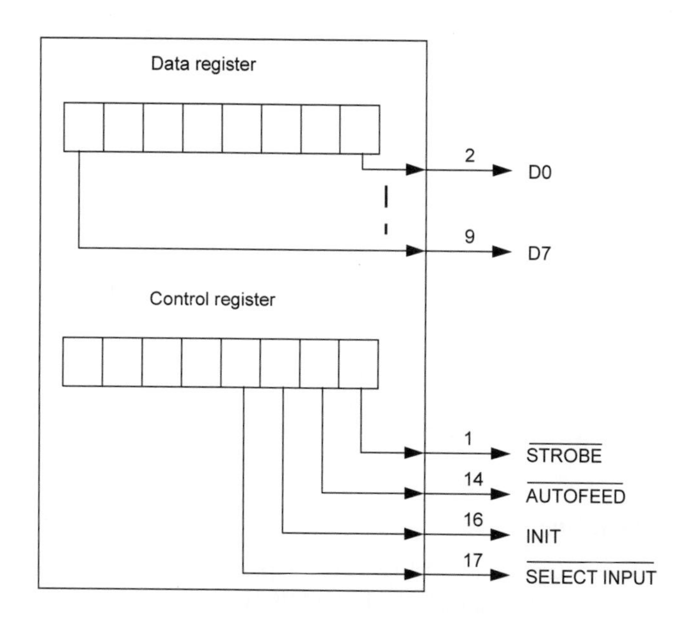

Figure 16.4 Output lines

16.4.3 Inputs

There are five inputs from the parallel port (BUSY, $\overline{\text{ACK}}$, PE, SELECT and $\overline{\text{ERROR}}$). The status of these lines can be found by simply reading the upper 5 bits of the status register, as illustrated in Figure 16.5.

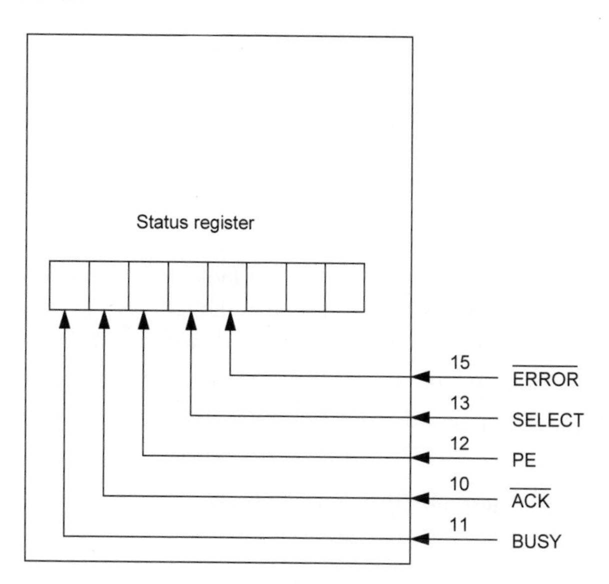

Figure 16.5 Input lines

Unfortunately, the BUSY line has an inverted status. Thus when a LOW is present on BUSY, the bit will actually be read as a 1. For example Program 16.3 reads the bits from the status register, inverts the BUSY bit and then shifts the bits three places to the right so that the five inputs bit are in the five least significant bits.

Program 16.3

```
#include          <stdio.h>
#define DATA      0x378
#define STATUS    DATA+1
int   main(void)
{
unsigned int in1;

      in1 = _inp(STATUS); /* read from status register */
      in1 = in1 ^ 0x80        /* invert  BUSY bit  */
      in1 = in1 >> 3;              /* move bits so that the inputs are the least
                                   significant bits */
      printf("Status bits are %d\n",in1);
      return(0);
}
```

16.4.4 Electrical interfacing

The output lines can be used to drive LEDs. Figure 16.6 shows an example circuit where a LOW output will cause the LED to be ON while a HIGH causes the output to be OFF. For an input an open push button causes a HIGH on the input.

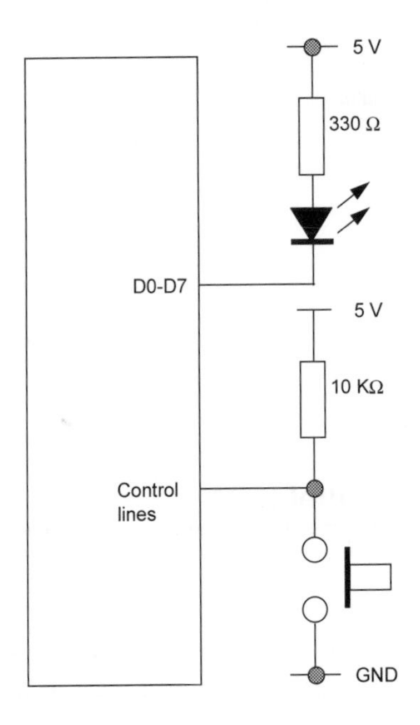

Figure 16.6 Interfacing to inputs and outputs

16.4.5 Simple example

Program 16.4 uses a push button connected to pin 11 (BUSY). When the button is open then the input to BUSY will be a HIGH and the most significant bit in the status register will thus be a 0 (as the BUSY signal is inverted). When the button is closed then this bit will be a 1.

This is tested with

```
if (in1&0x80)==1)
```

When this condition is TRUE (that is, when the button is closed) then the output data lines (D0–D7) will flash on and off with a delay of 1 second between flashes. An output of all 1s to the data lines causes the LEDs to be off, and all 0s cause the LEDs to be on.

Program 16.4

```
/*    Flash LEDs on and off when the push button connected to BUSY      */
/*    is closed                                                          */
#include <stdio.h>
#include <dos.h>

#define DATA       0x378
#define STATUS     DATA+1
#define CONTROL    DATA+2

int main(void)
{
int in1;
   do
   {
      in1 = _inp(STATUS);

      if (in1&0x80)==1)   /* if switch closed this is TRUE  */
      {
         _outp(DATA,0x00);    /* LEDs on */
         delay(1000);
         _outp(DATA, 0xff);    /* LEDs off    */
         delay(1000);
      }
      else
         _outp(DATA,0x01); /* switch open */
      } while (!kbhit());
      return(0);
}
```

16.5 Interrupt-driven parallel port

16.5.1 Introduction

The previous section discusses how the parallel port is used to output data. This chapter discusses how an external device can interrupt the processor. It does this by hooking onto the interrupt server routine for the interrupt that the port is attached to. Normally this interrupt routine serves as a printer interrupt (such as lack of paper, paper jam and so on). Thus, an external device can use the interrupt service routine to transmit data to or from the PC.

16.5.2 Interrupts

Each parallel port is hooked to an interrupt. Normally the primary parallel port is connected to IRQ7. It is assumed in this section that this is the case. As with the serial port this interrupt line must be enabled by setting the appropriate bit in the interrupt mask register (IMR), which is based at address 21h. The bit for IRQ7 is the most significant bit, and it must be set to a 0 to enable the interrupt. As with the serial port, the end of interrupt signal must be acknowledged by setting the EOI signal bit of the interrupt control register (ICR) to a 1. See Section 8.5.2 for more information on these operations.

The interrupt on the parallel port is caused by the $\overline{\text{ACK}}$ line (pin 10) going from a high to a low (just as a printer would acknowledge the reception of a character). For this interrupt to be passed to the PIC then bit 4 of the control port (IRQ Enable) must be set to a 1.

16.5.3 Example program

Program 16.5 is a simple interrupt-driven parallel port Borland C program. The program interrupts the program each time the $\overline{\text{ACK}}$ line is pulled LOW. When this happens the output value should change corresponding to a binary count (0000 0000 to 1111 1111, and then back again). The user can stop the program by pressing any key on the keyboard. Figure 16.7 shows a sample set-up with a push button connected to the $\overline{\text{ACK}}$ line and LEDs connected to the output data lines.

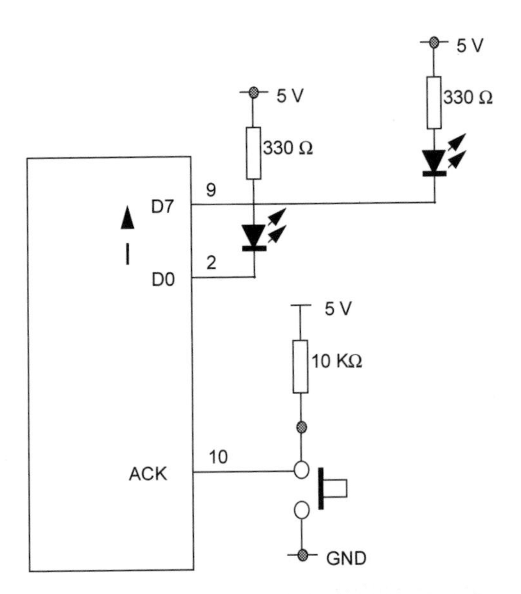

Figure 16.7 Example set-up for interrupt-driven parallel port

Program 16.5

```
/* Program to sample data from the parallel port   */
/* when the ACK line goes low                      */
#include <stdio.h>
#include <bios.h>
#include <conio.h>
#include <dos.h>
#define  TRUE     1
#define  FALSE    0
#define  DATA     0x378
#define  STATUS   DATA+1
#define  CONTROL  DATA+2
#define  IRQ7     0x7F  /* LPT1 interrupt */
```

```
#define  EOI      0x20  /* End of Interrupt */
#define  ICR      0x20  /* Interrupt Control Register */
#define  IMR      0x21  /* Interrupt Mask Register */

void  interrupt far pl_interrupt(void);
void  setup_parallel (void);
void  set_vectors(void);
void  enable_interrupts(void);
void  disable_interrupts(void);
void  reset_vectors(void);
void  interrupt far (*oldvect)();
int   int_flag = TRUE;
int   outval=0;

int main(void)
{
      set_vectors();
      setup_parallel();
      do
      {
         if (int_flag)
         {
            printf("New value sent\n");
            int_flag=FALSE;
         }
      } while (!kbhit());
      reset_vectors();
      return(0);
}

void  setup_parallel(void)
{

   outportb(CONTROL, inportb(CONTROL) | 0x10);
                  /* Set Bit 4 on control port to a 1 */
}

void interrupt far pl_interrupt(void)
{
      disable();
      outportb(DATA,outval);
      if (outval!=255) outval++; else outval=0;
      int_flag=TRUE;
      outportb(ICR,EOI);
      enable();
}

void set_vectors(void)
{
int int_mask;
      disable();                   /* disable all ints        */
      oldvect=getvect(0x0f);       /* save any old vector     */
      setvect (0x0f,pl_interrupt); /* set up for new int serv         */
}

void  enable_interrupts(void)
{
int ch;
      disable();
      ch=inportb(IMR);
      outportb(IMR, ch & IRQ7);
      enable();
```

```
}


void  disable_interrupts(void)
{
int ch;
       disable();
       outportb(IMR, ch & ~IRQ7);
       enable();
}

void  reset_vectors(void)
{
       setvect(0x0f,oldvect);
}
```

16.5.4 Program explanation

The initial part of the program enables the interrupt on the parallel port by setting bit 4 of the control register to 1:

```
void  setup_parallel(void)
{

   outportb(CONTROL, inportb(CONTROL) | 0x10); /* Set Bit 4 on control port*/
}
```

After the serial port has been initialized the interrupt service routine for the `IRQ7` line is set to point to a new 'user-defined' service routine. The primary parallel port `LPT1:` normally sets the `IRQ7` line active when the $\overline{\text{ACK}}$ line goes from a high to a low. The interrupt associated with `IRQ7` is 0Fh (15). The `getvect()` function gets the ISR address for this interrupt, which is then stored in the variable `oldvect` so that at the end of the program it can be restored. Finally, in the `set_vectors()` function, the interrupt assigns a new 'user-defined' ISR (in this case it is the function `pl_interrupt()`):

```
void set_vectors(void)
{
int int_mask;
      disable();  /* disable all ints */
      oldvect=getvect(0x0f);  /* save any old vector */
      setvect (0x0f,pl_interrupt);  /* set up for new int serv */
}
```

At the end of the program the ISR is restored with the following code:

```
void  reset_vectors(void)
{
      setvect(0x0f,oldvect);
}
```

To enable the `IRQ7` line on the PIC, bit 5 of the IMR (interrupt mask register) is to be set to a 0 (zero). The statement

```
      ch = inportb(IMR) & 0x7F;
```

achieves this as it bitwise ANDs all the bits, except for bit 7, with a 1. This is because any bit which is ANDed with a 0 results in a 0. The bit mask `0x7F` has been defined with the macro `IRQ7`:

```
void  enable_interrupts(void)
{
int ch;
      disable();
      ch=inportb(IMR);
      outportb(IMR, ch & IRQ7);
      enable();
}
```

At the end of the program the interrupt on the parallel port is disabled by setting bit 7 of the IMR to a 1; this disables `IRQ7` interrupts:

```
void  disable_interrupts(void)
{
int ch;
      disable();
      outportb(IMR, ch & ~IRQ7);
      enable();
}
```

The ISR for the `IRQ7` function is set to `pl_interrupt()`. It outputs the value of `outval`, which is incremented each time the interrupt is called (note that there is a roll-over statement which resets the value of `outval` back to zero when its value is 255). At the end of the ISR the end of interrupt flag is set in the interrupt control register with the statement `outportb(ICR, EOI);`, as follows:

```
void interrupt far pl_interrupt(void)
{
      disable();
      outportb(DATA,outval);
      if (outval!=255) outval++; else outval=0;
      int_flag=TRUE;
      outportb(ICR,EOI);
      enable();
}
```

The `main()` function calls the initialisation and the de-initialisation functions. It also contains a loop which continues until any key is pressed. Within this loop, the keyboard is tested to determine if a key has been pressed. The interrupt service routine sets `int_flag`. If the main routine detects that it is set it displays the message ‘New value sent’ and resets the flag:

```
int main(void)
{
   set_vectors();
   outportb(CONTROL, inportb(CONTROL) | 0x10);
                  /* set bit 4 on control port to logic one */
      do
      {
         if (int_flag)
         {
            printf("New value sent\n");
            int_flag=FALSE;
         }
   } while (!kbhit());
```

```
    reset_vectors();

    return(0);
}
```

16.6 Exercises

16.6.1 How many pins does a standard D-type parallel port connector have:

(a) 9 (b) 12
(c) 25 (d) 36

16.6.2 How many data bits can the parallel port transmit at a time:

(a) 8 (b) 12
(c) 16 (d) 32

16.6.3 What is the major limitation of a standard Centronics parallel port:

(a) It is only an output (b) It is not compatible with many printers
(c) Incompatibility of software (d) Limited cable types

16.6.4 What is the maximum data of a standard Centronics parallel port:

(a) 15 kbps (b) 150 kbps
(c) 1.5 Mbps (d) 10 Mbps

16.6.5 What is the standard I/O base address for a standard parallel port:

(a) 3F8h (b) 378h
(c) 2F8h (d) 278h

16.6.6 What is the standard I/O base address for a secondary parallel port:

(a) 3F8h (b) 378h
(c) 2F8h (d) 278h

16.6.7 What is the standard interrupt line for a standard parallel port:

(a) IRQ3 (b) IRQ4
(c) IRQ5 (d) IRQ7

16.6.8 Write a program that sends a 'walking-ones' code to the parallel port. The delay between changes should be 1 second. A 'walking-ones' code is as follows:

```
00000001
00000010
00000100
00001000
   : :
10000000
00000001
00000010
```
and so on.

Hint: Use a `do...while` loop with either the shift left operators (`<<`) or output the values 0x01, 0x02, 0x04, 0x08, 0x10, 0x20, 0x40, 0x80, 0x01, 0x02, and so on.

16.6.9 Write separate programs which output the patterns in (a) and (b). The sequences are as follows:

(a)
```
00000001
00000010
00000100
00001000
00010000
00100000
01000000
10000000
01000000
00100000
00010000
   ::
00000001
00000010
```
and so on.

(b)
```
10000001
01000010
00100100
00011000
00100100
01000010
10000001
01000010
00100100
00011000
00100100
```
and so on.

16.6.10 Write separate programs which output the following sequences:

(a)
```
1010 1010
0101 0101
1010 1010
0101 0101
```
and so on.

(b)
```
1111 1111
0000 0000
1111 1111
0000 0000
```
and so on.

(c)
```
0000 0001
0000 0011
0000 1111
0001 1111
0011 1111
0111 1111
1111 1111
0000 0001
0000 0011
0000 0111
0000 1111
0001 1111
```
and so on.

(d)
```
0000 0001
0000 0011
0000 0111
0000 1111
0001 1111
0011 1111
0111 1111
1111 1111
0111 1111
0011 1111
0001 1111
0000 1111
```
and so on.

(e) The inverse of (d) above.

16.6.11 Binary coded decimal (BCD) is used mainly in decimal displays and is equivalent to the decimal system where a 4-bit code represents each decimal number. The first 4 bits represent the units, the next 4 the tens, and so on. Write a program that outputs to the parallel port a BCD sequence with a 1-second delay between changes. A sample BCD table is given in Table 16.3. The output should count from 0 to 99.

Hint: One possible implementation is to use two variables to represent the units and tens. These would then be used in a nested loop. The resultant output value will then be `(tens << 4)+units`. An outline of the loop code is given next.

```
for (tens=0;tens<10;tens++)
   for (units=0;units<10;units++)
   {
   }
```

Table 16.3 BCD conversion

Digit	*BCD*
00	00000000
01	00000001
02	00000010
03	00000011
04	00000100
05	00000101
06	00000110
07	00000111
08	00001000
09	00001001
10	00010000
11	00010001
.	.
.	.
.	.
97	10010111
98	10011000
99	10011001

16.6.12 Write a program which interfaces to a 7-segment display and displays an incremented value every second. Each of the segments should be driven from one of the data lines on the parallel port. For example:

Value	Segment							Hex
	A	B	C	D	E	F	G	value
0	1	1	1	0	1	1	1	77h
1	0	0	1	0	0	1	0	12h
2	1	1	0	1	0	1	1	6Bh
:				:	:			
9	0	0	1	1	1	1	1	1Fh

Two ways of implementing this is either to determine the logic for each segment or to have a basic look-up table, such as:

```
int seg_val[8]={0x77, 0x12, 0x6B, ... 0x1F};
   val=seq_val(count % 10);
            /* mask-off the least-significant digit */
         outportb(0x378,seg_val[val]);
```

16.6.13 Write a program counts the number of pushes of a button. The display should show the value.

16.6.14 Modify the program developed in Exercise 16.6.13 so that it outputs the count value to the parallel port.

16.6.15 Modify the program developed in Exercise 16.6.14 so that the display is incremented when the user presses a button.

16.6.16 Write a program in which the user presses a button which causes the program to read from the parallel port.

16.6.17 Write a printer driver in which a string buffer is passed to it and this is then outputted to the printer. The driver should include all the correct error checking (such as out-of-paper, and so on).

16.7 Notes from the author

The parallel port is hardly the greatest piece of technology. In its truly standard form, it only allows for simplex communications, from the PC outwards. However, like the RS-232 port, it's a standard part of the PC, and its cheap. So, interface designers have worked under difficult circumstances to try and improving its specification, such as increasing its bit rate and allowing multiple devices to connect to it at the same time, but it still suffers from a lack of controllability. Anyone who has changed the interface of a device from the parallel port to the USB will know how much better the USB port is over the parallel port.

The parallel port and RS-232 are the two top requests that I get from users, especially related to project work. The Top 10 requests, in order of the most requests I have received, are:

1. RS-232.	*6. Interrupt-driven software.*
2. Parallel Port.	*7. PCMCIA.*
3. Converting a DOS program to Microsoft Windows.	*8. Network card design.*
4. Borland Delphi interfacing.	*9. Visual Basic interfacing*
5. ISA card design.	*10. Using buffered systems.*

One of the most amusing emails that I ever received related to an ISA card which I had drawn. In the card, I had drawn a few chips, to basically show that it had some electronics on it. So that the chips would not be confused with real chips I labelled one of them XYZ123. One user sent me an email saying:

'Thanks for ... Please could you tell me the function of the XYZ123 device. I have searched for this component, and cannot find any information on it. Please could you send me some'

I didn't really have the heart to write back to the user and say that it was a made-up chip, so I sent an email back saying that it was not available at the present time (which was true).

So why has the serial port become more popular than the parallel port. Well it's because of one reason: since PC's started, the serial port has always been a standard port and most manufacturers abide with it, whereas the parallel port was a quick fix so that the original PC could communicate with a printer. In its standard form, it can only send information in a single direction, and, even worse, only eight bits can be sent at a time. Nevertheless, it has survived, and now has several uses, especially with printers, scanners and external CD-ROMs. So it will hold the fort for a few years yet before the USB port takes over in creating a truly integrated bus system. But, you may say, the USB port is serial. So why transmit one bit at a time when you can transmit 8 or 16 or even 32 bits at a time. Well it's all to do with the number of wires that must be connected. A serial bus always has the advantage over a parallel bus, in that you only really need one signal line in a serial bus to transmit all the data. This saves space in both the connector, and in the cable. It is also cheaper to install.

Personally, I think that there is no better bus for a student to start to learn how to interface to external devices. It is relatively easy to build the interface electronics, and to connect a few LEDs. How great it is to see a student's face after they have written their first program to make a few LEDs flash on and off. I remember a third year student commented: 'I've been programming for three years, and finally, we're doing something real.' Whether you agree with this comment or not depends on the type of programming that you would like to do. Some of us like doing databases, some like writing user-interfaces, but there are lots who like to make computers sense things and make physical things happen. In the past, especially in the 1970s and 1980s, electronic engineers used breadboards and wires to prototypes circuits. Sometimes the circuits blew-up, or times they would stop working, but at least you knew where you were with the electronics. These days with massively integrated circuits, it is difficult to know one end of a microchip from another. They normally work first time, they're easy to connect to, and when they don't work you just throw them in the bin. Image the size of the bin that would have been required if someone had had build a Pentium processor from the discrete transistors (over 20 million of them). Image the heat that would have been generated. Assuming 15mW for each transistor, the total power would be 300kW, which is equivalent to the heat given of by 3000 100W light builds, or 300 1kW heaters. So it shows how far we have come in such a short time, as now we can touch the processor, and it just feels a little hot. Personally, I would have no problems in going back to the days when transistors had three legs and a tin hat, and you had to look up a data sheet to tell which of the legs was the base, and which was the collector.

So, as the technology has moved on, the parallel port seems like an old friend. It has watched the PC develop as the inners have become more integrated and faster, but it has never really been a high flier, preferring instead to quietly perform its duties without much bother. From CGA and EGA to VGA, from the serial port to the USB port, from 5.25inch floppy disks to 6550MB CD-ROMs, and so on. But, there's no way that the parallel port could be allowed to stay as it was in the original parallel specification. It has potential, but that potential is severely limited because it must always keep compatibility with previous ports. So how is it possible to connect a printer on the parallel port, and other devices, without the printer reading communications that are destined for another device. If it wasn't the PC, the designers would have simply ripped up the original specification, and started again. But, you don't do that with the PC, or you'll not sell. So, we'll see in the next chapter how the parallel port has been dragged into the modern age. But, as we'll see, it's more like a difficult toddler, than an enterprising businessman. The prize for the best upgrade goes to Ethernet, which has increased its transmission rate by a factor of 100 (10Mbps to 1Gbps).

17 Enhanced Parallel Port

17.1 Introduction

The Centronics parallel port only allows data to be sent from the host to a peripheral. To overcome this the IEEE published the 1284 standard, entitled 'Standard Signaling Method for a Bidirectional Parallel Peripheral Interface for Personal Computers'. It allows for bi-directional communication and high communication speeds, while being backwardly compatible with existing parallel ports.

The IEEE-1284 standard defines the following modes:

- Compatibility mode (forward direction only) – This mode defines the transfer of data between the PC and the printer (Centronics mode, as covered in the previous chapter).
- Nibble mode (reverse direction) – This mode defines how four bits are transferred, at a time, using status lines for the input data (sometimes known as Hewlett Packard Bitronics). The Nibble mode can thus be used for bidirectional communication, with the data lines being used as outputs. To input a byte, requires two nibble cycles.
- Byte mode (reverse direction) – This mode defines how eight bits are transferred at a time.
- Enhanced parallel port (EPP) – This mode defines a standard bidirectional communications and is used by many peripherals, such as CD-ROMs, tape drives, external hard disks, and so on.

In the IEEE 1284 standard the control and status signal for nibble, byte and EPP modes have been renamed. It also classifies the modes as forward (data goes from the PC), reverse (data is input into the PC) and bidirectional. Both the compatibility and nibble modes can be implemented with all parallel ports (as the nibble mode uses the status lines and the compatibility mode only outputs data). Some parallel ports support input and output on the data lines and thus support the byte mode. This is usually implemented by the addition of a direction bit on the control register.

17.2 Compatibility mode

The compatibility mode was discussed in Chapter 16. In this mode, the program sends data to the data lines and then sets the $\overline{\text{STROBE}}$ LOW and then HIGH. These then latch the data to the printer. The operations that the program does are:

1. Data is written to the data register.

2. The program reads from the status register to test to see if the BUSY signal is LOW (that is, the printer is not busy)
3. If the printer is not busy then the program sets the $\overline{\text{STROBE}}$ line active LOW.
4. Program then makes the $\overline{\text{STROBE}}$ line HIGH by de-asserting it.

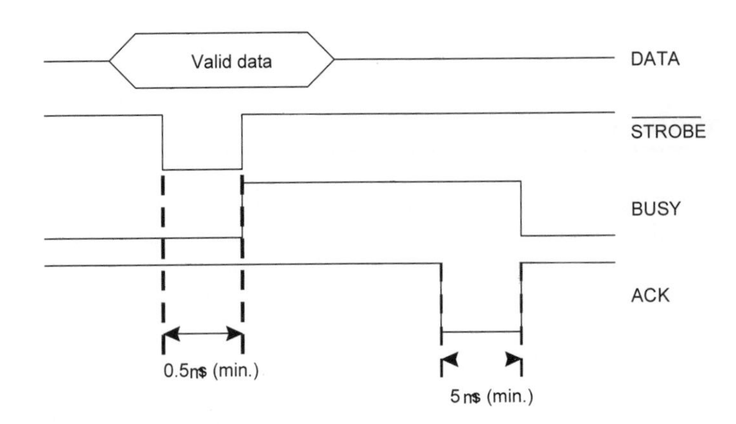

Figure 17.1 Compatibility mode transfer

17.3 Nibble mode

This mode defines how four bits are transferred, at a time, using status lines for the input data (sometimes known as Hewlett Packard Bi-tronics). The Nibble mode can thus be used for bi-directional communication, with the data lines being used as outputs. To input a byte, requires two nibble cycles.

As seen in Chapter 16 there are five inputs from the parallel port (BUSY, $\overline{\text{ACK}}$, PE, SELECT and $\overline{\text{ERROR}}$). The status of these lines can be found by simply reading the upper five bits of the status register. The BUSY, PE, SELECT and $\overline{\text{ERROR}}$ are normally used as $\overline{\text{ACK}}$ used to interrupt the processor.

Table 17.1 defines the names of the signal in the nibble mode and Figure 17.2 shows the handshaking for this mode.

The nibble mode has the following sequence:

1. Host (PC) indicates that it is ready to receive data by setting HostBusy LOW.
2. The peripheral then places the first nibble on the status lines.
3. The peripheral indicates that the data is valid on the status line by setting PtrClk low.
4. The host then reads from the status lines and sets HostBusy high to indicate that it has received the nibble, but it is not yet ready for another nibble.
5. The peripheral sets PtrClk HIGH as an acknowledgement to the host.
6. Repeat steps 1–5 for the second nibble.

Table 17.1 Nibble mode signals

Compatibility signal name	*Nibble mode name*	*In/out*	*Description*
$\overline{\text{STROBE}}$	$\overline{\text{STROBE}}$	Out	Not used.
$\overline{\text{AUTO FEED}}$	HostBusy	Out	Host nibble mode handshake signal. It is set LOW to indicate that the host is ready for nibble and set HIGH when the nibble has been received.
$\overline{\text{SELECT INPUT}}$	1284Active	Out	Set HIGH when the host is transferring data.
$\overline{\text{INIT}}$	$\overline{\text{INIT}}$	Out	Not used.
$\overline{\text{ACK}}$	PtrClk	In	Indicates valid data on the status lines. It is set low to indicate that there is valid data on the control lines and then set HIGH when the HostBusy going high.
BUSY	PtrBusy	In	Data bit 3 for one cycle then data bit 7.
PE	AckDataReq	In	Data bit 2 for one cycle then data bit 6.
SELECT	Xflag	In	Data bit 1 for one cycle then data bit 5.
$\overline{\text{ERROR}}$	$\overline{\text{DataAvail}}$	In	Data bit 0 for one cycle then data bit 4.
D0–D7	D0–D7		Not Used.

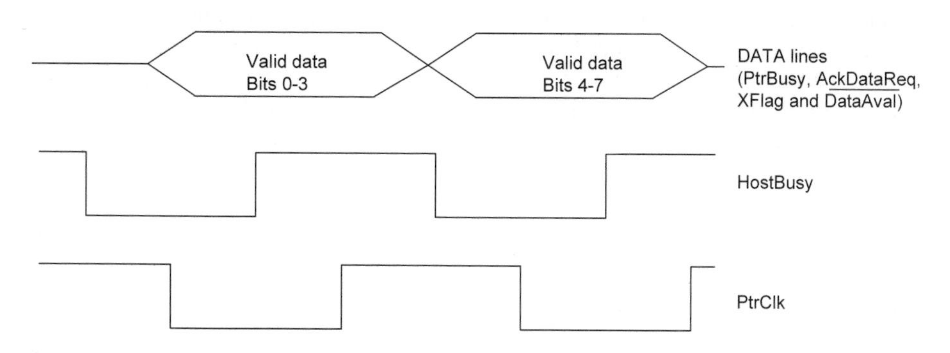

Figure 17.2 Nibble mode data transfer cycle

These operations are software intensive as the driver requires to set and read the handshaking lines. This limits transfer to about 50 kBps. Its main advantage is that it will work with all printer ports because it uses the standard Centronics set-up and is normally used in low-speed bi-directional operations, such as ADC adapters, reading data from switches, and so on.

Figure 17.3 illustrates the operation of the nibble mode, where four data bits are read into the parallel port using the four input handshaking lines. The status of these lines is then read by interrogating the upper four bits of the status register. This method is fine when there is no handshaking and when there are four, or less, data bits to be read in. If there are more, or if there is handshaking, then extra circuitry is required.

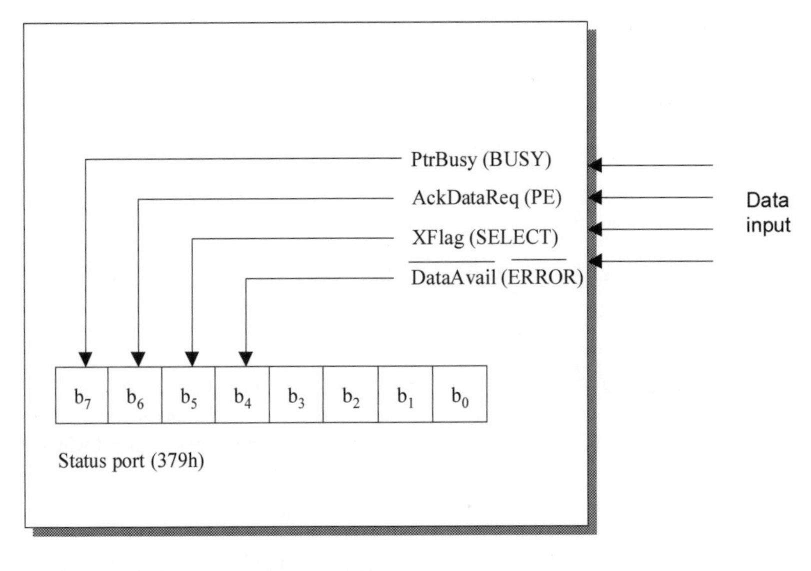


```
inval=(_inp(0x379) & 0xf0) >> 4
```

Figure 17.3 Nibble mode interfacing

Figure 17.4 shows how the nibble mode can be used to read-in eight bits at a time. For this one bit of the data output lines (D0) is used to select either the upper four bits or the lower four bits of the 8-bit data byte. If D0 is a low (0) then the lower four bits are selected, else if it is a high (1) then the upper four bits are selected. The D0 output line connects to a multiplexor which will select the lower or the upper four bits. If A is the multiplexor selector line, X[1:4] are the input data bits and Z[1:4] the output from the multiplexor, then the equation for the multiplexor is

$$Z[1] = AX[5] + \bar{A}X[1]$$

$$Z[2] = AX[6] + \bar{A}X[2]$$

$$Z[3] = AX[7] + \bar{A}X[3]$$

$$Z[4] = AX[8] + \bar{A}X[4]$$

A
X[1]
X[5]
Z[1]

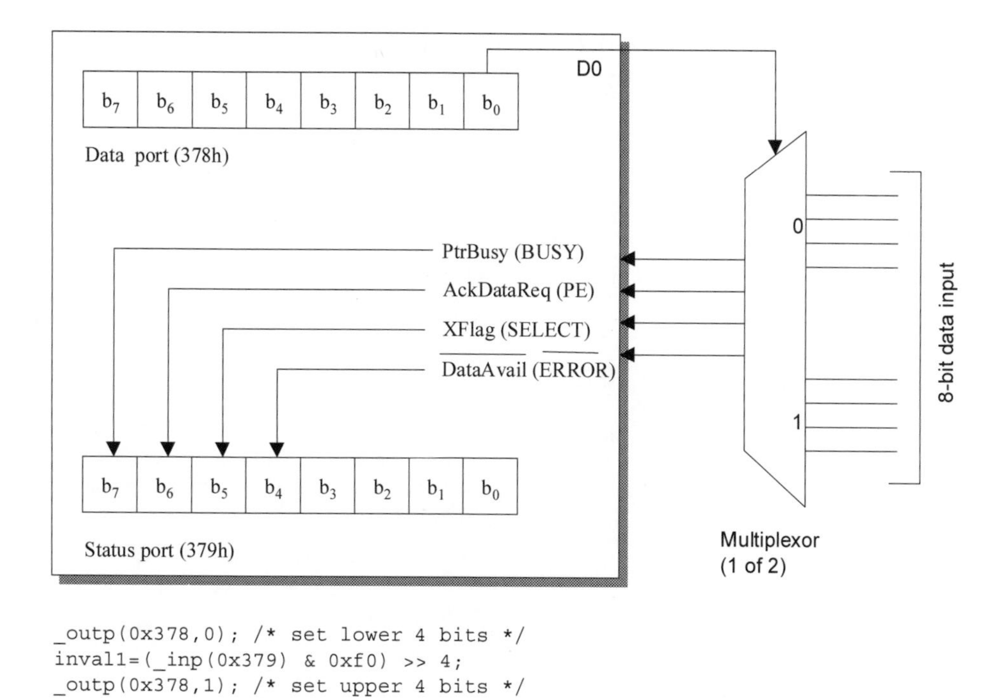


```
_outp(0x378,0); /* set lower 4 bits */
invall=(_inp(0x379) & 0xf0) >> 4;
_outp(0x378,1); /* set upper 4 bits */
inval=(_inp(0x379) & 0xf0) +invall;
```

Figure 17.4 Nibble mode for 8-bit input

17.4 Byte mode

The byte mode is often known as a bidirectional port and it uses bidirectional data lines. It has the advantage over nibble mode in that it only takes a single cycle to transfer a byte. Unfortunately, it is only compatible with newer ports. Table 17.2 defines the names of the signal in the nibble mode and Figure 17.5 shows the handshaking for this mode.

The byte mode has the following sequence:

1. Host (PC) indicates that it is ready to receive data by setting HostBusy LOW.
2. The peripheral then places the byte on the status lines.
3. The peripheral indicates that the data is valid on the status line by setting PtrClk LOW.
4. The host then reads from the data lines and sets HostBusy HIGH to indicate that it has received the nibble, but it is not yet ready for another nibble.
5. The peripheral sets PtrClk HIGH as an acknowledgement to the host.
6. Host then acknowledges the transfer by pulsing HostClk.

Table 17.2 Byte-mode signals

Compatibility signal name	*Byte-mode name*	*In/Out*	*Description*
$\overline{\text{STROBE}}$	HostClk	Out	Used as an acknowledgment signal. It is pulsed low after each transferred byte.
$\overline{\text{AUTO FEED}}$	HostBusy	Out	It is set LOW to indicate that the host is ready for nibble and set HIGH when the nibble has been received.
$\overline{\text{SELECT INPUT}}$	1284Active	Out	Set HIGH when the host is transferring data.
$\overline{\text{INIT}}$	$\overline{\text{INIT}}$	Out	Not used.
$\overline{\text{ACK}}$	PtrClk	In	Indicates valid data byte. It is set LOW to indicate that there is valid data on the data lines and then set HIGH when the HostBusy going high.
BUSY	PtrBusy	In	Busy status (for forward direction).
PE	AckDataReq	In	Same as $\overline{\text{DataAvail}}$.
SELECT	Xflag	In	Not used.
$\overline{\text{ERROR}}$	$\overline{\text{DataAvail}}$	In	Indicates that there is reverse data available.
D0–D7	D0–D7	In/Out	Input/output data lines.

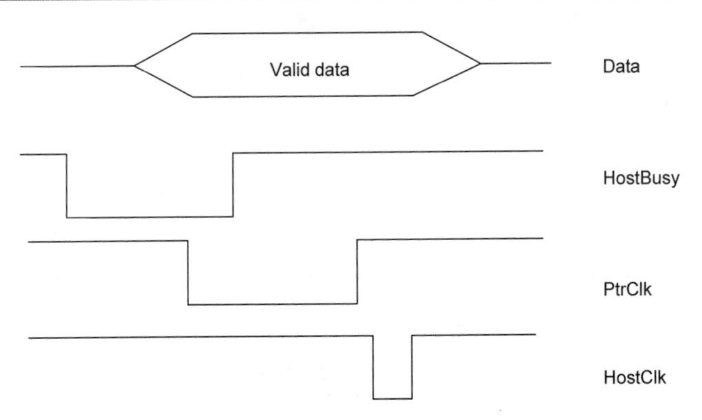

Figure 17.5 Byte mode data transfer cycle

17.5 EPP

The enhanced parallel port (EPP) mode defines a standard bidirectional communications mode and is used by many peripherals, such as CD-ROMs, tape drives, external hard disks and so on.

The EPP protocol provides four types of data transfer cycles:

1. Data read and write cycles – These involve transfers between the host and the peripheral.
2. Address read and write cycles – These pass address, channel, or command and control information.

Table 17.3 defines the names of the signal in the nibble mode. The $\overline{\text{WRITE}}$ occurs automatically when the host writes data to the output lines.

The data write cycle has the following sequence:

1. Program executes an I/O write cycle to the base address port + 4 (EPP data port), see Table 17.4. Then the following occur with hardware:
2. The $\overline{\text{WRITE}}$ line is set LOW, which puts the data on the data bus.
3. The $\overline{\text{DATASTB}}$ is then set LOW.
4. The host waits for peripheral to set the $\overline{\text{WAIT}}$ line HIGH.
5. The $\overline{\text{DATASTB}}$ and $\overline{\text{WRITE}}$ are then HIGH and the cycle ends.

The important parameter is that it takes just one memory-mapped I/O operation to transfer data. This gives transfer rates of up to 2 million bytes per second. Although it is not as fast as a peripheral transferring over the ISA, it has the advantage that the peripheral can transfer data at a rate that is determined by the peripheral.

Table 17.3 EPP mode signals

Compatibility signal name	*EPP mode name*	*In/out*	*Description*
$\overline{\text{STROBE}}$	$\overline{\text{WRITE}}$	Out	A LOW for a write operation while a HIGH indicates a read operation.
$\overline{\text{AUTO FEED}}$	$\overline{\text{DATASTB}}$	Out	Indicates a data read or write operation.
$\overline{\text{SELECT INPUT}}$	$\overline{\text{ADDRSTROBE}}$	Out	Indicates an address read or write operation.
$\overline{\text{INIT}}$	$\overline{\text{RESET}}$	Out	Peripheral reset when LOW.
$\overline{\text{ACK}}$	$\overline{\text{INTR}}$	In	Peripheral sets this line LOW when it wishes to interrupt to the host.
BUSY	$\overline{\text{WAIT}}$	In	When it is set LOW it indicates that it is valid to start a cycle, else if it is HIGH then it is valid to end the cycle.
PE	User defined	In	Can be set by each peripheral.
SELECT	User defined	In	Can be set by each peripheral.
$\overline{\text{ERROR}}$	User defined	In	Can be set by each peripheral.
D0–D7	AD0–AD7	In/out	Bidirectional address and data lines.

17.5.1 EPP registers

Several extra ports are defined, these are the EPP address register and EPP data register. The EPP address register has an offset of three bytes from the base address and the EPP data register is offset by four bytes. Table 17.4 defines the registers.

Table 17.4 EPP register definitions

Port Name	*I/O address*	*Read/ write*	*Description*
Data register	BASE_AD	W	
Status register	BASE_AD +1	R	
Control register	BASE_AD +2	W	
EPP address port	BASE_AD+3	R/W	Generates EPP address read or write cycle
EPP data port	BASE_AD+4	R/W	Generates EPP data read or write cycle

17.6 ECP

The extended capability port (ECP) protocol was proposed by Hewlett Packard and Microsoft as an advanced mode for communication with printer and scanner type peripherals. It provides a high performance bidirectional data transfer between a host and a peripheral.

The standard provides for two cycle types in both forward and reverse directions:

1. Data cycles.
2. Command cycles which can either be a run length count or a channel address.

It has many advantages over the EPP standard, including:

- Standard addresses – ECP has standard register addresses – Figure 17.6 shows that the addresses from 0778h to 077Ah have been defined for the extra functionality of ECP.
- Run length encoding (RLE) – RLE allows for compression. It allows high compression rates when there is a great deal of repetitive information in a file (typically with graphics files). A repetitive sequence is identified by a count followed by the repeated byte.
- FIFOs for both the forward and reverse channels.
- DMA as well as programmed I/O for the host register interface.
- Channel addressing – This allows multiple logical devices to be located within a single physical device. This channel address is passed in the command phase and can support up to 128 devices (addresses 0 to 127). For example, a single unit could have an integrated printer, fax and modem. ECP channel address allows them all to be accessed over a single connection. Within one physical package, having a single parallel port attached, there is a printer, fax and modem. This has the advantage that the printer can be busy printing while the modem can be accessed at the same time.

ECP redefines the SPP signals to be consistent with the ECP handshake. Table 17.5 describes these signals.

Figure 17.7 shows two forward data transfer cycles. It has data followed by a command phase. A high on the HostAck line indicates a data cycle, whereas a low indicates a command cycle. In the low state (command cycle) the data either represents an RLE count or a channel address. The most significant bit of the data byte indicates whether it is an RLE count or a channel address. If it is a 0, then bits the other 7 bits represent a RLE Count (from 0 to 127), else a 1 represents a channel address (from 0 to 127).

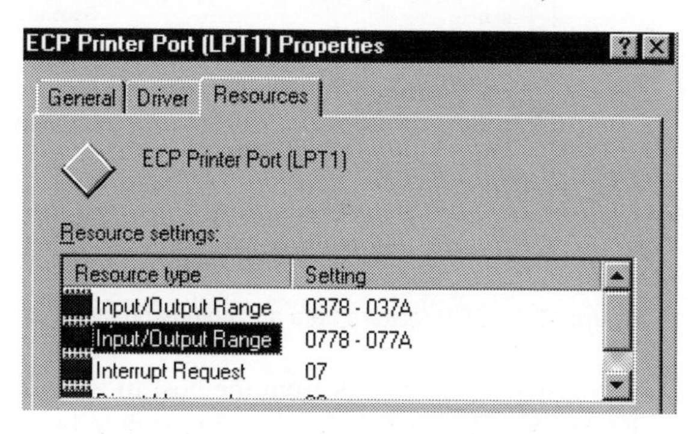

Figure 17.6 ECP input/output address ranges

Table 17.5 ECP mode signals

Compatibility signal name	*ECP mode name*	*In/out*	*Description*
$\overline{\text{STROBE}}$	HostClk	I	Transfers data or address information in the forward direction (along with PeriphAck).
$\overline{\text{AUTO FEED}}$	HostAck	O	Command/Data status in the forward direction. Data transfer in reverse direction (along with PeriphClk).
$\overline{\text{SELECT INPUT}}$	1284Active	O	Set high when host is in a 1284 transfer mode.
$\overline{\text{INIT}}$	$\overline{\text{ReverseRequest}}$	O	A low puts channnel in reverse direction.
$\overline{\text{ACK}}$	PeriphClk	I	Transfer data in the reverse direction (along with HostAck).
BUSY	PeriphAck	I	Transfer data or command information (along with HostClk).
PE	nAckReverse	I	Acknowledgement to nReverseRequest.
SELECT	Xflag	I	Extensibility flag.
$\overline{\text{ERROR}}$	nPeriphRequest	I	Set low by peripheral to indicate that reverse data is available.
D0–D7	Data[8:1]	I/O	Data lines.

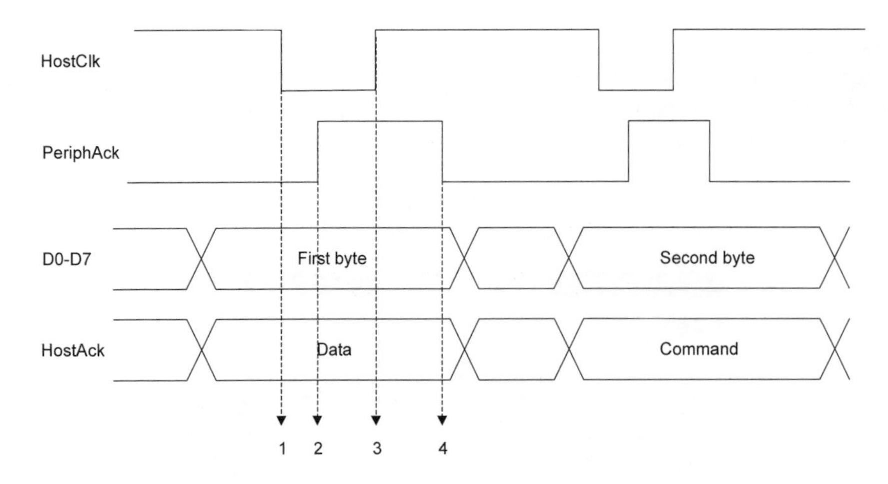

Figure 17.7 ECP forward data and command cycle

In the forward mode, the transfer of the data is from the host to the peripheral. Initially the host places its data on the data bus. It sets the HostAck line high to indicate a data cycle and sets HostClk low to indicate valid data. Next, the peripheral acknowledges the host by setting PeriphAck high. The host sets HostClk high which clocks the data into the peripheral. After this, the peripheral sets PeriphAck low to indicate that it is ready for the next byte.

Figure 17.7 illustrates an example of the reverse channel transfer where the peripheral transfers information to the host. As before, it shows a command cycle followed by a data cycle. It is similar to the forward phase except that the host requests a reverse channel by setting the nReverseRequest low. The peripheral then sets the nAckReverse line low to indicate that it is ready to transfer data, then it puts the data on the data bus. It then sets the PeriphAck high to indicate that it is a data cycle and set PeriphClk low to indicate valid data. After this the host sets HostAck high to acknowledge these events and the peripheral sets PeriphClk high. This clock edge then clocks the data into the host. Finally, the host sets HostAck low to indicate that it is ready for the next byte.

17.6.1 ECP software and register interface

The ECP specification ('The IEEE 1284 Extended Capabilities Port Protocol and ISA Interface Standard') defines a number of operational modes. These are defined in Table 17.6. The registers used to program ECP are based on the standard parallel port setting and uses an address which are offset by 1024 (400h) from the standard port address. Thus:

Standard port base address = 378h
ECP extended registers = 378h + 400h = 778h

There are six extra registers defined for ECP, these are given in Table 17.7. These six registers are mapped into three memory addresses and are shown in Figure 17.8 (778h, 779h and 77Ah). The ECR register used to set the current operational mode and can also be used to determine if an ECP-capable port is installed in the PC. Detection software can try to access any ECR registers by adding 402h to the base address of the LPT ports identified in the BIOS LPT port table.

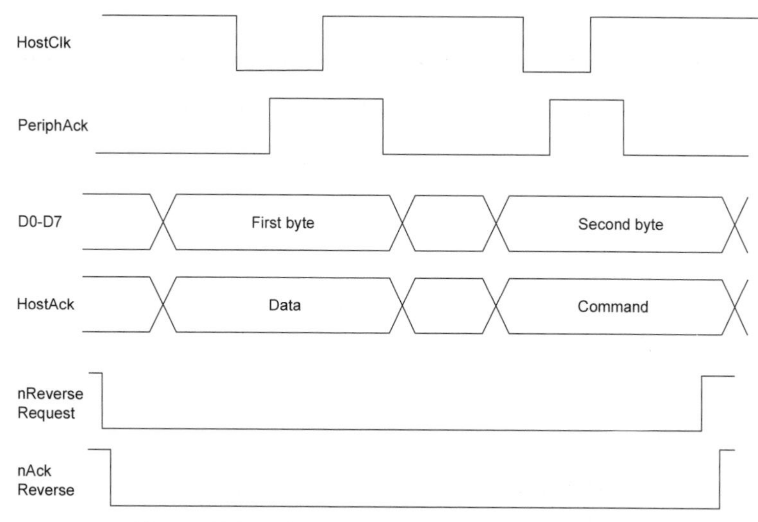

Figure 17.8 ECP Reverse data and command cycle

The operation of the ECP port is similar to the EPP port. The ECR register is used to set an operational mode, after which an I/O port is used to transfer data (the actual port depends on the mode). Handshaking is done automatically by the hardware and there is no need for the software to control it.

Table 17.6 ECR Register Modes

Mode	*Description*	*Mode*	*Description*
000	SPP mode	100	EPP parallel port mode (note 1)
001	Bidirectional mode (byte mode)	101	(reserved)
010	Fast Centronics	110	Test mode
011	ECP parallel port mode	111	Configuration mode

Table 17.7 ECP register description

Offset	*Name*	*Read/Write*	*ECP Mode*	*Function*
000	Data	R/W	000-001	Data register
000	ecpAfifo	R/W	011	ECP address FIFO
001	dsr	R/W	all	Status register
002	dcr	R/W	all	Control register
400	cFifo	R/W	010	Parallel port data FIFO
400	ecpDfifo	R/W	011	ECP data FIFO
400	tfifo	R/W	110	Test FIFO
400	cnfgA	R	111	Configuration register A
401	cnfgB	R/W	111	Configuration register B
402	ecr	R/W	all	Extended control register

17.7 Exercises

17.8.1 How many pins does a standard D-type parallel port connector have:

(a)	9	(b)	12
(c)	25	(d)	36

17.8.2 What is the maximum transfer rate for ECP/EPP mode:

(a)	100 kB/s	(b)	150 kB/s
(c)	1 MB/s	(d)	1.2 MB/s

17.8.3 Outline the operation of the nibble mode. How does the parallel port allow data to be inputted?

17.8.4 Design a circuit for nibble mode operation which will sample data bits. The design should include ground connections (GND), connector types and pin numbers. If possible, implement the design by adding switches to simulate input levels (power can be supplied by the parallel port connection).

17.8.5 Explain how several devices can be connected to the parallel port, and identify how the operating system identifies each of the devices.

17.8 Note from the author

The parallel port was never really been destined for glory. It is basically a legacy port, which, in the past, was only really useful in connecting printers. The future for printer connections is either with network connections, such as Ethernet, or with a USB connection. In its standard form, it has a large, bulky connector, which in many systems is never even used.

It has always struggled against the serial port, because it lacks the flexibility of RS-232 and, until recently, had no standards agency to support it. However, it's there and it has great potential for fast data transfers. RS-232 has always been a great success and has many of the large manufacturers supporting it, and all importantly, it is defined by several standards agencies. The key to its current success was due to the intervention of the NPA which brought together many of the leading computer and printer manufacturers. In these days, there are only a few major companies, such as Intel and Microsoft, who can lead the market and define new standards (such as the PCI bus, with Intel).

The main difficulties are how to keep compatibility with previous implementations and software, and also how to connect multiple devices on a bus system, and allow them to pass data back and forward without interfering with other devices. This has finally been achieved with ECP/EPP mode. It is a bit complex, but it works, and even supports data compression. At the present, my notebook connects to a CD-R drive, a scanner and a printer, all of the same parallel port (just like SCSI). This arrangement works well most of the time and is a relative cheap way of connecting devices, but it is in no way as flexible and as fast a SCSI.

18 Modbus

18.1 Modbus protocol

The Modbus protocol is an industry-standard protocol which allows programmable controllers to communicate over a network or local communications link. It defines a standard message structure that all Modbus-compatible controllers recognise and implement, regardless of the network type. It describes:

- The format of requests to Modbus-compatible devices.
- The format of responses from Modbus-compatible devices.
- The layout and contents of message fields for Modbus-compatible devices. The Modbus protocol provides the internal standard that the Modicon controllers use for parsing messages.
- How each controller knows its own device address and recognizes any messages addressed to it.
- The format of the data and other information contained in the message.

18.1.1 Transactions on Modbus networks

Standard Modbus controllers communicate using RS-232C and can be networked or connected via a modem. Each controller (such as a host processor) communicates with the connected devices (such as a PLC) using a master–slave technique (Figure 18.1). The controller (the master) initiates transactions (queries) which are sent to the other devices (the slaves). The addressed slave then responses to the request by sending back data or by implementing the required action. This addressing can be to an individual device, or can be broadcasted to all connected slaves. There are no responses from a broadcast query.

The query takes the form of:

- An address (either an individual address or a broadcast address).
- A function code, which defines the requested action.
- Sent data, the format of which depends on the function code. For example, a function code of 03 defines that the slave read from the started register defined in the data field and it also contains the number of registers to read.
- Error-checking field, to allow the slave to validate the message integrity.

The response message takes the form:

- Action confirmation – on error, this field contains an echo of the query function code. On an error, the function code is modified to indicate that the response is an error response, and the returned data field contains an error code.

- Returned data – this contains the data returned by the slave, either register values or a status.
- Error-checking field – this allows the master to validate the message integrity.

A standard Modbus network only contains masters and slaves. On a Modbus Plus network controllers can operate as a master or a slave. The Modbus protocol is still applied to the transaction. This typically occurs over a network.

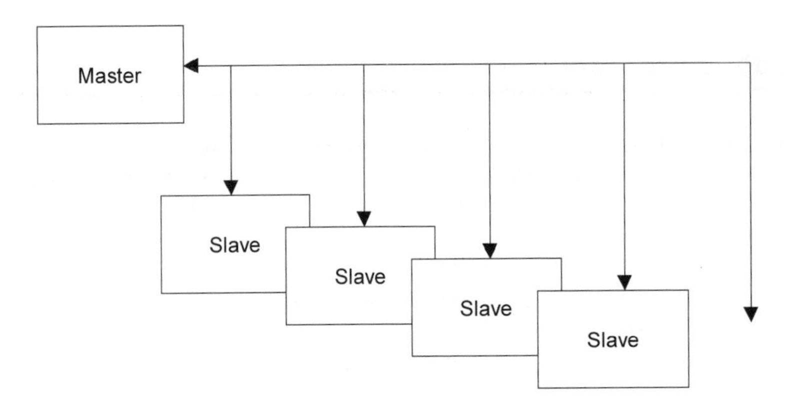

Figure 18.1 Master–slave

18.1.2 Transmission modes

Modbus transmits values from the master to the slave either using ASCII or RTU (remote terminal unit). All the devices on the network must be set to the same setting. These are:

- ASCII – Modbus transmits the bit values as ASCII characters which represent the hexadecimal of the transmitted bit values. The transmitted characters will range from '0' to '9' and 'A' to 'F'. For example, if the transmitted bit stream is to be:

  0110 1111 0001 0011 1100 1100 1011 0000

  this would be transmitted as the ASCII characters:

  '6' 'F' '1' '3' 'C' 'C' 'B' '0'

  In this mode, a start bit is transmitted, followed by a 7-bit ASCII character, an optional parity bit and then two stop bits. The least-significant bit of the ASCII character is sent first.

- RTU – an 8-bit value is sent as two hexadecimal values. For example:

  0110 1111 0001 0011 1100 1100 1011 0000

  this would be transmitted as the following:

'0110 1111' '0001 0011' '1100 1100' '1011 0000'

which allows for a faster transmission of values, and they can thus be decoded quicker than the ASCII mode. RTU will obviously be twice as fast as the ASCII method. It also allows continuous bit streams to be transmitted. In this mode, a start bit is transmitted, followed by an 8-bit binary value, an optional parity bit and then two stop bits.

In summary, the modes are:

	ASCII	*RTU*
Coding	Hexadecimal characters	8-bit binary
Start bits	1	1
No of bits/character	7	8
Parity	Optional	Optional
Stop bits	1 or 2	1 or 2
Error checking	LRC	CRC

18.1.3 Modbus message frame

The Modbus message has different formats, depending on the transmission mode. These are:

- ASCII framing – a colon ASCII character (:, or 3Ah) starts the message and the carriage return–line feed sequence ends the message (CRLF, or 0Dh and 0Ah). The characters within the message will then be '0' to '9' or 'A' to 'F'. On a network, devices continually listen for the colon character. The field after this is the address field. The maximum interval between characters is one second. Figure 18.2 shows the standard format.
- RTU framing – messages start with a silent interval of at least 3.5 character times. After this, the device address is transmitted. All devices on the network listen to the bus, and wait for a silent period, which must be at least 3.5 characters since the last message. It then transmits the message as a continuous stream. The first eight bits are the target address. Errors occur if there is a silent period of more than 1.5 character times or if a device transmits its message before 3.5 character delays after the previous message. Figure 18.2 shows the standard format.

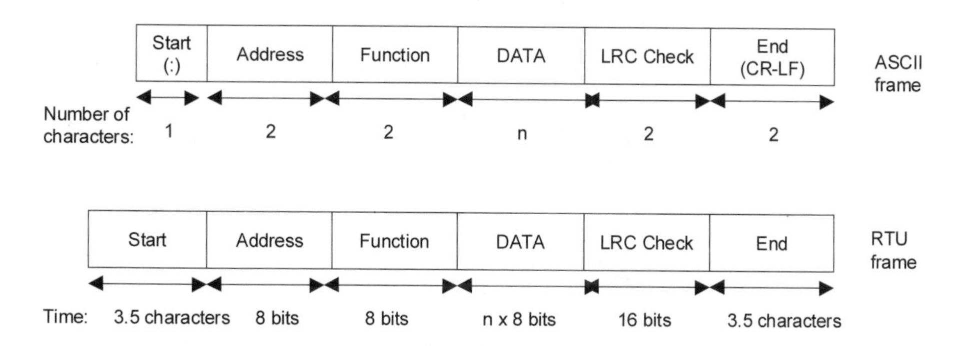

Figure 18.2 ASCII and RTU message frame

18.1.4 Address field

The address field contains either two ASCII characters (for ASCII mode) or eight bits (for RTU mode). Addresses range from 0 to 247 (00h to F7h), where 0 is the broadcast address and 1 to 247 are used for slaves addresses. A master communicating with the slave puts the slave's address in the address field, and the slave, when responding, puts its own address in the address field (to identify itself).

18.1.5 Function field

The function field contains either two ASCII characters (for ASCII mode) or eight bits (for RTU mode). Codes range from 1 to 255 (00h to FFh), and they are used by the master to inform the slave as to the action which requires to be performed. Typical codes (in decimal) are:

01	Read coil status	02 Read input status	03 Read holding registers
04	Read input registers	05 Force single coil	06 Preset single register
07	Read exception status	08 Diagnostics	11 Fetch comm. event counter
12	Fetch comm. event log	13 Program controller	14 Poll controller
15	Force multiple coils	16 Preset multiple registers	
17	Report Slave ID	18 Program 884/M84	19 Reset communication link
20	Read general reference	21 Write general reference	
22	Mask write 4x reference	23 Read/write 3x registers	24 Read FIFO queue

For example, the read coil status gives an ON/OFF status for discrete outputs. When there are no errors the slave sends back the original function code, else, on an error, the same code is sent back, expect the most-significant bit is set to a 1. For example, if the function code was 0000 1000, then, on an error, the return value will be 1000 1000. A status code is also added in the data field, these are outlined in Table 18.1.

Table 18.1 Exception codes

Code	*Name*	*Description*
01	Illegal function	The message function received is not an allowable action for the addressed salve.
02	Illegal data address	The address referenced in the data field is not an allowable address for the addressed slave location.
03	Illegal data value	The value referenced in the data field is not allowable in the addressed slave location.
04	Failure in associated device	The slave's subcontroller has failed to respond to a message or an abortive error occurred.
05	Acknowledge	The slave has accepted and is processing the long duration program command.
06	Busy, rejected message	The message was received without error, but the slave is currently busy.

For example, if the master sends the message

Address	*Function*	*Start address (hi)*	*Start address (lo)*	*No. (hi)*	*No. (low)*	*LRC*
:12	01	02	10	00	01	DA

Then on an error, the response would send back the function code of 81 (which sets the most significant bit of the function code (that is, 1000 0001). If the slave were busy then the exception code would be 06. Thus the code sent back will be:

Address	*Function*	*Exception code*	*LRC*
12	81	06	67

18.1.6 Data field

The data fields contains even multiples of hexadecimal digits (in ASCII mode) or an even number of binary values (in RTU mode). The format of the field depends on the function code, and contains information, such as register addresses, the number of values required and the number of bytes in the data field.

For example if the function code is 01 (read code status), then the format of the frame send from the master to the slave is:

- Slave address (*xx*).
- Function (01) – read coil status.
- Starting address high (*xx*) – most-significant byte of the starting register address.
- Starting address low (*xx*) – least-significant byte of the starting register address.
- Number of Points high (*xx*) – most-significant byte of the number of points to be sent.
- Number of Points low (*xx*) – least-significant byte of the number of points to be sent.
- Error check (*xx*).

If there are no errors, then the response is:

- Slave address (*xx*).
- Function (01) – read coil status.
- Byte count.
- Data (Coils 8 to 1) – data for the first eight coils, where a 1 value in a coil bit position represents ON, whereas a 0 represent OFF.
- Data (Coil 16 to 9) –data for the next eight coils.
- etc.
- Error check (*xx*).

Some data fields are empty, such as the communication event log function (12, or 0Bh).

18.1.7 Error checking field

The error checking method depends on the type of transmission, these are:

- ASCII – in this mode the error checking field contains two characters, which performs a longitudinal redundancy check (LRC) for all characters, excluding the start and end terminating characters (:, CR and LF).
- RTU – in this mode, the error-checking field contains a 16-bit value which performs a cyclical redundancy check (CRC). This field is added to the end of the message; the low-order byte of the field is appended first, followed by the high-order byte.

LRC

The ASCII mode uses the LRC method. It basically adds up the values of each of the 8-bit fields, apart from the starting colon and the end CRLF, and then takes the two's complement of the result (ignoring any carries). For example, from the previous example the transmitted values are

Start address (12)	0001 0010
Function (01)	0000 0001
Start address, high (02)	0000 0010
Start address, low (10)	0001 0000
Number, high (00)	0000 0000
Number, low (01)	0000 0001
Total	**0010 0110**

To convert to 2's complement, invert all the bits, to give

1101 1001

and then add 1, to give

1101 1010

which is DA, in hexadecimal. Thus the transmitted message would be:

`:120102100001DA`*<CRLF>*

CRC checking

The RTU mode uses CRC, which is a much stronger error checking method. This method is outlined in Appendix D. Its operation is as follows:

1. 16-bit register is preloaded with all bits set to 1.
2. The first eight-bit data character is exclusive ORed (XOR) with the higher order-byte in the register and the result is put in the register.
3. The register is then right shifted by one bit position and a zero filled into the most significant bit (MSB) position.
4. If the shifted bit out is a 1, XOR the generator polynomial 1010 0000 0000 0001 with the 16-bit register, else return to Step 3.
5. Repeat steps 3 and 4 for eight right shifts.
6. XOR the next 8-bit value with the 16-bit register.
7. Repeat Steps 3 to 6 until all the bytes in the message have been XOR with the 16-bit register and shifted eight times.
8. The resultant content of the 16-bit register is the CRC error check.

18.2 Function codes

Each value is addressed via a register. The first register address on the Modbus is referenced to zero. The following sections outline the main function codes.

18.2.1 Read coil status (01)

This function reads the ON/OFF status in Boolean logic. The query message specifies the starting coil and quantity of coils to be read. For example to read 12 values (0Ch) from device 18 (12h), starting at address 02DE, then the following is used:

Address	*Function*	*Start address (hi)*	*Start address (lo)*	*No. (hi)*	*No. (low)*	*LRC*
`:12`	`01`	`02`	`DE`	`00`	`0C`	`01`

The response contains the coil status, in which the data field is packed with bit values, one for each coil. A one represents ON, a zero represents OFF and the lsb of the first byte contains the first address coil. Other coil values follow this and, if the number of coil values is not a multiple of eight, then zeros are used to pad the end values. The byte count field precedes the coil values and specifies the quantity of complete bytes of data. An example response to the above query is

Address	*Function*	*Number of bytes*	*Data values (8 to 1)*	*Data values (12 to 9)*	*LRC*
`:12`	`01`	`02`	`BA`	`10`	`FB`

Thus, if the addressed coils are Coil 1 to Coil 12, then the Coils 8 to Coil 1 have the status of 1101 1100 (BAh), which means that Coil 8, Coil 7, Coil 5, Coil 4 and Coil 3 are ON, and Coil 6, Coil 2 and Coil 1 are OFF. The other four coils are 0001 for Coil 12 to Coil 9. Thus, Coil 9 is ON and Coil 12, Coil 11 and Coil 10 are OFF.

18.2.2 Read Input Status (02)

This function reads the ON/OFF status of discrete inputs from the slave device. This function reads the ON/OFF status of logic Boolean. It has the same format as the read coil Status function code. For example to read four values, starting at address 11FF, then the following is used:

Address	*Function*	*Start address (hi)*	*Start address (lo)*	*No. (hi)*	*No. (low)*	*LRC*
`:12`	`02`	`11`	`FF`	`00`	`04`	`D8`

The response is in the same format as the read coil status function. An example response to the above query is

Address	*Function*	*Number of bytes*	*Data values (4 to 1)*	*LRC*
`:12`	`02`	`01`	`02`	`CC`

which returns the status of the four inputs as

Input 4	(Address: 1202)	OFF	(0000 **0**010)
Input 3	(Address: 1201)	OFF	(0000 0**0**10)
Input 2	(Address: 1200)	ON	(0000 00**1**0)
Input 1	(Address: 11FF)	OFF	(0000 001**0**)

18.2.3 Read holding registers (03)

The function reads the binary contents of holding registers (4*x* references) in the slave. Holding registers are identified starting from 40001, which is addressed as register 0000. Register 40002 is addressed as register 0001, and so on.

For example, to read two values, starting at address 0E2 (register 40226), then the following is used:

Address	*Function*	*Start address (hi)*	*Start address (lo)*	*No. (hi)*	*No. (low)*	*LRC*
`:12`	`03`	`00`	`E1`	`00`	`02`	`05`

The response gives 16 bits for every register value. An example response is,

Address	*Function*	*Number of bytes*	*Data value (40226)*	*Data values (40227)*	*LRC*
`:12`	`03`	`04`	`BA A2`	`FF 10`	`7c`

18.2.4 Read input registers (04)

This function reads the binary contents of input registers (3*x* references) in the slave. Input registers are identified starting from 30001, which is addressed as register 0000. Register 30002 is addressed as register 0001, and so on. The response gives 16 bits for every register value.

18.2.5 Force single coil (05)

This function forces a single coil (0*x* reference) to either an ON or an OFF state. For example, to force coil at address 101 (65h) to be ON, then the following is used:

Address	*Fun.*	*Start add. (hi)*	*Start add. (lo)*	*Force data (hi)*	*Force data (low)*	*LRC*
`:12`	`05`	`00`	`65`	`FF`	`00`	`85`

A value of FF00 sets the coil ON, while a value of 0000 sets the coil OFF. The response should just be the echo of the query.

18.2.6 Preset Single Register (06)

This function presets a value into a single 16-bit holding register (4*x* reference). For example to preset register 40226 (address E1) to 021F then:

Address	*Function*	*Start address (hi)*	*Start address (lo)*	*No. (hi)*	*No. (low)*	*LRC*
`:12`	`06`	`00`	`E1`	`02`	`1F`	`05`

The response should just be the echo of the query.

18.2.7 Read Exception Status (07)

This function reads the current status of the slave. Normally the settings for the addresses and the bits within the addresses are normally manufacture defined. For example, for the Honeywell 2500 series chromatograph the returned status codes are:

Coil	*Assignment*	*Coil*	*Assignment*
1	Shutdown	2	Unknown fail
3	Power fail	4	Unacknowledged alarms
5	Starting	6	Running
7	Warm start	8	Cold start

Other set-ups (especially on newer equipment) allow access to the batteries status. An example query is

Address	*Function*	*LRC*
:12	07	E7

and an example response is

Address	*Function*	*Flag data*	*LRC*
:12	07	7D	6A

18.2.8 Fetch Communications event counter (11, 0Bh)

This function returns a status word and an event count for the slave's communications event counter. An example query is

Address	*Function*	*LRC*
:12	0B	E3

and an example response is

Address	*Function*	*Status (hi)*	*Status (lo)*	*Event count (hi)*	*Event count (lo)*	*LRC*
:12	0B	FF	FF	02	08	DB

A status of FFFFh indicates that the slave is still progressing a program function, else it will be 0000h. The event counter holds the number of events that have been counted by the controller.

18.2.9 Fetch communication event log (12, 0Ch)

This function returns a status word, event count, message count, and a field of event bytes from the slave. The status word and event count are identical to that returned by the fetch communications event counter function, but it is followed by a 16-bit value which defines the number of events stored. The events are then listed after this.

18.3 Modbus diagnostics

The 08 function is used to get slave diagnostics. This is used with a number of subfunctions. The format, and example, of a diagnostics function are

Address	*Function*	*Subfunction (hi)*	*Subfunction (lo)*	*Data (hi)*	*Data (lo)*	*LRC*
:12	08	00	FF	02	08	DB

The subfunctions are given in the following table.

Sub function	*Description*	*Query (data field)*	*Reply*
00 00	Return query data	Same as the query.	Same as query.
00 01	Restart communication option	00 00 (leave log as it was prior to restart) FF 00 (clear event log)	Same as query.
00 02	Return diagnostic register	00 00	Diagnostic 16-bit register contents. The contents depends on the Modbus type. An example is: Bit 0 Continue on error 1 Run light failed 2 T-Bus test failed 3 Asynchronous bus test failed 4 Force listen mode 7 ROM Chip 0 test failed 8 Continuous ROM checksum test in execution 9 ROM 1 test failed 10 ROM 2 test failed 11 ROM 3 test failed
00 03	Change ASCII input delimiter	'*char*' 00	Change end-of-message character. Return is the same as query.
00 04	Force listen only-mode	00 00	Slave goes into a listen-only mode and thus does not respond.
00 0A	Clear counters and diagnostic register	00 00	Return is the same as query.
00 0B	Return bus message count	00 00	Return is the same as query.
00 0C	Return bus communication error count	00 00	CRC Error count.
00 0D	Return bus exception error count	00 00	Exception error count.
00 0E	Return slave message count	00 00	Slave message count.
00 0F	Return slave no response count	00 00	Slave no response count.
00 10	Return slave NAK count	00 00	Slave NAK count.
00 11	Return slave busy count	00 00	Slave busy count.
00 12	Slave character overrun count	00 00	Slave character overrun count.
00 13	Return overrun error count	00 00	
00 14	Clear overrun counter and flag	00 00	
00 15	Get/clear Modbus Plus statistics		

18.4 Exercises

18.4.1 What is the basic topology of a Modbus network:

(a) One or many masters and one slave
(b) One or many master and one or many slaves
(c) One master and one slave
(d) One master and one or many slaves

18.4.2 How is the start of an ASCII message frame identified:

(a) 01111110 (b) ':'
(c) Start bit (a 1) (d) LFCR (line feed, carriage return)

18.4.3 What is the maximum number of nodes on a Modbus network:

(a) 8 (b) 256
(c) 1024 (d) No limit

18.4.4 What is the addressing range for the Modbus protocol:

(a) 00h–FFh (b) 0000h–FFFFh
(c) 000000h–FFFFFFh (d) No limit

18.4.5 Determine the LRC (in hex) that is to be added to the message transmission of 4F2A10h:

(a) 89h (b) 77h
(c) 88h (d) 76h

18.4.6 Determine the LRC (in binary) that is to be added to the message transmission of 1000 1100 0001 0110 1111 0110b:

(a) 01101000b (b) 10011000b
(c) 00000000b (d) 01100111b

18.4.7 What ASCII characters are transmitted for the data transmission of 1010 0011 1110 1010b:

(a) '10', '3', '15', 'A' (b) '1', '0', '3', '1', '5', 'A'
(c) 'A', '3', 'E', 'A' (d) '0', '1', '3', '5', '1', 'A'

18.5 Notes from the author

Modbus is an important protocol and has grown in its popularity because of its simplicity. It has a very basic structure, and is easy extremely easy to implement as it is based on a master–slave relationship where a master device sends commands and the addressed slave responses back with the required information. Its main advantages are its simplicity, its standardization and its robustness.

Modbus can be operated on a wide range of computers running any type of software, from a simple terminal-type connection, where the user can enter the required commands and views the responses, through to a graphical user interface, with the commands and response messages hidden from the user. The basic protocol is, of course, limited in its basic specification, such as the limited number of nodes (256, maximum) and the limited addressing range (0000h to FFFFh).

The basic communications link is also simple to implement (normally, RS-232), but newer Modbus implementations use network connections, such as Ethernet. Another change is to implement the Modbus protocol over a standard TCP/IP-based network. This will allow Modbus to be used over an Internet connection.

RS-232 does not have strong error checking, and only provides for basic parity check. Modbus using ASCII-based transmission of the Modbus protocol adds a simple checksum to provide an improved error detection technique (LRC). For more powerful error detection the data can be transmitted in RTU format, which uses the more powerful technique (CRC).

The Modbus Plus protocol now allows for devices to be either a master or a slave. This allows for distributed instrumentation, where any device can request data from any other device, at a given time.

19 Fieldbus

19.1 Introduction

Field buses are special local area networks that are dedicated to data acquisition and the control of sensors and actuators. They typically run over low-cost twisted pair cable. They differ from many traditional LANs (such as Ethernet) in that they are optimised for the exchange of short point-to-point status and command messages. There are many Fieldbus standards that exist, each developed for a specific purpose.

The potential market for Fieldbus equipment is enormous. Figure 19.1 shows an estimate of sales over time. It can be seen that the expected market in 2003 is over 50%. Instrumentation interfaces have evolved from 3–15 PSI transmitters, to 4–20 mA analogue interfaces, now to serial interfaces (typically either RS-232 or RS-485) and now to Fieldbus interfaces. This evolution over time is illustrated in Figure 19.2.

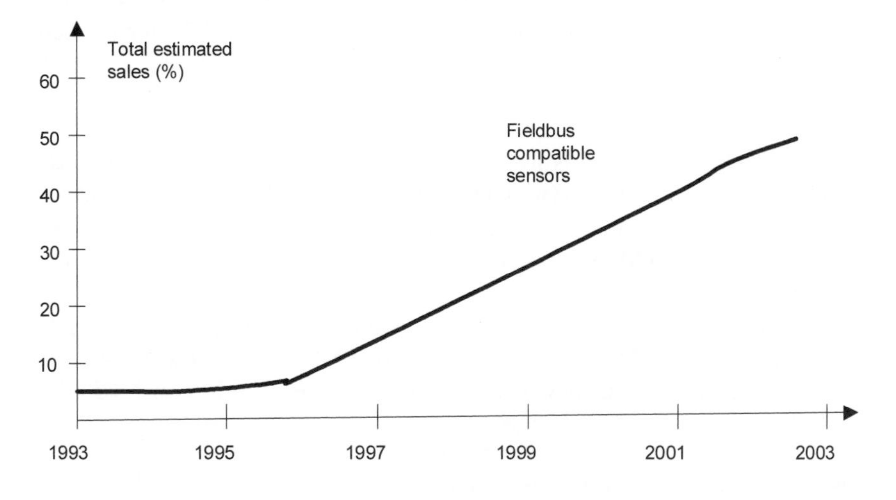

Figure 19.1 Market for Fieldbus

19.2 Fieldbus types

The main Fieldbus types are outlined in this section, but most of the chapter is devoted to the FOUNDATION Fieldbus, which is a truly open standard, and the WorldFIB, which are supported by many major vendors.

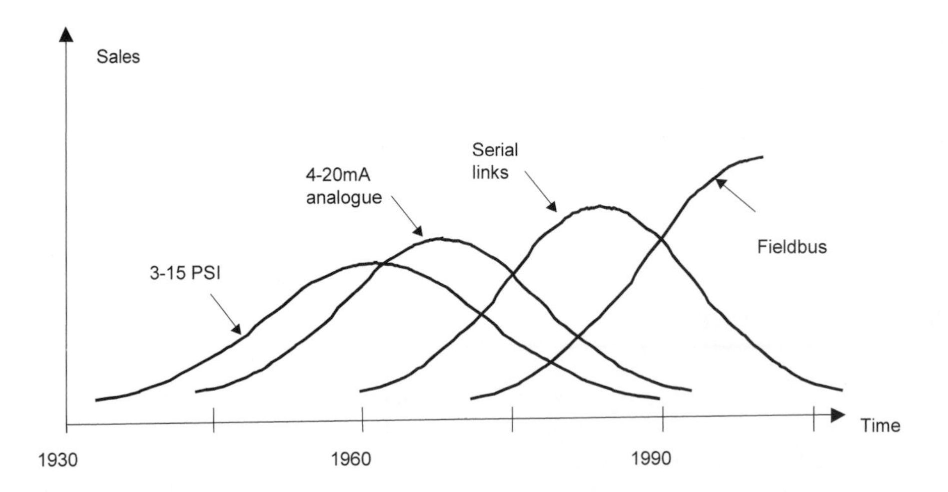

Figure 19.2 Changes in the market for instrumentation parameter transmissions

There are three main categories for Fieldbus installations, these are Fieldbus standard, Other domain standards and non-Fieldbus. The main Fieldbus standard are WorldFIP (standard in France), Profibus (standard in Germany) and P-Net (standard in Denmark), and are all part of the CENELEC European standard EN 50170. WorldFIP has the advantage over the others in that it uses the IEC physical layer (the IEC 1158-2). The Fieldbus Foundation is an initiative of mainly USA-based vendors. Its main aim is to standardise Fieldbus for the petrochemical/chemical industries. One of its aims is not to replace traditional DCSs (distributed control systems), but to integrate with them. The main standards for Fieldbus products are:

- Fieldbus Foundation – this was formed 1994 have defined low speed 31.25 kbps transmission (H1). The H2 standard (which is equivalent to WorldFIP) will operate at 1Mbps.
- WorldFIP (or WorldFIP Europe) – this standard has been incorporated into many products and supports a 1 Mbps transmission rate. WorldFIP contributes to the Fieldbus Foundation, in their standardisation process.
- Profibus – this has three main types: FMS (flexible manufacturing systems), DP (distributed peripherals) and PA (process automation). FMS and DP use RS-485 signalling, whereas PA uses the IEC physical layer at low speeds. FMS and DP are part of EN50170.

Other busses, such as the CAN bus, use only the lower layers of functionality, especially for remote I/O.

19.2.1 BITBUS

Intel introduced the BITBUS for remote I/O capability to multibus systems. It allows programs to be downloaded and executed in a remote node for truly distributed system configurations. Its outline specification is

Speed	375 kbps
Maximum nodes with/without repeaters	250/32
Maximum distance with/without repeaters	13.2 km max/1.2 km max
Arbitration	master/slave

Cable type	twisted-pair
Header/data size	1 to 13 or 52 bytes
Major benefits	large number of users
Nodes programmable	intelligent I/O modules
Primary applications	process control

19.2.2 WorldFIP

WorldFIP operates at 1 Mbps over twisted-pair cables, and is a reliable method of transmitting variables (from sensors and to actuators) and messages (such as events, configuration commands). It uses a bus arbitrator that broadcasts variable identifiers to all the nodes on the network. This triggers the required node which produces the node to respond with the required value. All modules that need this value must then read it. Its main characteristics are:

- It supports a distributed, decentralised database of variables.
- It does not require node addresses as messages are broadcasted by a bus arbitrator, and then the response is from the node which contains the processor parameter.

Its outline specification is

Speed	1 Mbps
Maximum nodes with/without repeaters	256/64
Maximum distance with/without repeaters	Greater than 10 km/2 km
Arbitration	bus arbiter
Cable type	twisted-pair
Header/Data size	1 to 128 bytes
Major benefits	distributed data base/ very deterministic
Primary applications	Real-time control/process/machine

19.2.3 CAN

Controller area network (CAN) was developed mainly for the automobile industry, and is now popular in factory automation. It transmits at 1Mbps and uses twisted-pair cable for up to 40 devices. Its main features are:

- Nodes can communicate when there are no nodes communicating on the bus.
- It uses a non-destructive bit-wise arbitration which allows fast detection of multiple accesses. This allows full use of the bandwidth. This differs from bus-topology LAN technologies, such as Ethernet, which detects collisions over long distances, and suffers from propagation delays, and nodes may transmit many bits before they can determine if two or more nodes are communicating at a time. In Ethernet, nodes back off from the network when a collision occurs.
- Message priority system, which is based on an 11-bit packet identifier.
- Architecture can be many masters, and involves peer-to-peer communications or multicast transmissions.
- Automatic error detection, signalling and retries.
- Short data packets of eight bytes.

Its outline specification is

Speed	1 Mbps
Maximum nodes with (/without repeaters)	N/A (30)
Maximum distance with (/without repeaters)	N/A (40 m at 1Mbps, 1 km at 20 kbps)
Arbitration	CSMA
Cable type	twisted-pair
Header/data size	8 bytes fixed
Major benefits	Low cost/efficient for short messages
Primary applications	Automotive

19.3 FOUNDATION Fieldbus

FOUNDATION Fieldbus, is an open specification for sensors, actuators, analysers, and so on. It allows:

- The control functionality actually resides in field devices.
- The support of other diagnostic, process operation and maintenance functions within field devices.

In the past, 4–20 mA standards have been used to transmit plant information to controllers. This has in some places, been replaced by transmitter vendors providing their own digital protocol to allow bidirectional communication between the control system and smart transmitters. Fieldbus has finally allowed a standardised method for process control to move from being centralized to become distributed, and the control to actual reside in field devices, such as transmitters, values and analysers. It provides a digital communications channel and a user layer to provide intercommunications. Its benefits are:

- Interoperability – this allows different suppliers to be used for devices.
- Wiring cost savings – one communication channel can transmit many digital signals.
- Flexible control implementations.
- Increased field information – this includes processed data, averages, minimas, maxima, diagnostic information and operational information.

Fieldbus was initially defined by the ISA's SP50 fieldbus standards committee, which outlined a two-way, multidrop, digital communications standard for the interconnection of sensors, actuators, instruments and control systems. The Fieldbus Foundation has since set out to commercialise it as the FOUNDATION Fieldbus.

19.3.1 Fieldbus topology

Most analogue transmission methods, and many digital field communications methods, require a single twisted-pair wire for the transmission of a single process variable. Fieldbus differs from this in that it can connect using point-to-point, with buses with spurs, as a daisy chain, as a tree, or as a combination of any of these. The methods are:

- Bus with spurs – all the devices connect to a common bus and they connect through junction boxes, as illustrated in Figure 19.3.

- Daisy-chained – all the devices are chained to each other, one-by-one. It is similar to the bus with spurs, but does not use junction boxes. It is a useful method of connecting devices as new devices can be added by simply daisy-chaining from a close device. The disadvantage is devices must be disconnected in order to connect a nearby device, unless a special connector can be used that allows a connected device to be connected.
- Tree – type of topology uses a single junction box, with the devices connecting directly to the junction box. Typically, it is used when devices are added and deleted from the network on a regular basis.

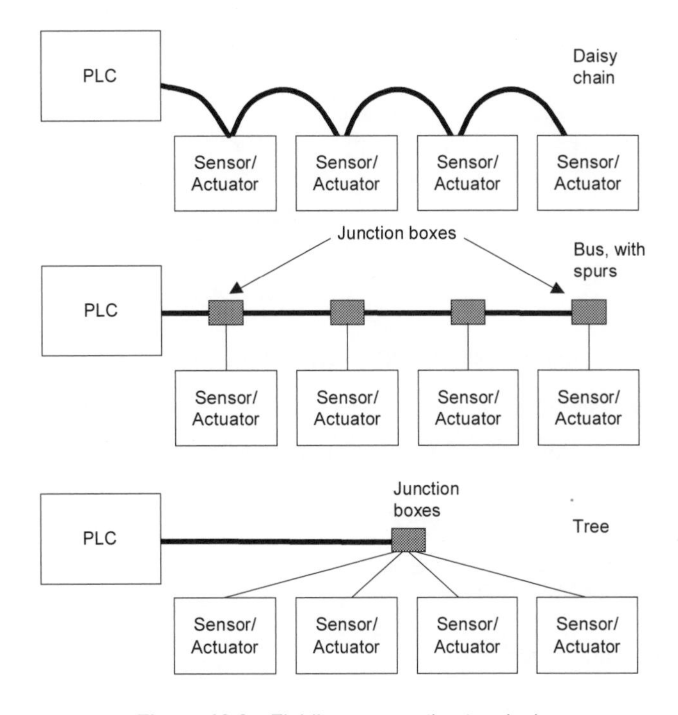

Figure 19.3 Fieldbus connection topologies

19.3.2 FOUNDATION Fieldbus layers

Process control communications can be group into three different levels:

- Hardware-address buses – this type of bus uses hardware addresses and registers to store values. Examples are I/O buses, PLCs, SCADA protocols with RTUs (remote termination units).
- Symbolically addressed buses – this type of bus uses addresses that actually have a symbolic name. This works at a higher-level than the hardware-address bus.
- Comprehensive user-layer functionality buses – this type of bus operates at a higher level than hardware and symbolic addressing. It is used in the FOUNDATION Fieldbus and supports function blocks, standardised parameters, operational modes, cascade initialisation sequences, antiwindup mechanisms, quality-of-data propagation and response, fail-state initiation, alarm reporting and control mechanisms, process control data structures, and so on.

The FOUNDATION Fieldbus consists of two main layers: the communications layer and the user layer. The components in these layers are illustrated in Figure 19.4. The user layer operates above the communications layer and includes function blocks, resource blocks, transducer blocks and alarm notifications.

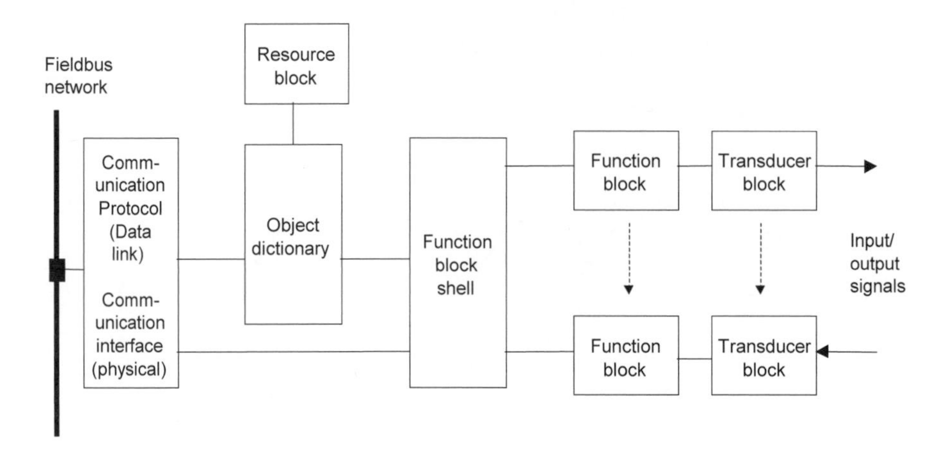

Figure 19.4 FOUNDATION Fieldbus architecture

19.3.3 Function blocks (FB)

The user layer supports device configuration, and uses function blocks. A device can have any number of function blocks. These are used for control, diagnostic, safety and production accounting purposes, and define such things as:

- Standardised parameter names.
- Data types.
- A cascade initialisation mechanism.
- Status propagation.
- An antiwindup mechanism.
- Trend collection mechanism.
- An execution scheduling mechanism.
- Block modes and behaviors in response to mode changes.
- Status of process variables.
- Rules for propagation of status.
- Behaviours in response to status changes.

These include:

- Standardised function blocks.
- Vendor-enhanced function blocks and vendor-custom function blocks.

19.3.4 Resource blocks (RB)

Each device also has an RB, which contain parameters relating to the physical device, such as:

- Manufacturer ID.
- Device type of device.
- Revision.
- Memory usage and free space.
- Computational time.
- Device state (on-line/off-line/standby/fault condition/etc).

19.3.5 Transducer block (TB)

Each device has a TB, which the named entity that stores the parameters associated with the sensor or actuator.

19.3.6 Alarm notifications

Each block (functional, resource or transducer) can produce an alarm notification which is associated with that particular block, such as:

- Process problems with function blocks.
- Sensor/actuator problems with transducer blocks.
- Overall device problems with resource blocks.

The devices each have parameters, which are structured using an object dictionary, which is a standardised method of interrogating and referencing parameters over the communications link.

In FOUNDATION Fieldbus the trip-point values are low, low-low (LO_LO_LIM), high (HI_LIM), high-high (HI_HI_LIM), deviation-low (DEV_LO) and deviation-high (DEV_HI). These trip values can generate alarms, which have certain priority levels. These are:

- Priority 0 – disables alarms including the setting of the alarm condition status flag.
- Priority 1 – disable report, but causes the status flag to be set.
- Priority 2 to 7 – advisory alarms.
- Priority 8 to 15 – critical alarms.

19.3.7 Device description language (DDL)

The FBs, TBs and RBs are not just limited to a standardized set of parameters, a new DDL allows manufactures to specify additional parameters in a standardised manner. This includes names, data types, enumerations, units, valid ranges, user entry limits, entry conditions (such as out-of service or manual mode), connection properties, presentation information and help text. Updates can be easily installed with DDs (device descriptions), which is a compiled form of DDL. This allows easy updates and bug fixes on equipment, as updates can be downloaded onto the equipment.

19.3.8 Control

The two main control advantages of the Fieldbus are that it truly distributes control and that control processing can be done concurrently (rather than being centralised in a controller). The devices can implement many of the control functions that a traditional DCS would do. Most of the control functions are implemented by the following:

- Two input blocks (analogue and digital).
- Two output blocks (analogue and digital).
- Six control blocks, such as PID (proportional-integral-derivative).

Other, less well used, control functions include pulse input, arithmetic, dead time, splitter and signal characterisation.

Most current control systems use a DCS (distributed control system) to control and the transmitter simply determines the process variable. Smart transmitters will change this as basic regulatory control can be moved to the transmitter. This reduces the loading on the controller, and as the functions become more complex may eliminate the controller all together.

The control with Fieldbus is relatively easy if the devices are located on the same bus, and are located in near proximity. Control in the field devices works well if the elements of control are located relatively close to each other. This allows function blocks to be linked without having to span different bus segments, and thus reduces delays.

Messages on the bus are divided into two classes:

- Cyclic – these messages involve process data which is transferred between linked function blocks and can be made part of a network. The 31.25 kbps Fieldbus supports approximately 30 messages per second
- Acyclic – these are single transfers of data. The scheduling of these is determined by the control equipment and is flexible in its approach, thus allowing the bandwidth to be used effectively.

19.3.9 Diagnostic information

Maintenance methods differ from plant to plant. These are:

- Preventative maintenance (PM) – this is where plant is inspected and, if necessary, replaced, before faults occur. In some cases, PM can cause more problems than it is worth, because when a piece of plant is disturbed it can often lead to faults that would not have happened.
- Unfortunately, to operate PM properly, it requires a great deal of information about the operation of the plant for its previous history.
- Deferred maintenance (DM) – this is where maintenance is deferred to save costs. Unfortunately, deferred maintenance can often lead to long-term costs, typically causing plant shutdowns, complete rebuilds for expensive equipment or leading to an unsafe plant.

In the past manufacturers have built in diagnostic information to microprocessor-based devices. Unfortunately, the method of implementation has been non-standard. Typically, each diagnostic signal required an additional 4–20mA signal to be sent to a host or DCS. In some cases, a proprietary digital protocol allowed the transmission of multiple diagnostic signals over the same pair of wires. This all required extra control system programming and alarm handling.

The Fieldbus overcomes this with a standardised comprehensive alarm-reporting mechanism and the DDL. A host or DCS supporting Fieldbus does not require any special configuration or programming to accept the manufacturer-specific predictive diagnostic information.

The diagnostic information can be used to determine when a device needs to be maintained or replaced. For example, an instrument may have a battery backup. The microprocessor can then monitor the voltage level of the battery. If it falls to a given value, the microprocessor can report an alarm that the battery requires maintenance. The intelligent plant that warns its operators when it is about to fail is one step closer.

The Fieldbus allows for peer-to-peer communications. Thus intelligent sensors can talk to each other, and allow the interaction of devices, typically to make calculations from process measurements, that allow instruments to determine if a fault is localised or due to a process upset.

19.3.10 Operational Information

In the transmission of 4–20 mA, analogue signals reduces the resolution of any measured value. As the Fieldbus uses digital technology, the resolution of the measured signals does not depend on the transmission channel, and only on the sensor and its associated analogue-to-digital conversion. The TB passes transducer passes the measured signal, which has been compensated for operating conditions, to the FB which converts the value into the required units. The Fieldbus allows for 6.5 decimal places of precision with the desired engineering units. Typical operation information includes:

- Range values, engineering units, secondary variables, serial number or tag name, calibration information, calibration date, materials of construction, and time stamping.
- Trending/calculations – from these measurements, statistics such as minimum, maximum, rms and average values can be logged and trended, over any time interval. The operation can be controlled by certain events, such as process start-up, process changes, time events, and so on. System time is automatically synchronised between devices. These values can be stored and automatically reset without the host or DCS having to control the operation. In custody transactions, such as gas/oil flow rates, values can be automatically stored and not tampered with by external parties. The calculation of trend releases the processing of a centralised system and also reduces network traffic, as the host or DCS only has to communicate with the Fieldbus device when it requires information.
- Device-related statistics – these include device identification (such as ID, model number, and so) and operation parameters, such as operational time, number of alarm conditions, number of power-ups, and so on, and are typically used in maintenance records.

19.3.11 New installations and upgrades

Fieldbus has the greater advantage in new installations in that it can significantly reduce the amount of cabling on the plant, and provide an increased amount of information than many other digital protocols. It also allows different venders equipment to be used (as long as they abide with a Fieldbus standard). On an existing site it is often difficult to justify the complete upgrade of a plant, as plant upsets can lead to financial losses. Thus many venders allow existing equipment to be upgraded. This normally requires a change of hardware, but is difficult in hazardous environments. A more typical situation is to replace failed devices with new smart transmitters, but this obviously requires a change of host software to be able to communicate with the device.

There has been considerable investment in DCSs. For Fieldbus to be a success it will have to be integrated with existing DCSs. The DCS will then change its functionality from low-level/high-level control to implementing high-level control, leaving the low-level control to the Fieldbus devices, as illustrated in Figure 19.5. Functions that will move from the

DCS to the Fieldbus are:

- Low-level control.
- Process parameters.
- Alarm generation.
- Calibration information.
- Device status.
- Area control.

The DCS will still be responsible for the high-level control functionality and the interoperatiablily between areas. Thus, the functions that will stay on the DCS are:

- Advanced control.
- Interarea control (bring together the control of areas).
- Production co-ordination.
- Centralised configuration.
- Alarm filtering.
- Network administration.
- Communicate with devices and service their requirements.
- Maintain an overall database.

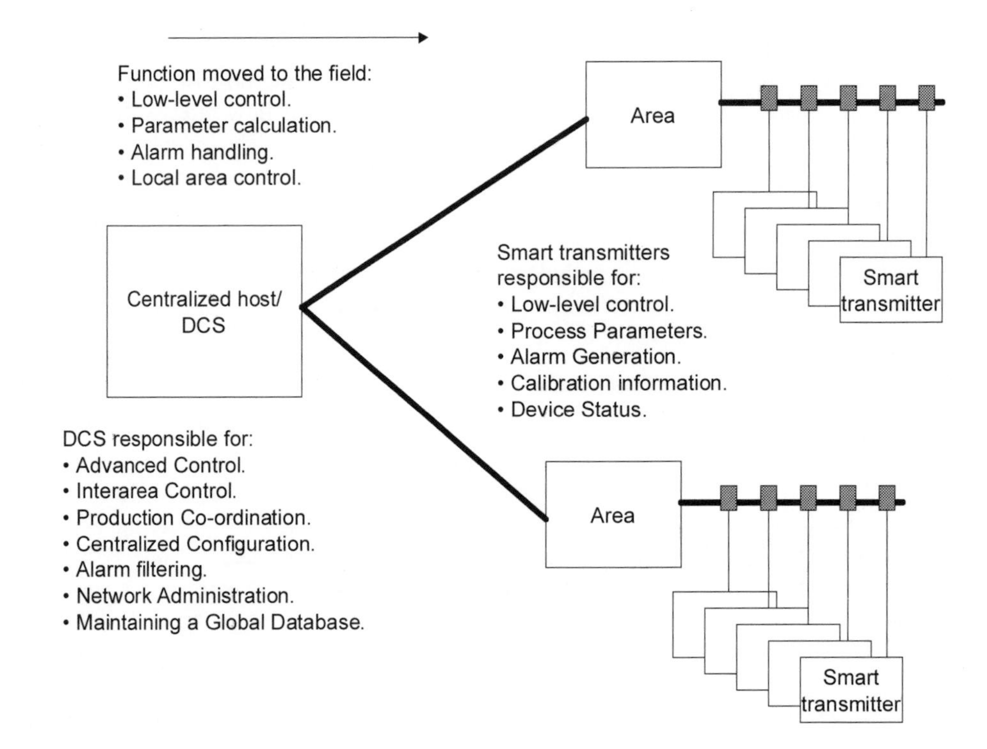

Figure 19.5 Functionality of a DCS/Fieldbus system

19.4 Exercises

19.4.1 The evolution of instrumentation transmission has generally progressed:

(a) 3–15 psi → 4–20 mA → serial link → Fieldbus
(b) 4–20 mA → 3–15 psi → serial link → Fieldbus
(c) 3–15 psi → serial link → 4–20 mA → Fieldbus
(d) serial link → 3–15 psi → 4–20 mA → Fieldbus

19.4.2 A 500 Ω resistor at the end of a 4-20mA transmission signal will give which voltage range:

(a) 0–5 V (b) 0–10 V
(c) 1–5 V (d) 2–10 V

19.4.3 Which field bus is used in the automotive industry:

(a) CAN bus (b) Profibus
(a) FOUNDATION (b) WorldFIP

19.4.4 Outline the advantages of Fieldbus compared with traditional methods. This should include topology flexibility, interoperatability, wiring costs, parameter resolution, and so on.

19.4.5 Discuss how Fieldbus will change the operation of a large instrumentation network. Outline the functions that are likely to be controlled by the DCS and which are likely to be controlled locally, in the field (refer to Figure 19.5).

19.5 Notes from the author

The instumentation industry has moved over the years from instrumentation networks made from dumb instruments which reported back to a central controller, to smart instruments with distributed control. Fieldbus is really the first true implementer of totally distributed systems, but as the scope of the Fieldbus is limited to areas, there is still a need for a global control system (such as a DCS). The Fieldbus is excellent at allowing local control and parameter conversion, but is not so good at providing a global control strategy. This disadvantage is outweighed by reduced cabling costs, as Fieldbus connects onto a bus, and devices easily connect to the bus.

Future instrumentation networks have no need to involve a complex main controller, as the instruments themselves have the power to locally control. The function of the main controller will change from getting involved with the low-level operations of instrumentation to the high-level functions of advanced control, interarea control, centralised configuration, alarm filtering and maintaining a global database.

Serial communication, such as RS-485, has allowed for multidrop serial communication networks, and has proved to be an excellent method of providing a highly reliable, fast communications channel. Unfortunately, it is still basically a communication channel and most of the higher-level protocols are vendor specific. The Fieldbus is an attempt to overcome this and to provide a standard method that is well matched for control and data acquisition.

The days of manufacturers creating a virtual monopoly with vendor-specific components is now, thankfully, receding. At one time organisations were generally tied by the vendor of the main control system, this was the only way that they could guarantee compatibility. International standards overcome this problem by forcing manufacturers to conform to the standard. Any vendor who does not conform will quickly lose market share, unless they are a true market leader, and have the power to force the whole industry in a certain direction. Today even the market leaders, such as Honeywell, have to conform to standards and become involved with other companies to develop industry standard, which are then developed as international standards by the relevant standards agency.

20 WorldFIP

20.1 Introduction

The WorldFIP protocol is an open system, international fieldbus standard (EN50170) and is used to interface to level zero (sensors and actuators) and level one (PLCs, controllers, and so on) devices. It can be used in many different architectures, such as centralised, decentralised and master–slave. The control algorithm can be located within a single processor or can be distributed. Figure 20.1 shows the layers of the WorldFIP standard.

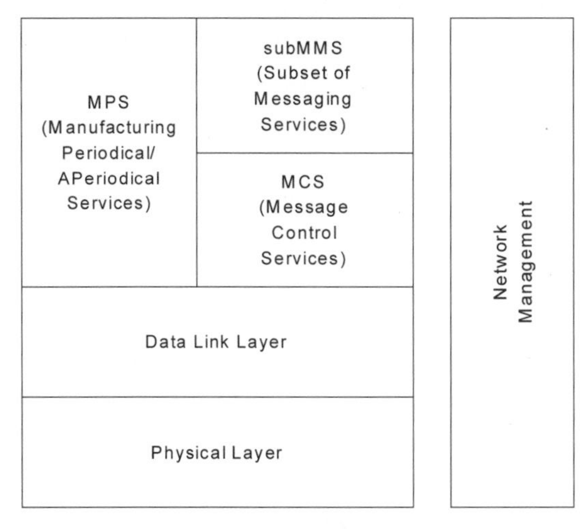

Figure 20.1 WorldFIP layers

20.2 Physical layer

The physical layer ensures the transfer of bits from one device to another. In the main specification the transmission rate is 1 Mbps over shielded twisted-pair (STP) or optical fibre cable. The three defined rates are S1 (31.2 kbps), S2 (1 Mbps) and S3 (2.5 Mbps) and an additional speed of 5 Mbps (fibre optic).

As Ethernet, WorldFIP uses Manchester coding. This codes a '1' as a high to a low transition, and a '0' as a low to a high transition, as illustrated in Figure 20.2. A constant high or a constant low level is a violation to the coding. A high level is a V^+ volition and a low level is a V^- violation. Manchester coding has the advantage of embedding the clock within the transmitted signal.

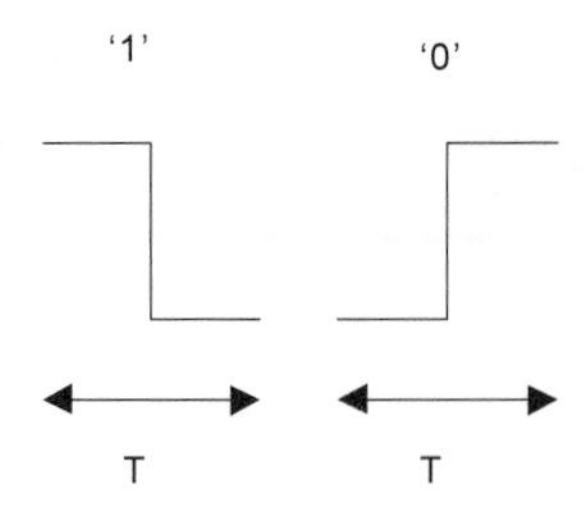

Figure 20.2 WorldFIP bit coding

The frame start sequence (FSS) is illustrated in Figure 20.3, and contains the following fields:

- Preamble (PRE) – apredefined pattern of 10101010. This is used to synchronise the receiving clocks.
- Frame start delimiter (FSD) – a predefined pattern of $1V^+V^-10V^-V^+0$, which defines the start of the CAD field. Note that violations (V^+ and V^-) cannot occur within this field.
- Control and data (CAD) – contains information on the data link layer.
- Frame end delimiter (FED) – a predefined pattern of $1V^+V^-V^+V^-101$, which defines the end of the CAD field.

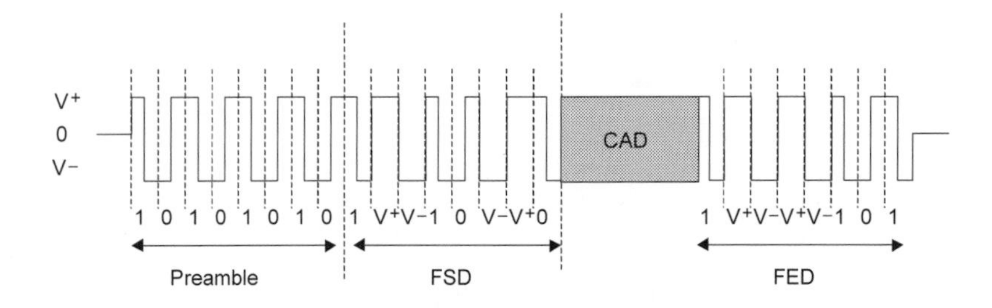

Figure 20.3 WorldFIP frame

20.3 Data link layer

The data link layer supports two types of service:

- Exchanges of variables.
- Message transfers.

These can either be cyclic or an explicit user request. A cyclic message is when the system configures object names. These exchanges are automatically sent without the user requesting them. An explicit user request involves requesting variables and the related response.

20.3.1 Addressing

WorldFIP has two addressing modes:

- Variable addressing – this is a global addressing scheme, where each variable in the distributed system has an associated identifier, which uniquely identifies the variable. Each identifier is a 16-bit integer value. A device requesting the variable does not need to know the location of the variable, and uses broadcasting to all connected devices.
- Message addressing – this is an addressing scheme which uses a 24-bit address for each device on the segment. Each address identifies the network segment and the node address on that segment.

20.3.2 Application layer–physical layer interfaces

The data link layer provides an interface between the application layer and the physical layer. It consists of a number of produced and consumed buffers, which contain the latest values updated by the user or by the network. These buffers are overwritten when the value is updated, and are automatically created on the initial configuration of a station.

Transactions involve passing an ID_DAT frame, which is followed by an RP_DAT frame. The format of these is illustrated in Figure 20.4. Frames begin with a control byte which allows network devices to determine the frame type. It is used to identify variable transfer requests, acknowledgement frames, and so on. Frames end with a frame check sequence (FCS) which is used to provide error detection.

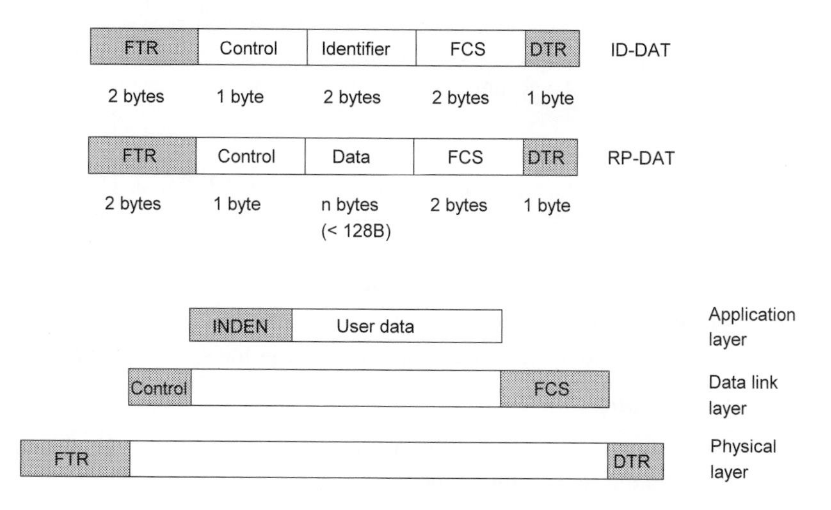

Figure 20.4 WorldFIP frames

Question frames are ID_DAT types, and are ID_RQ (ID request) and ID_MSG (ID message). These types are identified by the bits set in the control field. The responses are:

- **RP_RQ** (response request transfer). When a station that has made a variable transfer request receives an ID_RQ frame it responds with an RP_RQ frame coded as follows. The DATA field contains a list of identifiers, each of 16 bits (it can thus store up to 64 identifiers).
- **RP_MSG_xxx** (response message transfer). When a station that has made a message transfer request receives an ID_MSG frame it responds with an RP_MSG_NOACK or an RP_MSG_ACK frame coded as follows. The Data field contains a 3-byte destination address, a 3-byte source address and, up to, 256 message bytes. A bit in the control field indicates if the message transfer is acknowledged or unacknowledged. The destination and source fields show the addresses of the communicating entities.
- **RP_ACK** (response acknowledgement transfer). When a destination station receives a message with a request for acknowledgement it transmits an acknowledgement frame. No data is transmitted in the DATA field. This frame is very short, as the acknowledgement information is contained in the control field.
- **RP_FIN** (end-of message transaction response frame). When a message has been transmitted the sender, after waiting for an acknowledgement if necessary, transmits an end of message transaction frame. No data is transmitted in the DATA field. This frame is very short, as the transaction finished information is contained in the control field.

WorldFIP involves the sending of an ID_DAT, followed by an RP_DAT. The RP_DAT frame must be send within a given time interval. This is called the turnaround time. Typically the turnaround time is between 10 bit and 70 bit transmissions. Thus at 1 Mbps the time to transmit a single bit will be 1 μs, and the turnaround time will be between 10 μs and 70 μs.

20.3.3 Medium allocation mechanism

The WorldFIP network automatically achieves bus arbitration and production/consumption functions. The bus arbitrator (BA) contains the resources required to scan for variables when they are required. This involves a scanning table with a list of identifiers to circulate on the bus. It does this by:

- Broadcasting the name of the variable by sending an ID_DAT frame. This is read by all the devices on the bus. One of these identifies itself as being the producer of the variable. One or more other stations can recognise that they are consumers of the variable.
- Next the single producer of the variable transmits the value of the variable within a response frame (RP_DAT). All the consumers then read this.
- The arbitrator next goes to the next identifier, and follows the same sequence.

20.3.4 Bus arbitrating tables

The bus arbitrating table contains the variable identifier, the periodicity, the scanning time period, the data type and the conversation time:

Variable	*Period(ms)*	*Type*	*Time (μs)*
TEMP_1	5	INT_8	170
TEMP_2	10	INT_16	178
PRES_1	15	OSTR_32	418
PRES_2	20	SFPOINT	194

The bus arbitrator then repeats the table, indefinitely. The *period* is the scan period (in ms), the *type* defines the data type (such as INT_8 for an 8-bit integer, INT_16 for a 16-bit integer, OSTR_32 for a 32-character string) and *time* represent the total transaction time. In the case, the initial sequence would be:

(0 ms)	*(5 ms)*	*(10 ms)*	*(15 ms)*	*(20 ms)*	*(25 ms)*
TEMP_1	TEMP_1	TEMP_1	TEMP_1	TEMP_1	TEMP_2
TEMP_2		TEMP_2		TEMP_2	
PRES_1			PRES_1		
PRES_2				PRES_2	

The total time scanning must not exceed the repetition period, and the time which is not used with a periodic scan can be used by an aperiod transfer. A macrocycle goes through the scan sequence, and repeats. For example, in the case above the macrocycle will be:

```
1st Macrocycle        2nd Macrocycle
111111111111          111111111111
2 2 2 2 2 2           2 2 2 2 2 2
3   3   3   3         3   3   3   3
4     4     4         4     4     4
```

On a WorldFIP network there may be one or more bus arbitrators, but only one bus arbitrator can be active at any time. When a bus arbitrator is active, the others are silent to the traffic on the bus, and if a fault occurs on the currently active arbitrator elect a new arbitrator. This election takes place without consultation.

Within a WorldFIP network, each station has a physical station address of between 0 and 255. The mechanism of electing the new arbitrator is a function of the stations address and also for a time period (T3). When a dormant arbitrator detects a silence on the bus it waits for the time period T3, and then elects itself as the bus arbitrator (if another arbitrator has not elected itself). The time period (T3) is calculated as:

$$T_3 = 4(n+1)T_o$$

where n is the station address and T_0 is basic time filler (110 μs by default).

Thus the lower the address of a potential arbitrator the higher the chance it has to become the arbitrator. After election, the bus arbitrator begins scanning. The new arbitrators must be set up with the same elementary cycles and macrocycles as the previous arbitrator. The new bus arbitrator can then change these by transmitting a bus arbitrator synchronisation variable, which contains an elementary cycle number and a macrocycle number. All other dormant bus arbitrators read these values and change their values for the macrocycles and elementary cycles.

20.3.5 Aperiodic transfer

Variables that are not in the bus arbitrators cyclic scanning table can also be transmitted using an aperiodic transfer. A station that can request an aperiodic transfer can be a producer of a variable, a consumer, or both. It involves:

- The bus arbitrator broadcasting a question frame for the required parameter, in a time that is used not used for periodic traffic. Next the producer of the parameter responds with the parameter and sets an aperiodic request bit in the control field of the response frame (RQ). Aperiodic transfers have the two priority levels of urgent or normal.
- The bus arbitrator sending an identification request frame (ID_RQ) to ask the producer of the required parameter to transmit its request. The producer of the parameter responds with an RP_RQ frame. This frame contains a list of identifiers (between 1 and 64 identifier).

20.4 Exercises

20.4.1 In WorldFIP, Manchester coding is used as a line code. How does this code the bits:

(a) 0 is a positive voltage, 1 is a negative voltage
(b) 0 is a negative voltage, 1 is a positive voltage
(c) 0 is a transition from low to high, 1 is a transition from high to low
(d) 0 is a zero voltage, 1 is a positive voltage

20.4.2 In WorldFIP, what is defined as a bit volition over a single bit period:

(a) A constant low or a constant high voltage level:
(b) A change from low to high voltage
(c) A change from high to low voltage
(d) A transition which is the same as the previous transition

20.4.3 How is the start and end of a frame determined in WorldFIP:

(a) They are a constant voltage level (b) They contain violations
(c) They contain no violations (d) They contain no transitions

20.4.4 What is the purpose of the preamble in WorldFIP:

(a) It reduces power dissipation
(b) It allows all the connected devices to synchronise their receiving clocks
(c) It increases the bit rate
(d) It contain information

20.4.5 Discuss how parameters on a WorldFIP network can be assigned a unique ID. What is the maximum number of IDs that can be allocated. Also, what is the maximum number of devices that can be connected to a WorldFIP network.

20.4.6 Outline the main method that WorldFIP uses to request and broadcast data on the network.

20.4.7 Discuss how WorldFIP uses bus arbitrating tables.

20.5 Notes from the author

WorldFIP is an excellent example of a well-designed bus that is simple to set up and use. It uses many of the techniques developed in computer networks, such as the use of Manchester coding and collision detection. It is also based on a layer approach, such as having a physical layer, a data link layer, a management layer, and so on. This fits in well with the OSI seven-layered model that is used in computer networks (see Chapter 25), and allows manufacturers of different systems to interconnect their equipment through standard interfaces. It also allows software and hardware to integrate well and be portable on differing systems.

The layered approach also allows for different implementations on each layer. In its current form it supports bitrates of 31.5 kbps, 1 Mbps, 2.5 Mbps and 5 Mbps, over copper and fibre optic cables. The polling of data on a WorldFIP network is also extremely flexible where messages can either be sent periodically or aperiodically.

Another great advantage of WorldFIP is that each parameter on the network can be assigned a unique ID (a tag). As it is a 16-bit field, up to 65,636 of these tags can be used. The addressing of the devices is also powerful, and over 1 million addressable devices is possible (24-bit address).

21 CAN bus

21.1 Introduction

The Controller Area Network (CAN) protocol is an ISO-defined standard (ISO 11898) for serial data communication at bit rates up to 1 Mbps. It was initially developed for the automotive industry, and has the great advantage that it uses a common bus which reduces the need for wiring harnesses. It has since outgrown this application. The standard includes a physical layer and a data-link layer, which defines different message types, arbitration rules for bus access and methods for fault detection and fault confinement.

Its basic features are:

- Differential transmission using twisted-pair cables.
- Arbitration – access to the bus is controlled by a non-destructive bitwise arbitration technique. In arbitration, every transmitter compares the level of the bit transmitted with the level that is monitored on the bus. If these levels are the same then the unit will continue to send.
- Small messages (only up to 8 bytes in length), each with an associated checksum. These have fixed format messages, which transmit data and other information. Nodes only transmit messages when the bus is free. The content of a message is identified by the IDENTIFIER, which describes the meaning of the data, but does not provide for the destination of the information. All the nodes on the network can decide whether they need to read the data or not (multicast).
- No addressing structure – messages are broadcast on the common bus with a message with a priority level and identification.
- Prioritisation of messages – each message has a defined priority. On a free bus, any node can transmit their message. Two or more units which transmit a message at the same time produce an error condition. The unit with a message of higher priority always gains access to the bus over a lower priority node.
- Multicast reception with time synchronisation – all nodes are able to receive transmitted data and can also quickly synchronise their clocks to the transmitted data.
- Error detection and signalling – powerful error handling scheme that allows for the retransmission of messages when they are not properly received. There is also automatic retransmission of corrupted messages as soon as the bus is idle again. The recovery time from detecting an error until the start of the next message is at most 29 bit times, if there is no further error.
- Multimaster – any device can gain control of the bus.
- Enhanced fault finding and fault isolation. Implementation of methods of fault finding and removal of faulty nodes from the bus. There is also a distinction between temporary

errors and permanent failures of nodes. Defective nodes are switched off.

- Unlimited number of connections – any number of units can connect to the bus (without causing a disruption) and they can also be easily disconnected.
- Guaranteed latency times.
- System-wide data consistency.
- System flexibility – nodes can be added to the CAN network without requiring any change in the software or hardware of any node and application layer.
- Sleep mode/wake-up – a CAN device may be set into sleep mode without any internal activity and with disconnected bus drivers. This helps to save power. The sleep mode is finished with a wake-up by any activity or by internal conditions of the system.
- Acknowledgements – all receivers check the consistency of received messages, acknowledge a consistent message and flag an inconsistent message.

The CAN protocol defines the two layer layers of the OSI seven-layered model, the physical layer and the data-link layer. It does not contain specifications on higher level protocols, such as flow control, transportation of larger data packets, node addresses, communication establishments, and so on. These are implemented by a HLP (higher level protocol), which:

- Standardises the start-up procedures, such as setting the bitrate.
- Creates logical addresses for nodes.
- Formats messages.
- Organises system error handling.

The CAN bus is a truly distributed control system as it does not need a controller to control the flow of data between nodes.

The CAN bus splits into three main layers, as shown in Figure 21.1, and consists of:

- Object layer – implements part of the data link layer. It involves finding which messages are to be transmitted, deciding which messages that are received by the transfer layer are actually to be used and also provides an interface to the application layer.
- Transfer layer – implements the other part of the data link layer. It involves controlling the framing, performing arbitration, error checking, error signalling and fault confinement. This layer decides whether the bus is free for starting a new transmission or whether reception is just starting. It also provides other features, such as bit timing.
- Physical layer – the layer involves the definition of the electrical (signal levels and bit representations) and mechanical aspects (cable/connector type) of the physical connection.

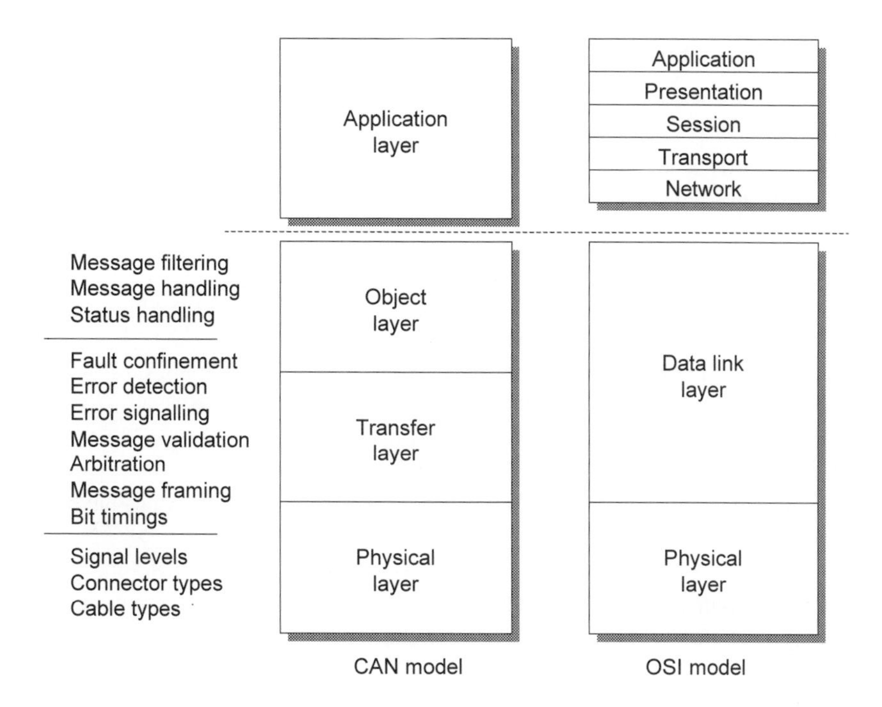

Figure 21.1 CAN/OSI models

21.2 CAN physical

The CAN bus uses non-return to zero (NRZ) with bit-stuffing for the physical layer. A bit can either be *dominant* (a logical 0) or *recessive* (logical 1), which corresponds to certain electrical levels which depend on the physical layer used. Modules are connected to the bus in a wired AND, thus if one node puts the line to a dominant level, then the whole line goes to this state, regardless of the other levels on the line.

Several physical layers can be used, such as:

- ISO 11898 – this uses a two-wire balanced signalling scheme, which can be shielded or unshielded. Implemented by the 82C250 transceiver device; the cable impedance is nominally 120 Ω.
- ISO 11519 – lower speed applications for two-wire balanced signalling scheme.
- Proprietary physical layers.
- RS-485 standard connection.
- SAE J2411 – single-wire cable.

The maximum speed of a CAN bus is 1 Mbps. At this maximum speed the maximum cable length of is 40 m (130 ft). This is a limited length, as the arbitration scheme requires that the first pulse propagates to the most remote node and back again before the bit is sampled.

Typical maximum lengths are:

- 500 kbps (100 m).
- 250 kbps (200 m).
- 125 kbps (500 m).
- 10 kbps (6 km).

There are no standard CAN connectors, typical types, as illustrated in Figure 21.2, are:

- 9-pole DSUB. The main connections are CAN_L (CAN_L bus line), CAN_GND (CAN ground), CAN_H (CAN_H bus line) and CAN_V+ (power, between +7 V and 13 V, 100 mA). Modules have a male connector (plug) and cables have a female connector (socket). Pins 3 and 6 are connected internally.
- 5-pole Mini-C (DeviceNet and SDS). Power supply is 24 V and modules have male connectors.
- 6-pole Deutch DT04-6P connector (proposed by CANHUG for hydraulic systems). It uses a male connector on the module and has six pins. The main connections are CAN_H (pin 2), power (pin 5), GND (pin 1) and CAN_L (pin 6).

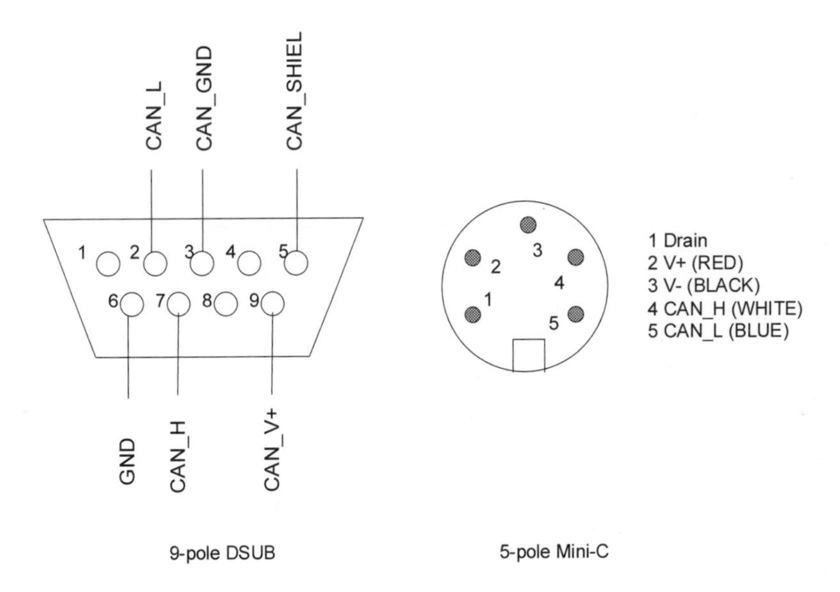

Figure 21.2 Typical CAN connectors

21.3 CAN bus basics

Many of the basics of the CAN bus were outlined in Section 21.1.

21.3.1 Bus values

The CAN bus defines two levels for bits on the bus, these are:

- Dominant level – all nodes on the bus connect to the common bus. A dominant bit will always overrule a recessive bit. In a wired-AND implementation of the bus, a dominant level is represented by a logical 0 and the recessive level by a logical 1.
- Recessive level – a recessive level on the bus is always overruled by a dominant level. If all levels are at a recessive level, the resulting level will also be recessive. In a wired-OR connection a recessive level will be a logical 0 (as a single logical 1 level will make the output also a logical 1).

Message bit streams uses a non-return-to zero (NRZ) technique. The frame segments start of frame, arbitration field, control field, data field and CRC sequence are coded using bit stuffing. When the transmitter detects five consecutive bits of identical values in the bit stream, it automatically adds an extra complementary bit in the actually transmitted bit stream. When the receiver detects this is automatically deletes the inserted bit.

In data frames and remote frames, the CRC delimiter, ACK field and end of frame are not stuffed. Error frames and overload frames are not bit stuffed.

21.3.2 Error detection

The main error detection methods are:

- Monitoring bit levels – transmitters check the transmitted level with the level on the bus.
- Cyclic redundancy check.
- Bit stuffing.
- Message frame check.

The error detection scheme detects the following:

- All global errors.
- All local errors at transmitters.
- Up to five randomly distributed errors in a message.
- Burst errors of length less that 15 in a message.

The resultant probability of undeleted errors is less than 4.7×10^{-11}.

21.4 Message transfer

Message transfer involves four different frame types:

- Data frame – contains data from a transmitter to a number of receivers.
- Remote frame – transmitted by a unit to request the transmission of the data frame with the same identifier.

- Error frame – transmitted by a unit on a bus error.
- Overload frame– provides for an extra delay between the preceding and the succeeding data or remote frames.

Data frames and remote frames are separated from preceding frames by an interframespace.

There is no explicit address in the messages. Instead, messages are *contents-addressed*, so that their contents implicitly determine their address.

In order to wake up other nodes of the system, which are in sleep-mode, a special wake-up message with the dedicated, lowest possible identifier (rrr rrrd rrrr; r= recessive, d = dominant) may be used.

21.4.1 Data frame

Data frames have seven different bit fields, as illustrated in Figure 21.3, these are:

- Start of frame – this defines the start of a data frame or a remote frame. It consists of a single dominant bit. All units on the bus synchronise to the leading edge of the bit.
- Arbitration field – this field consists of the identifier and the RTR bit. The identifier length is 11 bits (from ID-10 to ID-0). The seven most-signification bits (ID-10 to ID-4) must not be all recessive. The RTR bit (remote transmission request bit) is dominant in a dataframe, and recessive in a remote frame. Note that CAN 2.0B (extended CAN), uses a 29-bit Identifier (which also contains two recessive bits: SRR and IDE) and the RTR bit.
- Control field – this field has six bits and includes a 4-bit data length code(DLC) and two bits which are reserved for future expansion (dominant, at present). The codes for the DLC are *dddd* (for 0 bytes), *dddr* (for 1 byte), *ddrd* (for 2 bytes), up to *rddd* (for 8 bytes). Thus up to 8 bytes can be defined in the data field.
- Data field – this field consists of up to eight bytes of data (MSB first).
- CRC field – this field contains the CRC sequence, followed by a CRC delimiter (which is a single recessive bit).
- ACK slot – a receiver which correctly receives a message, reports an acknowledgement by sending a message back to the transmitter with a dominant bit in the ACK slot. After this field is the ACK delimiter field which is a single recessive bit.
- End of frame – this delimits a data frame and remote frame and consists of seven recessive bits.

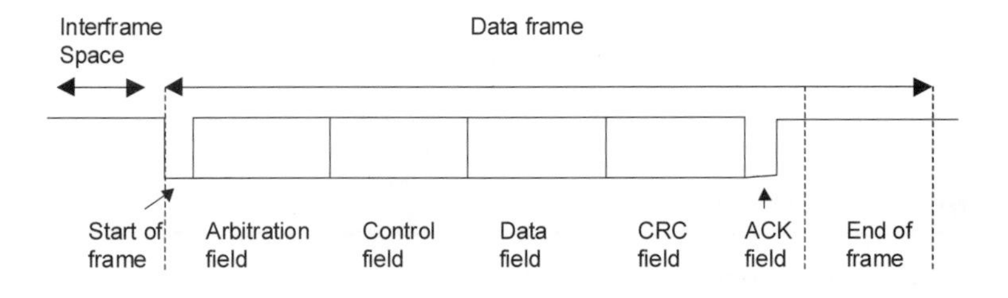

Figure 21.3 Data frame format

The 15-bit CRC calculation uses the message fields as a polynomial which is divided by a defined CRC generator polynomial (using modulo-2 division), which is:

$$x^{15}+x^{14}+x^{10}+x^{8}+x^{7}+x^{4}+x^{3}+1$$

The remainder of this polynomial division is the CRC sequence. The algorithm uses a 15-bit shift register buffer(14:0).

```
buffer=0;
repeat
   nextbit= nextbit ⊕ buffer(14);
   buffer(14:1) = buffer(13:0);
   buffer(0)=0;
   if nextbit then
      buffer(14:0) = buffer(14:0) ⊕  (100010110011001b);
   endif
until (the start of the CRC SEQUENCE field);
```

where nextbit denotes the next bit of the bit stream. The resulting buffer will then be stored in the CRC sequence field.

21.4.2 Remote frame

The receiver for certain data can initiate the transmission of the required data by sending a remote frame. It has the same fields as a data frame, but does not have a data field. It is identified with a recessive bit in the RTR bit (a dominant bit identifies a data frame).

For example, if node A transmits a remote frame with the arbitration field set to 123, then node B could respond with a data frame with the arbitration field also set to 123. This type of frame is used to implement a request–response type of bus traffic management.

21.4.3 Error frame

Error frames have two different fields, error flags and an error delimiter (eight recessive bits). There are two types of error flag, these are an active error flag and a passive error flag. An active error flag has six consecutive dominant bits, whereas the passive error flag has six consecutive recessive bits (unless they are overwritten by dominant bits from other nodes).

After transmission of an error flag, each node sends recessive bits and then monitors the bus until it detects a recessive bit. Afterwards it starts transmitting seven more recessive bits. The error frame thus violates the bit stuffing rules of a CAN message. It is transmitted when a node detects a fault and causes all other nodes to detect a fault (and they will also send error frames). After this the transmitter automatically resends the message.

The error frame consists of an error flag, which is 6 bits of the same value (thus violating the bit-stuffing rule) and an error delimiter, which is 8 recessive bits.

21.4.4 Overload frame

An overload frame contains the two bit fields: overload flag (six dominant bits) and overload delimiter (eight recessive bits). At the most, two overload frames may be generated to delay the next data or remote frame. The overload flags form destroys the fixed form of the intermission field. Thus, all other stations detect an overload condition. The conditions that cause an overload frame are:

- Receiver internal conditions, which requires a delay of the next data frame or remote frame.
- Detection of a dominant bit during intermission.

After transmission of an overload flag, nodes monitor the bus until they detect a transition from a dominant to a recessive bit. Then every node start transmission of seven more recessive bits.

21.4.5 Interframe spacing

Data frames and remote frames are separated from preceding frames by an interframe space. It contains intermission (three recessive bits) and bus idle bit fields. During intermission, no station can start to transmit data. The period of bus idle is of any length (a free bus condition). After this, the detection of a dominant bit on the bus is interpreted as a start of frame.

21.5 Fault confinement

With respect to fault confinement a unit may be in one of three states:

- Error active – these nodes can normally take part in bus communications and sends an active errorr flag when an error has been detected.
- Error passive – these nodes must not send an active error flag. They take part in bus communication but when an error has been detected they only send a passive error flag.
- Bus off – these nodes are not allowed to have any influence on the bus.

For fault confinement, two counts are implemented in every bus unit. The first is the transmit error count. For example:

- A transmitter sending an error flag, increases it by 8.
- A transmitter detecting a bit error while sending an active error flag or an overload Tflag, increases it by 8.
- Successful transmission of a message, decreases by 1 (unless it is already 0).

The second is the receive error count. For example:

- A receiver detects an error, increase it by one.
- A receiver detecting a dominant bit as the first bit after sending an error flag, increased by eight.
- A receiver detects a bit error while sending an active error flag or an overload flag, increased by eight.
- After the successful reception of a message, decreased by one.

Nodes are initially error active. An error active node transmits active error flags when it detects errors. A node becomes error passive when the transmit error count equals or exceeds 128, or when the receive error count equals or exceeds 128. An error passive node transmits passive error flags when it detects errors. A node is bus off when the transmit error count is greater that or equal to 256. A node which is bus off will not transmit anything on the bus at all.

21.6 Bit timing

Each bit on the CAN bus is, for timing purposes, divided into at least 4 *quanta*. The quanta are logically divided into four groups or *segments*:

- Synchronisation segment – this is one quantum long and is used for synchronisation of the clocks. A bit edge is expected to take place here when the data changes on the bus.
- Propagation segment – this is required to compensate for the delay in the bus lines.
- Phase segment 1 – this may be shortened (Phase segment 1) or lengthened (Phase segment 2), if necessary, to keep the clocks in synchronisation. The bus levels are sampled at the border between Phase segment 1 and Phase segment 2.
- Phase segment 2.

Figure 21.4 shows a schematic of the bit. Most CAN controllers also provide an option to sample three times during a bit. In this case, the sampling occurs on the borders of the two quanta that precedes the sampling point, and the result is subject to majority decoding.

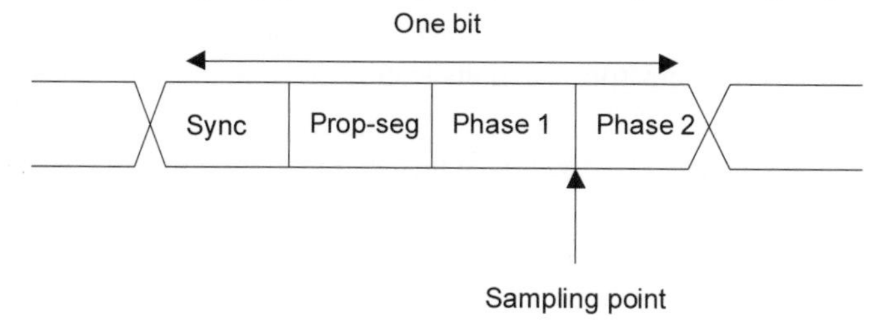

Figure 21.4 Bit timing

21.6.1 Clock synchronisation

In order to adjust the on-chip bus clock, a CAN controller can either shorten or lengthen a bit by a whole number of quanta. The maximum number of quanta is defined as the synchronisation jump width:

- Hard synchronisation – occurs on the recessive-to-dominant transition of the start bit. The bit time is restarted from that edge.
- Resynchronisation – occurs when a bit edge does not occur within the synchronisation segment in a message. For this one of the phase segments is shortened or lengthened with an amount that depends on the phase error in the signal (the maximum value is defined by the synchronisation jump width).

21.6.2 Bus failure modes

The ISO 11898 standard defines several fault modes on a CAN bus cable, these are:

1. CAN_H interrupted.
2. CAN_L interrupted.
3. CAN_H shorted to battery voltage.

4. CAN_L shorted to ground.
5. CAN_H shorted to ground.
6. CAN_L shorted to battery voltage.
7. CAN_L shorted to CAN_H wire.
8. CAN_H and CAN_L interrupted at the same location.
9. Loss of connection to termination network.

A fault tolerant network will be able to survive these faults, and still transmit data (although the SNR will be reduced). An 82C250-type transceiver may not be able to survive many of the conditions. Fault-tolerant drivers, such as the TJA1053, can handle these failures (though at a reduced maximum speed).

21.7 CAN open

The CAN Application Layer (CAL) was originally developed by CiA and involves:

- CMS (CAN-based message specification) – defines protocols for transferring data between CAN modules.
- NMT (network management service) – defines the protocols for system start-up and shutdown and error logging.
- DBT (distributor service) – defines a protocol for distributing identifiers to the different modules in a system.

21.8 Exercises

21.8.1 In the CAN bus, what is the basic bit rate:

(a) 125 kbps (b) 1 Mbps
(c) 10 Mbps (d) 100 Mbps

21.8.2 Which of the following best describes a dominant bit:

(a) A bit, that when transmitted will be overruled by a recessive bit
(b) A bit, that when transmitted will overrule a recessive bit
(c) A bit that is always a high level
(d) A bit that is always a low level

21.8.3 Which of the following best describes a recessive bit:

(a) A bit, that when transmitted will be overruled by a dominant bit
(b) A bit, that when transmitted will overrule a dominant bit
(c) A bit that is always a high level
(d) A bit that is always a low level

21.8.4 Explain the concept of dominant and recessive levels, and how these can be used to determine if two or more devices are communicating at the same time.

21.9 Notes from the author

As with the WorldFIP bus, the CAN bus is a well-designed network, based on techniques learned from computer networks. It is a serially connected bus, where all nodes have access to the network, and collisions between nodes are detected within a very short time. This allows devices to have a relatively equal share of the bandwidth of the bus. As automobiles are noisy environments, the CAN bus is a rugged bus which copes well with errors, and also devices which are not operating correctly.

The relatively high bit rates of the CAN bus allows a great deal of information to be passed between instruments and their controllers. To prevent major problems, the bus can be organized into segments, so that a major fault in one are a does not greatly affect other areas. A failure of any of the controllers can lead to major problems, so secondary controllers can be made to monitor the operation of the primary, and can remove the primary controller from the bus if they are not operating correctly. Another method is to allow localized control when the primary control is not functioning properly.

Power dissipation is also a major factor in cars as devices must be ready to respond quickly to events, but not to dissipate much power when they are idle. Thus, the CAN bus has methods to allow devices to sleep if they are inactive and then is awoken when a specific event occurs.

The car of the future, based on the CAN bus, would have little need for complex wiring harnesses, and would simply require the daisy chaining of devices onto the common bus. The connector used can be matched to the environment, such as heavy-duty connector for robust situations, or a light connector for ease of connection/disconnection.

As the CAN bus has been designed with a thought for the seven-layered OSI model, which is used in computer networks, there is great potential for using standard computer network protocols, such as TCP/IP. Thus will allow CAN busses to connect straight into the Internet, and allow for remote control and remote data acquisition over the Internet, or over a local or wide area network. The data could be protected using data encryption techniques. So, maybe one day you could log into the Internet and switch on the air conditioning in your car before you even leave your house.

22 IEEE-488, VME and VXI

22.1 Introduction

The IEEE-488, VME and VXI busses have all been used to interface to programmable instrumentation and controllers. They were at one time very popular, but with the increasing development of PCI, SCSI and RS-232 they have been replaced in many applications. IEEE-488 has the advantage of being very robust and simple to use, but suffers from a lack of speed. The VME bus has the opposite problem. It has fast data transfers and is powerful, but it is also difficult to use.

22.2 IEEE-488 bus

The IEEE-488 bus was developed in the 1970s as an answer to problems in interfacing with programmable instruments and controllers. Its history is as follows:

- 1975 – the IEEE published the IEEE-488 standard, which was based on work done by Gerald Nelson and David Ricci at Hewlett-Packard.
- 1978 – the IEEE published a revised specification known as the ANSI/IEEE standard 488-1978. IEC (International Electrotechnical Commission) then adopted the specification as an international standard (IEC 625-i).
- 1987 – revised standard, known as ANSI/IEEE standard 488.l-l987. This specification deals with the mechanical interconnection and the electrical protocol.
- 1987 – ANSI adopted the ANSI/IEEE standard. 488.2-1987 which standardises the software interfaces in terms of codes, formats, protocols and common commands.

The IEEE-488 bus is an excellent interface to programmable instruments as it is relatively simple to add to, operates at reasonable speeds and is available from various manufacturers. It is also know as the general-purpose interface bus (GPIB).

The IEEE-488 bus allows for the interconnection of instruments using a standard cable and a standard interface. It supports different data-transfer rates and also different message lengths. The maximum transmission path length is 20 m, and it is recommended the maximum interconnection length between instruments is limited to 2 m, each. A maximum of 15 instruments can be connected to the bus, although bus extender can increase this number. All devices connect to a common bus (a party-line) using 16 lines. These are shown in Figure 22.1 and are:

- Eight data lines (DIO1…DIO8). These allow eight bits to be transmitted, at a time. If more than eight bits are to be transmitted then they are sent one at a time. This gives a maximum throughput of 1 MB/s.
- Three handshaking lines ($\overline{\text{DAV}}$, $\overline{\text{NRFD}}$ and $\overline{\text{NDAC}}$). These handshake the data between a talker and a listener.
- Five bus management lines (IFC, ATN, SRQ, REN and EOI). These provide status information on the bus.

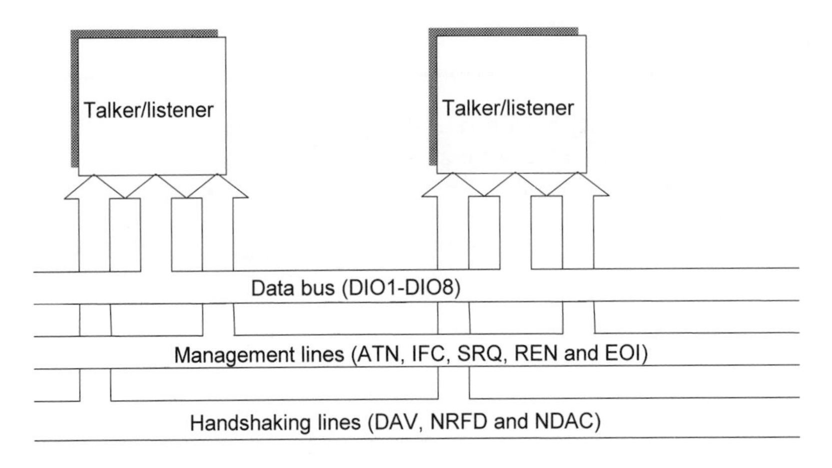

Figure 22.1 IEEE-488 bus lines

The IEEE-488 bus is a bidirectional bus where data can flow in either direction. Devices are set up with their functionality. These are either a talker, a listener or a controller, or any combination of these. The function types are:

- Talkers – these are devices which can send data to other devices or to the control. These include multi-meters and oscilloscopes. Only one device is allowed to talk at a time.
- Listeners – these are devices which only receive data. A device can also be a talker and a listener. For example, a multimeter can be a talker when it is sending voltage samples, but a listener when it is receiving data on range changes.
- Controller – these are devices which are responsible for information management on the bus, and include setting up tasks, monitoring the progress of measurements and interpreting results. To avoid a conflict, only one controller may be active at any time, but there may be more than one controller connected to the bus. Thus a controller can pass control to another controller.

It is possible to construct a system which does not have a controller and only has a talker and a listener. For example, a talker could be a digital multimeter and the listener could be a printer. The multimeter is then hardwired as a talker (using switches) and the printer as a listener.

22.2.1 Signal lines

Figure 22.2 shows the connections on the GPIB connector. Each instrument on the bus has an assignable and unique address on the bus. These addresses are selected by switches, although many new types are configured using configuration menus. The addresses range from 0 to 31 (although address 31 is often used by some manufacturers for self-diagnostics). Normally only 15 address can be used, unless a bus extender is used. The assigned address is the primary address of the instrument. If the instrument is a host for other instruments, each of the connected instruments can be assigned a unique secondary address.

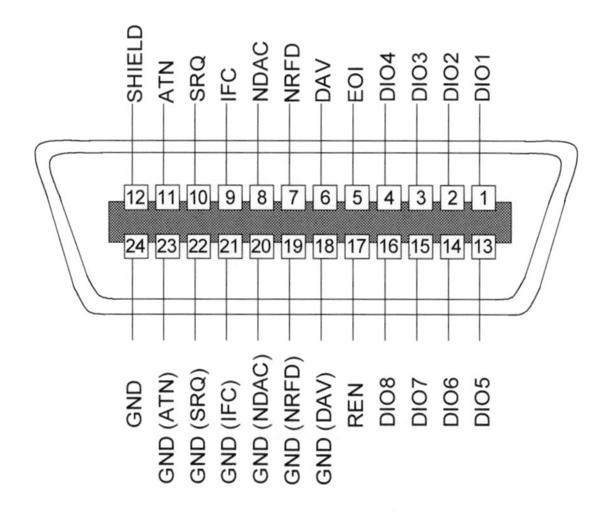

Figure 22.2 GPIB connector

Handshaking lines

The handshaking signal lines are:

- $\overline{\text{DAV}}$ (data valid) – a talker sets this line to a LOW, after it detects a HIGH on the $\overline{\text{NRFD}}$ line, and when the data on its I/O are settled and valid.
- $\overline{\text{NRFD}}$ (not ready for data) – a listener sets this line LOW when it is not ready for data. A HIGH level thus indicates that it is ready for data. This line will be held LOW. The $\overline{\text{NRFD}}$ line only goes HIGH when all the addressed listeners are ready to accept data.
- $\overline{\text{NDAC}}$ (not data accepted) – the listener sets this line LOW when it has not accepted the data. When it accepts the data, it releases the $\overline{\text{NDAC}}$ line. The $\overline{\text{NDAC}}$ line does not go HIGH until all the listeners have accepted the data.

Every byte transferred must be handshaked, and more than one device can receive data at a time. The three handshaking signals are active-LOW and can be pulled LOW by any device. Figure 22.3 shows a basic handshaking operation. Initially, with the first byte, the talker puts the data byte of the data bus and then waits for the $\overline{\text{NRFD}}$ line to be set HIGH. It then sets the $\overline{\text{DAV}}$ line active LOW. The devices then read the data and set their $\overline{\text{NDAC}}$ lines HIGH, but as long as one device as not read the data it will be set to an active LOW. Once the slowest device has read the data it sets the $\overline{\text{NDAC}}$ line HIGH. The talker then knows that the data has been read and can then put more data on the bus.

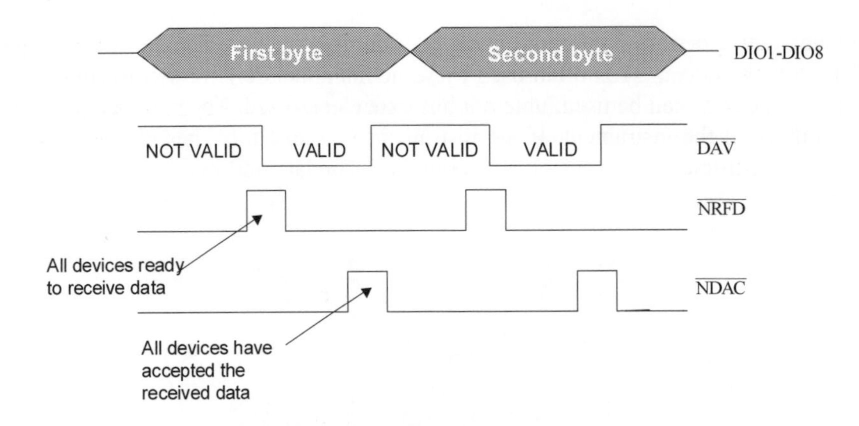

Figure 22.3 GPIB handshaking

Interface management lines

The interface management lines are:

- ATN (attention) – this causes all the devices on the bus to interpret the data, either as a controller command or as an address.
- IFC (interface clear) – bus reset.
- SRQ (service request) – used by a device to alert the controller that it requires to communicate.
- REN (remote enable) – enables devices to respond to remote program control.
- EOI (end or identify) – indicates the last transferred data byte of data.

22.3 VME bus

The VME computer bus is based on the IEEE-1014-1987 standard. Its main features are:

- A high-speed asynchronous data bus to transfer 8, 16 or 32 bits at a time.
- Four buses: data-transfer, arbitration, priority interrupt and utilities.
- Supports several bus controllers, such as the CPU, DMA, I/O controllers and any other device that needs to control the bus. The arbitration bus avoids conflicts.
- Priority-interrupt buses where devices can request service from a VME interrupt handler (similar to the service request [SRQ] line in IEEE-488).
- A utilities bus which provides power distribution, clocks, initialisation and failure detection.

The VXI (VME extension for instrumentation) is an extension to the VME bus, and is similar in its approach to the IEEE-488 bus. It is made up of several buses: VME bus, clock and sync bus, the star bus, the trigger bus, the local bus, the analogue sum bus, the module identification bus and the power distribution bus.

The VXI bus is intended for rack-mounted devices; each slot can take a module that is 30 mm (larger modules can take up more than one slot). Figure 22.4 shows the different sizes of modules that can be used with the VXI bus. Modules stand on edge, with cooling holes at the top and bottom edge of each module.

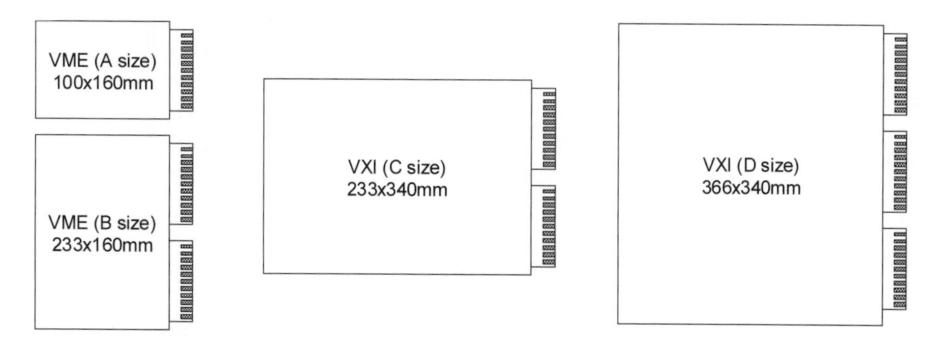

Figure 22.4 VMI module sizes

22.4 VXI bus

The VXI has several buses. These buses are global, unique or private. A global bus is a common bus that connects to all of the cards. A private bus is used for local communications between a set of cards, and a unique bus provides additional, named signal lines. The buses can be grouped as follows:

- Global – VME computer bus, trigger bus, analogue sum bus and power distribution bus.
- Private – local bus.
- Unique – star bus.

22.4.1 Clock and synchronisation bus

This bus contains two clocks and a clock synchronisation signal. One clock operates at 10 MHz, and the other is at 100 MHz, and these are accompanied by a sync signal. Each of the clocks uses ECL (emitter-coupled logic) and are buffered on the backplane, as illustrated in Figure 22.5.

22.4.2 Star bus

The star bus allows high-serial communications between each of the modules. It uses two high-performance ECL lines, named STARX and STARY. The bus is designed so that the path length between slot 0 and the other 12 slots is the same, and gives a maximum delay of 5 ns between slot 0 and any other module, as illustrated in Figure 22.6.

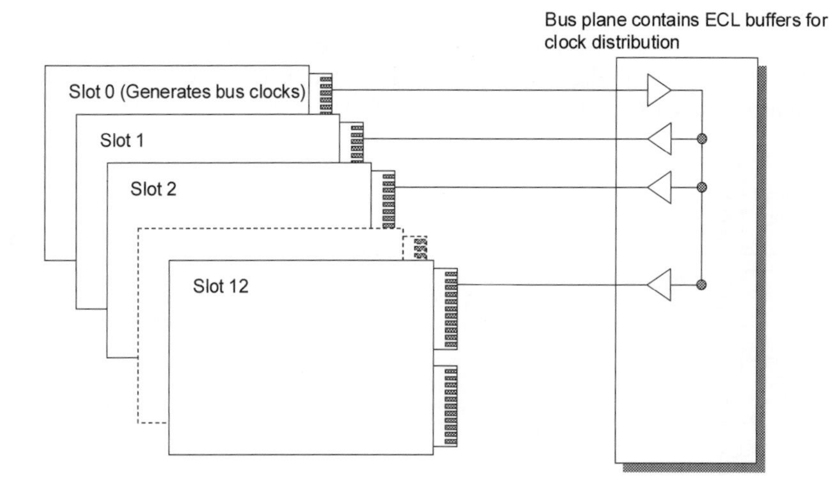

Figure 22.5 VMI clock distribution

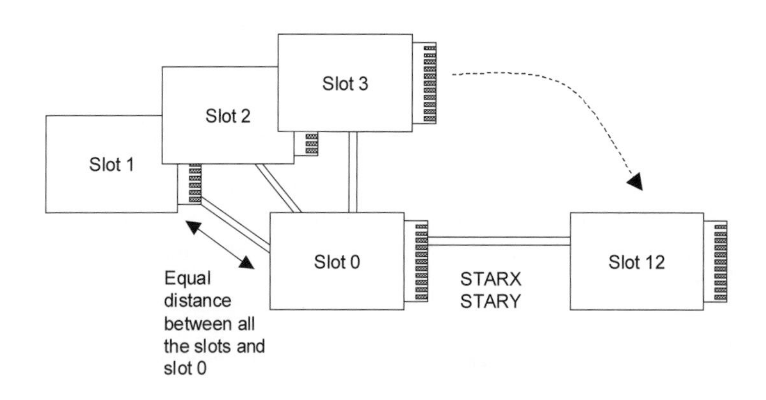

Figure 22.6 VMI clock distribution

22.4.3 Trigger bus

The trigger bus is a general-purpose logic bus that can be used for triggering, handshaking, clocking or data transmission. It uses eight TTL trigger lines and six ECL trigger lines.

22.4.4 Local bus

The local bus is used for intercommunication between two or more modules that does not use a global bus. It has 72 lines on each module, which are partitioned into 36 lines on each side of the module, as illustrated in Figure 22.7. The purpose of the bus is to decrease the need for jumpers between modules and to provide local communications between two or more modules without using a global bus. For example, two cards could be placed in consecutive sockets and communicate with each other using the right-hand side pins on left-most card and the left-hand side pins on the right-most card.

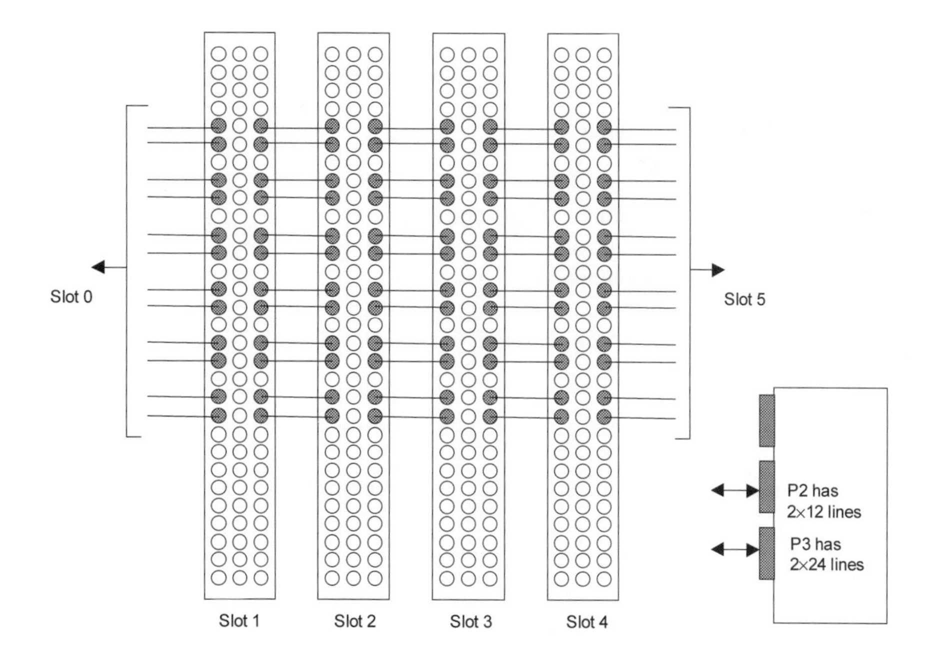

Figure 22.7 Local bus backplane

22.4.5 Analogue SUMBUS

The Analogue SUMBUS provides for an analogue-summing node for all connected modules. It is possible for each module to drive the summing line using an analogue current source. Every module can also read from this line, using a high-impedance receiver. It is typically used to generate a complex waveform, where one or more modules can each provide a part of the waveform. For example, three modules could each provide a single frequency which are then summated onto the SUMBUS.

22.4.6 Module identification bus

This bus provides for an identification of the connected modules. It uses 12 identification wires which all connect to slot 0, and then one connection to each of the other slots. If a module is plugged into a slot, then it provides a ground connection for the module identification line, otherwise it will be an open circuit. This allows for quick system configuration at start-up and aids in diagnostics. Typically the module in slot 0 will scan the bus when it is started, if a module that is connected fails to communicate, the device in slot 0 may display an error for that device.

22.4.7 Power bus

The power bus provides:

- Up to 268 W for each module.
- Voltages of +5 V, +12 V, –12 V and +5V standby (battery backup) from the VME bus.
- Voltages of +24 V and –24 V for analogue circuits for the VXI bus.
- ECL voltages of –5.2 V and –2 V for ECL.

22.5 Exercises

22.5.1 What does GPIB standard for:

(a) General-purpose interrupt bus
(b) General-programmable interface bus
(c) General-purpose interface bus
(d) General-programmable interrupt bus

22.5.2 Which international specification defines the GPIB:

(a) IEEE-488 (b) RS-486
(c) IEE-488 (d) GP-IB

22.5.3 How many data bits can be transferred over the IEEE-488 bus, at a time:

(a) 8 (b) 16
(c) 32 (d) 64

22.5.4 An oscilloscope is likely to be which of the following GPIB function types:

(a) Talker (b) Listener
(c) Priority generator (d) Controller

22.5.5 A temperature instrument device is likely to be which of the following GPIB function types:

(a) Talker (b) Listener
(c) Priority generator (d) Controller

22.5.6 A data acquisition/controller PC is likely to be which of the following GPIB function types:

(a) Talker (b) Listener
(c) Priority generator (d) Controller

22.5.7 What is the function of the $\overline{\text{DAV}}$ handshaking line:

(a) It defines that all the devices have received the data
(b) It identifies when all the devices are ready
(c) It resets the bus
(d) It defines valid data on the bus

22.5.8 What is the function of the $\overline{\text{NDAC}}$ handshaking line:

(a) It defines that all the devices have received the data
(b) It identifies when all the devices are ready
(c) It resets the bus

(d) It defines valid data on the bus

22.5.9 What is the function of the $\overline{\text{NDFD}}$ handshaking line:

(a) It defines that all the devices have received the data
(b) It identifies when all the devices are ready
(c) It resets the bus
(d) It defines valid data on the bus

22.5.10 Define the operation of the handshaking lines on the IEEE-488 bus.

22.5.11 Define the actions of talkers, listeners and controllers on an IEEE-488 bus. Give examples of each type of device.

22.5.12 The IEEE-488 bus is a common bus, where all the devices connect to each of the signal lines. Why are most of the handshaking lines active low?

22.6 Notes from the author

This chapter has covered one of the most simple busses, the IEEE-488 bus, and one of the most complex ones, the VME bus. So, why did I include them in a single chapter. Well, I cheated a little, because they didn't really merit a chapter of their own, so I merged them (I'm sure I'll receive lots of e-mails complaining about this, so I'll give them both a chapter of their own in the second edition, if they both still exist and if I'm allowed a second edition).

The IEEE-488 is a beautifully designed bus, which is well supported by software vendors, and is easy to set up. It will basically run quietly for many years without requiring any intervention by the user. The connector and cable are very well designed and can stand a great deal of abuse. It has typically been used a standard interface for instrumentation, but the growth of the serial busses is likely to reduce its importance.

And what can I say about the VME bus. Oh boy, it's complex. Its little brother, the VXI is a little less complex, but still is an extremely powerful and flexible bus for building modular instrumentation systems. Unfortunately it suffers from being too flexible and can be complex to write software for. The popularity of the PCI bus, and especially the CompactPCI bus (the PCI bus for modular systems) is overtaking the VXI bus.

23 TCP/IP

23.1 Introduction

Networking technologies such as Ethernet, Token Ring and FDDI provide a data link layer function; that is, they allow a reliable connection between one node and another on the same network. They do not provide internetworking where data can be transferred from one network to another or from one network segment to another. For data to be transmitted across a network requires an addressing structure which is read by a bridge, gateway and router. The interconnection of networks is known as internetworking (or an internet). Each part of an internet is a subnetwork (or subnet). Transmission control protocol (TCP) and Internet protocol (IP) are a pair of protocols that allow one subnet to communicate with another. A protocol is a set of rules that allows the orderly exchange of information. The IP part corresponds to the network layer of the OSI model and the TCP part to the transport layer. Their operation is transparent to the physical and data link layers and can thus be used on ethernet, FDDI or token ring networks. This is illustrated in Figure 23.1. The address of the data link layer corresponds to the physical address of the node, such as the MAC address (in Ethernet and Token Ring) or the telephone number (for a modem connection). The IP address is assigned to each node on the Internet. It is used to identify the location of the network and any subnets.

TCP/IP was originally developed by the US Defense Advanced Research Projects Agency (DARPA). Their objective was to connect a number of universities and other research establishments to DARPA. The resultant internet is now known as the Internet. It has since outgrown this application and many commercial organizations now connect to the Internet. The Internet uses TCP/IP to transfer data. Each node on the Internet is assigned a unique network address, called an IP address. Note that any organisation can have its own internets, but if it is to connect to the Internet then the addresses must conform to the Internet addressing format.

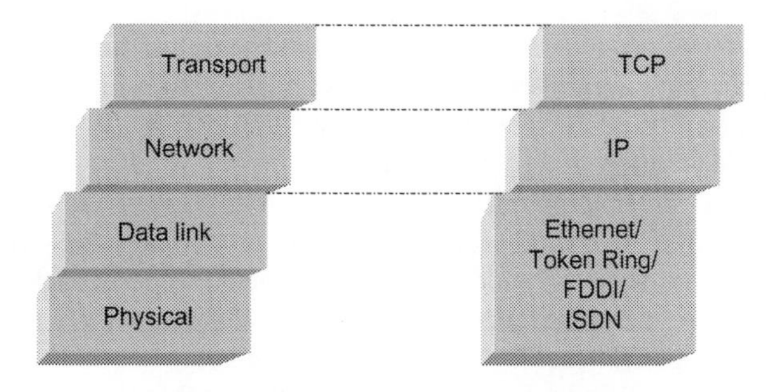

Figure 23.1 TCP/IP and the OSI model

The ISO have adopted TCP/IP as the basis for the standards relating to the network and transport layers of the OSI model. This standard is known as ISO-IP. Most currently available systems conform to the IP addressing standard.

Common applications that use TCP/IP communications are remote login and file transfer. Typical programs used in file transfer and login over TCP communication are `ftp` for file transfer program and `telnet` which allows remote log into another computer. The `ping` program determines if a node is responding to TCP/IP communications.

23.2 TCP/IP gateways and hosts

TCP/IP hosts are nodes which communicate over interconnected networks using TCP/IP communications. A TCP/IP gateway node connects one type of network to another. It contains hardware to provide the physical link between the different networks and the hardware and software to convert frames from one network to the other. Typically, it converts a Token Ring MAC layer to an equivalent Ethernet MAC layer, and vice versa.

A router connects a network of a similar type to another of the same kind through a point-to-point link. The main operational difference between a gateway, a router, and a bridge is that for a Token Ring and Ethernet network, the bridge uses the 48-bit MAC address to route frames, whereas the gateway and router use the IP network address. As an analogy to the public telephone system, the MAC address would be equivalent to a randomly assigned telephone number, whereas the IP address would contain the information on where the telephone is logically located, such as which country, area code, and so on.

Figure 23.2 shows how a gateway (or router) routes information. It reads the frame from the computer on network A, and reads the IP address contained in the frame and makes a decision whether it is routed out of network A to network B. If it does then it relays the frame to network B.

23.3 Function of the IP protocol

The main functions of the IP protocol are to:

- Route IP data frames (which are called Internet datagrams) around an internet. The IP protocol program running on each node knows the location of the gateway on the network. The gateway must then be able to locate the interconnected network. Data then passes from node to gateway through the Internet.
- Fragment the data into smaller units, if it is greater than a given amount (64 kB).
- Report errors – when a datagram is being routed or is being reassembled an error can occur. If this happens then the node that detects the error reports back to the source node. Datagrams are deleted from the network if they travel through the network for more than a set time. Again, an error message is returned to the source node to inform it that the internet routing could not find a route for the datagram or that the destination node, or network, does not exist.

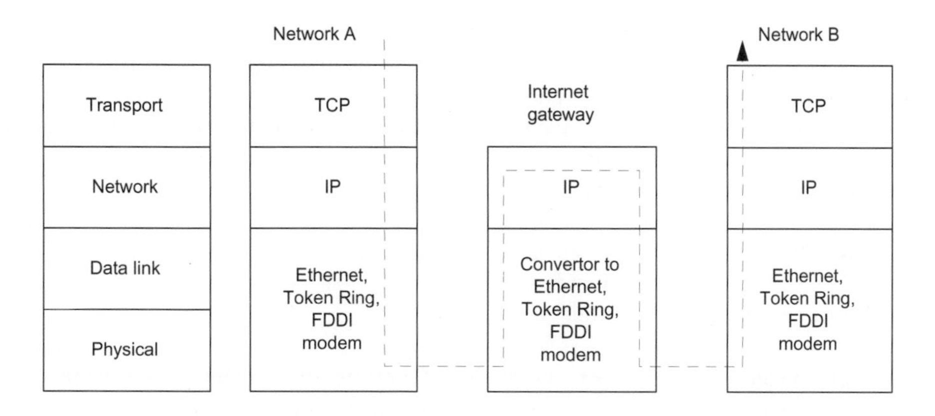

Figure 23.2 Internet gateway layers

23.4 Internet datagram

The IP protocol is an implementation of the network layer of the OSI model. It adds a data header onto the information passed from the transport layer, the resultant data packet is known as an internet datagram. The header contains information such as the destination and source IP addresses, the version number of the IP protocol and so on. Figure 23.3 shows its format.

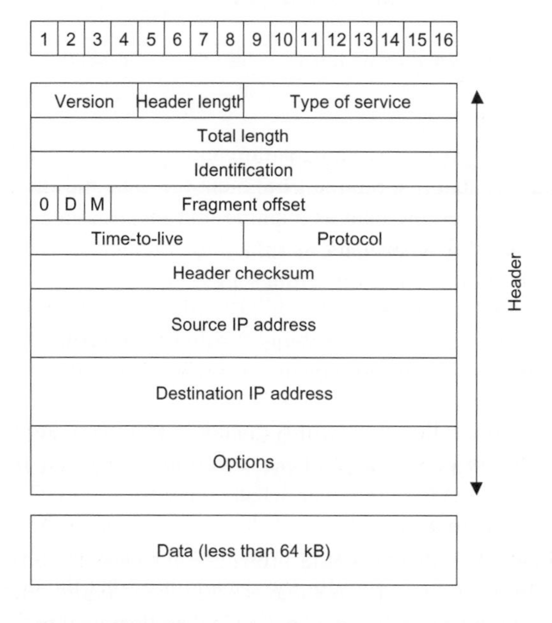

Figure 23.3 Internet datagram format and contents

The datagram can contain up to 65 536 bytes (64 kB) of data. If the data to be transmitted is less than, or equal to, 64 kB, then it is sent as one datagram. If it is more than this then the sender splits the data into fragments and sends multiple datagrams. When transmitted from the source each datagram is routed separately through the internet and the received fragments are finally reassembled at the destination.

The fields in the IP datagram are:

- **Version**. The TCP/IP `version number` helps gateways and nodes interpret the data unit correctly. Differing versions may have a different format. Most current implementations will have a version number of four (IPv4).

- **Type of service**. The `type of service` bit field is an 8-bit bit pattern in the form `PPPDTRXX`, where `PPP` defines the priority of the datagram (from 0 to 7). The precedence levels are:

111 (network control)	110 (Internetwork control)	101 (CRITIC/ECP)
100 (flash override)	011 (flash)	010 (immediate)
001 (priority)	000 (routine)	

  `D` sets a low delay service (0 = normal delay, 1 = low delay).
  `T` sets high throughput (0 = normal throughput, 1 = high throughput).
  `R` sets high reliability (0 = normal reliability, 1 = high reliability).
  The `XX` bits are currently not used (and set to 00).

- **Header length** (4 bits). The `header length` defines the size of the data unit in multiplies of four bytes (32 bits). The minimum length is five bytes and the maximum is 65 536 bytes. Padding bytes fill any unused spaces.

- **Identification** (16 bits). A value which is assigned by the sender to aid the assembly of the frames of a datagram.

- **D** and **M** bits. A gateway may route a datagram and split it into smaller fragments. The `D` bit informs the gateway that it should not fragment the data and thus it signifies that a receiving node should receive the data as a single unit or not at all. The `M` bit is the 'more fragments' bit and is used when data is split into fragments. The `fragment offset` contains the fragment number. The bit settings are:

  D (don't fragment) – 0 = may fragment, 1 = don't fragment.
  M (last fragment) – 0 = last fragment, 1 = more fragments.

- **Fragment offset** (13 bits). Indicates which datagram this fragment belongs to. The fragment offset is measured in units of eight bytes (64 bits). The first fragment has an offset of zero.

- **Time-to-live** (8 bits). A datagram could propagate through the internet indefinitely. To prevent this, the 8-bit `time-to-live` value is set to the maximum transit time in seconds and is set initially by the source IP. Each gateway then decrements this value by a defined amount. When it becomes zero the datagram is discarded. It also defines the maximum amount of time that a destination IP node should wait for the next datagram fragment.

- **Protocol** (8 bits). Different IP protocols can be used on the datagram. The 8-bit `protocol` field defines the type to be used. A full list is given later in Table 23.5 (Section 23.16.1). Typical values are: 1 = ICMP and 6 = TCP.

- **Header checksum** (16 bits). The `header checksum` contains a 16-bit pattern for error detection. As values within the header change from gateway to gateway (such as the time-to-live field), it must be recomputed every time the IP header is processed. The algorithm is:
  The 16-bit 1's complement of the 1's complement sum of all the 16-bit words in the header. When calculating the checksum the header checksum field is assumed to be set to a zero.

- **Source and destination IP addresses** (32 bits). The `source` and `destination IP addresses` are stored in the 32-bit source and destination IP address fields.

- **Options**. The `options` field contains information such as debugging, error control and routing information. Section 23.16.2 gives further information.

23.5 ICMP

Messages, such as control data, information data and error recovery data, are carried between Internet hosts using the Internet control message protocol (ICMP). These messages are sent with a standard IP header. Typical messages are:

- Destination unreachable (message type 3) – sent by a host on the network to say that a host is unreachable. The message can also include the reason the host cannot be reached.
- Echo request/echo reply (message type 8 or 0) – used to check the connectivity between two hosts. The `ping` command uses this message, where it sends an ICMP 'echo request' message to the target host and waits for the destination host to reply with an 'echo reply' message.
- Redirection (message type 5) – sent by a router to a host that is requesting its routing services. This helps to find the shortest path to a desired host.
- Source quench (message type 4) – used when a host cannot receive any more IP packets at the present (or reduce the flow).

An ICMP message is sent within an IP header, with the version field, source and destination IP addresses, and so on. The type of service field is set to a 0 and the protocol field is set to a 1 (which identifies ICMP). After the IP header, follows the ICMP message, which starts with three fields, as per Figure 23.4. The message type has eight bits and identifies the type of message, as given in Table 23.1. The code fields are also eight bits long and a checksum field is 16 bits long. The checksum is the 1's complement of the 1's complement sum of all 16-bit words in the header (the checksum field is assumed to be zero in the addition).

The information after this field depends on the type of message, such as:

- For echo request and reply, the message header is followed by an 8-bit identifier, then an 8-bit sequence number followed by the original IP header.
- For destination unreachable, source quelch and time, the message header is followed by 32 bits which are unused and then the original IP header.
- For timestamp request, the message header is followed by a 16-bit identifier, then by a 16-bit sequence number, followed by a 32-bit originating timestamp.

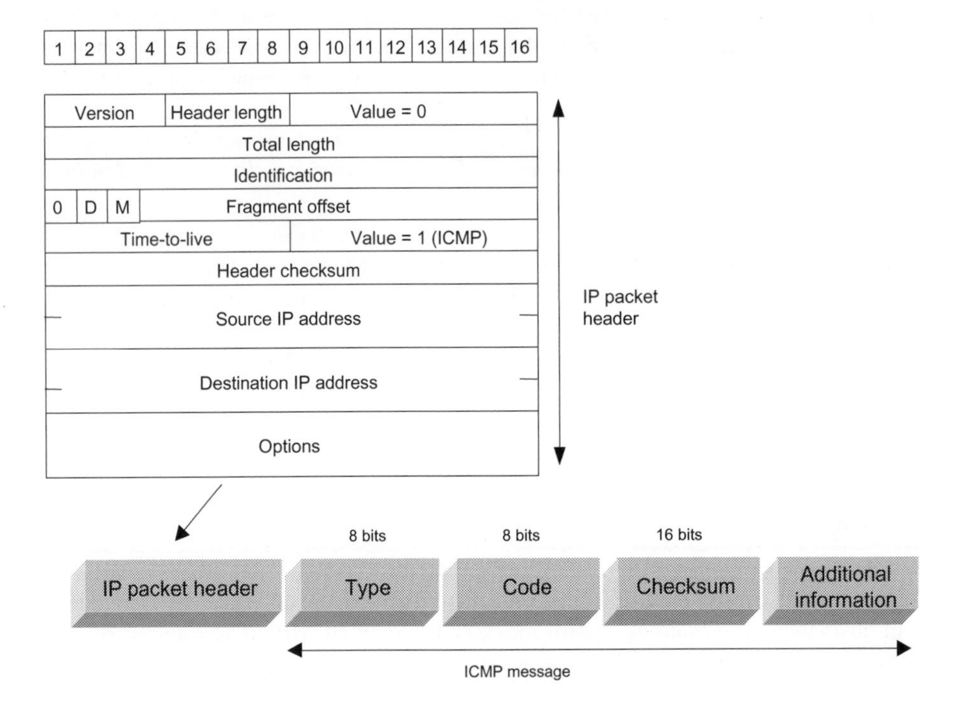

Figure 23.4 ICMP message format

Where:

- Pointer (8-bit) – identifies the byte location of the parameter error in the original IP header. For example, a value of 9 would identify the protocol field, and 12 would identify the source IP address field.
- Identifier (16-bit) – helps the matching of requests and replies (possibly set to zero). It can be used to identify a unique connection.
- Sequence number (16-bit) – helps in matching request and replies (possibly set to zero).
- Timestamps (32-bit) –this is the time in milliseconds since midnight UT (universal time). If this is not possible then it is anytime, as long as the high-order bit of the timestamp is set to a 1 to indicate that it is non-standard time.
- Gateway address (32-bit) – the address of the gateway to which network traffic specified in the original datagram should be sent to.
- Internet header + 64 bits of data datagram – this is the original IP header and the first 64 byte of the data part. It is used by the host to match the match to the required high-level application (such as TCP port values).

Table 23.1 Message type field value

Value	*Description*	*Code field*	*Additional information*
0	Echo reply message	0	16-bit identifier 16-bit sequence number
3	Destination Unreachable	0 = net unreachable 1 = host unreachable 2 = protocol unreachable 3 = port unreachable 4 = fragmentation needed and D bit set 5 = source route failed	32 bits unused. Internet header + 64 bits of original data datagram
4	Source quench message	0	32 bits unused. Internet header + 64 bits of original data datagram
5	Redirect message	0 = redirect datagram for the network 1 = redirect datagram for the host 2 = redirect datagram for the type of service and network 3 = redirect datagram for the type of service and host	32 bits gateway address. Internet header + 64 bits of original data datagram
8	Echo request	0	
11	Time-to-live exceeded	0 = time-to-live exceeded in transit 1 = fragment reassembly time exceeded	32 bits unused. Internet header + 64 bits of original data datagram
12	Parameter problem	0 = pointer indicates the error	8-bit pointer. 24 bits unused. Internet header + 64 bits of original data datagram
13	Timestamp request	0	16-bit identifier 16-bit sequence number 32-bit originate timestamp 32-bit receive timestamp 32-bit transmit timestamp
14	Timestamp reply	0	As above
15	Information request	0	16-bit identifier 16-bit sequence number
16	Information reply	0	As above

The descriptions of the messages and replies are:

- Source quench message (4) – sent by a gateway or a destination host when it discards a datagram (possibly through lack of buffer memory), and identifies that the sender should reduce the flow of traffic transmission. The host should then reduce the flow, and gradually increase it, as long as it does not receive any more source quench messages.
- Time exceeded message (11) – this is sent either by a gateway when a datagram has a Time-to-Live field which is zero and has been deleted, or when a host cannot reassemble a fragmented datagram due to missing fragments, within a certain time limit.
- Parameter problem message (12) – sent by a gateway or a host when they encounter a problem with one of the parameters in an IP header.
- Destination unreachable message (3) – sent by a gateway to identify that a host cannot be reached or a TCP port process does not exist.
- Redirected message (5) – sent by a gateway to inform other gateways that there is a better route to a given network destination address.
- Information reply message (15)– sent in reply to an information request.(see information request (16) for a typical usage).
- Information request (16) – this request can be sent with a fully specified source IP address, and a zero destination IP address. The replying IP gateway then replies with an information reply message with its fully specified IP address. In this way the host can determine the network address that it is connected to.
- Echo message (8) – requests an echo. (see echo reply message (0)).
- Echo reply message (0) – the data received in the echo message (8) must be returned in this message.

23.6 TCP/IP internets

Figure 23.5 illustrates a sample TCP/IP implementation. A gateway `MERCURY` provides a link between a Token Ring network (NETWORK A) and the Ethernet network (ETHER C). Another gateway `PLUTO` connects NETWORK B to ETHER C. The TCP/IP protocol allows a host on NETWORK A to communicate with `VAX01`.

23.6.1 Selecting internet addresses

Each node using TCP/IP communications requires an IP address which is then matched to its Token Ring or Ethernet MAC address. The MAC address allows nodes on the same segment to communicate with each other. In order for nodes on a different network to communicate, each must be configured with an IP address.

Nodes on a TCP/IP network are either hosts or gateways. Any nodes that run application software or are terminals are hosts. Any node that routes TCP/IP packets between networks is called a TCP/IP gateway node. This node must have the necessary network controller boards to physically interface to other networks it connects with.

23.6.2 Format of the IP address

A typical IP address consists of two fields: the left field (or the network number) identifies the network, and the right number (or the host number) identifies the particular host within that network. Figure 23.6 illustrates this.

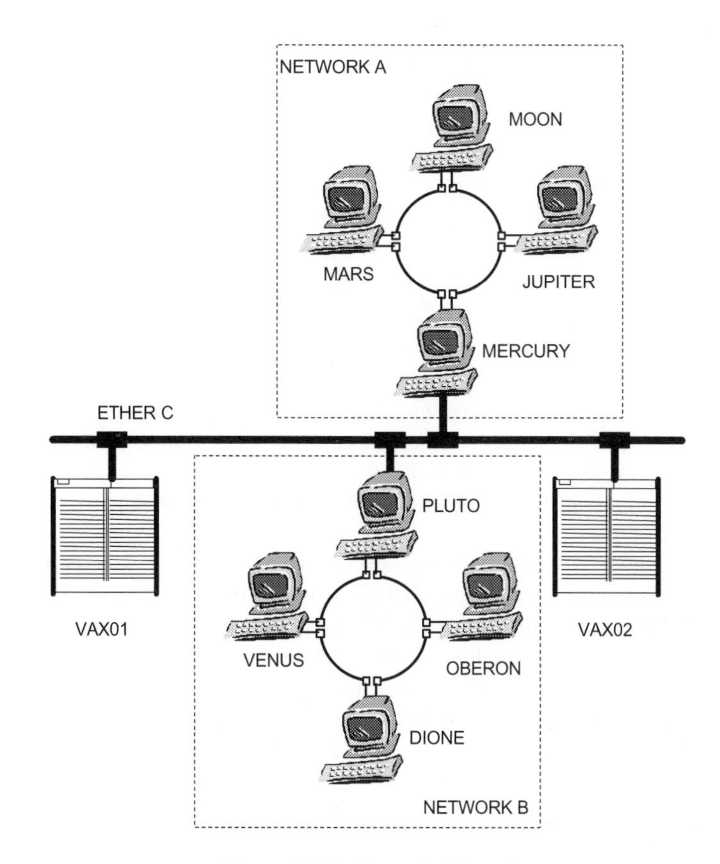

Figure 23.5 Example internet

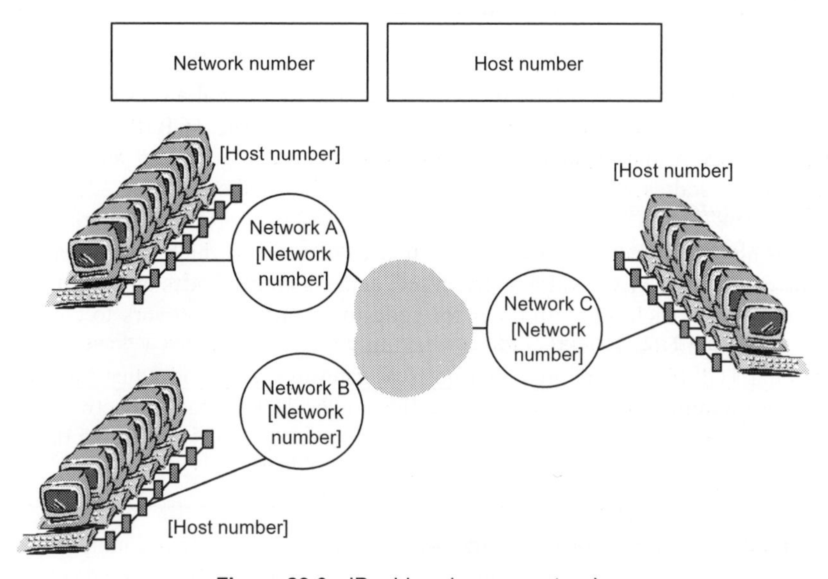

Figure 23.6 IP addressing over networks

The IP address is 32 bits long and can address over four billion physical addresses (2^{32} or 4294967296 hosts). There are three main address formats and these are shown in Figure 23.7.

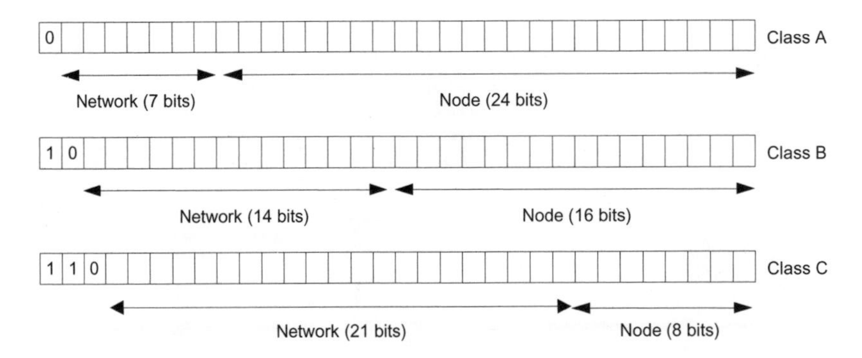

Figure 23.7 Type A, B and C IP address classes

Each of these types is applicable to certain types of networks. Class A allows up to 128 (2^7) different networks and up to 16777216 (2^{24}) hosts on each network. Class B allows up to 16384 (2^{14}) networks and up to 65536 (2^{16}) hosts on each network. Class C allows up to 2097152 (2^{21}) networks each with up to 256 (2^8) hosts.

The class A address is thus useful where there are a small number of networks with a large number of hosts connected to them. Class C is useful where there are many networks with a relatively small number of hosts connected to each network. Class B addressing gives a good compromise of networks and connected hosts.

When selecting internet addresses for the network, the address can be specified simply with decimal numbers within a specific range. The standard DARPA IP addressing format is of the form:

`W.X.Y.Z`

where `W`, `X`, `Y` and `Z` represent 1 byte of the IP address. As decimal numbers they range from 0 to 255. The 4 bytes together represent both the network and host address.

The valid range of the different IP addresses is given in Figure 23.7 and Table 23.2 defines the valid IP addresses. Thus for a class A type address there can be 127 networks and 16711680 (256×256×255) hosts. Class B can have 16320 (64×255) networks and class C can have 2088960 (32×256×255) networks and 255 hosts.

Addresses above `223.255.254` are reserved, as are addresses with groups of zeros.

Table 23.2 Ranges of addresses for type A, B and C internet address

Type	*Network portion*	*Host portion*
A	`1 - 126`	`0.0.1 - 255.255.254`
B	`128.1 - 191.254`	`0.1 - 255.254`
C	`192.0.1 - 223.255.254`	`1 - 254`

23.6.3 Creating IP addresses with subnet numbers

Besides selecting IP addresses of internets and host numbers, it is also possible to designate an intermediate number called a subnet number. Subnets extend the network field of the IP address beyond the limit defined by the type A, B, C scheme. They allow a hierarchy of internets within a network. For example, it is possible to have one network number for a network attached to the internet, and various subnet numbers for each subnet within the network. This is illustrated in Figure 23.8.

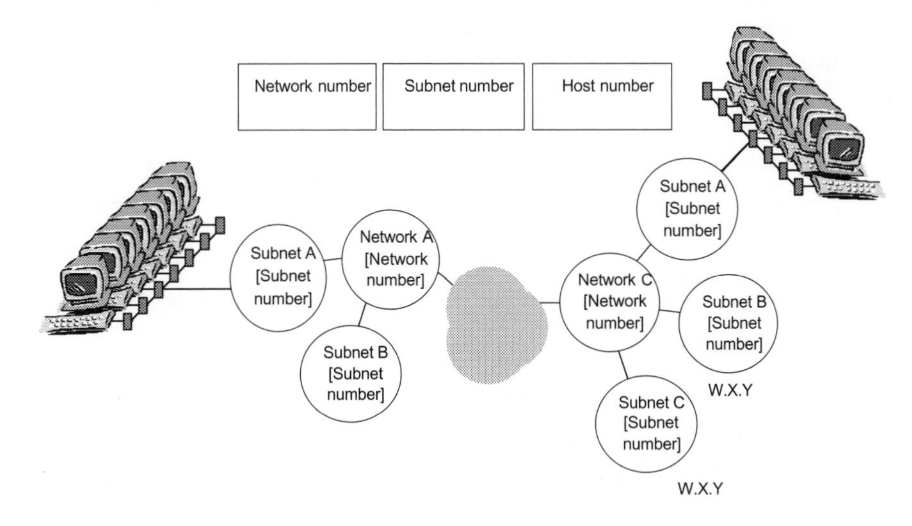

Figure 23.8 IP addresses with subnets

For an address `W.X.Y.Z` for a type A address `W` specifies the network and `X` the subnet. For type B the `Y` field specifies the subnet, as illustrated in Figure 23.9.

To connect to a global network a number is normally assigned by a central authority. For the Internet network it is assigned by the network information center (NIC). Typically, on the Internet an organisation is assigned a type B network address. The first two fields of the address specify the organisation network, the third specifies the subnet within the organization and the final value specifies the host.

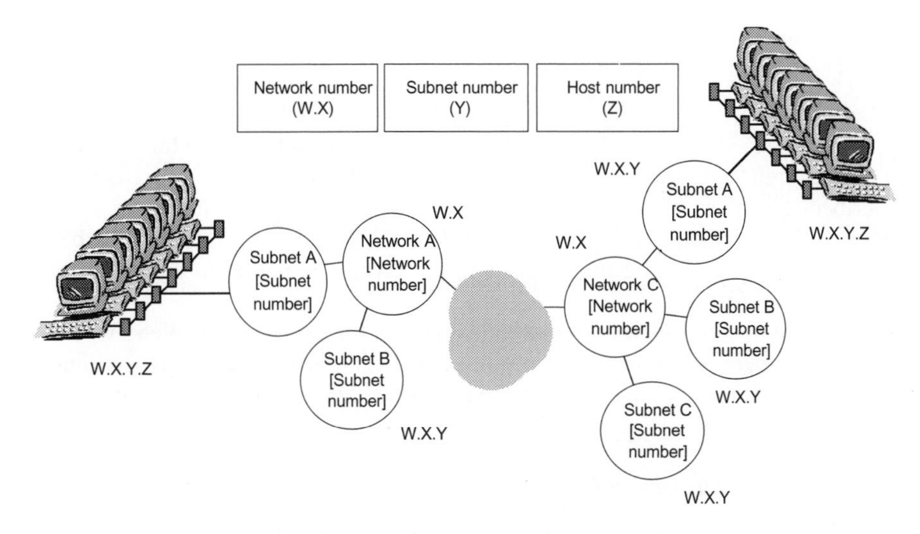

Figure 23.9 Internet addresses with subnets

23.6.4 Specifying subnet masks

If a subnet is used then a bit mask, or subnet mask, must be specified to show which part of the address is the network part and which is the host.

The subnet mask is a 32-bit number that has 1's for bit positions specifying the network and subnet parts and 0's for the host part. A text file called `hosts` is normally used to set up the subnet mask. Table 23.3 shows example subnet masks.

Table 23.3 Default subnet mask for type A, B and C IP addresses

Address Type	*Default mask*
Class A	`255.0.0.0`
Class B	`255.255.0.0`
Class C and Class B with a subnet	`255.255.255.0`

To set up the default mask the following line is added to the `hosts` file.

```
Hosts file
255.255.255.0defaultmask
```

23.7 Domain name system

An IP address can be defined in the form `WWW.XXX.YYY.ZZZ`, where `XXX`, `YYY`, `ZZZ` and `WWW` are integer values in the range 0 to 255. On the Internet, with Class B IP addresses, it is `WWW.XXX.YYY` that normally defines the subnet and `ZZZ` that defines the host. Such names may be difficult to remember. A better method is to use symbolic names rather than IP addresses.

Users and application programs can then use symbolic names rather than the IP addresses. The directory network service on the Internet determines the IP address of the named destination user or application program. This has the advantage that users and application programs can move around the Internet and are not fixed to an IP address.

An analogy relates to the public telephone service. A telephone directory contains a list of subscribers and their associated telephone numbers. If someone looks for a telephone number, first the user name is looked up and then the associated telephone number found. The telephone directory listing maps a user name (symbolic name) to an actual telephone number (the actual address).

Table 23.4 lists some Internet domain assignments for World Wide Web (WWW) servers. Note that domain assignments are not fixed and can change their corresponding IP addresses, if required. The binding between the symbolic name and its address can thus change at any time.

Table 23.4 Internet domain assignments for web servers

Web server	*Internet domain name*	*Internet IP address*
NEC	`web.nec.com`	`143.101.112.6`
Sony	`www.sony.com`	`198.83.178.11`
Intel	`www.intel.com`	`134.134.214.1`
IEEE	`www.ieee.com`	`140.98.1.1`
University of Bath	`www.bath.ac.uk`	`136.38.32.1`
University of Edinburgh	`www.ed.ac.uk`	`129.218.128.43`
IEE	`www.iee.org.uk`	`193.130.181.10`
University of Manchester	`www.man.ac.uk`	`130.88.203.16`

23.8 Internet naming structure

The Internet naming structure uses labels separated by periods; an example is `eece.napier.ac.uk`. It uses a hierarchical structure where organisations are grouped into primary domain names. These are `com` (for commercial organisations), `edu` (for educational organisations), `gov` (for government organisations), `mil` (for military organisations), `net` (Internet network support centers) or `org` (other organisations). The primary domain name may also define the country in which the host is located, such as `uk` (United Kingdom), `fr` (France), and so on. All hosts on the Internet must be registered to one of these primary domain names.

The labels after the primary field describe the subnets within the network. For example in the address `eece.napier.ac.uk`, the `ac` label relates to an academic institution within the `uk`, `napier` to the name of the institution and `eece` the subnet within that organisation. An example structure is illustrated in Figure 23.10.

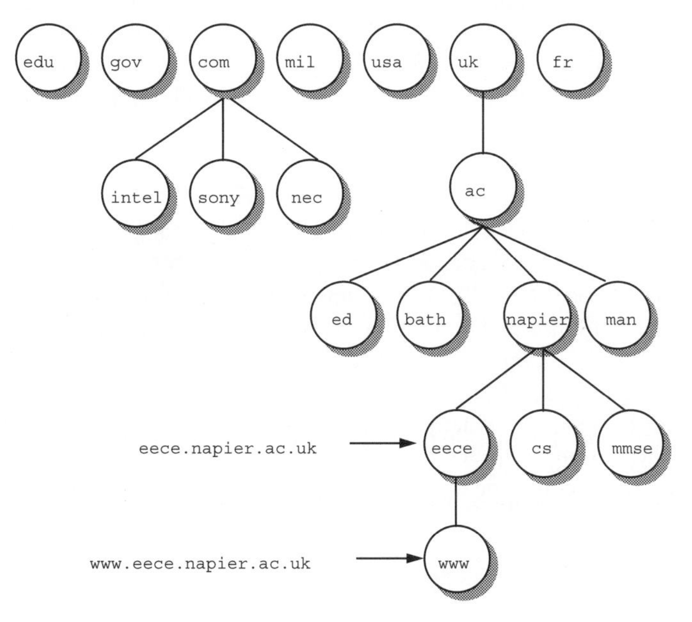

Figure 23.10 Example domain naming

23.9 Domain name server

Each institution on the Internet has a host that runs a process called the domain name server (DNS). The DNS maintains a database called the directory information base (DIB) which contains directory information for that institution. When a new host is added, the system manager adds its name and its IP address. It can then access the Internet.

23.9.1 DNS program

The DNS program is typically run on a Linux-based PC with a program called `named` (located in `/usr/sbin`) with an information file of `named.boot`. To run the program the following is used:

```
/usr/bin/named -b /usr/local/adm/named/named.boot
```

The following shows that the DNS program is currently running.

```
$ ps -ax
  PID TTY STAT  TIME COMMAND
  295 con S     0:00 bootpd
   35 con S     0:00 /usr/sbin/lpd
  272 con S     0:00 /usr/sbin/named -b /usr/local/adm/named/named.boot
  264 p 1 S     0:01 bash
  306 pp0 R     0:00 ps -ax
```

In this case the data file `named.boot` is located in the `/usr/local/adm/named` directory. A sample `named.boot` file is

```
/usr/local/adm/named - soabasefile
              eece.napier.ac.uk -main record of computer names
              net/net144   -reverse look-up database
              net/net145      "      "
              net/net146      "      "
              net/net147      "      "
              net/net150      "      "
              net/net151      "      "
```

This file specifies that the reverse look-up information on computers on the subnets 144, 145, 146, 147, 150 and 151 is contained in the `net144`, `net145`, `net146`, `net147`, `net150` and `net151` files, respectively. These are stored in the `net` subdirectory. The main file which contains the DNS information is, in this case, `eece.napier.ac.uk`.

Whenever a new computer is added onto a network, in this case, the `eece.napier.ac.uk` file and the `net/net1**` (where ** is the relevant subnet name) are updated to reflect the changes. Finally, the serial number at the top of these data files is updated to reflect the current date, such as 19970321 (for 21st March 1997).

The DNS program can then be tested using `nslookup;` For example,

```
$ nslookup
Default Server:  ees99.eece.napier.ac.uk
Address:  146.176.151.99
> src.doc.ic.ac.uk
Server:  ees99.eece.napier.ac.uk
```

```
Address:  146.176.151.99
Non-authoritative answer:
Name:     swallow.doc.ic.ac.uk
Address:  193.63.255.4
Aliases:  src.doc.ic.ac.uk
```

23.10 Bootp protocol

The bootp protocol allocates IP addresses to computers based on a table of network card MAC addresses. When a computer is first booted, the bootp server interrogates its MAC address and then looks up the bootp table for its entry. It then grants the corresponding IP address to the computer. The computer then uses it for connections.

23.10.1 Bootp program

The bootp program is typically run on a Linux-based PC with the `bootp` program. The following shows that the `bootp` program is currently running on a computer:

```
$ ps -ax

  PID TTY STAT  TIME COMMAND
      1 con S    0:06 init
     31 con S    0:01 /usr/sbin/inetd
  14142 con S    0:00 bootpd -d 1
     35 con S    0:00 /usr/sbin/lpd
     49 p 3 S    0:00 /sbin/agetty 38400 tty3
  14155 pp0 R    0:00 ps -ax
  10762 con S    0:18 /usr/sbin/named -b /usr/local/adm/named/named.boot
```

For the bootp system to operate then a table is required that reconciles the MAC addresses of the card to an IP address. In the previous example this table is contained in the `bootptab` file which is located in the `/etc` directory. The following file gives an example `bootptab`:

Contents of bootptab file

```
# /etc/bootptab: database for bootp server
# Blank lines and lines beginning with '#' are ignored.
# Legend:
#     first field -- hostname
#           (may be full domain name and probably should be)
#     hd -- home directory
#     bf -- bootfile
#     cs -- cookie servers
#     ds -- domain name servers
#     gw -- gateways
#     ha -- hardware address
#     ht -- hardware type
#     im -- impress servers
#     ip -- host IP address
#     lg -- log servers
#     lp -- LPR servers
#     ns -- IEN-116 name servers
#     rl -- resource location protocol servers
#     sm -- subnet mask
```

```
#      tc -- template host (points to similar host entry)
#      to -- time offset (seconds)
#      ts -- time servers
#
#hostname:ht=1:ha=ether_addr_in_hex:ip=ip_addr_in_dec:tc=allhost:
.default150:\
       :hd=/tmp:bf=null:\
       :ds=146.176.151.99 146.176.150.62 146.176.1.5:\
       :sm=255.255.255.0:gw=146.176.150.253:\
       :hn:vm=auto:to=0:
.default151:\
       :hd=/tmp:bf=null:\
       :ds=146.176.151.99 146.176.150.62 146.176.1.5:\
       :sm=255.255.255.0:gw=146.176.151.254:\
       :hn:vm=auto:to=0:

pc345: ht=ethernet: ha=0080C8226BE2:   ip=146.176.150.2: tc=.default150:
pc307: ht=ethernet: ha=0080C822CD4E:   ip=146.176.150.3: tc=.default150:
pc320: ht=ethernet: ha=0080C823114C:   ip=146.176.150.4: tc=.default150:
pc331: ht=ethernet: ha=0080C823124B:   ip=146.176.150.5: tc=.default150:
pc401: ht=ethernet: ha=0080C82379F7:   ip=146.176.150.6: tc=.default150:
pc404: ht=ethernet: ha=0080C8238369:   ip=146.176.150.7: tc=.default150:
pc402: ht=ethernet: ha=0080C8238467:   ip=146.176.150.8: tc=.default150:
   :       :
pc460: ht=ethernet: ha=0000E8C7BB63:   ip=146.176.151.142: tc=.default151:
pc414: ht=ethernet: ha=0080C8246A84:   ip=146.176.151.143: tc=.default151:
pc405: ht=ethernet: ha=0080C82382EE:   ip=146.176.151.145: tc=.default151:
```

The format of the file is:

```
#hostname:ht=1:ha=ether_addr_in_hex:ip=ip_addr_in_dec:tc=allhost:
```

where `hostname` is the hostname, the value defined after `ha=` is the Ethernet MAC address, the value after `ip=` is the IP address and the name after the `tc=` field defines the host information script. For example:

```
pc345:  ht=ethernet: ha=0080C8226BE2:   ip=146.176.150.2:
tc=.default150;
```

defines the hostname of `pc345`, `ethernet` indicates it is on an Ethernet network, and shows its IP address is `146.176.150.2`. The MAC address of the computer is `00:80:C8:22:6B:E2` and it is defined by the script `.default150`. This file defines a subnet of 255.255.255.0 and has associated DNS of

```
146.176.151.99 146.176.150.62 146.176.1.5
```

and uses the gateway at

```
146.176.150.253
```

23.11 Example network

A university network is shown in Figure 23.11. The connection to the outside global Internet is via the Janet gateway node and its IP address is 146.176.1.3. Three subnets, 146.176.160, 146.176.129 and 146.176.151, connect the gateway to departmental bridges. The Computer Studies bridge address is 146.176.160.1 and the Electrical Department bridge has an address 146.176.151.254.

The Electrical Department bridge links, through other bridges, to the subnets 146.176.144, 146.176.145, 146.176.147, 146.176.150 and 146.176.151. The main bridge (Figure 23.12) into the department connects to two Ethernet networks of PCs (subnets 146.176.150 and 146.176.151) and to another bridge (Bridge 1). Bridge 1 connects to the subnet 146.176.144. Subnet 146.176.144 connects to workstations and X-terminals. It also connects to the gateway Moon that links the Token Ring subnet 146.176.145 with the Ethernet subnet 146.176.144. The gateway Oberon, on the 146.176.145 subnet, connects to an Ethernet link 146.176.146. This then connects to the gateway Dione that is also connected to the Token Ring subnet 146.176.147.

The topology of the Electrical Department network is shown in Figure 23.12. Each node on the network is assigned an IP address. The *hosts* file for the set-up in Figure 23.12 is shown next. For example the IP address of Mimas is 146.176.145.21 and for miranda it is 146.176.144.14. Notice that the gateway nodes, Oberon, Moon and Dione, all have two IP addresses.

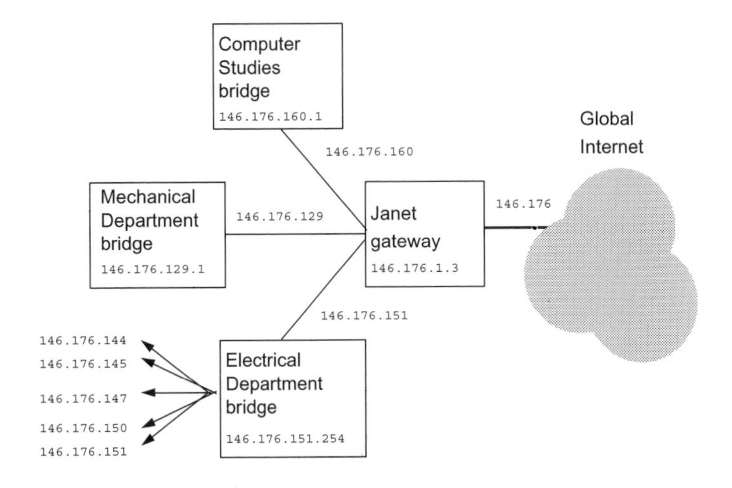

Figure 23.11 A university network

Contents of host file

```
146.176.1.3         janet
146.176.144.10      hp
146.176.145.21      mimas
146.176.144.11      mwave
146.176.144.13      vax
146.176.144.14      miranda
146.176.144.20      triton
146.176.146.23      oberon
146.176.145.23      oberon
146.176.145.24      moon
```

```
146.176.144.24      moon
146.176.147.25      uranus
146.176.146.30      dione
146.176.147.30      dione
146.176.147.31      saturn
146.176.147.32      mercury
146.176.147.33      earth
146.176.147.34      deimos
146.176.147.35      ariel
146.176.147.36      neptune
146.176.147.37      phobos
146.176.147.39      io
146.176.147.40      titan
146.176.147.41      venus
146.176.147.42      pluto
146.176.147.43      mars
146.176.147.44      rhea
146.176.147.22      jupiter
146.176.144.54      leda
146.176.144.55      castor
146.176.144.56      pollux
146.176.144.57      rigel
146.176.144.58      spica
146.176.151.254     cubridge
146.176.151.99      bridge_1
146.176.151.98      pc2
                    :::::
146.176.151.71      pc29
146.176.151.70      pc30
146.176.151.99      ees99
146.176.150.61      eepc01
146.176.150.62      eepc02
255.255.255.0       defaultmask
```

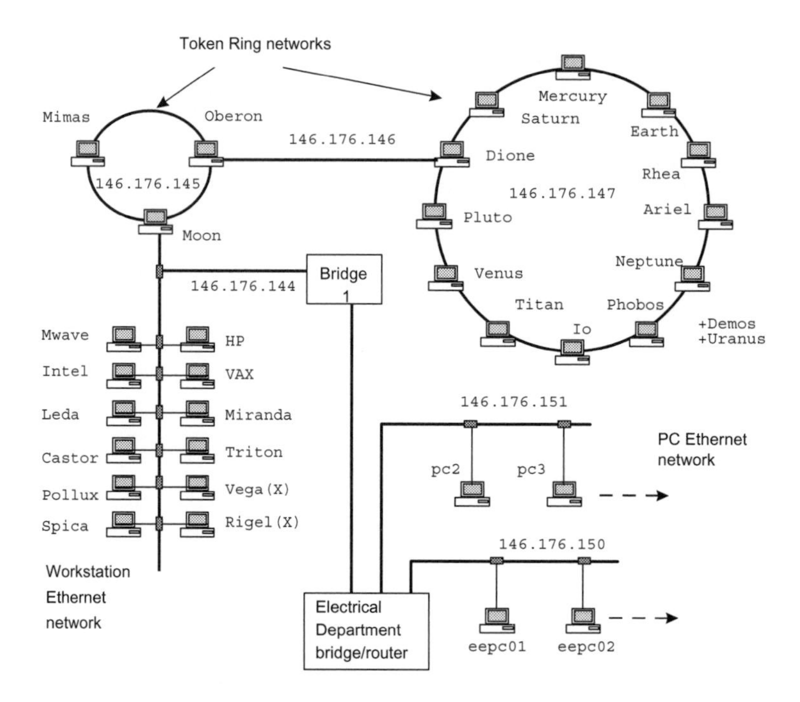

Figure 23.12 Network topology for the Electrical Department network

23.12 ARP

ARP (address resolution protocol) translates IP addresses to Ethernet addresses. This is used when IP packets are sent from a computer, and the Ethernet address is added to the Ethernet frame. A table look-up, called the ARP table, is used to translate the addresses. One column has the IP address and the other has the Ethernet address. The following is an example ARP table:

IP address	Ethernet address
146.176.150.2	00-80-C8-22-6BE2
146.176.150.3	00-80-C8-22-CD4E
146.176.150.4	00-80-C8-23-114C

A typical conversation is as follows:

1. Application sends an application message to TCP.
2. TCP sends the corresponding TCP message to the IP module. The destination IP address is known by the application, the TCP module, and the IP module.
3. At this point the IP packet has been constructed and is ready to be given to the Ethernet driver, but first the destination Ethernet address must be determined.
4. The ARP table is used to look up the destination Ethernet address.

The sequence of determining the Ethernet address is as follows:

1. An ARP request packet with a broadcast Ethernet address (`FF-FF-FF-FF-FF-FF`) is sent out on the network to every computer. Other typical Ethernet broadcast addresses are given in Section 23.16.3.
2. The outgoing IP packet is queued.
3. All the computers on the network segment read the broadcast Ethernet frame, and examine the Type field to determine if it is an ARP packet. If it is then it is passed to the ARP module.
4. If the IP address of a receiving station matches the IP address in the IP packet then it sends a response directly to the source Ethernet address.
5. The originator then receives the Ethernet frame and checks the Type field to determine if it an ARP packet. If it is then it adds the sender's IP address and Ethernet address to its ARP table.
6. The IP packet can now be sent with the correct Ethernet address.

Each computer has a separate ARP table for each of its Ethernet interfaces.

23.13 IP multicasting

Many applications of modern communications require the transmission of IP datagrams to multiple hosts. Typical applications are video conferencing, remote teaching and so on. This is supported by IP multicasting, where a host group is identified by a single IP address. The main parameters of IP multicasting are:

- The group membership is dynamic.
- Hosts may join and leave the group at any time.
- There is also no limit to the location or number of members in a host group.
- A host may be a member of more than one group at a time.
- A host group may be permanent or transient. Permanent groups are well-known and are administratively assigned a permanent IP address. The group is then dynamically associated with this IP address. IP multicast addresses that are not reserved to permanent groups are available for dynamic assignment to transient groups.
- Multicast routers forward IP multicast datagrams into the Internet.

23.13.1 Group addresses

A special group of addresses are assigned to multicasting. These are known as Class D addresses, and they begin with 1110 as their starting 4 bits (Class E addresses with the upper bits of 1111 are reserved for future uses). The Class D addresses thus range from

224.0.0.0 (11100000 00000000 00000000 00000000)

239.255.255.255 (11101111 11111111 11111111 11111111)

The address 224.0.0.0 is reserved. 224.0.0.1 is also assigned to the permanent group of all IP hosts (including gateways), and is used to address all multicast hosts on the directly connected network. Reserved and allocated addresses are:

224.0.0.0	Reserved
224.0.0.1	All systems on current subnet
224.0.0.2	All routers on current subnet
224.0.0.3	Unassigned
224.0.0.4	DVMRP routers
224.0.0.5	OSPFIGP all routers
224.0.0.6	OSPFIGP designated routers
224.0.0.7	ST routers
224.0.0.8	ST hosts
224.0.0.9	RIP2 routers
224.0.0.10–224.0.0.255	Unassigned
224.0.1.0	VMTP managers group
224.0.1.1	NTP network time protocol
224.0.1.2	SGI-dogfight
224.0.1.3	Rwhod
224.0.1.4	VNP
224.0.1.5	Artificial horizons–aviator
224.0.1.6	NSS–name service server
224.0.1.7	AUDIONEWS–audio news multicast
224.0.1.8	SUN NIS+ information service
224.0.1.9	MTP multicast transport protocol
224.0.1.10–224.0.1.255	Unassigned
224.0.2.1	rwho group (BSD) (unofficial)
224.0.2.2	SUN RPC PMAPPROC_CALLIT
224.0.3.0–224.0.3.255	RFE generic service
224.0.4.0–224.0.4.255	RFE individual conferences

224.1.0.0–224.1.255.255	ST multicast groups
224.2.0.0–224.2.255.255	Multimedia conference calls
232.*x.x.x*	VMTP transient groups

All the above addresses are listed in the domain name service under MCAST.NET and 224.IN-ADDR.ARPA. On an Ethernet or IEEE 802 network, the 23 low-order bits of the IP multicast address are placed in the low-order 23 bits of the Ethernet or IEEE 802 net multicast address.

23.13.2 Conformance

There are three levels of conformance:

- Level 0, no IP multicasting support – in this, a Level 0 host ignores, or deletes, all Class D addressed datagrams.
- Level 1, sending support, but no receiving – in this, a Level 1 host can send multicast datagrams, but cannot receive them.
- Level 2, Full multicasting support – in this, a Level 2 host can send and receive IP multicasting. It also requires the implementation of the Internet group management protocol (IGMP).

23.14 Exercises

23.14.1 Which OSI layer does the IP layer correspond to:

(a)	Data link	(b)	Network
(c)	Transport	(d)	Session

23.14.2 Which OSI layer does the TCP layer correspond to:

(a)	Data link	(b)	Network
(c)	Transport	(d)	Session

23.14.3 Which IP version do most TCP/IP hosts use:

(a)	Version 2	(b)	Version 4
(c)	Version 5	(d)	Version 6

23.14.4 How much data can be carried within an IP datagram:

(a)	64 kB	(b)	128 kB
(c)	256 kB	(d)	Unlimited

23.14.5 How many IP addresses are possible:

(a)	1 048 576	(b)	16 777 216
(c)	4 294 967 296	(d)	$3.402\,823\,669 \times 10^{38}$

23.14.6 How are IP datagrams deleted from the network:

(a) They are deleted when the time-to-live field becomes zero.
(b) They are never deleted, and will always be delivered.
(c) They are buffered on intermediate systems, and then deleted after a given time.
(d) They are returned to the originator if they are not deleted, and the originator either resends them or deletes them.

23.14.7 Which of the following is a Class A IP address:

(a) 12.1.14.12 (b) 146.176.151.130
(c) 194.50.100.1 (d) 224.50.50.1

23.14.8 Which of the following is a Class D IP address:

(a) 12.1.14.12 (b) 146.176.151.130
(c) 194.50.100.1 (d) 224.50.50.1

23.14.9 What are Class D IP addresses used for:

(a) Dynamic IP addressing (b) Testing networks
(c) Static IP addressing (d) Multicasting

23.14.10 Which of the following is the country domain for Germany:

(a) ge (b) de (c) dr (d) gy

23.14.11 Which service allows hosts to determine the IP address for a given domain name:

(a) TCP (b) ICMP
(c) ARP (d) DNS

23.14.12 Which protocol is used by a node to determine the Ethernet address to a host with a given IP address:

(a) TCP (b) ICMP
(c) ARP (d) DNS

23.14.13 Which Ethernet address is used for broadcast messages:

(a) FF-FF-FF-FF-FF-FF (b) 11-11-11-11-11-11-11
(c) 00-00-00-00-00-00 (d) AA-AA-AA-AA-AA-AA

23.14.14 Outline how ARP uses the broadcast address and the type field to identify that an ARP request is being transmitted. Also, discuss a typical ARP conversation.

23.14.15 Outline how the protocol is identified in the IP header. Discuss how the format of the data after the header differs with different protocols (such as TCP and ICMP).

23.14.16 Explain how ICMP and the options field would be used to determine the following information:

(i) Whether a destination node is responding to TCP/IP communications.
(ii) The route to a destination node.
(iii) The route to a destination node, with the time delay between each gateway.

23.14.17 Explain how the Options field can be used to set the route that a datagram can take.

23.14.18 Determine the IP addresses, and their type (i.e. class A, B or C), of the following 32-bit addresses:

(i) `10001100.01110001.00000001.00001001`
(ii) `01000000.01111101.01000001.11101001`
(iii) `10101110.01110001.00011101.00111001`

23.14.19 Determine the countries which use the following primary domain names:

(a) `de` (b) `nl` (c) `it` (d) `se` (e) `dk` (f) `sg`
(g) `ca` (h) `ch` (i) `tr` (j) `jp` (k) `au`

Determine some other domain names.

23.14.20 For a known TCP/IP network determine the names of the nodes and their Internet addresses.

23.14.21 For a known TCP/IP network determine how the DNS is implemented and how IP addresses are granted.

23.14.22 If a subnet mask on a Class B network is 255.255.240.0, show that there can be 16 connected networks, each with 4095 nodes on a Class B network.

23.15 Notes from the author

Which two people have done more for world unity than anyone else? Well, Prof. TCP and Dr. IP must be somewhere in the Top 10. They have done more to unify the world than all the diplomats in the world have. They do not respect national borders, time zones, cultures, industrial conglomerates or anything like that. They allow the sharing of information around the world, and are totally open for anyone to use. Top marks to Prof. TCP and Dr. IP, the true champions of freedom and democracy.

Many of the great inventions/developments of our time were things that were not really predicted, such as CD-ROMs, RADAR, silicon transistors, fibre optic cables, and, of course, the Internet. The Internet itself is basically an infrastructure of interconnected networks which run a common protocol. The nightmare of interfacing the many computer systems around the world was solved because of two simple protocols: TCP and IP. Without them the

Internet would not have evolved so quickly and possibly would not have occurred at all. TCP and IP are excellent protocols as they are simple and can be run over any type of network, on any type of computer system.

The Internet is often confused with the World Wide Web (WWW), but the WWW is only one application of the Internet. Others include electronic mail (the No.1 application), file transfer, remote login and so on.

The amount of information transmitted over networks increases by a large factor every year. This is due to local area networks, wide area networks and of course, traffic over the Internet. It is currently estimated that traffic on the Internet doubles every 100 days and that three people join the Internet every second. This means an eightfold increase in traffic over a whole year. It is hard to imagine such growth in any other technological area. Imagine if cars were eight times faster each year, or could carry eight times the number of passengers each year (and of course roads and driveways would have to be eight times larger each year).

23.16 Additional material

23.16.1 Assigned Internet protocol numbers

Table 23.5 outlines the values that are used in the protocol field of the IP header.

Table 23.5 Assigned Internet protocol numbers

Value	*Protocol*	*Value*	*Protocol*
0	Reserved	18	Multiplexing
1	ICMP	19	DCN
2	IGMP (Internet group management)	20	TAC monitoring
3	Gateway-to-gateway	21–62	
4	CMCC gateway monitoring message	63	Any local network
5	ST	64	SATNET and backroom EXPAK
6	TCP	65	MIT subnet support
7	UCL	66–68	Unassigned
8	EGP (exterior gateway protocol)	69	SATNET monitoring
9	Secure	70	Unassigned
10	BBN RCC monitoring	71	Internet Packet core utility
11	NVP	72–75	Unassigned
12	PUP	76	Backroom SATNET monitoring
13	Pluribus	77	Unassigned
14	Telenet	78	WIDEBAND monitoring
15	XNET	79	WIDEBAND EXPAK
16	Chaos	80–254	Unassigned
17	User datagram	255	Reserved

23.16.2 Options field in an IP header

The options field in an IP header is an optional field which may or may not appear in the header, and is also variable in length. It is a field which must be implemented by all hosts and gateways. There are two classes of option:

- An option-type byte.
- An option-type byte, followed by an option-length byte, and then the actual option-data bytes. The option-length byte counts all the bytes in the options field.

The option-type byte is the first byte and has three fields, as illustrated in Figure 23.13. The copied flag indicates that this option is (or is not) copied into all fragments on fragmentation.

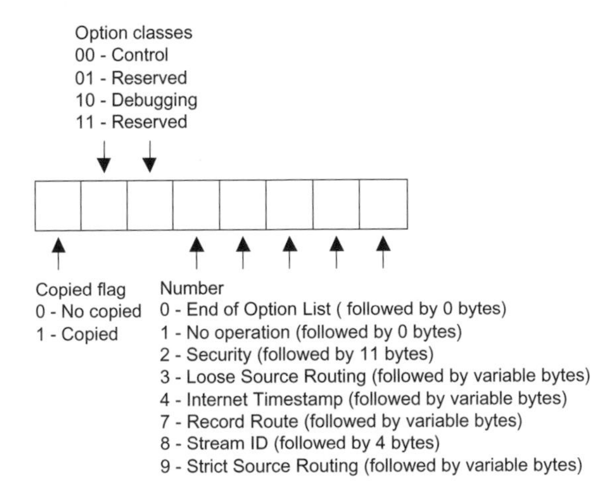

Figure 23.13 Options-type byte

End of option list (Type = 0)

This option indicates the end of the option list, but does not necessarily need to coincide with the end of the IP header according to the internet header length. It is used at the end of all options, but not the end of each option. It may be copied, introduced, or deleted on fragmentation, or for any other reason.

No operation (type = 1)

This option may be used between options, and can be used to align the beginning of a subsequent option on a 32-bit boundary. It may be copied, introduced, or deleted on fragmentation, or for any other reason.

Security (type = 130)

This option allows hosts to send security, compartmentation, handling restrictions, and TCC (closed user group) parameters. In this option, the Type field is a 2, and the Class field is also a 2. Thus the option-type byte has a value of 130 (0100 0010), and has 11 bytes in total. Its format is

```
+--------+--------+---...---+---...---+---...---+---...---+
|10000010|00001011|SSS   SSS|CCC   CCC|HHH   HHH|   TCC   |
+--------+--------+---...---+---...---+---...---+---...---+
```

The fields are

- SSS...SSS, security (16 bits) – These specify one of 16 levels of security, such as

00000000 00000000 – Unclassified	11110001 00110101 – Confidential
01111000 10011010 – EFTO	10111100 01001101 – MMMM
01011110 00100110 – PROG	10101111 00010011 – Restricted
11010111 10001000 – Secret	01101011 11000101 – Top Secret
00110101 11100010 – Reserved	10011010 11110001 – Reserved
01001101 01111000 – Reserved	00100100 10111101 – Reserved
00010011 01011110 – Reserved	10001001 10101111 – Reserved
11000100 11010110 – Reserved	11100010 01101011 – Reserved

- CCC...CCC, compartments (16 bits) – When this field contains all zero values then the transmitted information is not compartmented, other values can be obtained from the Defense Intelligence Agency.

- HHH...HHH, handling restrictions (16 bits) – This field is defined in the Defense Intelligence Agency Manual DIAM 65-19.

- TCC, transmission control code (24 bits) – This field allows the segregation of traffic and to define controlled communities of interest among subscribers (available from HQ DCA Code 530). Must be copied on fragmentation.

Loose source and record route (Type = 131)

Loose source and record route (LSRR) allows for the source of an internet datagram to supply routing information to be used by the gateways in forwarding the datagram to the destination. It can also be used to record routing information.

When routing the source host adds the IP addresses of the route to the route data, and each gateway routes the datagram using the recorded route, and not with its own internal routing table. This allows datagrams to take alternative routes through the Internet. Its format is

```
+--------+--------+--------+---------...--------+
|10000011| Length | Pointer|     Route data     |
+--------+--------+--------+---------...--------+
```

where

- Length – this is a single byte which contains the number of bytes in the option field.
- Pointer – this is a pointer, which is relative to this option, into the route data which indicates the byte which begins the next source address to be processed. The smallest value is 4.
- Route data – this is constructed with a number of internet addresses, each of 4 bytes in length. If the pointer is greater than the length, the source route is empty (and the recorded route full) and the routing is to be based on the destination address field.

When reaching the address in the destination address field, and when the pointer is not greater than the length in the route data, then the next address in the source route data replaces the address in the destination field. The pointer is also incremented by 4, to point to the next address. It is loose as the gateways are allowed to use any route to get to the next specified address in the routing table.

It must be copied on fragmentation and occurs, at the most, once in a datagram.

Strict source and record route (Type = 137)

The SSRR is similar to the LSRR, but the routing must follow, exactly, the addresses in the routing table. It thus cannot use any intermediate routes to get to these addresses. Its format is

```
+--------+--------+--------+---------...--------+
|10001001| Length | Pointer|     Route data     |
+--------+--------+--------+---------...--------+
```

Record route (type = 7)

The record route option records the route of an internet datagram. It can thus be used by such utilities as Traceroute. Its format is

```
+--------+--------+--------+---------...--------+
|00000111| Length | Pointer|     Route data     |
+--------+--------+--------+---------...--------+
```

where

- Length – this is a single byte which contains the number of bytes in the option field.
- Pointer – this is a pointer, which is relative to this option, into the route data which indicates the byte at which the next address should be added to. The smallest value is 4.
- Route data – contains a list of the route which a datagram has taken. Each entry has 4 bytes. The originating host must reserve enough area for the total number of addresses in the routing table, as the size of this option does not change as it transverses over the Internet. If there is a problem adding the address then an ICMP Parameter Problem can be sent back to the source host.

It is not copied on fragmentation, and goes in the first fragment only. In addition, it occurs, at the most, once in a datagram.

Internet timestamp (type = 68)

The Internet timestamp option records a timestamp for each gateway along the route of a datagram. It allows the source host to trace the time that each part of the route takes. Its format is

```
+--------+--------+--------+--------+
|01000100| Length | Pointer|Ov  |Flg|
+--------+--------+--------+--------+
|         internet address          |
+--------+--------+--------+--------+
|             timestamp             |
+--------+--------+--------+--------+
|                 .                 |
                  .
```

where

- Length. – is a single byte which contains the number of bytes in the option field (maximum is 40).
- Pointer – this is a pointer, which is relative to this option, into the route data which indicates the byte at which the next timestamp should be added to. The smallest value is 5.
- Overflow (Ov) – this has four bits and holds the number of IP modules that cannot register timestamps due to lack of space.
- Flag (Flg) – this has four bits and defines the format of the timestamp. Valid values are:

  0 – Store only the time stamps as 32-bit words.
  1 – Store IP address followed by a time stamp.
  3 – In this mode the IP addresses are specified in a table. A gateway only adds its timestamp if its IP address is in this table.

- Timestamp – this is a 32-bit value for the number of milliseconds since midnight UT (universal time). If this is not possible then it is any time, as long as the high-order bit of the timestamp is set to a 1 to indicate that it is non-standard time.

The originating host must reserve enough area for the total number of timestamps, as the size of this option does not change as it transverses over the Internet. If there is a problem adding the address then an ICMP parameter problem can be sent back to the source host. Initially the contents of the timestamp data area is either zero, or has IP addresses with zero time stamps. The timestamp area is full when the pointer is greater than the length.

It is not copied on fragmentation, and goes in the first fragment only. Also, it occurs, at the most once in a datagram.

Stream identifier (type =136)

This option allows for a 16-bit SATNET stream identifier to be carried through networks that do not support the stream concept. Its format is

```
+--------+--------+--------+--------+
|10001000|00000010|    Stream ID    |
+--------+--------+--------+--------+
```

23.16.3 *Ethernet multicast/broadcast addresses*

The following is a list of typical Ethernet multicast addresses:

Ethernet address	*Type field*	*Usage*
01-00-5E-00-00-00	0800	Internet multicast (RFC-1112)
01-80-C2-00-00-00	0802	Spanning tree (for bridges)
09-00-09-00-00-01	8005	HP probe
09-00-09-00-00-04	8005	HP DTC
09-00-1E-00-00-00	8019	Apollo DOMAIN
09-00-2B-00-00-03	8038	DEC lanbridge traffic monitor (LTM)
09-00-4E-00-00-02	8137	Novell IPX
CF-00-00-00-00-00	9000	Ethernet configuration test protocol

The following is a list of typical Ethernet broadcast addresses:

Ethernet address	*Type field*	*Usage*
FF-FF-FF-FF-FF-FF	0600	XNS packets, hello or gateway search.

FF-FF-FF-FF-FF-FF	0800	IP (such as RWHOD with UDP)
FF-FF-FF-FF-FF-FF	0804	CHAOS
FF-FF-FF-FF-FF-FF	0806	ARP (for IP and CHAOS) as needed
FF-FF-FF-FF-FF-FF	0BAD	Banyan
FF-FF-FF-FF-FF-FF	1600	VALID packets, hello or gateway search.
FF-FF-FF-FF-FF-FF	8035	Reverse ARP
FF-FF-FF-FF-FF-FF	807C	Merit Internodal (INP)
FF-FF-FF-FF-FF-FF	809B	EtherTalk

24 TCP and UDP

24.1 Introduction

TCP, ICMP and IP are extremely important protocols as they allow hosts to communicate over the Internet in a reliable way. The TCP layer is defined by RFC793 and RFC1122, ICMP by RFC792 and the IP layer by RFC791. TCP provides a connection between two hosts and supports error handling. This chapter discusses TCP in more detail and shows how a connection is established and then maintained. An important concept of TCP/IP communications is the usage of ports and sockets. A port identifies the process type (such as FTP, TELNET and so on) and the socket identifies a unique connection number. In this way, TCP/IP can support multiple simultaneous connections of applications over a network.

The IP header is added to higher-level data. This header contains a 32-bit IP address of the destination node. Unfortunately, the standard 32-bit IP address is not large enough to support the growth in nodes connecting to the Internet. Thus a new standard, IP Version 6, has been developed to support a 128-bit address, as well as additional enhancements.

24.2 Transmission control protocol

In the OSI model, TCP fits into the transport layer and IP fits into the network layer. TCP thus sits above IP, which means that the IP header is added onto the higher-level information (such as transport, session, presentation and application). The main function of TCP is to provide a robust and reliable transport protocol. It is characterised as a reliable, connection-oriented, acknowledged and datastream-oriented server. IP, itself, does not support the connection of two nodes, whereas TCP does. With TCP, a connection is initially established and is then maintained for the length of the transmission.

The main aspects of TCP are:

- Data transfer – data is transmitted between two applications by packaging the data within TCP packets. This data is buffered and forwarded whenever necessary. A push function can be used when the data is required to be sent immediately.
- Reliability – TCP uses sequence numbers and positive acknowledgements (ACK) to keep track of transmitted packets. Thus, it can recover from data that is damaged, lost, duplicated, or delivered out of order, such as:
  - Time-outs – the transmitter waits for a given time (the timeout interval), and if it does not receive an ACK, the data is retransmitted.
  - Sequence numbers – the sequence numbers are used at the receiver to correctly order the packets and to delete duplicates.

- Error detection and recovery – each packet has a checksum, which is checked by the receiver. If it is incorrect the receiver discards it, and can use the acknowledgements to indicate the retransmission of the packets.

- Flow control – TCP returns a window with every ACK. This window indicates a range of acceptable sequence numbers beyond the last segment successfully received. This window also indicates the number of bytes that the sender can transmit before receiving further acknowledgements.
- Multiplexing – to support multiple connections to a single host, TCP provides a set of ports within each host. This, along with the IP addresses of the source and destination, makes a socket. Each connection is uniquely identified by a pair of sockets. Ports are normally associated with various services and allow service programs to listen for defined port numbers.
- Connections – a connection is defined by the sockets, sequence numbers and window sizes. Each host must maintain this information for the length of the connection. When the connection is closed, all associated resources are freed. As TCP connections can be made with unreliable hosts and over unreliable communication channels, TCP uses a handshake mechanism with clock-based sequence numbers to avoid inaccurate connection initialisation.
- Precedence and security – TCP allows for different security and precedence levels.

TCP information contains simple acknowledgement messages and a set of sequential numbers. It also supports multiple simultaneous connections using destination and source port numbers, and manages them for both transmission and reception. As with IP, it supports data fragmentation and reassembly, and data multiplexing/demultiplexing.

The set-up and operation of TCP is as follows:

1. When a host wishes to make a connection, TCP sends out a request message to the destination machine that contains unique numbers called a socket number, and a port number. The port number has a value which is associated with the application (for example a TELNET connection has the port number 23 and an FTP connection has the port number 21). The message is then passed to the IP layer, which assembles a datagram for transmission to the destination.
2. When the destination host receives the connection request, it returns a message containing its own unique socket number and a port number. The socket number and port number thus identify the virtual connection between the two hosts.
3. After the connection has been made the data can flow between the two hosts (called a data stream).

After TCP receives the stream of data, it assembles the data into packets, called TCP segments. After the segment has been constructed, TCP adds a header (called the protocol data unit) to the front of the segment. This header contains information such as a checksum, the port number, the destination and source socket numbers, the socket number of both machines and segment sequence numbers. The TCP layer then sends the packaged segment down to the IP layer, which encapsulates it and sends it over the network as a datagram.

24.2.1 Ports and sockets

As previously mentioned, TCP adds a port number and socket number for each host. The port number identifies the required service, whereas the socket number is a unique number for that connection. Thus, a node can have several TELNET connections with the same port number but each connection will have a different socket number. A port number can be any value but there is a standard convention that most systems adopt. Table 24.1 defines some of the most common values. Standard applications normally use port values from 0 to 255, while unspecified applications can use values above 255. Section 24.12 outlines the main ports.

Table 24.1 Typical TCP port numbers

Port	*Process name*	*Notes*
20	FTP-DATA	File transfer protocol (data)
21	FTP	File transfer protocol (control)
23	TELNET	Telnet
25	SMTP	Simple mail transfer protocol
49	LOGIN	Login protocol
53	DOMAIN	Domain name server
79	FINGER	Finger
161	SNMP	Simple network management protocol

24.2.2 TCP header format

The sender's TCP layer communicates with the receiver's TCP layer using the TCP protocol data unit. It defines parameters such as the source port, destination port, and so on, and is illustrated in Figure 24.1. The fields are:

- Source and destination port number – 16-bit values that identify the local port number (source number and destination port number or destination port).
- Sequence number – identifies the current sequence number of the data segment. This allows the receiver to keep track of the data segments received. Any segments that are missing can be easily identified. The sequence number of the first data byte in this segment (except when SYN is present). If SYN is present the sequence number is the initial sequence number (ISN) and the first data octet is ISN+1.
- Acknowledgement number – when the ACK bit is set, it contains the value of the next sequence number the sender of the packet is expecting to receive. This is always set after the connection is made.
- Data offset – a 32-bit value that identifies the start of the data. It is defined as the number of 32-bit words in the header (as the TCP header always has a multiple number of 32-bit words).
- Flags – the flag field is defined as `UAPRSF`, where `U` is the urgent flag (URG), `A` the acknowledgement flag (ACK), `P` the push function (PSH), `R` the reset flag (RST), `S` the sequence synchronise flag (SYN) and `F` the end-of-transmission flag (FIN).
- Window – a 16-bit value that gives the number of data bytes that the receiving host can accept at a time, beginning with the one indicated in the acknowledgement field of this segment.
- Checksum – a 16-bit checksum for the data and header. It is the 1's complement of all the 1's complement sum of all the 16-bit words in the TCP header and text. The checksum is assumed to be a zero when calculating the checksum.
- UrgPtr – the urgent pointer used to identify an important area of data (most systems do not support this facility). It is only used when the URG bit is set. This field communicates the current value of the urgent pointer as a positive offset from the sequence number in this segment.

- Options (discussed in Section 24.2.3).
- Padding (variable) – The TCP header padding is used to ensure that the TCP header ends and data begins on a 32-bit boundary. The padding is composed of zeros.

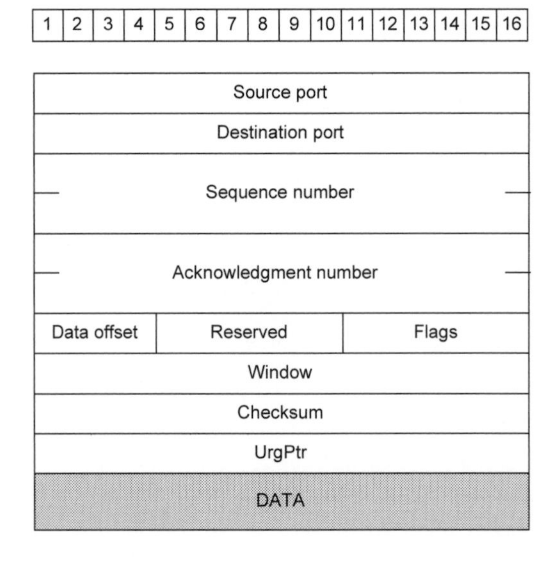

Figure 24.1 TCP header format

In TCP, a packet is termed as the complete TCP unit; that is, the header and the data. A segment is a logical unit of data, which is transferred between two TCP hosts. Thus a packet is made up of a header and a segment.

24.2.3 Options

Like IP, the option field can precede the data. It is variable in length and the contents of the header beyond the end-of-option must be header padding. It must be implemented by all hosts and gateways. There are two classes of option:

- An option-type byte.
- An option-type byte, followed by an option-length byte, and then the actual option-data bytes. The option-length byte counts all the bytes in the options field.

Supported types are:

Type	*Length*	*Description*
0		End of option list
1		No operation
2	4	Maximum segment size

24.2.4 End of option list (Type=0)

The end of option list indicates the end of all the options, not just the end of each option. It may not necessarily coincide with the end of the TCP header (according to the data offset field). It is only needed if the end of the options would not otherwise coincide with the end of the TCP header. Its format is

```
+--------+
|00000000|
+--------+
```

24.2.5 No-operation (type=1)

The no-operation can be used between options. A typical application is to align the beginning of a subsequent option, so that it is on a 32-bit word boundary. Its format is

```
+--------+
|00000001|
+--------+
```

24.2.6 Maximum segment size (type=2, length=4)

In this option the maximum receive segment size is defined, and is preceeded by a 16-bit maximum segment size. It is only sent in an initial connection request, that is, when the SYN control bit is set. If it is not included, then any segment size is allowed. Its format is

```
+--------+--------+---------+--------+
|00000010|00000100|   max seg size   |
+--------+--------+---------+--------+
```

24.3 UDP

TCP allows for a reliable connection-based transfer of data. The User datagram protocol (UDP) is an unreliable connection-less approach, where datagrams are sent into the network without an acknowledgement or connections. It is defined in RFC768 and uses IP as its underlying protocol. It has the advantage over TCP in that it has a minimal protocol mechanism, but does not guarantee delivery of any of the data. Figure 24.2 shows its format. The fields are:

- Source port – this is an optional field and is set to a zero if not used. It identifies the local port number which should be used when the destination host requires to contact the originator.
- Destination – port to connect to on the destination.
- Length – number of bytes in the datagram, including the UDP header and the data.
- Checksum – it is the 16-bit 1's complement of all 1's complement sum of the IP header, the UDP header and the data (which, if necessary, is padded with zero bytes at the end, to make an even number of bytes).

When used with IP the UDP/IP header is shown in Figure 24.3. The protocol field is set to 17 to identify UDP.

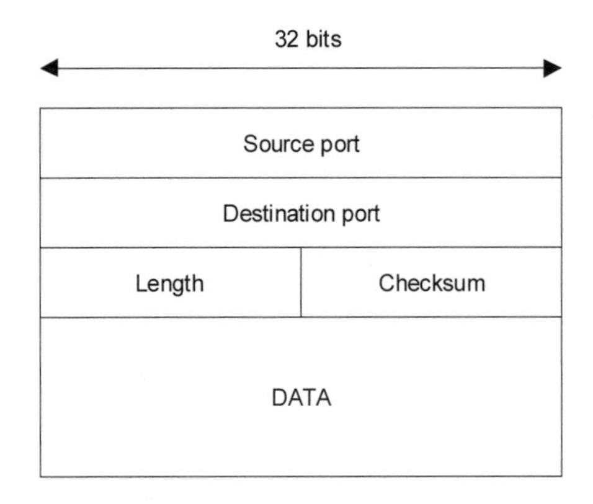

Figure 24.2 UDP header format

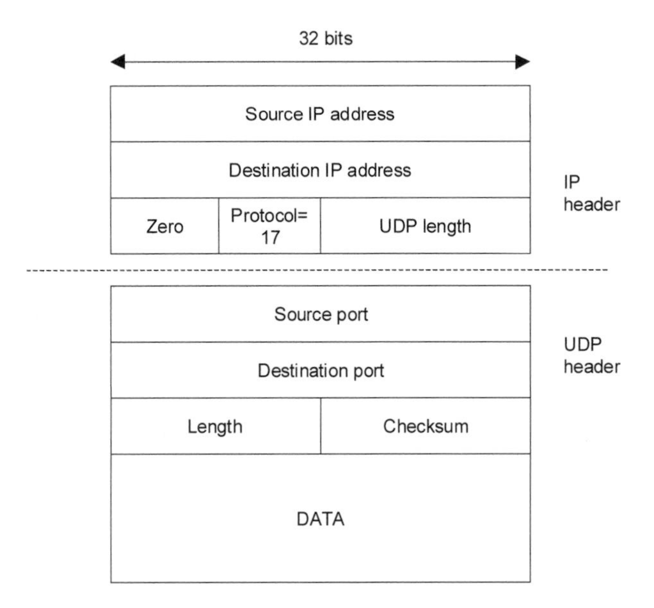

Figure 24.3 UDP/IP header format

24.4 TCP specification

TCP is made reliable with the following:

- Sequence numbers – each TCP packet is sent with a sequence number. Theoretically, each data byte is assigned a sequence number. The sequence number of the first data byte in the segment is transmitted with that segment and is called the segment sequence number (SSN).

- Acknowledgements – packets contain an acknowledgement number, which is the sequence number of the next expected transmitted data byte in the reverse direction. On sending, a host stores the transmitted data in a storage buffer, and starts a timer. If the packet is acknowledged then this data is deleted, else, if no acknowledgement is received before the timer runs out, the packet is retransmitted.
- Window – with this, a host sends a window value which specifies the number of bytes, starting with the acknowledgement number, that the host can receive.

24.4.1 Connection establishment, clearing and data transmission

The main interfaces in TCP are shown in Figure 24.4. The calls from the application program to TCP include:

- OPEN and CLOSE – to open and close a connection.
- SEND and RECEIVE – to send and receive.
- STATUS – to receive status information.

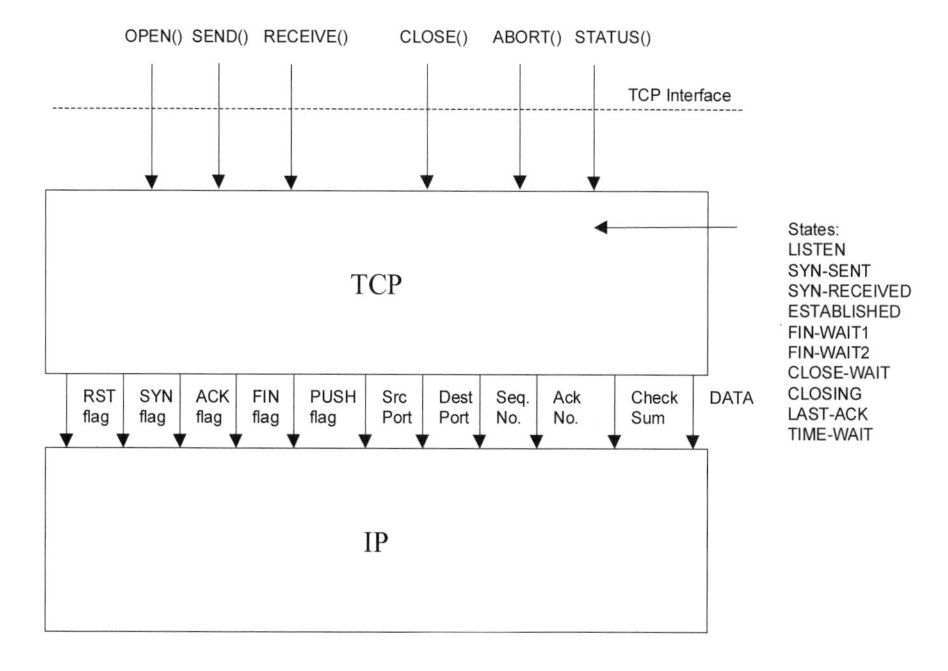

Figure 24.4 TCP interface

The OPEN call initiates a connection with a local port and foreign socket arguments. A transmission control block (TCB) stores the information on the connection. After a successful connection, TCP adds a local connection name by which the application program refers to the connection in subsequent calls.

The OPEN call supports two different types of call, as illustrated in Figure 24.5. These are:

- Passive OPEN – TCP waits for a connection from a foreign host, such as from an active OPEN. In this case, the foreign socket is defined by a zero. This is typically used by servers, such as TELNET and FTP servers. The connection can either be from a fully specified or an unspecified socket.

- Active OPEN – TCP actively connects to a foreign host, typically a server (which is opened with a passive OPEN). Two processes which issue active OPENs to each other, at the same time, will also be connected.

A connection is established with the transmission of TCP packets with the SYN control flag set and uses a three-way handshake (see Section 24.6). A connection is cleared by the exchange of packets with the FIN control flag set. Data flows in a stream using the SEND call to send data and RECEIVE to receive data.

The PUSH flag is used to send data in the SEND immediately to the recipient. This is required as a sending TCP is allowed to collect data from the sending application program and sends the data in segments when convenient. Thus, the PUSH flag forces it to be sent. When the receiving TCP sees the PUSH flag, it does not wait for any more data from the sending TCP before passing the data to the receiving process.

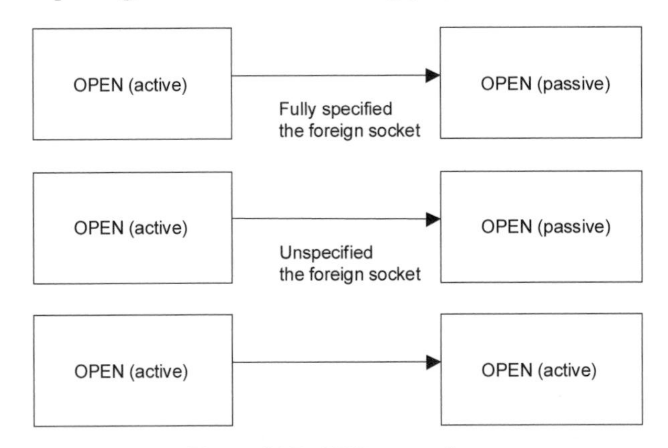

Figure 24.5 TCP connections

24.5 TCB parameters

Table 24.2 outlines the send and receive packet parameters, as well as the current segment parameter, which are stored in the TCB. Along with this, the local and remote port number require to be stored.

24.6 Connection states

Figure 24.6 outlines the states the connection goes into, and the events that cause them. The events from applications programs are OPEN, SEND, RECEIVE, CLOSE, ABORT and STATUS, and the events from the incoming TCP packets include the SYN, ACK, RST and FIN flags. The definition of each of the connection states are:

Table 24.2 TCB parameters

Send sequence variables	*Receive sequence variables*	*Current packet variable*
SND.UNA	RCV.NXT Receive next	SEG.SEQ Segment sequence number
SND.NXT Send next	RCV.WND Receive window	SEG.ACK Segment acknowledgement number
SND.WND Send window	RCV.UP Receive urgent pointer	SEG.LEN Segment length
SND.UP Send urgent pointer	IRS Initial receive sequence number	SEG.WND Segment window
SND.WL1 Segment sequence number used for last window update		SEG.UP Segment urgent pointer
SND.WL2 Segment acknowledgement number used for last window update		SEG.PRC Segment precedence value
ISS Initial send sequence number		

- LISTEN – this is the state in which TCP is waiting for a remote connection on a given port.
- SYN-SENT– this is the state where TCP is waiting for a matching connection request after it has sent a connection request.
- SYN-RECEIVED – is the state where TCP is waiting for a confirming connection request acknowledgement after having both received and sent a connection request.
- ESTABLISHED – this is the state that represents an open connection. Any data received can be delivered to the application program. This is the normal state for data to be transmitted.
- FIN-WAIT-1– this is the state in which TCP is waiting for a connection termination request, or an acknowledgement of a connection termination, from the remote TCP.
- FIN-WAIT-2 – this is the state in which TCP is waiting for a connection termination request from the remote TCP.
- CLOSE-WAIT – this is the state where TCP is waiting for a connection termination request from the local application.
- CLOSING – this is the state where TCP is waiting for a connection termination request acknowledgement from the remote TCP.
- LAST-ACK – this is the state where TCP is waiting for an acknowledgement of the connection termination request previously sent to the remote TCP.
- TIME-WAIT – this is the state in which TCP is waiting for enough time to pass to be sure the remote TCP received the acknowledgement of its connection termination request.
- CLOSED – this is the notational state, which occurs after the connection has been closed.

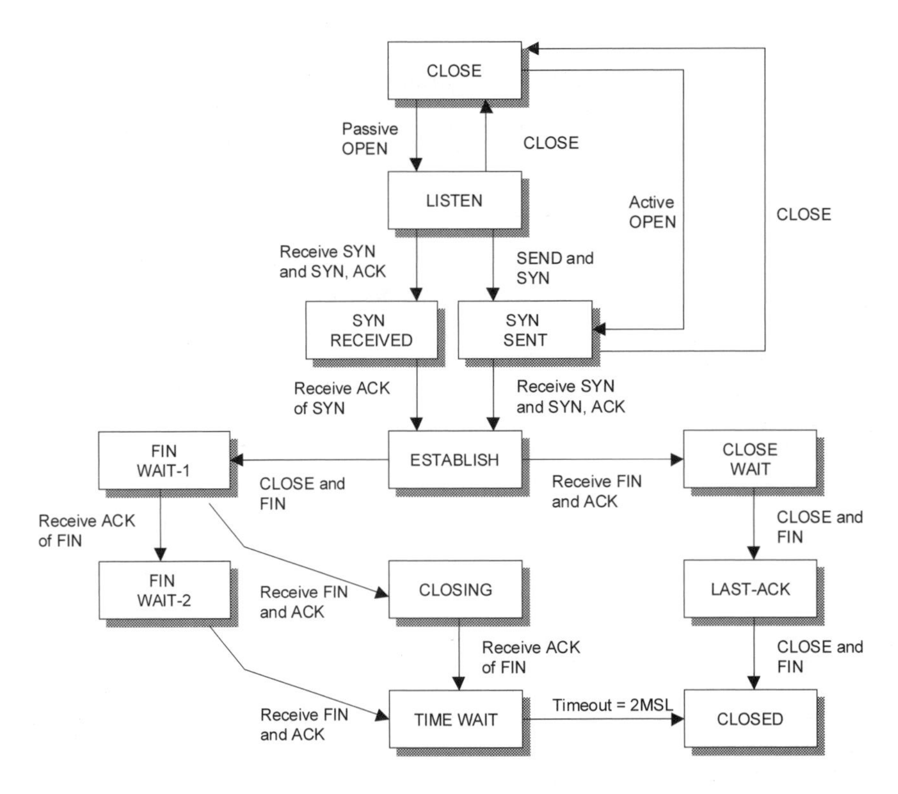

Figure 24.6 TCP connection states

24.6.1 Sequence numbers

TCP packets contain a 32-bit sequence number (0 to 4294967295), which relates to every byte sent. It uses a cumulative acknowledgement scheme, where an acknowledgement with a value of VAL, validates all bytes up to, but not including, byte VAL. Each byte at which the packet starts is numbered consecutively, at the first byte.

When sending data, TCP should receive acknowledgements for the transmitted data. The required TCB parameters will be:

SND.UNA	Oldest unacknowledged sequence number.
SND.NXT	Next sequence number to send.
SEG.ACK	Acknowledgement from the receiving TCP (next sequence number expected by the receiving TCP).
SEG.SEQ	First sequence number of a segment.
SEG.LEN	Number of bytes in the TCP packet.
SEG.SEQ+SEG.LEN–1	Last sequence number of a segment.

On receiving data, the following TCB parameters are required:

RCV.NXT	Next sequence number expected on an incoming segment, and is the left or lower edge of the receive window.
RCV.NXT+RCV.WND–1	Last sequence number expected on an incoming segment, and is

	the right or upper edge of the receive window.
SEG.SEQ	First sequence number occupied by the incoming segment.
SEG.SEQ+SEG.LEN–1	Last sequence number occupied by the incoming segment.

24.6.2 ISN selection

The initial sequence number (ISN) is selected so that previous sockets are not confused with new sockets. Typically, this can happen when a host application crashes and then quickly re-establishes the connection before the other side can time-out the connection. To avoid this a 32-bit initial sequence number (ISN) generator is created when the connection is made. The number is generated by a 32-bit clock, which is incremented approximately every 4 μs (giving an ISN cycle of 4.55 hours). Thus, within 4.55 hours, each ISN will be unique.

As each connection has a send and receive sequence number, these are an initial send sequence number (ISS) and an initial receive sequence number (IRS). When establishing a connection, the two TCPs synchronise their initial sequence numbers. This is done by exchanging connection establishing packets, with the SYN bit set and with the initial sequence numbers (these packets are typically called SYNs). Thus four packets must be initially exchanged:

- A sends to B. SYN with A_{SEQ}.
- B sends to A. ACK of the sequence number (A_{SEQ}).
- B sends to A. SYN with B_{SEQ}.
- A sends to B. ACK of the sequence number (B_{SEQ}).

Note that the two intermediate steps can be combined into a single message. This is sometimes knows as a three-way handshake. This handshake is necessary as the sequence numbers are not tied to a global clock, only to local clocks, and has many advantages, including the fact that old packets will be discarded as they occurred in a previous time.

To make sure that a sequence number is not duplicated, a host must wait for a maximum segment lifetime (MSL) before starting to retransmit packets (segments) after start-up or when recovering from a crash. An example MSL is 2 minutes. However, if it is recovering, and it has a memory of the previous sequence numbers, it may not need to wait for the MSL, as it can use sequence numbers which are much greater than the previously used sequence numbers.

24.7 Opening and closing a connection

Figure 24.7 shows a basic three-way handshake. The steps are:

1. The initial state on the initiator is CLOSED and, on the recipient, it is LISTEN (the recipient is waiting for a connection see figure 24.7).
2. The initiator goes into the SYN-SENT state and sends a packet with the SYN bit set and then indicates that the starting sequence number will be 999 (the current sequence number, thus the next number sent will be 1000). When this is received the recipient goes into the SYN-RECEIVED state.
3. The recipient sends back a TCP packet with the SYN and ACK bits set (which identifies that it is a SYN packet and also that it is acknowledging the previous SYN packet). In

this case, the recipient tells the originator that it will start transmitting at a sequence number of 100. The acknowledgement number is 1000, which is the sequence number that the recipient expects to receive next. When this is received, the originator goes into the ESTABLISHED state.

4. The originator sends back a TCP packet with the SYN and ACK bits set and the acknowledgement number is 101, which is the sequence number it expects to see next.
5. The originator transmits data with the sequence number of 1000.

	Originator					**Recipient**
1.	CLOSED					LISTEN
2.	SYN-SENT	→	<SEQ=999><CTL=SYN>			SYN-RECEIVED
3.	ESTABLISHED		<SEQ=100><ACK=1000>	<CTL=SYN,ACK>	←	SYN-RECEIVED
4.	ESTABLISHED	→	<SEQ=1000><ACK=101>	<CTL=ACK>		ESTABLISHED
5.	ESTABLISHED	→	<SEQ=1000><ACK=101>	<CTL=ACK><DATA>		ESTABLISHED

Figure 24.7 TCP connection

Note that the acknowledgement number acknowledges every sequence number up to but not including the acknowledgement number.

Figure 24.8 shows how the three-way handshake prevents old duplicate connection initiations from causing confusion. In state 3, a duplicate SYN has been received, which is from a previous connection. The recipient sends back an acknowledgement for this (4), but when this is received by the originator, the originator sends back a RST (reset) packet. This causes the recipient to go back into a LISTEN state. It will then receive the SYN packet sent in 2, and after acknowledging it, a connection is made.

TCP connections are half-open if one of the TCPs has closed or aborted, and the other end is still connected. They can also occur if the two connections have become desynchronised because of a system crash. This connection is automatically reset if data is sent in either direction. This is because the sequence numbers will be incorrect, otherwise the connection will time-out.

A connection is normally closed with the CLOSE call. A host who has closed cannot continue to send, but can continue to RECEIVE until it is told to close by the other side. Figure 24.9 shows a typical sequence for closing a connection. Normally the application program sends a CLOSE call for the given connection. Next, a TCP packet is sent with the FIN bit set, the originator enters into the FIN-WAIT-1 state. When the other TCP has acknowledged the FIN and sent a FIN of its own, the first TCP can ACK this FIN.

	Originator				**Recipient**
1.	CLOSED				LISTEN
2.	SYN-SENT	→	<SEQ=999><CTL=SYN>		
3.	(duplicate)	→	<SEQ=900><CTL=SYN>		
4.	SYN-SENT		<SEQ=100><ACK=901> <CTL=SYN,ACK>	←	SYN-RECEIVED
5.	SYN-SENT	→	<SEQ=901><CTL=RST>		LISTEN
6.			(packet 2 received)	→	
7.	SYN-SENT		<SEQ=100><ACK=1000><CTL=SYN,ACK>	←	SYN-RECEIVED
8.	ESTABLISHED	→	<SEQ=1000><ACK=101><CTL=ACK><DATA>		ESTABLISHED

Figure 24.8 TCP connection with duplicate connections

	Originator		**Recipient**
1.	ESTABLISHED (*CLOSE call*)		ESTABLISHED
2.	FIN-WAIT-1	→ <SEQ=1000><ACK=99> <CTL=SFIN,ACK>	CLOSE-WAIT
3.	FIN-WAIT-2	<SEQ=99><ACK=1001> <CTL=ACK> ←	CLOSE-WAIT
4.	TIME-WAIT	<SEQ=99><ACK=101><CTL=FIN,ACK> ←	LAST-ACK
5.	TIME-WAIT	→ <SEQ=1001><ACK=102><CTL=ACK>	CLOSED

Figure 24.9 TCP close connection

24.8 TCP user commands

The commands in this section characterise the interface between TCP and the application program. Their actual implementation depends on the operating system. Section 24.9 discusses the WinSock implementation.

24.8.1 OPEN

The OPEN call initiates an active or a passive TCP connection. The basic parameters passed and returned from the call are given next. Parameters in brackets are optional.

Parameters passed:	local port, foreign socket, active/passive [, timeout] [, precedence] [, security/compartment] [, options])
Parameters returned:	local connection name

These parameters are defined as:

- Local port – the local port to be used.
- Foreign socket – the definition of the foreign socket.
- Active/passive – a passive flag causes TCP to LISTEN, else it will actively seek a connection.
- Timeout – if present, this parameter allows the caller to set up a timeout for all data submitted to TCP. If the data is not transmitted successfully within the timeout period, the connection is aborted.
- Security/compartment – specifies the security of the connection.
- Local connection name – a unique connection name is returned which identifies the socket.

24.8.2 SEND

The SEND call causes the data in the output buffer to be sent to the indicated connection. Most implementations return immediately from the SEND call, even if the data has not been sent, although some implementations will not return until either there is a timeout or the data has been sent. The basic parameters passed and returned from the call are given next. Parameters in brackets are optional.

Parameters passed:	local connection name, buffer address, byte count, PUSH flag, URGENT flag [,timeout]

These parameters are defined as:

- Local connection name – a unique connection name which identifies the socket.
- Buffer address – address of data buffer.
- Byte count – number of bytes in the buffer.
- PUSH flag – if this flag is set then the data will be transmitted immediately, else the TCP may wait until it has enough data.
- URGENT flag – sets the urgent pointer.
- Timeout – sets a new timeout for the connection.

24.8.3 RECEIVE

The RECEIVE call allocates a receiving buffer for the specified connection. Most implementations return immediately from the RECEIVE call, even if the data has not been received, although some implementation will not return until either there is a timeout or the data has been received. The basic parameters passed and returned from the call are given next. Parameters in brackets are optional.

Parameters passed:	local connection name, buffer address, byte count
Parameters returned:	byte count, URGENT flag, PUSH flag

These parameters are defined as:

- Local connection name – a unique connection name which identifies the socket.
- Buffer address – address of the receive data buffer.
- Byte count – number of bytes received in the buffer.
- PUSH flag – if this flag is set then the PUSH flag has been set on the received data.
- URGENT flag – if this flag is set then the URGENT flag has been set on the received data.

24.8.4 CLOSE

The CLOSE call closes the connections and releases associated resources. All pending SENDs will be transmitted, but after the CLOSE call has been implemented, no further SENDs can occur. RECEIVEs can occur until the other host has also closed the connection. The basic parameters passed and returned from the call are given next.

Parameters passed:	local connection name

24.8.5 STATUS

The STATUS call determines the current status of a connection, typically listing the TCBs. The basic parameters passed and returned from the call are given next.

Parameters passed:	local connection name
Parameters returned:	status data

The returned information should include status information on the following:

- local socket, foreign socket, local connection name;

- receive window, send window, connection state;
- number of buffers awaiting acknowledgement, number of buffers pending receipt;
- urgent state, precedence, security/compartment;
- transmission timeout.

24.8.6 ABORT

The ABORT call causes all pending SENDs and RECEIVEs to be aborted. All TCBs are also removed and a RESET message sent to the other TCP. The basic parameters passed and returned from the call are given next. Parameters in brackets are optional.

Parameters passed: local connection name

24.9 WinSock

24.9.1 Introduction

The Windows sockets specification describes a common interface for networked Windows programs. WinSock uses TCP/IP communications and provides for binary and source code compatibility for different network types.

The Windows sockets API (WinSock application programming interface or WSA) is a library of functions that implement the socket interface by the Berkley Software distribution of UNIX. WinSock augments the Berkley socket implementation by adding Windows-specific extensions to support the message-driven nature of Windows system.

The basic implementation normally involves:

- Opening a socket – this allows for multiple connections with multiple hosts. Each socket has a unique identifier. It normally involves defining the protocol suite, the socket type and the protocol name. The API call used for this is `socket()`.
- Naming a socket – this involves assigning location and identity attributes to a socket. The API call used for this is `bind()`.
- Associate with another socket – this involves either listening for a connection or actively seeking a connection. The API calls used in this are `listen()`, `connect()` and `accept()`.
- Send and receive between sockets – the API calls used in this are `send()`, `sendto()`, `recv()` and `recvfrom()`.
- Close the socket – the API calls used in this are `close()` and `shutdown()`.

24.9.2 Windows sockets

The main WinSock API calls are:

`socket()`	Creates a socket.
`accept()`	Accepts a connection on a socket.
`connect()`	Establishes a connection to a peer.
`bind()`	Associates a local address with a socket.
`listen()`	Establishes a socket to listen for incoming connection.

`send()`	Sends data on a connected socket.
`sendto()`	Sends data on an unconnected socket.
`recv()`	Receives data from a connected socket.
`recvfrom()`	Receives data from an unconnected socket.
`shutdown()`	Disables send or receive operations on a socket.
`closesocket()`	Closes a socket.

Figure 24.10 shows the operation of a connection of a client to a server. The server is defined as the computer which waits for a connection, the client is the computer which initially makes contact with the server.

On the server the computer initially creates a socket with the `socket()` function, and this is bound to a name with the `bind()` function. After this, the server listens for a connection with the `listen()` function. When the client calls the `connect()` function the server then accepts the connection with `accept()`. After this the server and client can send and receive data with the `send()` or `recv()` functions. When the data transfer is complete the `closesocket()` is used to close the socket.

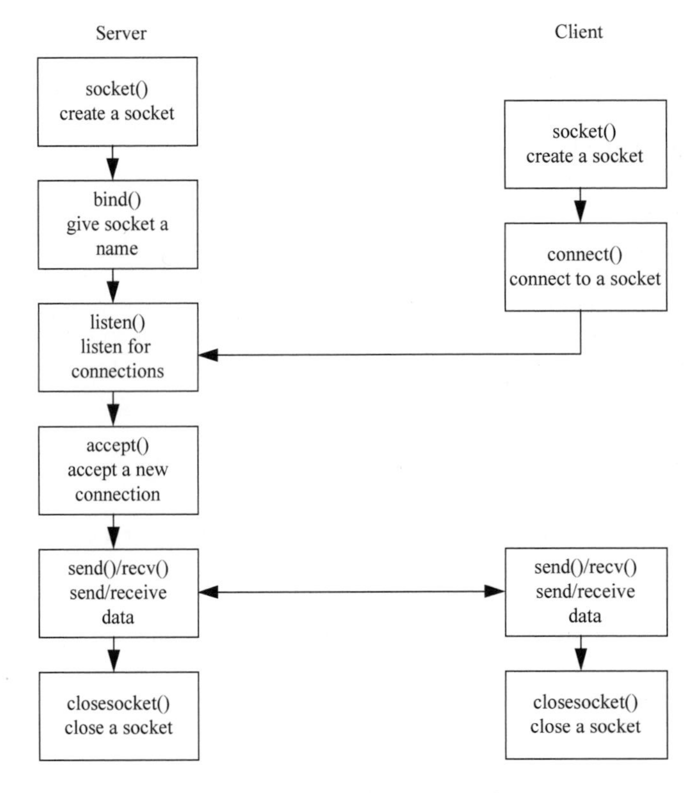

Figure 24.10 WinSock connection

socket()

The `socket()` function creates a socket. Its syntax is

```
SOCKET socket (int af,  int type,   int protocol)
```

where

`af`	A value of `PF_INET` specifies the ARPA Internet address format specification (others include `AF_IPX` for SPX/IPX and `AF_APPLETALK` for AppleTalk).
`type`	Socket specification, which is typically either `SOCK_STREAM` or `SOCK_DGRAM`. The `SOCK_STREAM` uses TCP and provides a sequenced, reliable, two-way, connection-based stream. `SOCK_DGRAM` uses UDP and provides for connectionless datagrams. This type of connection is not recommended. A third type is `SOCK_RAW`, for types other than UDP or TCP, such as for ICMP.
`protocol`	Defines the protocol to be used with the socket. If it is zero then the caller does not wish to specify a protocol.

If the `socket` function succeeds then the return value is a descriptor referencing the new socket. Otherwise, it returns `SOCKET_ERROR`, and the specific error code can be tested with `WSAGetLastError`. An example creation of a socket is given next:

```
SOCKET s;

      s=socket(PF_INET,SOCK_STREAM,0);
      if (s == INVALID_SOCKET)
      {
         cout << "Socket error"
      }
```

bind()

The `bind()` function associates a local address with a socket. It is called before the `connect` or `listen` function. When a socket is created with `socket`, it exists in a name space (address family), but it has no name assigned. The `bind` function gives the socket a local association (host address/port number). Its syntax is:

```
int bind(SOCKET s, const struct sockaddr FAR * addr, int namelen)
```

where

`s`	A descriptor identifying an unbound socket.
`namelen`	The length of `addr`.
`addr`	The address to assign to the socket. The `sockaddr` structure is defined as follows:

```
      struct sockaddr
      {
         u_short    sa_family;
         char       sa_data[14];
      };
```

In the Internet address family, the `sockadd_in` structure is used by Windows Sockets to specify a local or remote endpoint address to which to connect a socket. This is the form of the `sockaddr` structure specific to the Internet address family and can be cast to `sockaddr`. This structure can be filled with the `sockaddr_in` structure which has the following form:

```
struct SOCKADDR_IN
{
    short               sin_family;
    unsigned short      sin_port;
    struct              in_addr  sin_addr;
    char                sin_zero[8];
}
```

where

`sin_family`	must be set to `AF_INET`.
`sin_port`	IP port.
`sin_addr`	IP address.
`sin_zero`	padding to make structure the same size as `sockaddr`.

If an application does not care what address is assigned to it, it may specify an Internet address equal to `INADDR_ANY`, a port equal to 0, or both. An Internet address equal to `INADDR_ANY` causes any appropriate network interface be used. A port value of 0 causes the Windows sockets implementation to assign a unique port to the application with a value between 1024 and 5000.

If no error occurs then it returns a zero value. Otherwise, it returns `INVALID_SOCKET`, and the specific error code can be tested with `WSAGetLastError`. If an application needs to bind to an arbitrary port outside of the range 1024 to 5000 then the following outline code can be used:

```
#include <windows.h>
#include <winsock.h>
int main(void)
{

SOCKADDR_IN    sin;
SOCKET         s;
s = socket(AF_INET,SOCK_STREAM,0);

if (s == INVALID_SOCKET)
{
   // Socket failed
}

sin.sin_family = AF_INET;
sin.sin_addr.s_addr = 0;

sin.sin_port = htons(100); // port=100

if (bind(s, (LPSOCKADDR)&sin, sizeof (sin)) == 0)
{
 // Bind failed
}
return(0);
}
```

The Windows sockets `htons` function converts an unsigned short (`u_short`) from host byte order to network byte order.

connect()

The `connect()` function establishes a connection with a peer. If the specified socket is unbound then unique values are assigned to the local association by the system and the socket is marked as bound. Its syntax is

```
int connect (SOCKET s, const struct sockaddr FAR * name,   int namelen)
```

where

`s`	Descriptor identifying an unconnected socket.
`name`	Name of the peer to which the socket is to be connected.
`namelen`	Name length.

If no error occurs then it returns a zero value. Otherwise, it returns `SOCKET_ERROR`, and the specific error code can be tested with `WSAGetLastError`.

listen()

The `listen()` function establishes a socket which listens for an incoming connection. The sequence to create and accept a socket is:

- `socket()`- Creates a socket.
- `listen()`- this creates a queue for incoming connections and is typically used by a server that can have more than one connection at a time.
- `accept()`- these connections are then accepted with accept.

The syntax of `listen()` is

```
int listen (SOCKET s, int backlog)
```

where

`s`	Describes a bound, unconnected socket.
`backlog`	Defines the queue size for the maximum number of pending connections may grow (typically a maximum of 5).

If no error occurs then it returns a zero value. Otherwise, it returns `SOCKET_ERROR`, and the specific error code can be tested with `WSAGetLastError`.

```
#include <windows.h>
#include <winsock.h>

int main(void)
{

SOCKADDR_IN    sin;
SOCKET         s;

      s = socket(AF_INET,SOCK_STREAM,0);
      if (s == INVALID_SOCKET)
      {
```

```
            // Socket failed
        }

        sin.sin_family = AF_INET;
        sin.sin_addr.s_addr = 0;

        sin.sin_port = htons(100); // port=100

        if (bind(s, (struct sockaddr FAR *)&sin, sizeof (sin))==SOCKET_ERROR)
        {
         // Bind failed
      }

        if (listen(s,4)==SOCKET_ERROR)
        {
            // Listen failed
        }
        return(0);
}
```

accept()

The `accept()` function accepts a connection on a socket. It extracts any pending connections from the queue and creates a new socket with the same properties as the specified socket. Finally, it returns a handle to the new socket. Its syntax is

```
SOCKET accept(SOCKET s, struct sockaddr FAR *addr, int FAR  *addrlen )
```

where

`s`	Descriptor identifying a socket that is in listen mode.
`addr`	Pointer to a buffer that receives the address of the connecting entity, as known to the communications layer.
`addrlen`	Pointer to an integer which contains the length of the address `addr`.

If no error occurs then it returns a zero value. Otherwise, it returns `INVALID_SOCKET`, and the specific error code can be tested with `WSAGetLastError`.

```
#include <windows.h>
#include <winsock.h>

int main(void)
{

SOCKADDR_IN     sin;
SOCKET          s;
int             sin_len;

        s = socket(AF_INET,SOCK_STREAM,0);
        if (s == INVALID_SOCKET)
        {
            // Socket failed
        }

        sin.sin_family = AF_INET;
        sin.sin_addr.s_addr = 0;
        sin.sin_port = htons(100); // port=100
```

```
      if (bind(s, (struct sockaddr FAR *)&sin, sizeof (sin))==SOCKET_ERROR)
      {
       // Bind failed
      }

      if (listen(s,4)<0)
      {
         // Listen failed
      }
      sin_len = sizeof(sin);
      s=accept(s,(struct sockaddr FAR *) & sin,(int FAR *) &sin_len);
      if (s==INVALID_SOCKET)
      {
         // Accept failed
      }
      return(0);
}
```

send()

The `send()` function sends data to a connected socket. Its syntax is:

```
int send (SOCKET s, const char FAR *buf, int len, int flags)
```

where

`s`	Connected socket descriptor.
`buf`	Transmission data buffer.
`len`	Buffer length.
`flags`	Calling flag.

The *flags* parameter influences the behaviour of the function. These can be

`MSG_DONTROUTE`	Specifies that the data should not be subject to routing.
`MSG_OOB`	Send out-of-band data.

If `send()` succeeds then the return value is the number of characters sent (which can be less than the number indicated by `len`). Otherwise, it returns `SOCKET_ERROR`, and the specific error code can be tested with `WSAGetLastError`.

```
#include <windows.h>
#include <winsock.h>
#include <string.h>
#define  STRLENGTH 100

int main(void)
{

SOCKADDR_IN     sin;
SOCKET          s;
int sin_len;
char  sendbuf[STRLENGTH];

      s = socket(AF_INET,SOCK_STREAM,0);
      if (s == INVALID_SOCKET)
```

```
        {
           // Socket failed
        }
        sin.sin_family = AF_INET;
        sin.sin_addr.s_addr = 0;
        sin.sin_port = htons(100); // port=100
        if (bind(s, (struct sockaddr FAR *)&sin, sizeof (sin))==SOCKET_ERROR)
        {
         // Bind failed
        }

        if (listen(s,4)<0)
        {
           // Listen failed
        }
        sin_len = sizeof(sin);

        s=accept(s,(struct sockaddr FAR *) & sin,(int FAR *) &sin_len);

        if (s<0)
        {
           // Accept failed
        }

        while (1)
        {
           // get message to send and put into sendbuff
           send(s,sendbuf,strlen(sendbuf),80);
        }
        return(0);
}
```

recv()

The `recv()` function receives data from a socket. It waits until data arrives and its syntax is:

```
int recv(SOCKET s, char FAR *buf, int len, int flags)
```

where

`s`	Connected socket descriptor.
`buf`	Incoming data buffer.
`len`	Buffer length.
`flags`	Specifies the method by which the data is received.

If `recv()` succeeds then the return value is the number of bytes received (a zero identifies that the connection has been closed). Otherwise, it returns `SOCKET_ERROR`, and the specific error code can be tested with `WSAGetLastError`.

The flags parameter may have one of the following values:

`MSG_PEEK`	Peek at the incoming data. Any received data is copied into the buffer, but not removed from the input queue.
`MSG_OOB`	Process out-of-band data.

```
#include <windows.h>
#include <winsock.h>
```

```
#define  STRLENGTH 100

int main(void)
{

SOCKADDR_IN     sin;
SOCKET          s;
int             sin_len,status;
char            recmsg[STRLENGTH];

      s = socket(AF_INET,SOCK_STREAM,0);

      if (s == INVALID_SOCKET)
      {
         // Socket failed
      }

      sin.sin_family = AF_INET;
      sin.sin_addr.s_addr = 0;

      sin.sin_port = htons(100); // port=100

      if (bind(s, (struct sockaddr FAR *)&sin, sizeof (sin))==SOCKET_ERROR)
      {
       // Bind failed
      }

      if (listen(s,4)<0)
      {
         // Listen failed
      }
      sin_len = sizeof(sin);

      s=accept(s,(struct sockaddr FAR *) & sin,(int FAR *) &sin_len);

      if (s<0)
      {
         // Accept failed
      }
      while (1)
      {
         status=recv(s,recmsg,STRLENGTH,80);

         if (status==SOCKET_ERROR)
         {
            // no socket
            break;
         }
         recmsg[status]=NULL; // terminate string
         if (status)
         {
            // szMsg contains received string
         }
         else
         {
            break;
            // connection broken
         }
      }
      return(0);
}
```

shutdown()

The `shutdown()` function disables send or receive operations on a socket and does not close any opened sockets. Its syntax is

```
int shutdown(SOCKET s, int how)
```

where

`s`	Socket descriptor.
`how`	Flag that identifies operation types that will no longer be allowed. These are: 0 – Disallows subsequent receives. 1 – Disallows subsequent sends. 2 – Disables send and receive.

If no error occurs then it returns a zero value. Otherwise, it returns `INVALID_SOCKET`, and the specific error code can be tested with `WSAGetLastError`.

closesocket()

The `closesocket()` function closes a socket. Its syntax is:

```
int closesocket (SOCKET s);
```

where

s	Socket descriptor.

If no error occurs then it returns a zero value. Otherwise, it returns `INVALID_SOCKET`, and the specific error code can be tested with `WSAGetLastError`.

24.10 Visual Basic socket implementation

Visual Basic supports a WinSock control and allows the connection of hosts over a network. It supports both UDP and TCP. Figure 24.11 shows a sample Visual Basic screen with a WinSock object (in this case, it is named Winsock1). To set the protocol used then either select the properties window on the WinSock object, click protocol and select sckTCPProtocol, or sckUDPProtocol. Otherwise, within the code it can be set to TCP with:

```
Winsock1.Protocol = sckTCPProtocol
```

The WinSock object has various properties, such as:

obj.`RemoteHost`	Defines the IP address or domain name of the remote host.
obj.`LocalPort`	Defines the local port number.

The methods that are used with the WinSock object are:

obj.`Connect`	Connects to a remote host (client invoked).
obj.`Listen`	Listens for a connection (server invoked).
obj.`GetData`	Reads data from the input steam.
obj.`SendData`	Sends data to an output stream.

The main events are:

`ConnectionRequest`	Occurs when a remote host wants to make a connection with a server.
`DataArrival`	Occurs when data has arrived from a connection (data is then read with GetData).

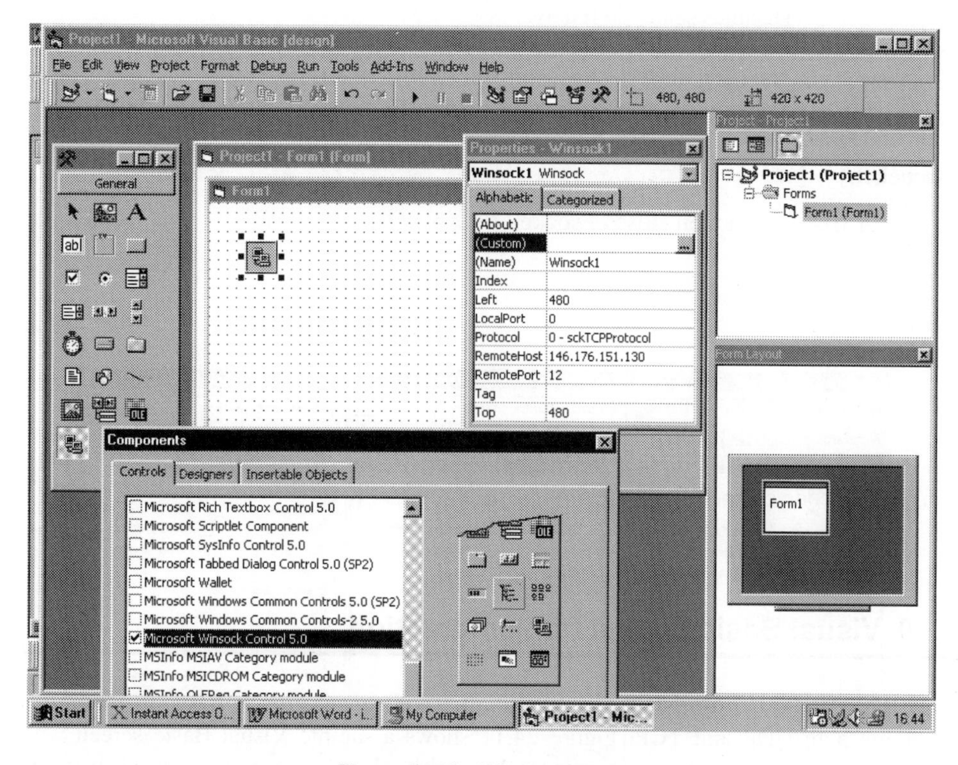

Figure 24.11 WinSock object

24.10.1 Creating a server

A server must listen for connection. To do this, do the following:

1. Create a new standard EXE project.
2. Change the name of the default form to myServer.
3. Change the caption of the form to 'Server Application' (see Figure 24.12).
4. Put a WinSock control on the main form and change its name to myTCPServer.
5. Add two TextBox controls to the form. Name the first SendTextData, and the second ShowText (see Figure 24.12).

6 Add the code given below to the form.

```
Private Sub Form_Load()
   ' Set the local port to 1001 and listen for a connection
   myTCPServer.LocalPort = 1001
   myTCPServer.Listen
   myClient.Show
End Sub

Private Sub myTCPServer_ConnectionRequest (ByVal requestID As Long)
   ' Check state of socket, if it is not closed then close it.
   If myTCPServer.State <> sckClosed Then myTCPServer.Close
   ' Accept the request with the requestID parameter.
   myTCPServer.Accept  requestID
End Sub

Private Sub SendTextData_Change()
   ' SendTextData contains the data to be sent.
   ' This data is setn using the SendData method
   myTCPServer.SendData = SendTextData.Text
End Sub

Private Sub myTCPServer_DataArrival (ByVal bytesTotal As Long)
   ' Read incoming data into the str variable,
   ' then display it to ShowText
   Dim str As String
   myTCPServer.GetData = str
   ShowText.Text = str
End Sub
```

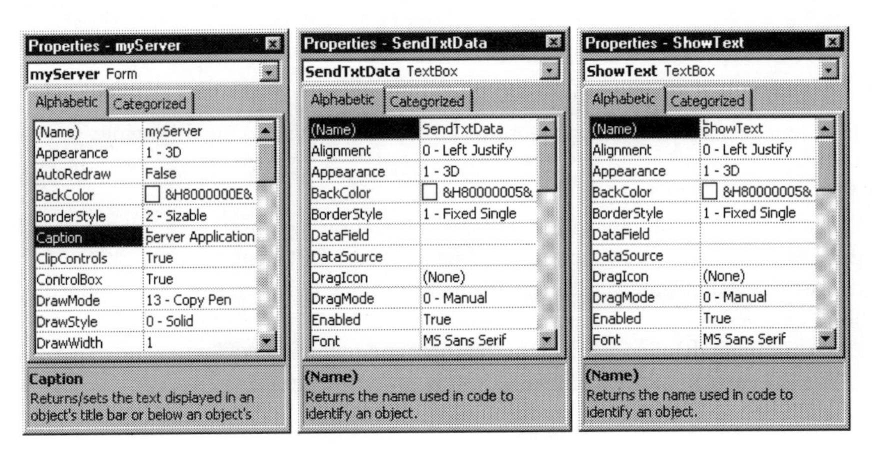

Figure 24.12 Server set-ups

Figure 24.13 shows the server setup.

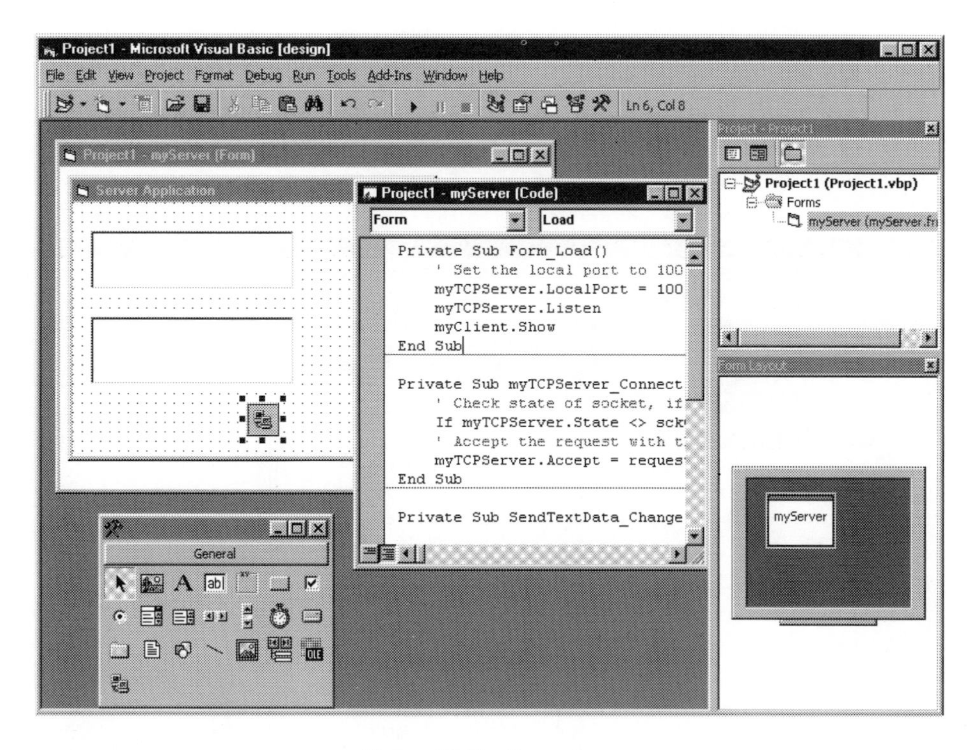

Figure 24.13 Server form

24.10.2 Creating a client

The client must actively seek a connection. To create a client, do the following:

1. Add a new form to the project, and name it myClient.
2. Change the caption of the form to 'Client Application'.
3. Add a WinSock control to the form and name it myTCPClient.
4. Add two TextBox controls to the form. Name the first SendTextData, and the second ShowText.
5. Draw a CommandButton control on the form and name it cmdConnect.
6. Change the caption of the CommandButton control to Connect.
7. Add the code given below to the form.

```
Private Sub Form_Load()
   ' In this case it will connect to 146.176.151.130
   ' change this to the local IP address or DNS of the local computer
   myTCPClient.RemoteHost = "146.176.151.130"
   myTCPClient.RemotePort = 1001
End Sub

Private Sub cmdConnect_Click()
   ' Connect to the server
   myTCPClient.Connect
End Sub

Private Sub SendTextData_Change()
   tcpClient.SendData txtSend.Text
```

```
End Sub

Private Sub tcpClient_DataArrival (ByVal bytesTotal As Long)
   Dim str As String
   myTCPClient.GetData str
   ShowText.Text = str
End Sub
```

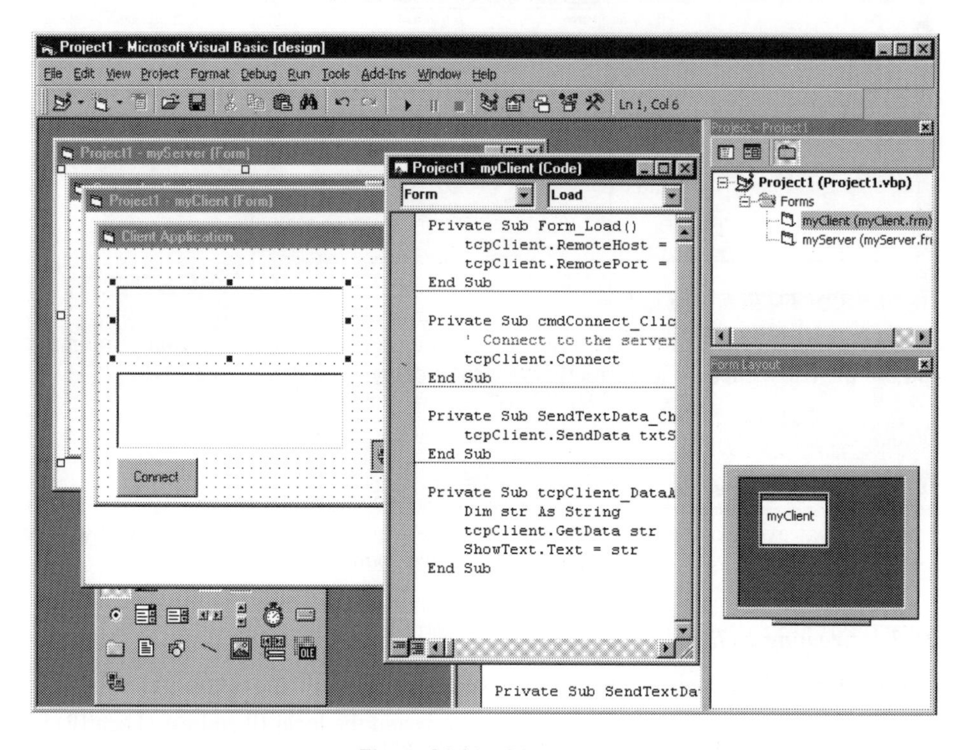

Figure 24.14 Client form

The program, when it is run, will act as a client and a server. Any text typed in the SendTxtData TextBox will be sent to the ShowText TextBox on the other form.

24.10.3 Multiple connections

In Visual Basic, it is also possible to create multiple connections to a server. This is done by creating multiple occurrences of the server object. A new one is created every time there is a new connection (with the Connection_Request event). Each new server accepts the incoming connection. The following code, which has a WinSock control on a form called multServer, is given below.

```
Private ConnectNo As Long

Private Sub Form_Load()
   ConnectNo = 0
   multServer(0).LocalPort = 1001
   multServer(0).Listen
End Sub

Private Sub multServer_ConnectionRequest _
```

```
                              (Index As Integer, ByVal requestID As Long)
   If Index = 0 Then
      ConnectNo = ConnectNo + 1
      Load multServer(ConnectNo)
      multServer(ConnectNo).LocalPort = 0
      multServer(ConnectNo).Accept requestID
      Load txtData(ConnectNo)
   End If
End Sub
```

24.10.4 Connect event

The Connect event connects to a server. If an error occurs then a flag (ErrorOccurred) is set to True, else it is False. Its syntax is

Private Sub *object*.Connect(ErrorOccurred As Boolean)

24.10.5 Close event

The Close event occurs when the remote computer closes the connection. Applications should use the Close method to correctly close their connection. Its syntax is

*object*_Close()

24.10.6 DataArrival event

The DataArrival event occurs when new data arrives, and returns the number of bytes read (bytesTotal). Its syntax is

*object*_DataArrival (bytesTotal As Long)

24.10.7 Bind method

The Bind method specifies the local port (LocalPort) and the local IP address (LocalIP) to be used for TCP connections. Its syntax is

object.Bind *LocalPort, LocalIP*

24.10.8 Listen method

The Listen method creates a socket and goes into listen mode (for server applications). Its stays in this mode until a ConnectionRequest event occurs, which indicates an incoming connection. After this, the Accept method should be used to accept the connection. Its syntax is:

object.Listen

24.10.9 Accept method

The Accept method accepts incoming connections after a ConnectionRequest event. Its syntax is

object.Accept requestID

The requestID parameter is passed into the ConnectionRequest event and is used with the

Accept method.

24.10.10 Close method

The Close method closes a TCP connection. Its syntax is

object.`Close`

24.10.11 SendData method

The SendData methods sends data (`Data`) to a remote computer. Its syntax is

object.`SendData` *Data*

24.10.12 GetData method

The GetData method gets data (Data) from an object. Its syntax is

object.`GetData` *data*, [*type*,] [*maxLen*]

24.11 Exercises

24.11.1 Which of the following is not part of a TCP header:

(a) Host IP address (b) Time-to-live field
(c) Host port number (d) Acknowledgement number

24.11.2 Which port does a TELNET server listen to:

(a) 21 (b) 25
(c) 25 (d) 80

24.11.3 Which port does an Email server (using SMTP) listen to:

(a) 21 (b) 25
(c) 25 (d) 80

24.11.4 Which port does a WWW server (using HTTP) listen to:

(a) 21 (b) 25
(c) 25 (d) 80

24.11.5 Which port does an FTP server listen to:

(a) 21 (b) 25
(c) 25 (d) 80

24.11.6 What is the main difference between UDP and TCP:

(a) TCP uses sequence numbers, makes connections and uses acknowledge-

ments.
(b) They use different addressing schemes.
(c) They use different port allocations.
(d) UDP only supports one-way traffic, while TCP supports multiplexed traffic.

24.11.7 What is the main method that TCP uses to create a reliable connection:

(a) Enhanced error correction
(b) Specially coded data
(c) Encrypted data
(d) Sequence numbers and acknowledgements

24.11.8 How is the initial sequence number of a TCP packet generated:

(a) Randomly
(b) From a 32-bit clock which is updated every 4 μs
(c) From a universal Internet-based clock
(d) From the system clock

24.11.9 How many packets are exchanged in setting up an established TCP connection:

(a) 1 (b) 2
(c) 3 (d) 4

24.11.10 Outline the operation of the three-way handshaking.

24.11.11 What advantages does TCP have over UDP. Investigate server applications which use UDP.

24.11.12 If possible, implement a basic client/server application with either C++ or Visual Basic. As a test, run the client and the server on the same computer. (Note the IP address of the computer as this is required by the client.)

24.11.13 Change the program in Exercise 24.11.12 so that the client and the server run on different computers (note the IP address of the server as this is required by the client). If possible, run the program on different network segments.

For the following questions, download a program from the WWW which connects to a specified port on a specified server.

24.11.14 Connect to a WWW server using port 13. This port should return the current date and time.

24.11.15 Connect to a WWW server using port 19.

24.11.16 Connect two computers over a network and set up a chat connection. One of the computers should be the chat server and the other the chat client. Modify it so that the server accepts calls from one or many clients.

24.12 TCP/IP services reference

Port	*Service*	*Comment*	*Port*	*Service*	*Comment*
1	TCPmux		7	echo	
9	discard	Null	11	systat	Users
13	daytime		15	netstat	
17	qotd	Quote	18	msp	Message send protocol
19	chargen	ttytst source	21	ftp	
23	telnet		25	smtp	Mail
37	time	Timserver	39	rlp	Resource location
42	nameserver	IEN 116	43	whois	Nicname
53	domain	DNS	57	mtp	Deprecated
67	bootps	BOOTP server	67	bootps	
68	bootpc	BOOTP client	69	tftp	
70	gopher	Internet Gopher	77	rje	Netrjs
79	finger		80	www	WWW HTTP
87	link	Ttylink	88	kerberos	Kerberos v5
95	supdup		101	hostnames	
102	iso-tsap	ISODE	105	csnet-ns	CSO name server
107	rtelnet	Remote Telnet	109	pop2	POP version 2
110	pop3	POP version 3	111	sunrpc	
113	auth	Rap ID	115	sftp	
117	uucp-path		119	nntp	USENET
123	ntp	Network Timel	137	netbios-ns	NETBIOS name service
138	netbios-dgm	NETBIOS	139	netbios-ssn	NETBIOS session
143	imap2		161	snmp	SNMP
162	snmp-trap	SNMP trap	163	cmip-man	ISO management over IP
164	cmip-agent		177	xdmcp	X display manager
178	nextstep	NeXTStep	179	bgp	BGP
191	prospero		194	irc	Internet relay chat
199	smux	SNMP multiplexor	201	at-rtmp	AppleTalk routing
202	at-nbp	AppleTalk name binding	204	at-echo	AppleTalk echo
206	at-zis	AppleTalk zone information	210	z3950	NISO Z39.50 database
213	ipx	IPX	220	imap3	Interactive mail access
372	ulistserv	UNIX Listserv	512	exec	Comsat 513 login
513	who	Whod	514	shell	No passwords used
514	syslog		515	printer	Line printer spooler
517	talk		518	ntalk	
520	route	RIP	525	timed	Timeserver
526	tempo	Newdate	530	courier	Rpc
531	conference	Chat	532	netnews	Readnews
533	netwall	Emergency broadcasts	540	uucp	Uucp daemon
543	klogin	Kerberized 'rlogin' (v5)	544	kshell	Kerberized 'rsh' (v5)

24.13 Notes from the author

In this chapter I have presented the two opposite ends of code development for TCP/IP communications. The C++ code is complex, but very powerful, and allows for a great deal of flexibility. On the other hand, the Visual Basic code is simple to implement but is difficult to implement for non-typical applications. Thus, the code used tends to reflect the type of application. In many cases Visual Basic gives an easy-to-implement package, with the required functionality. I've seen many a student wilt at the prospect of implementing a Microsoft Windows program in C++. 'Where do I start', is always the first comment, and then 'How do I do text input', and so on. Visual Basic, on the other hand, has matured into an excellent development system which hides much of the complexity of Microsoft Windows away from the developer. So, don't worry about computer language snobbery. Pick the best language to implement the specification.

UDP transmission can be likened to sending electronic mail. In most electronic mail packages the user can request that a receipt is sent back to the originator when the electronic mail has been opened. This is equivalent to TCP, where data is acknowledged after a certain amount of data has been sent. If the user does not receive a receipt for their electronic mail then they will send another one, until it is receipted or until there is a reply. UDP is equivalent to a user sending an electronic mail without asking for a receipt, thus the originator has no idea if the data has been received, or not.

TCP/IP is an excellent method for networked communications, as IP provides the routing of the data, and TCP allows acknowledgements for the data. Thus, the data can always be guaranteed to be correct. Unfortunately there is an overhead in the connection of the TCP socket, where the two communicating stations must exchange parameters before the connection is made, then they must maintain and acknowledge received TCP packets. UDP has the advantage that it is connectionless. So there is no need for a connection to be made, and data is simply thrown in the network, without the requirement for acknowledgments. Thus UDP packets are much less reliable in their operation, and a sending station cannot guarantee that the data is going to be received. UDP is thus useful for remote data acquisition where data can be simply transmitted without it being requested or without a TCP/IP connection being made.

The concept of ports and sockets is important in TCP/IP. Servers wait and listen on a given port number. They only read packets which have the correct port number. For example, a WWW server listens for data on port 80, and an FTP server listens for port 21. Thus a properly set up communication network requires a knowledge of the ports which are accessed. An excellent method for virus writers and hackers to get into a network is to install a program which responds to a given port which the hacker uses to connect to. Once into the system they can do a great deal of damage. Programming languages such as Java have built-in security to reduce this problem.

25 Networks

25.1 Introduction

Most computers in organisations connect to a network using a LAN. These networks normally consist of a backbone, which is the common link to all the networks within the organization. This backbone allows users on different network segments to communicate and allows data into and out of the local network. Figure 25.1 shows a local area network which contains various segments: LAN A, LAN B, LAN C, LAN D, LAN E and LAN F. These are connected to the local network via the BACKBONE 1. Thus, if LAN A talks to LAN E then the data must travel out of LAN A, onto BACKBONE 1, then into LAN C and through onto LAN E.

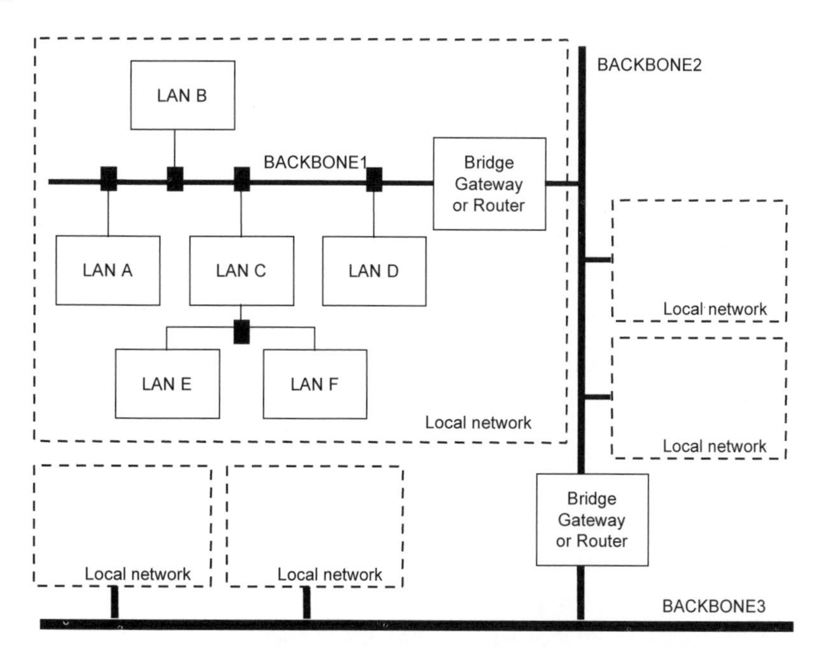

Figure 25.1 Interconnection of local networks

Networks are partitioned from other networks with a bridge, a gateway or a router. A bridge links a network of one type to an identical type, such as Ethernet to Ethernet, or Token Ring to Token Ring. A gateway connects two dissimilar types of networks and routers operate in a similar way to gateways and can either connect to two similar or dissimilar networks. The essential operation of a gateway, bridge or router is that they only allow data traffic through that is intended for another network, which is outside the connected network.

This filters traffic and stops traffic, not intended for the network, from clogging-up the backbone. Most modern bridges, gateways and routers are intelligent and can automatically determine the topology of the network.

Spanning-tree bridges have built-in intelligence and can communicate with other bridges. They can then build up a picture of the interconnected networks. So, if more than one path exists between individual segments, the bridge automatically finds alternate routes. This is useful when a fault develops on a route or a route becomes too heavily loaded. Conventional bridges can cause frames to loop around forever.

25.1.1 Peer-to-peer and client/server

An important concept is the differentiation between a peer-to-peer connection and a client/server connection. A peer-to-peer connection allows users on a local network access to a local computer. Typically, this might be access to:

Local printers – Printers, local to a computer, can be accessed by other users if the printer is shareable. This can be password protected, or not. Shareable printers on a Microsoft network have a small hand under the icon.

Local disk drives and folders – The disk drives, such as the hard disk or CD-ROM drives can be accessed if they are shareable. Normally the drives must be shareable. On a Microsoft network a drive can be made shareable by selecting the drive and selecting the right-hand mouse button, then selecting the Sharing option. User names and passwords can be set-up locally or can be accessed from a network server. Typically, only the local computer grants access to certain folders, whereas others are not shared.

These shared resources can also be mounted as disk drives to the remote computer. Thus, the user of the remote computer can simply access resources on the other computers as if they were mounted locally. This option is often the best when there is a small local network, as it requires the minimum of set-up and does not need any complicated server set-ups.

A client/server network has a central server which is typically used to:

- Store usernames, group names and passwords.
- Run print queues for networked printers.
- Allocate IP addresses for Internet accesses.
- Provide centralised file services, such as hard disks or networked CD-ROM drives.
- Provide system back-up facilities, such as CD-R disk drives and DAT tape drives.
- Centralise computer settings.
- Provide access to other centralised peripherals, such as networked faxes, dial-in network connections, and so on.
- Provide WWW and TCP/IP services, such as remote login and file transfer.

If in doubt, a peer-to-peer network is normally the best for a small office environment. Care must be taken, though, when setting up the attributes of the shared resources. Figure 25.2 shows an example of the sharing setting for a disk drive. It can be seen that the main attributes are:

- Read only – this should be used when the remote user only requires to copy or execute files. The remote user cannot modify any of the files.
- Full – this option should only be used when the remote user has full access to the files and can copy, erase or modify the files.
- Depends on password – in this mode the remote user must provide a password for either read-only access or full access.

If the peer-to-peer network has a local server, such as Novell NetWare or Windows NT/2000 then access can be provided for certain users and/or groups, if they provide the correct password.

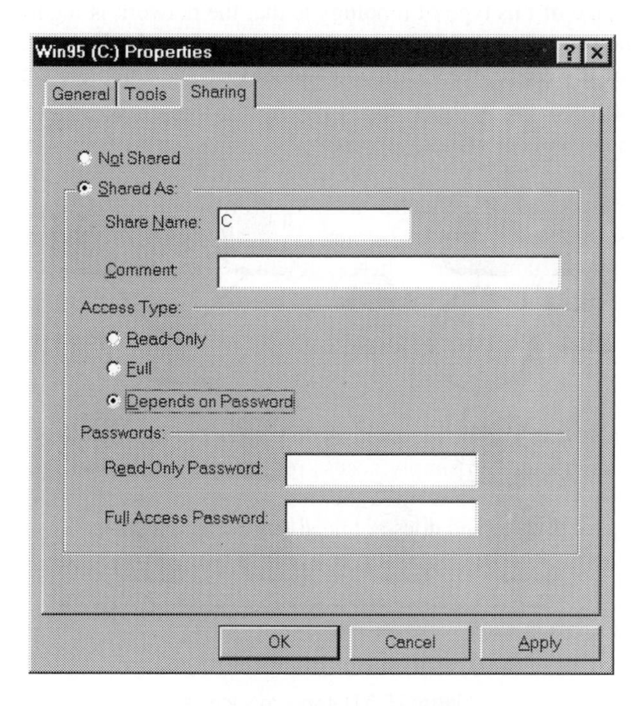

Figure 25.2 File access rights

25.2 Network topologies

There are three basic topologies for LANs, which are shown in Figure 25.3. These are

a star network, a ring network and a bus network.

There are other topologies which are either a combination of two or more topologies or are derivatives of the main types. A typical topology is a tree topology, that is essentially a combined star and a bus network, as illustrated in Figure 25.4. A concentrator (or hub) is used to connect the nodes to the network.

25.2.1 Star network

In a star topology, a central server switches data around the network. Data traffic between nodes and the server will thus be relatively low. Its main advantages are:

- As the data rate is relatively low between central server and the node, a low-specification twisted-pair cable can be used to connect the nodes to the server.
- A fault on one of the nodes will not affect the rest of the network. Typically, mainframe computers use a central server with terminals connected to it.

The main disadvantage of this type of topology is that the network is highly dependent upon the operation of the central server. If it were to slow significantly then the network becomes slow. In addition, if it were to become unoperational then the complete network would shut down.

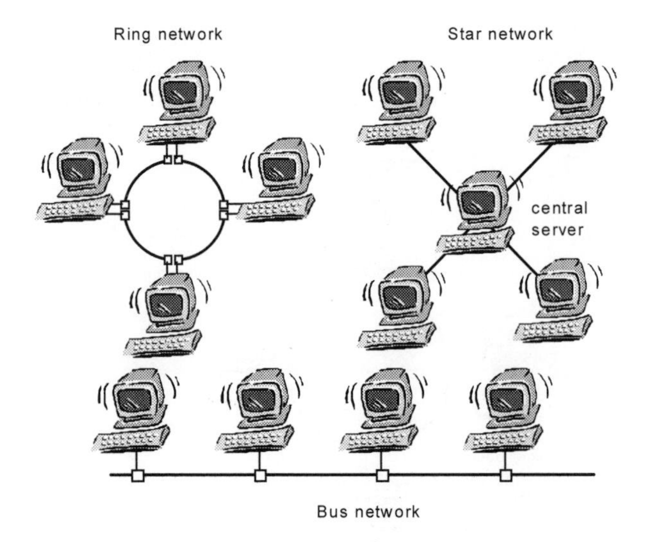

Figure 25.3 Network topologies

25.2.2 Ring network

In a ring network, computers link together to form a ring. To allow an orderly access to the ring, a single electronic token passes from one computer to the next around the ring, as illustrated in Figure 25.6. A computer can only transmit data when it captures the token. In a manner similar to the star network, each link between nodes is a point-to-point link and allows the usage of almost any type of transmission medium. Typically, twisted-pair cables to allow a bit rate of up to 16Mbps, but coaxial and fibre optic cables are normally used for extra reliability and higher data rates.

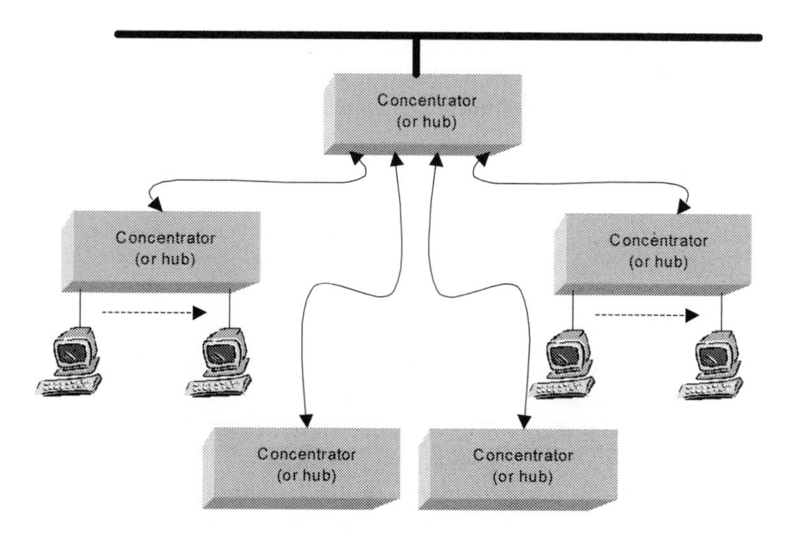

Figure 25.4 Tree topology

A typical ring network is IBM Token Ring. The main advantage of token ring networks is that all nodes on the network have an equal chance of transmitting data. Unfortunately it suffers from several problems; the most severe is that if one of the nodes goes down then the whole network may go down.

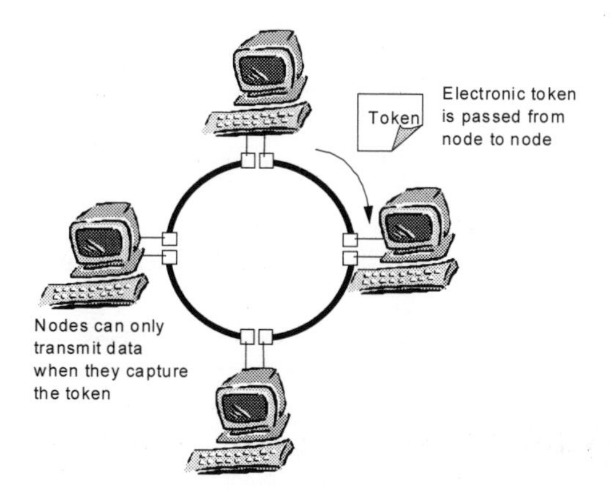

Figure 25.5 Token passing ring network

25.2.3 Bus network

A bus network uses a multidrop transmission medium, as shown in Figure 25.6. All nodes on the network share a common bus and all share communications. This allows only one device to communicate at a time. A distributed medium access protocol determines which station is to transmit. As with the ring network, data frames contain source and destination addresses, where each station monitors the bus and copies frames addressed to itself.

Twisted-pair cables give data rates up to 100 Mbps, whereas, coaxial and fibre optic cables give higher bit rates and longer transmission distances. A bus network is a good compromise over the other two topologies as it allows relatively high data rates. Also, if a node goes down, it does not affect the rest of the network. The main disadvantage of this topology is that it requires a network protocol to detect when two nodes are transmitting at the same time. It also does not cope well with heavy traffic rates. A typical bus network is Ethernet 2.0.

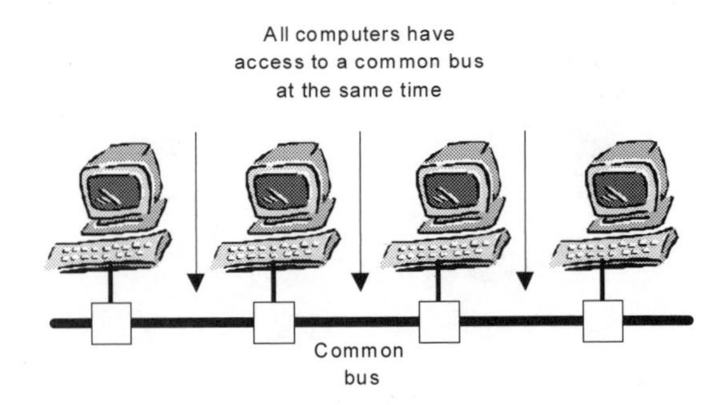

Figure 25.6 Bus topology

25.3 OSI model

A major problem in the electronics industry is the interconnection of equipment and software compatibility. Other problems can occur in the connection of electronic equipment in one part of the world to another, in another part. For these reasons, the International Standards Organization (ISO) developed a model known as the OSI (open systems interconnection) model. Its main objects were to:

- Allow manufacturers of different systems to interconnect their equipment through standard interfaces.
- Allow software and hardware to integrate well and be portable on differing systems.
- Create a model which all the countries of the world use.

Figure 25.7 shows the OSI model. Data passes from the top layer of the sender to the bottom and then up from the bottom layer to the top on the recipient. Each layer on the sender, though, communicates directly the recipient's corresponding layer. This creates a virtual data flow between layers.

The top layer (the application layer) initially gets data from an application and appends it with data that the recipients application layer reads. This appended data passes to the next layer (the presentation layer). Again, it appends it with its own data, and so on, down to the physical layer. The physical layer is then responsible for transmitting the data to the recipient. The data sent can be termed as a data frame, whereas data sent by the network or transport layer is typically referred to as a data packet.

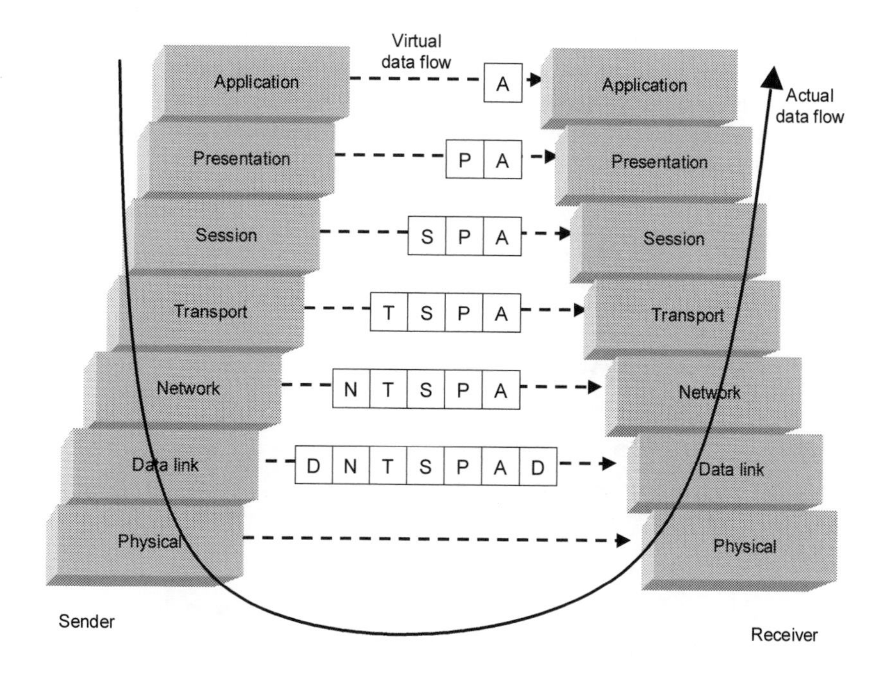

Figure 25.7 Seven-layer OSI model

The basic function of each of the layers are:

1. **Physical**. Defines the electrical characteristics of the communications channel and the transmitted signals, such as voltage levels, connector types, cabling, and so on.
2. **Data link**. Ensures that the transmitted bits are received in a reliable way, such as adding extra bits to define the start and end of a data frame, adding extra error detection/correction bits and ensuring that multiple nodes do not try to access a common communication channel at the same time.
3. **Network**. Routes data frames through a network. If data packets require to go out of a network then the transport layer routes them through interconnected networks. Its task may involve, for example, splitting data for transmission and re-assembling it upon reception. The IP part of TCP/IP is involved with the network layer.
4. **Transport**. Network transparent data transfer and transmission protocol. It supports the transmission of multiple streams from a single computer. The TCP part of TCP/IP is involved with the transport layer.
5. **Session**. Provides an open communications path with the other system. It involves the setting up, maintaining and closing down of a session. The communication channel and the internetworking of the data should be transparent to the session layer. A typical session protocol is telnet, which allows for the remote login over a network.
6. **Presentation**. Uses a set of translations that allows the data to be interpreted properly. It may have to translate between two systems if they use different presentation standards, such as different character sets or differing character codes. The presentation layer can also add data encryption for security purposes.
7. **Application**. Provides network services to application programs, such as file transfer and electronic mail.

Figure 25.8 shows how typical networking systems fit into the OSI model. The data link and physical layers are covered by networking technologies such as Ethernet, Token Ring and FDDI. The networking layer is covered by IP (internet protocol) and transport by TCP (transport control protocol).

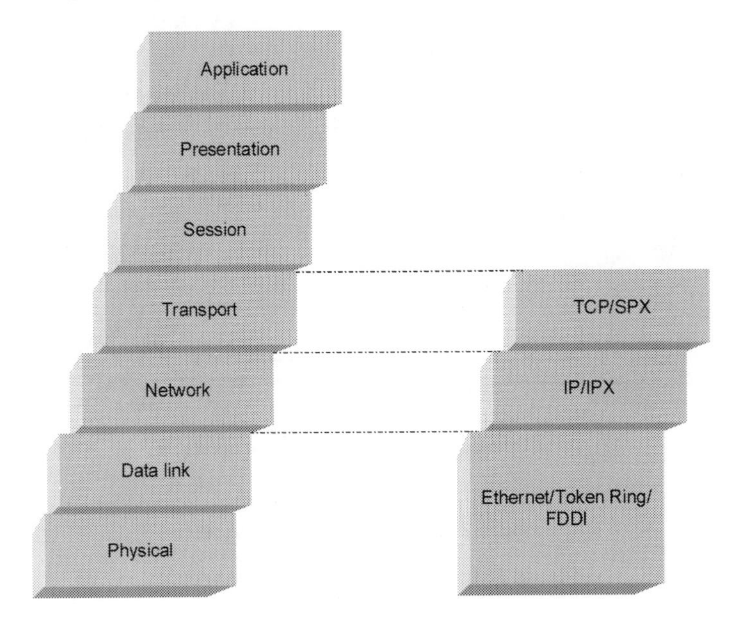

Figure 25.8 Typical technologies used in network communications

25.4 Routers, bridges and repeaters

Networks connect to other networks through repeaters, bridges or routers. A repeater corresponds to the physical layer of the OSI model and routes data from one network segment to another. Bridges route data using the data link layer (with the media access control address (MAC) address), and routers route data using the network layer (that is, using a network address, such as an IP address). Normally, at the data-link layer, the transmitted data is known as a data frame, while at the network layer it is referred to as a data packet. Figure 25.9 illustrates the three interconnection types.

25.4.1 Repeaters

All network connections suffer from a reduction in signal strength (attenuation) and digital pulse distortion. Thus, for a given cable specification and bit rate, each connection will have a maximum length of cable that can be used to transmit the data reliably. Repeaters can be used to increase the maximum interconnection length, and may do the following:

- Clean signal pulses.

- Pass all signals between attached segments.
- Boost signal power.
- Possibly translate between two different media types (such as fibre to twisted-pair cable).

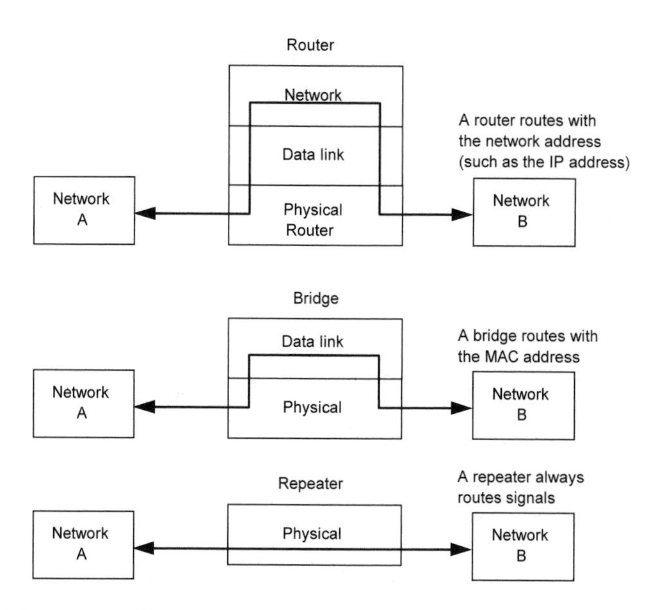

Figure 25.9 Repeaters, bridges and routers

25.4.2 Bridges

Bridges filter input and output traffic so that only data frames distended for a network are actually routed into that network and only data frames destined for the outside are allowed out of the network.

The performance of a bridge is governed by two main factors:

- **The filtering rate**. A bridge reads the MAC address of the Ethernet/Token ring/FDDI node and then decides if it should forward the frames into the network. Filter rates for bridges range from around 5000 – 70 000 pps (packets per second).
- **The forward rate**. Once the bridge has decided to route the frame into the internetwork, the bridge must forward the frame onto the destination network. Forwarding rates range from 500 to 140 000 pps and a typical forwarding rate is 90 000 pps.

A typical Ethernet bridge has the following specifications:

Bit rate	10 Mbps
Filtering rate	17 500 pps
Forwarding rate	11 000 pps
Connectors	Two DB15 AUI (female), one DB9 male console port, two BNC (for 10BASE2) or two RJ-45 (for 10BASE-T).

Algorithm Spanning tree protocol. This automatically learns the addresses of all devices on both interconnected networks and builds a separate table for each network.

25.4.3 Spanning tree architecture (STA) bridges

The IEEE 802.1 standard has defined the spanning tree algorithm. It is normally implemented as software on STA-compliant bridges. On power-up they automatically learn the addresses of all the nodes on both interconnected networks and build up a separate table for each network.

They can also support two connections between two LANs so that when the primary path becomes disabled, the spanning tree algorithm re-enables the previously disabled redundant link, as illustrated in Figure 25.10.

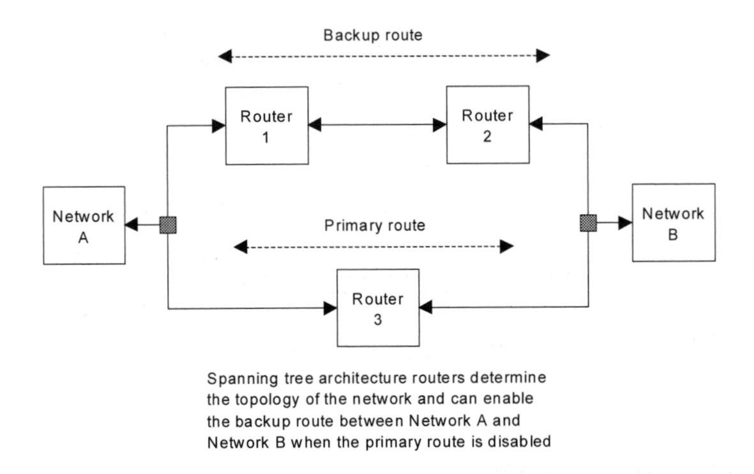
Backup route
Router 1
Router 2
Network A
Primary route
Network B
Router 3
Spanning tree architecture routers determine the topology of the network and can enable the backup route between Network A and Network B when the primary route is disabled

Figure 25.10 Spanning tree routers

25.4.4 Source route bridging

With source route bridging a source device, not the bridge, is used to send special explorer packets, these are then used to determine the best path to the destination. Explorer packets are sent out from the source routing bridges until they reach their destination workstation. Each source routing bridge along the route enters its address in the routing information field (RIF) of the explorer packet. The destination node then sends back the completed RIF field to the source node. When the source device has determined the best path to the destination, it sends the data message along with the path instructions to the local bridge. It then forwards the data message according to the received path instructions.

25.4.5 Routers

Routers examine the network address field and determine the best route for the packet. They have the great advantage in that they normally support several different types of network layer protocols.

Routers need to communicate with other routers so that they can exchange routing information. Most network operating systems have associated routing protocols which support the transfer of routing information. Typical routing protocols using Internet communications are:

- BGP (border gateway protocol).
- EGP (exterior gateway protocol).
- OSPF (open shortest path first).
- RIP (routing information protocol).

Most routers support RIP and EGP. In the past, RIP was the most popular router protocol standard. Its widespread use is due, in no small part, to the fact that it was distributed along with the Berkeley Software Distribution (BSD) of UNIX (from which most commercial versions of UNIX are derived). It suffers from several disadvantages and has been largely replaced by OSFP and EGB. These newer protocols have the advantage over RIP in that they can handle large internetworks, as well as reducing routing table update traffic.

RIP uses a distance vector algorithm which measures the number of network jumps (known as hops), up to a maximum of 16, to the destination router. This has the disadvantage that the smallest number of hops may not be the best route from source to destination. The OSPF and EGB protocol uses a link state algorithm that can decide between multiple paths to the destination router. These are based, not only on hops, but on other parameters such as delay capacity, reliability and throughput.

With distance vector routing each router maintains tables by communicating with neighbouring routers. The number of hops in its own table are then computed as it knows the number of hops to local routers, as illustrated in Figure 25.11. Unfortunately, the routing table can take some time to be updated when changes occur, because it takes time for all the routers to communicate with each other (known as slow convergence).

25.5 Network cable types

The cable type used on a network depends on several parameters, including:

- The data bit rate.
- The reliability of the cable.
- The maximum length between nodes.
- The possibility of electrical hazards and tolerance to harsh conditions.
- Power loss in the cables.
- Expense and general availability of the cable.
- Ease of connection, maintenance and ease of running cables.

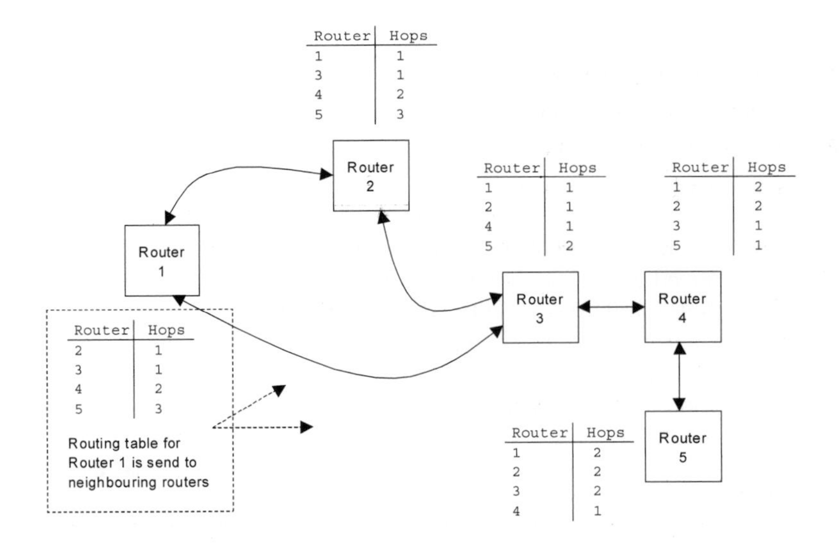

Figure 25.11 Routing tables with number of hops

The main types of cables used in networks are twisted-pair, coaxial and fibre optic; these are illustrated in Figure 25.12. Twisted-pair and coaxial cables transmit electrical signals, whereas fibre optic cables transmit light pulses.

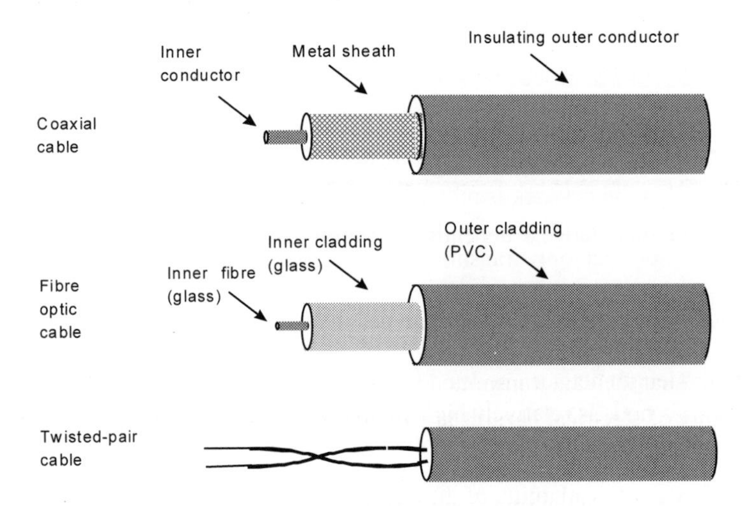

Figure 25.12 Typical network cable types

The basic specification for the cable types are:

- Twisted-pair cables – unshielded twisted-pair (UTP) cables are not shielded and thus interfere with nearby cables, whereas, shielded twisted-pair (STP) cables have a less effect on nearby cables. Public telephone lines generally use UTP cables. In LANs, they are generally used up to bit rates of 100 Mbps and with maximum lengths of 100 m.

- Coaxial cables – these have a grounded metal sheath around the signal conductor. This limits the amount of interference between cables and thus allows higher data rates. Typically, they are used at bit rates of 100 Mbps for maximum lengths of 1 km.
- Fibre optic cables – having the highest specifications of the three types, they allow extremely high bit rates over long distances. Fibre optic cables do not interfere with nearby cables and give greater security, give more protection from electrical damage from by external equipment, are more resistance to harsh environments and are safer in hazardous environments. A typical bit rate for a LAN using fibre optic cables is 100 Mbps, in other applications this can reach several gigabits/per second. The maximum length of the fiber optic cable depends on the transmitter and receiver electronics, but a single length of 20 km is possible.

25.6 Exercises

The following questions are multiple choice. Please select from a to d.

25.6.1 The cable type which offers the highest bit rate is:

(a) Fibre optic cable (b) Twisted pair cable
(c) Coaxial cable (d) Untwisted pair cable

25.6.2 Which of the following is the main disadvantage of a star network:

(a) That the data transmitted between the central server and the node is relatively high compared to other network topologies
(b) That the network is reliant on a central server
(c) All nodes compete for the network
(d) Nodes can only transmit data once they have a token

25.6.3 Which of the following is the main disadvantage of a ring network:

(a) That the data transmitted between the central server and the node is relatively high compared to other network topologies
(b) That the network is reliant on a central server
(c) All nodes compete for the network
(d) A break in the ring stops data from being transmitted

25.6.4 Which of the following is the main disadvantage of a bus network:

(a) Nodes can only transmit data once they have a token
(b) That the network is reliant on a central server
(c) All nodes compete for the network
(d) A break in the ring stops data from being transmitted

25.6.5 On a network which address does a bridge route with:

(a) IP address
(b) Interrupt address
(c) MAC address
(d) Source address

25.6.6 On a network which address does a router route with:

(a) IP address
(b) Interrupt address
(c) MAC address
(d) Source address

25.6.7 Which of the following best describes a peer-to-peer network:

(a) Resources are centralised on a server
(b) Local resources, such as memory and processor, are shared between users over the network
(c) Local resources, such as disk drives and printers, are shared between users over the network
(d) Internet connections are allocated centrally

25.6.8 Which of the following best describes a client/server network:

(a) Resources are centralised on a server
(b) Local resources, such as memory and processor, are shared between users
(c) Local resources, such as disk drives and printers, are shared between users
(d) Internet connections are allocated centrally

25.6.9 Explain how peer-to-peer networks differ from server-based networks. When might peer-to-peer networks be used and how must they be carefully set up.

25.6.10 If possible, set up a peer-to-peer connection between two computers and share some folders.

25.6.11 Locate a LAN within an organisation, such as a college or university network, and determine the cables that are used.

25.7 Notes from the author

Many of the great inventions/developments of our time were things that were not really predicted, such as CD-ROMs, RADAR, silicon transistors, fibre optic cables, and, of course, the Internet. The Internet itself is basically an infrastructure of interconnected networks which run a common protocol. The nightmare of interfacing the many computer systems around the world was solved because of two simple protocols: TCP and IP. Without them the Internet would not have evolved so quickly and possibly would not have occurred at all. TCP and IP are excellent protocols as they are simple and can be run over any type of network, on any type of computer system.

The Internet is often confused with the World Wide Web (WWW), but the WWW is only one application of the Internet. Others include electronic mail (the No.1 application), file transfer, remote login, and so on.

The amount of information transmitted over networks increases by a large factor every year. This is due to local area networks, wide area networks and traffic over the Internet. It is currently estimated that traffic on the Internet doubles every 100 days and that three people join the Internet every second. This means an eight-fold increase in traffic over a whole year. It is hard to imagine such growth in any other technological area. Imagine if cars were eight times faster each year, or could carry eight times the number of passengers each year (and of course roads and driveways would have to be eight times larger each year).

Networks have grown vastly since the 1970s ,and most companies now have some form of network. At the beginning of the 1980s, PCs were relatively complex machines to use, and required application programs to be installed locally to their disk drives. Many modern computers now run their application programs over a network, which makes the administration of the application software must simpler, and also allows users to share their resources.

The topology of a network is all-important, as it can severely effect the performance of the network, and can also be used to find network faults. I have run a network for many years and know the problems that can occur if a network grows without any long-term strategy. Many users (especially managers) perceive that a network can be expanded to an infinite degree. Many also think that new users can simply be added to the network without a thought on the amount of traffic that they are likely to generate, and its effect on other users. It is thus important for network managers to have a short-term, a medium-term and a long-term plan for the network.

So, what are the basic elements of a network. I would say:

- *IP addresses/domain names (but only if the network connects to the Internet or uses TCP/IP).*
- *A network operating system (such as Microsoft Windows, Novell NetWare, UNIX and Linux). Many companies run more than one type of network operating system, which causes many problems, but has the advantage of being able to migrate from one network operating system to another. One type of network operating system can also have advantages over other types. For example, UNIX is a very robust networking operating system which has good network security and directly supports TCP/IP for all network traffic.*
- *The cables (twisted-pair, fibre optic or coaxial cables). These directly affect the bit rate of the network, its reliability and the ease of upgrade of the network.*
- *Network servers, client/server connections and peer-to-peer connections.*
- *Bridges, routers and repeaters. These help to isolate traffic from one network segment to another. Routers and bridges are always a good long-term investment and help to isolate network traffic and can also isolate segment faults.*

The networking topology of the future is likely to evolve around a client/server architecture. With this, server machines run special programs which wait for connections from client machines. These server programs typically respond to networked applications, such as electronic mail, WWW, file transfer, remote login, date/time servers, and so on.

Many application programs are currently run over local area networks, but in the future many could be run over wide area networks, or even over the Internet. This means that computers would require the minimum amount of configuration and allows the standardisation of programs at a single point (this also helps with bug fixes and updates). There may also be a time when software licensing is charged by the amount of time that a user actually uses the package. This requires applications to be run from a central source (the server).

26 Ethernet

26.1 Introduction

Most of the computers in business now connect through a LAN and the most commonly used LAN is Ethernet. DEC, Intel and the Xerox Corporation initially developed Ethernet and the IEEE 802 committee has since defined standards for it, the most common of which are Ethernet 2.0 and IEEE 802.3. This section discusses Ethernet technology and the different types of Ethernet.

In itself Ethernet cannot make a network and needs some other protocol such as TCP/IP to allow nodes to communicate. Unfortunately, Ethernet in its standard form does not cope well with heavy traffic, but this is offset by the following:

- Ethernet networks are easy to plan and cheap to install.
- Ethernet network components, such as network cards and connectors, are cheap and well supported.
- It is a well-proven technology, which is fairly robust and reliable.
- It is simple to add and delete computers on the network.
- It is supported by most software and hardware systems.

A major problem with Ethernet is that, because computers must contend to get access to the network, there is no guarantee that they will get access within a given time. This contention also causes problems when two computers try to communicate at the same time – they must both back off and no data can be transmitted. In its standard form Ethernet allows a bit rate of 10 Mbps, but new standards for fast Ethernet systems minimise the problems of contention and also increase the bit rate to 100 Mbps (and even 1 Gbps). Ethernet uses coaxial, fibre optic or twisted-pair cable.

Ethernet uses a shared-media, bus-type network topology where all nodes share a common bus. These nodes must then contend for access to the network as only one node can communicate at a time. Data is then transmitted in frames which contain the MAC (media access control) source and destination addresses of the sending and receiving node, respectively. The local shared media is known as a segment. Each node on the network monitors the segment and copies any frames addressed to it.

Ethernet uses carrier sense, multiple access with collision detection (CSMA/CD). On a CSMA/CD network, nodes monitor the bus (or Ether) to determine if it is busy. A node wishing to send data waits for an idle condition then transmits its message. Unfortunately, collisions can occur when two nodes transmit at the same time, thus nodes must monitor the cable when they transmit. When a collision occurs, both nodes stop transmitting frames and transmit a jamming signal. This informs all nodes on the network that a collision has occurred. Each of the nodes involved in the collision then waits a random period of time before attempting a retransmission. As each node has a random delay time then there can be a prioritisation of the nodes on the network.

Each node on the network must be able to detect collisions and be capable of transmitting and receiving simultaneously. These nodes either connect onto a common Ethernet connection or can connect to an Ethernet hub. Nodes thus contend for the network and are not guaranteed access to it, as illustrated in Figure 26.1. Collisions generally slow the network.

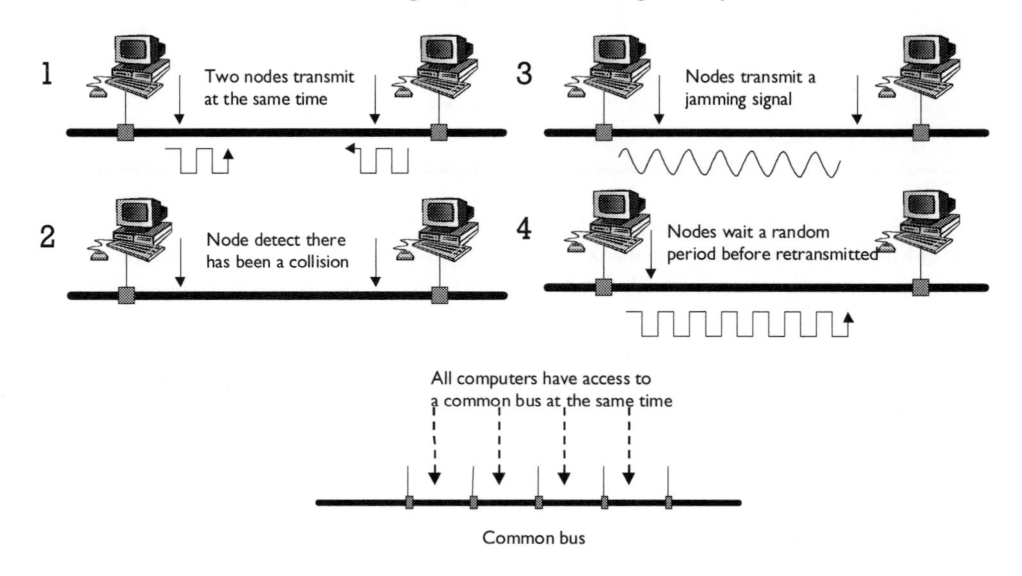

Figure 26.1 Ethernet transmission

26.2 IEEE standards

The IEEE are the main standards organization for LANs and they refer to the standard for Ethernet as CSMA/CD. Figure 26.2 shows how the IEEE standards for CSMA/CD fit into the OSI model. The two layers of the IEEE standards correspond to the physical and data link layers of the OSI model. On Ethernet networks, most hardware complies with IEEE 802.3 standard. The MAC layer allows many nodes to share a single communication channel. It also adds the start and end frame delimiters, error detection bits, access control information, and source and destination addresses. Each frame also has an error detection scheme known as cyclic redundancy check (CRC).

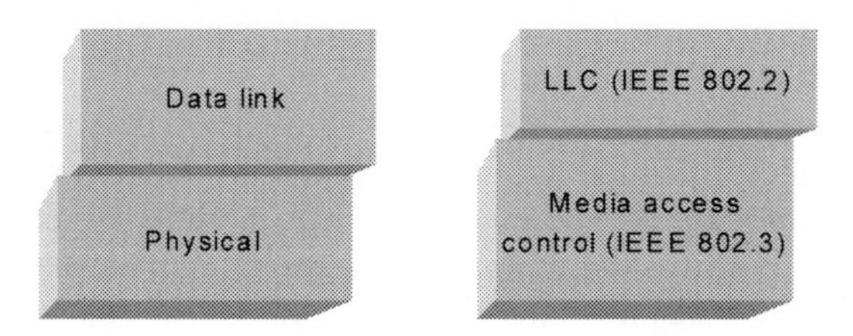

Figure 26.2 Standards for IEEE 802 LANs

26.3 Ethernet – media access control (MAC) layer

When sending data the MAC layer takes the information from the LLC link layer. Figure 26.3 shows the IEEE 802.3 frame format. It contains 2 or 6 bytes for the source and destination addresses (16 or 48 bits each), 4 bytes for the CRC (32 bits) and 2 bytes for the LLC length (16 bits). The LLC part may be up to 1500 bytes long. The preamble and delay components define the start and end of the frame. The initial preamble and start delimiter are, in total, 8 bytes long and the delay component is a minimum of 96 bytes long.

A seven-byte preamble precedes the Ethernet 802.3 frame. Each byte of the preamble has a fixed binary pattern of `10101010` and each node on the network uses it to synchronise their clock and transmission timings. It also informs nodes that a frame is to be sent and for them to check the destination address in the frame.

The end of the frame there is a 96-bit delay period, which provides the minimum delay between two frames. This slot time delay allows for the worst-case network propagation delay. The start delimiter field (SDF) is a single byte (or octet) of `10101011`. It follows the preamble and identifies that there is a valid frame being transmitted. Most Ethernet systems use a 48-bit MAC address for the sending and receiving node. Each Ethernet node has a unique MAC address, which is normally defined as hexadecimal digits, such as:

4C - 31 - 22 - 10 - F1 - 32 or 4C31 : 2210: F132.

A 48-bit address field allows 2^{48} different addresses (or approximately 281 474 976 710 000 different addresses). The LLC length field defines whether the frame contains information or it can be used to define the number of bytes in the logical link field. The logical link field can contain up to 1500 bytes of information and has a minimum of 46 bytes; its format is given in Figure 26.3. If the information is greater than this upper limit then multiple frames are sent. Also, if the field is less than the lower limit then it is padded with extra redundant bits.

The 32-bit frame check sequence (or FCS) is an error detection scheme. It is used to determine transmission errors and is often referred to as a cyclic redundancy check (CRC) or simply as a checksum.

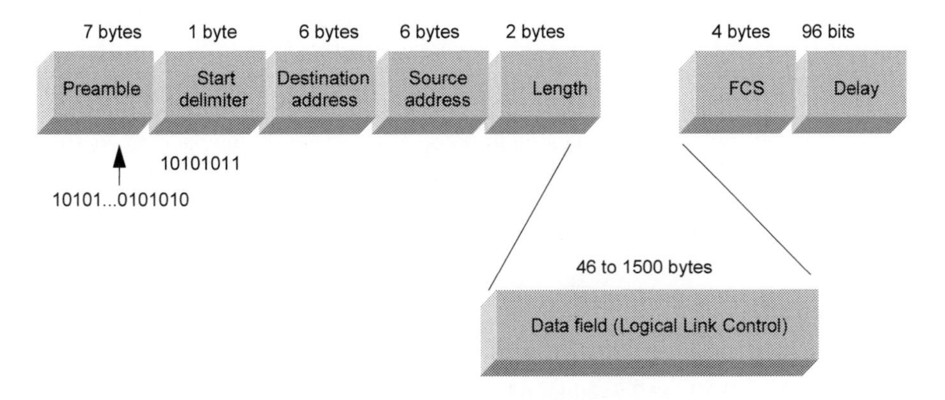

Figure 26.3 IEEE 802.3 frame format

If the transmission rate is 10Mbps, the time for one bit to be transmitted will be:

$$T = \frac{1}{\text{bit rate}} = \frac{1}{10\times10^6}\,\text{s} = 100\,\text{ns}$$

Thus the maximum and minimum times to transmit a frame will be:

$$T_{\max} = (7+1+6+6+2+1500+4+12)\times8\times100\,\text{ns} = 1.2\,\text{ms}$$
$$T_{\min} = (7+1+6+6+2+46+4+12)\times8\times100\,\text{ns} = 0.067\,\mu\text{s}$$

It may be assumed that an electrical signal propagates at about half the speed of light ($c = 3\times10^8$m/s). Thus, the time for a bit to propagate a distance of 500 m is:

$$T_{500m} = \frac{dist}{speed} = \frac{500}{1.5\times10^8} = 3.33\mu s$$

by which time, the number of bits transmitted will be:

$$\text{Number of bits transmitted} = \frac{T_{500m}}{T_{bit}} = \frac{3.33\mu s}{100ns} = 33.33$$

Thus, if two nodes are separated by 500 m then it will take more than 33 bits to be transmitted before a node can determine if there has been a collision of the line, as illustrated in Figure 26.4. If the propagation speed is less that this, it will take even longer. This shows the need for the preamble and the requirement for a maximum segment length.

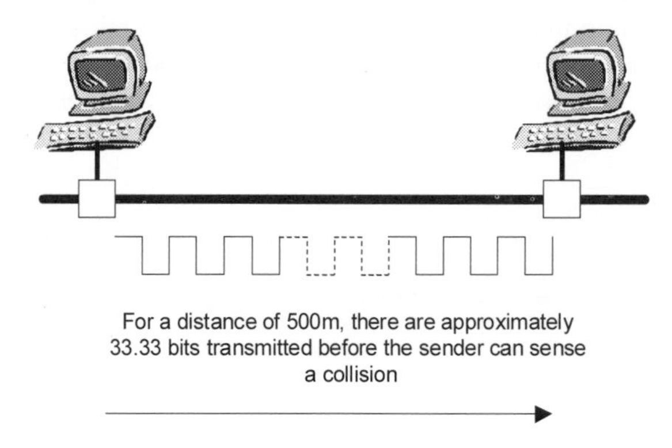

For a distance of 500m, there are approximately
33.33 bits transmitted before the sender can sense
a collision

Figure 26.4 Bits transmitted before a collision is detected

26.3.1 Ethernet II

The first standard for Ethernet was Ethernet I. Most currently available systems implement either Ethernet II or IEEE 802.3 (although most networks are now defined as being IEEE 802.3 compliant). An Ethernet II frame is similar to the IEEE 802.3 frame; it consists of eight bytes of preamble, six bytes of destination address, six bytes of source address, two bytes of frame type, between 46 and 1500 bytes of data, and four bytes of the frame check sequence field.

When the protocol is IPX/SPX the frame type field contains the bit pattern 1000 0001 0011 0111, but when the protocol is TCP/IP the type field contains 0000 1000 0000 0000.

26.4 IEEE 802.2 and Ethernet SNAP

The LLC is embedded in the Ethernet frame and is defined by the IEEE 802.2 standard. Figure 26.5 illustrates how the LLC field is inserted into the IEEE 802.3 frame. The DSAP and SSAP fields define the types of network protocol used. A SAP code of 1110 0000 identifies the network operating system layer as NetWare, whereas 0000 0110 identifies the TCP/IP protocol. These SAP numbers are issued by the IEEE. The control field is, among other things, for the sequencing of frames.

In some cases, it was difficult to modify networks to be IEEE 802 compliant. Thus, an alternative method was to identify the network protocol, known as Ethernet SNAP (subnetwork access protocol). This was defined to ease the transition to the IEEE 802.2 standard and is illustrated in Figure 26.6. It simply adds an extra two fields to the LLC field to define an organisation ID and a network layer identifier. NetWare allows for either Ethernet SNAP or Ethernet 802.2 (as Novell used Ethernet SNAP to translate to Ethernet 802.2).

Non-compliant protocols are identified with the DSAP and SSAP code of 1010 1010, and a control code of 0000 0011. After these fields come

- Organization ID – which indicates where the company that developed the embedded protocol belongs. If this field contains all zeros it indicates a non company-specific generic Ethernet frame.
- EtherType field – which defines the networking protocol. A TCP/IP protocol uses 0000 1000 0000 0000 for TCP/IP, whereas NetWare uses 1000 0001 0011 0111. NetWare frames adhering to this specification are known as NetWare 802.2 SNAP.

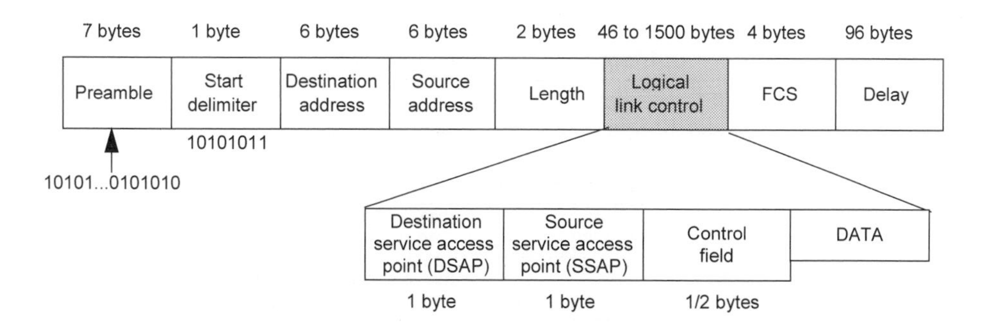

Figure 26.5 Ethernet IEEE 802.3 frame with LLC

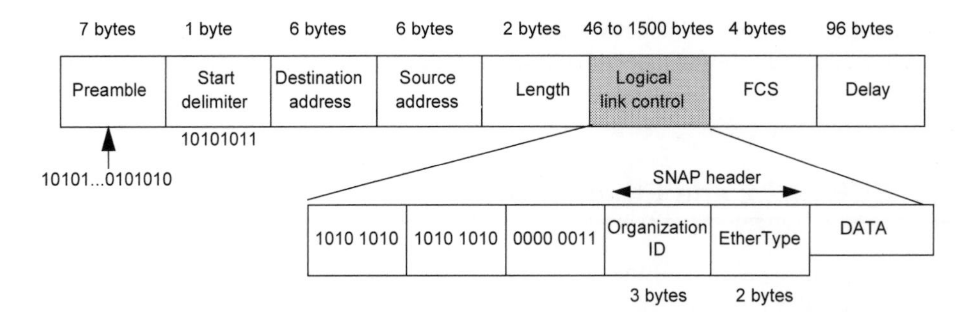

Figure 26.6 Ethernet IEEE 802.3 frame with LLC containing SNAP header

26.4.1 LLC protocol

The 802.3 frame provides some of the data link layer functions, such as node addressing (source and destination MAC addresses), the addition of framing bits (the preamble) and error control (the FCS). The rest of the functions of the data link layer are performed with the control field of the LLC field; these functions are:

- Flow and error control – each data frame sent has a frame number. A control frame is sent from the destination to a source node informing that it has or has not received the frames correctly.
- Sequencing of data – large amounts of data are sliced and sent with frame numbers. The spliced data is then reassembled at the destination node.

Figure 26.7 shows the basic format of the LLC frame. There are three principal types of frame: information, supervisory and unnumbered. An information frame contains data, a supervisory frame is used for acknowledgement and flow control, and an unnumbered frame is used for control purposes. The first two bits of the control field determine which type of frame it is. If they are 0X (where X is a don't care) then it is an information frame, 10 specifies a supervisory frame and 11 specifies an unnumbered frame.

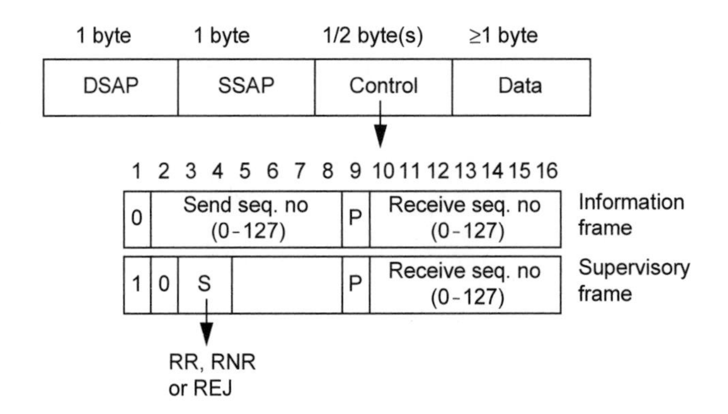

Figure 26.7 LLC frame format

An information frame contains a send sequence number in the control field which ranges from 0 to 127. Each information frame has a consecutive number, N(S) (note that there is a roll-over from frame 127 to frame 0). The destination node acknowledges that it has received the frames by sending a supervisory frame. The function of the supervisory frame is specified by the 2-bit S field. This can either be set to receiver ready (RR), receiver not ready (RNR) or reject (REJ). If an RNR function is set then the destination node acknowledges that all frames up to the number stored in the receive sequence number N(R) field were received correctly. An RNR function also acknowledges the frames up to the number N(R), but informs the source node that the destination node wishes to stop communicating. The REJ function specifies that frame N(R) has been rejected and all other frames up to N(R) are acknowledged.

26.5 OSI and the IEEE 802.3 standard

Ethernet fits into the data link and the physical layer of the OSI model. These two layers only deal with the hardware of the network. The data link layer splits into two parts: the LLC and the MAC layer.

The IEEE 802.3 standard splits into three sublayers:

- MAC (media access control).
- Physical signalling (PLS).
- Physical media attachment (PMA).

The interface between PLS and PMA is called the attachment unit interface (AUI) and the interface between PMA and the transmission media is called the media dependent interface (MDI). This grouping into modules allows Ethernet to be very flexible and to support a number of bit rates, signalling methods and media types. Figure 26.8 illustrates how the layers interconnect.

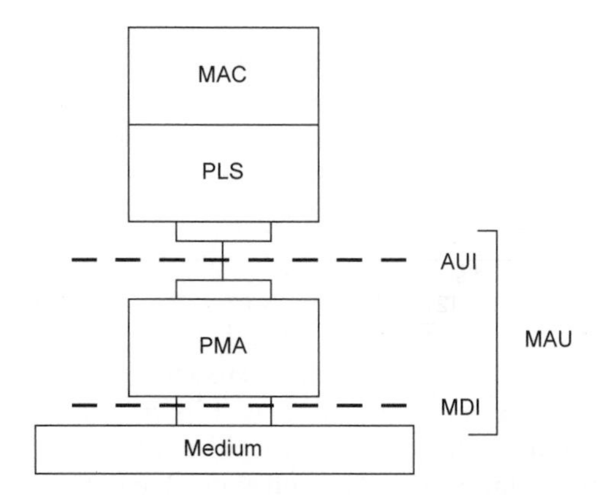

Figure 26.8 Organisation of the IEEE 802.3 standard

26.5.1 Media access control (MAC)

The CSMA/CD function is implemented in the MAC layer. The functions of the MAC layers are:

- When sending frames – receive frames from LLC; control whether the data fills the LLC data field, if not add redundant bits; make the number of bytes an integer, and calculate the FCS; add the preamble, SFD and address fields to the frame; send the frame to the PLS in a serial bit stream.
- When receiving frames – receive one frame at a time from the PLS in a serial bit stream; check whether the destination address is the same as the local node; ensure the frame contains an integer number of bytes and the FCS is correct; remove the preamble, SFD, address fields, FCS and remove redundant bits from the LLC data field; send the data to the LLC.
- Avoid collisions when transmitting frames and keep the right distance between frames by not sending when another node is sending; when the medium is free, wait a specified period before starting to transmit.
- Handle any collision that appears by sending a jam signal; generate a random number and back off from sending during that random time.

26.5.2 Physical signalling (PLS) and physical medium attachment (PMA)

PLS defines transmission rates, types of encoding/decoding and signalling methods. In PMA a further definition of the transmission media is accomplished, such as coaxial, fibre or twisted-pair. PMA and MDI together form the media attachment unit (MAU), often known as the transceiver.

26.6 Ethernet transceivers

Ethernet requires a minimal amount of hardware. The cables used to connect it are typically either unshielded twisted-pair cable (UTP) or coaxial cables. These cables must be terminated with their characteristic impedance, which is 50 Ω for coaxial cables and 100 Ω for UTP cables.

Each node has transmission and reception hardware to control access to the cable and also to monitor network traffic. The transmission/reception hardware is called a transceiver (short for *trans*mitter/re*ceiver*) and a controller builds up and strips down the frame. For 10 Mbps Ethernet, the transceiver builds the transmitted bits at a rate of 10 Mbps – thus the time for one bit is $1/10\times10^6$, which is 0.1 μs (100 ns).

The Ethernet transceiver transmits onto a single ether. When there are no nodes transmitting, the voltage on the line is +0.7 V. This provides a carrier sense signal for all nodes on the network, it is also known as the heartbeat. If a node detects this voltage then it knows that the network is active and there are no nodes currently transmitting.

Thus, when a node wishes to transmit a message it listens for a quiet period. Then, if two or more transmitters transmit at the same time, a collision results. When they detect a collision, each node transmits a 'jam' signal. The nodes involved in the collision then wait for a random period of time (ranging from 10 to 90 ms) before attempting to transmit again. Each node on the network also awaits for a retransmission. Thus, collisions are inefficient in networks as they stop nodes from transmitting. Transceivers normally detect collisions by monitoring the DC (or average) voltage on the line.

When transmitting, a transceiver unit transmits the preamble of consecutive 1s and 0s. The coding used is a Manchester coding, which represents a 0 as a high to a low voltage transition and a 1 as a low to high voltage transition. A low voltage is represented by –0.7 V and a high is +0.7 V. Thus, when the preamble is transmitted the voltage changes between +0.7V and –0.7 V; as illustrated in Figure 26.9. If after the transmission of the preamble no collisions are detected then the rest of the frame is sent.

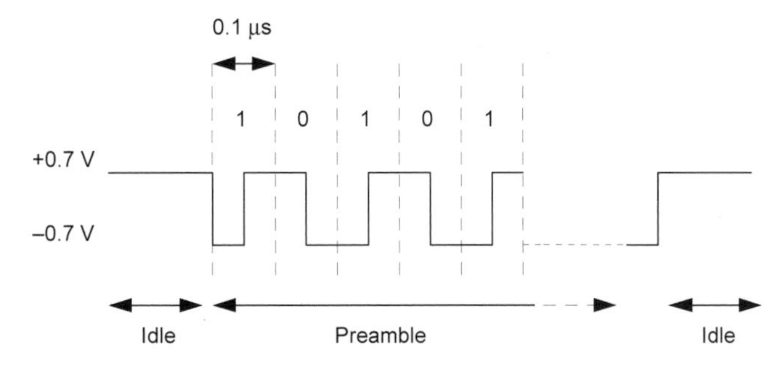

Figure 26.9 Ethernet digital signal

26.7 Ethernet types

The six main types of standard Ethernet are:

- Standard, or thick-wire, Ethernet (10BASE5).
- Thinnet, or thin-wire Ethernet, or Cheapernet (10BASE2).
- Twisted-pair Ethernet (10BASE-T).
- Optical fibre Ethernet (10BASE-FL).
- Fast Ethernet (100BASE-TX and 100VG-Any LAN).
- Gigabit Ethernet (1000BASE-SX, 1000BASE-T, 1000BASE-LX and 1000BASE-CX).

The thin- and thick-wire types connect directly to an Ethernet segment; these are shown in Figure 26.10 and Figure 26.11. Standard Ethernet, 10BASE5, uses a high specification cable (RG-50) and N-type plugs to connect the transceiver to the Ethernet segment. A node connects to the transceiver using a 9-pin D-type connector and a vampire (or bee-sting) connector can be used to clamp the transceiver to the backbone cable.

Thin-wire, or Cheapernet, uses a lower specification cable (it has a lower inner conductor diameter). The cable connector required is also of a lower specification, that is, BNC rather than N-type connectors. In standard Ethernet the transceiver unit is connected directly onto the backbone tap. On a Cheapernet network the transceiver is integrated into the node.

Most modern Ethernet connections are to a 10BASE-T hub, which connects UTP cables to the Ethernet segment. A RJ-45 connector is used for 10BASE-T. The fibre optic type, 10BASE-FL, allows long lengths of interconnected lines, typically up to 2 km. They use either SMA connectors or ST connectors. SMA connectors are screw-on types whereas ST connectors are push-on. Table 26.1 shows the basic specifications for the different types.

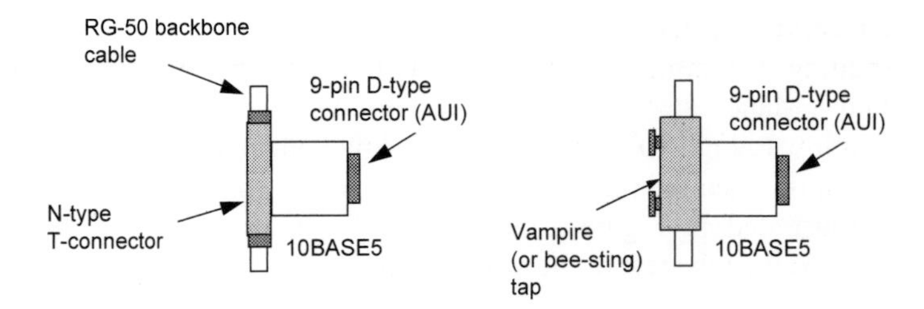

Figure 26.10 Ethernet connections for thick Ethernet

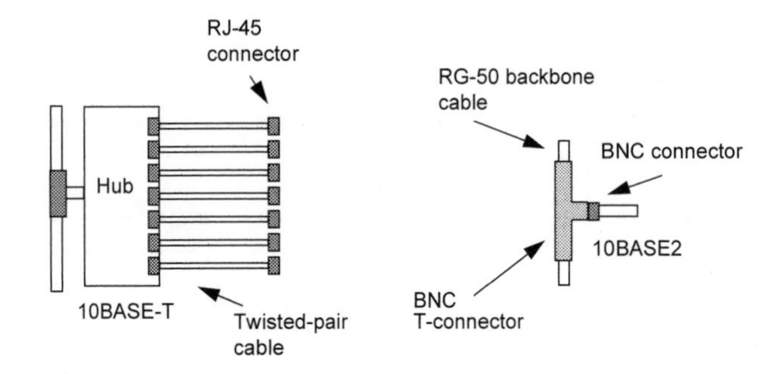

Figure 26.11 Ethernet connections for thin Ethernet and 10BASE-T

Table 26.1 10BASE network parameters

Parameter	*10BASE5*	*10BASE2*	*10BASE-T*
Common name	Standard or thick-wire Ethernet	Thinnet or thin-wire Ethernet	Twisted-pair Ethernet
Data rate	10 Mbps	10 Mbps	10 Mbps
Maximum segment length	500 m	200 m	100 m
Maximum nodes on a segment	100	30	3
Maximum number of repeaters	2	4	4
Maximum nodes per network	1024	1024	
Minimum node spacing	2.5 m	0.5 m	no limit
Location of transceiver electronics	located at the cable connection	integrated within the node	in a hub
Typical cable type	RG-50 (0.5 in diameter)	RG-6 (0.25 in diameter)	UTP cables
Connectors	N-type	BNC	RJ-45/ Telco
Cable impedance	50 Ω	50 Ω	100 Ω

26.8 Twisted-pair hubs

Twisted-pair Ethernet (10BASE-T) nodes normally connect to the backbone using a hub, as illustrated in Figure 26.12. Connection to the twisted-pair cable is via an RJ-45 connector. The connection to the backbone can either be to thin or thick Ethernet. Hubs are also stackable, with one hub connected to another. This leads to concentrated area networks (CANs) and limits the amount of traffic on the backbone. Twisted-pair hubs normally improve network performance.

10BASE-T uses two twisted-pair cables, one for transmit and one for receive. A collision occurs when the node (or hub) detects that it is receiving data when it is currently transmitting data.

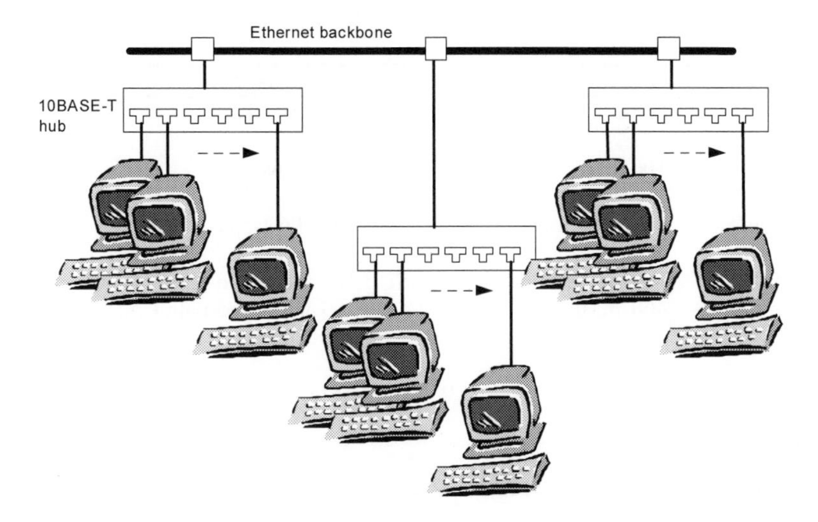

Figure 26.12 10BASE-T connection

26.9 100 Mbps Ethernet

Standard 10 Mbps Ethernet does not perform well when many users are running multimedia applications. Two improvements to the standard are Fast Ethernet and 100VG-AnyLAN. The IEEE has defined standards for both of them, IEEE 802.3u for Fast Ethernet and 802.12 for 100VG-AnyLAN. They are supported by many manufacturers and use bit rates of 100 Mbps. This gives at least 10 times the performance of standard Ethernet.

New standards relating to 100 Mbps Ethernet are now becoming popular:

- 100BASE-TX (twisted-pair) – which uses 100 Mbps over two pairs of Cat-5 UTP cable or two pairs of Type 1 STP cable.
- 100BASE-T4 (twisted-pair) – which is the physical layer standard for 100 Mbps over Cat-3, Cat-4 or Cat-5 UTP.
- 100VG-AnyLAN (twisted-pair) – which uses 100 Mbps over two pairs of Cat-5 UTP cable or two pairs of Type 1 STP cable.

- 100BASE-FX (fiber-optic cable) – which is the physical layer standard for 100 Mbps over fiber-optic cables.

Fast Ethernet, or 100BASE-T, is simply 10BASE-T running at 10 times the bit rate. It is a natural progression from standard Ethernet and thus allows existing Ethernet networks to be easily upgraded. Unfortunately, as with standard Ethernet, nodes contend for the network, reducing the network efficiency when there are high traffic rates. Also, as it uses collision detect, the maximum segment length is limited by the amount of time for the farthest nodes on a network to properly detect collisions. On a Fast Ethernet network with twisted-pair copper cables this distance is 100 m, and for a fibre-optic link, it is 400 m. Table 26.2 outlines the main network parameters for Fast Ethernet.

Table 26.2 Fast Ethernet network parameters

	100BASE-TX	*100VG-AnyLAN*
Standard	IEEE 802.3u	IEEE 802.12
Bit rate	100 Mbps	100 Mbps
Actual throughput	up to 50 Mbps	up to 96 Mbps
Maximum distance (hub to node)	100 m (twisted-pair, CAT-5) 400 m (fiber)	100 m (twisted-pair, CAT-3) 200 m (twisted-pair, CAT-5) 2 km (fibre)
Scaleability	none	up to 400 Mbps
Advantages	Easy migration from 10BASE-T	greater throughput, greater distance

As 100BASE-TX standards are compatible with 10BASE-TX networks then the network allows both 10 Mbps and 100 Mbps bit rates on the line. This makes upgrading simple, as the only additions to the network are dual-speed interface adapters. Nodes with the 100 Mbps capabilities can communicate at 100 Mbps, but they can also communicate with slower nodes, at 10 Mbps.

The basic rules of a 100BASE-TX network are:

- The network topology is a star network and there must be no loops.
- Cat-5 cable is used.
- Up to two hubs can be cascaded in a network.
- Each hub is the equivalent of 5 m in latency.
- Segment length is limited to 100 m.
- Network diameter must not exceed 205 m.

26.9.1 100BASE-T4

100BASE-T4 allows the use of standard Cat-3 cables. These contain eight wires made up of four twisted-pairs. 100BASE-4T uses all of the pairs to transmit at 100 Mbps. This differs from 10BASE-T in that 10BASE-T uses only two pairs, one to transmit and one to receive.

100BASE-T allows compatibility with 10BASE-T in that the first two pairs (Pair 1 and Pair 2) are used in the same way as 10BASE-T connections. 100BASE-T then uses the other two pairs (Pair 3 and Pair 4) with half-duplex links between the hub and the node. The connections are illustrated in Figure 26.13.

26.9.2 Line code

100BASE-4T uses four separate Cat-3 twisted-pair wires. The maximum clock rate that can be applied to Cat-3 cable is 30 Mbps. Thus, some mechanism must be devised which reduces the line bit rate to under 30 Mbps but gives a symbol rate of 100 Mbps. This is achieved with a three-level code (+, – and 0) and is known as 8B6T. This code converts eight binary digits into six ternary symbols.

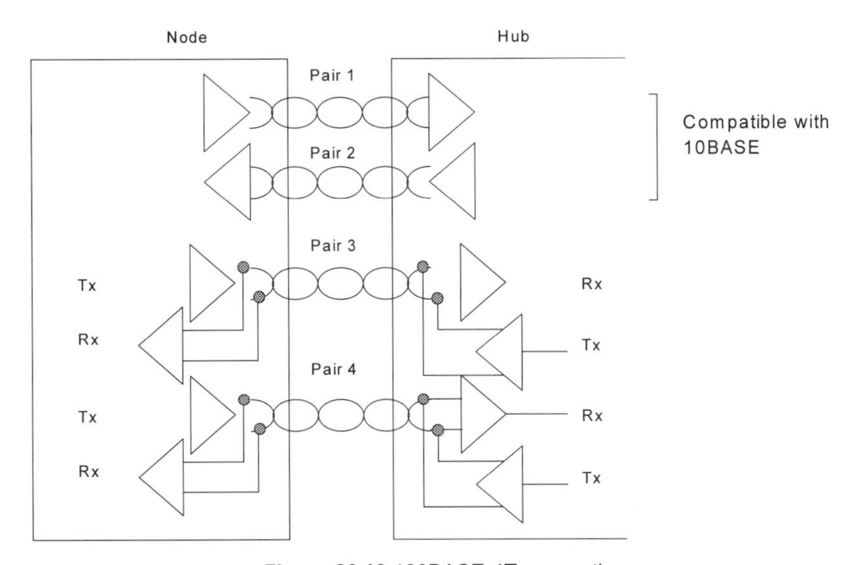

Figure 26.13 100BASE-4T connections

The first six codes are:

Data byte	*Code*	*Data byte*	*Code*	*Data byte*	*Code*
00000000	-+0 0-+	00000001	0-+ -+0	00000010	0-+ 0-+
00000011	0-+ +0-	00000100	-+0 +0-	00001001	+0- -+0

Thus, the bit sequence 00000000 will be coded as a negative voltage, a positive voltage, a zero voltage, a zero voltage, a negative voltage and a positive voltage.

The maximum base frequency for a 100 Mbps signal will be produced when the input bitstream is 010101010…01010. As each bit lasts 10 ns then the period between consecutive levels is 20 ns. Thus, the minimum frequency contained will be 50 MHz. This is greater than the bandwidth of Cat-3 cable, so it would not pass through the cable.

Apart from reducing the frequencies with the digital signal, the 8B6T code has the advantage of reducing the DC content of the signal. Most of the codes contain the same number of positive and negative voltages. This is because only 256 of the possible 729 (3^6) codes are actually used. The codes are also chosen to have at least two transitions in every code word, thus the clock information is embedded into the signal.

Unfortunately, it is not possible to have all codes with the same number of negative voltages as positive voltages. Thus, there are some codes that have a different number of negatives and positives, these include:

0100 0001+ 0 − 0 0 +
0111 1001+ + + − 0 −

Most transceiver circuits use a transformer to isolate the external equipment from the computer equipment. These transformers do not allow the passage of DC current. Thus if the line code has a sequence which consecutively has more positives than negatives, the DC current will move away from its zero value. As this does not pass across the transformer, the receive bitstream on the output of the transformer can reduce the amplitude of the received signal (and may thus cause errors). This phenomenon is known as DC wander. A code that has one more positive level than the negative levels is defined as having a weighing of +1.

The technique used to overcome this is to invert consecutive codes that have a weighing of +1. For example, suppose the line code were

+0++−− ++0+−− +++−−0 +++−−0

it would actually be coded as

+0++−− −−0−++ +++−−0 −−−++0

The receiver detects the –1 weighted codes as an inverted pattern.

26.9.3 100VG-AnyLAN

The 100VG-AnyLAN standard (IEEE 802.12) was developed mainly by Hewlett Packard and overcomes the contention problem by using a priority-based round-robin arbitration method, known as demand priority access method (DPAM). Unlike Fast Ethernet, nodes always connect to a hub which regularly scans its input ports to determine whether any nodes have requests pending.

100VG-AnyLAN has the great advantage over 100BASE in that it supports both IEEE 802.3 (Ethernet) and IEEE 802.5 (Token Ring) frames and can thus integrate well with existing 10BASET and Token Ring networks.

100VG-AnyLAN has an in-built priority mechanism with two priority levels: a high priority request and a normal priority request. A normal priority request is used for non real-time data, such as data files, and so on. High priority requests are used for real-time data, such as speech or video data. At present, there is limited usage of this feature and there is no support mechanism for this facility after the data has left the hub.

100VG-AnyLAN allows up to seven levels of hubs (i.e. one root and six cascaded hubs) with a maximum distance of 150 m between nodes. Unlike other forms of Ethernet, it allows any number of nodes to be connected to a segment.

Connections

100BASE-TX, 100BASE-T4 and 100VG-AnyLAN use the RJ-45 connector, which has eight connections. 100BASE-TX uses pairs 2 and 3, whereas 100BASE-T4 and 100VG-AnyLAN use pairs 1, 2, 3 and 4. The connections for the cables are defined in Table 26.3. The white/orange colour identifies the cable which is white with an orange stripe, whereas orange/white identifies an orange cable with a white stripe.

Table 26.3 Cable connections for 100BASE-TX

Pin	*Cable colour*	*Cable colour*	*Pair*
1	white/orange	white/orange	Pair 4
2	orange/white	orange/white	Pair 4
3	white/green	white/green	Pair 3
4	blue/white	blue/white	Pair 3
5	white/blue	white/blue	Pair 1
6	green/white	green/white	Pair 1
7	white/brown	white/brown	Pair 2
8	brown/white	brown/white	Pair 2

Migration to fast Ethernet

If an existing network is based on standard Ethernet then, in most cases, the best network upgrade is either to fast Ethernet or 100VG-AnyLAN. As the protocols and access methods are the same, there is no need to change any of the network management software or application programs. The upgrade path for Fast Ethernet is simple and could be:

- Upgrade high data rate nodes, such as servers or high-powered workstations to Fast Ethernet.
- Gradually upgrade NICs (network interface cards) on Ethernet segments to cards which support both 10BASE-T and 100BASE-T. These cards automatically detect the transmission rate to give either 10 or 100 Mbps.

The upgrade path to 100VG-AnyLAN is less easy as it relies on hubs and, unlike Fast Ethernet, most NICs have different network connectors, one for 10BASE-T and the other for 100VG-AnyLAN (although it is likely that more NICs will have automatic detection). A possible path could be:

- Upgrade high data rate nodes, such as servers or high-powered workstations to 100VG-AnyLAN.
- Install 100VG-AnyLAN hubs.
- Connect nodes to 100VG-AnyLAN hubs and change over connectors.

It is difficult to assess the performance differences between Fast Ethernet and 100VG-AnyLAN. Fast Ethernet uses a well-proven technology, but suffers from network contention. 100VG-AnyLAN is a relatively new technology and the handshaking with the hub increases delay time. The maximum data throughput of a 100BASE-TX network is limited to around 50 Mbps, whereas 100VG-AnyLAN allows rates up to 96 Mbps. 100VG-AnyLAN allows possible upgrades to 400 Mbps.

26.10 Comparison of fast Ethernet other technologies

Table 26.4 compares fast Ethernet with other types of networking technologies.

Table 26.4 Comparison of fast Ethernet with other networking technologies

Feature	*100VG-AnyLAN (Cat 3, 4, or 5)*	*100BASE-T (TX/FX/T4)*	*FDDI*	*ATM*	*Gigabit Ethernet (802.3z)*
Maximum segment length	100 m	100 m (Cat-5) 412 m (Fibre)	2000 m	200 m (Cat-5) 2000 m (Fibre)	100 m (Cat 5) 1k m (Fibre)
Maximum network diameter with repeater(s)	6000 m	320 m	100 km	N/A	To be determined by the standard
Bit rate	100 Mbps	100 Mbps	100 Mbps	155 Mbps	1 Gbps
Media access method	Demand priority	CSMA/CD	Token passing	PVC/SVC	CSMA/CD
Maximum nodes on each domain	1024	Limited by hub	500	N/A	To be determined
Frame type	Ethernet and Token Ring	Ethernet	802.5	53-byte cell	Ethernet
Multimedia support	✓	☒	FDDI-I (☒) FDDI-II (✓)	✓	YES (with 802.1p)
Integration with 10BASE2	Yes with bridges, switches and routers	Yes with switches	Yes with routers and switches	Yes with routers or switches	Yes with 10/100 Mbps switching
Relative cost	Low	Low	Medium	High	Medium
Relative complexity	Low	Low	Medium	High	Low

26.10.1 Switching technology

A switch uses store-and-forward packets to switch between ports. The main technologies used are:

- Shared bus – this method uses a high-speed backplane to interconnect the switched ports. It is frequently used to build modular switches that give a large number of ports, and to interconnect multiple LAN technologies, such as FDDI, 100VG-AnyLAN, 100BASE-T, and ATM.

- Shared memory – these use a common memory area (several megabytes) in which data is passed between the ports. It is very common in low-cost, small-scale switches and has the advantage that it can cope with different types of network, which are operating at different speeds. The main types of memory allocation are:

  - Pooled memory – memory is allocated as it is needed by the ports from a common memory pool.
  - Dedicated shared memory – memory is fixed and shared by a single pair of I/O ports.
  - Distributed memory – memory is fixed and dedicated to each port.

26.11 Switches and switching hubs

A switch is a very fast, low-latency, multiport bridge that is used to segment LANs. They are typically also used to increase communication rates between segments with multiple parallel conversations and also communication between technologies (such as between FDDI and 100BASE-TX).

A 4-port switching hub is a repeater that contains four distinct network segments (as if there were four hubs in one device). Through software, any of the ports on the hub can directly connect to any of the four segments at any time. This allows for a maximum capacity of 40 Mbps in a single hub.

Ethernet switches overcome the contention problem on normal CSMA/CD networks. They segment traffic by giving each connect a guaranteed bandwidth allocation. Figure 26.14 and Figure 26.15 show the two types of switches; their main features are:

- Desktop switch (or workgroup switch) – These connect directly to nodes. They are economical with fixed configurations for end-node connections and are designed for standalone networks or distributed workgroups in a larger network.
- Segment switch – These connect both 10 Mbps workgroup switches and 100 Mbps interconnect (backbone) switches that are used to interconnect hubs and desktop switches. They are modular, high-performance switches for interconnecting workgroups in mid- to large-size networks.

26.11.1 Segment switch

A segment switch allows simultaneous communication between any client and any server. A segment switch can simply replace existing Ethernet hubs. Figure 26.15 shows a switch with five ports each transmitting at 10 Mbps; this allows up to five simultaneous connections giving a maximum aggregated bandwidth of 50 Mbps. If the nodes support 100 Mbps communication then the maximum aggregated bandwidth will be 500 Mbps. To optimise the network, nodes should be connected to the switch that connects to the server with which it most often communicates. This allows for a direct connection with that server.

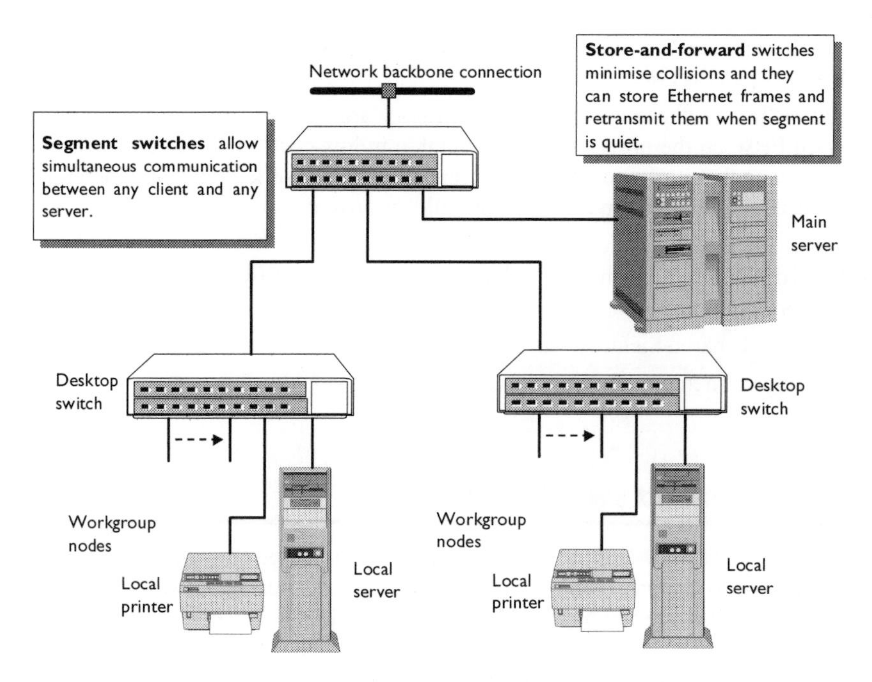

Figure 26.14 Desktop switch

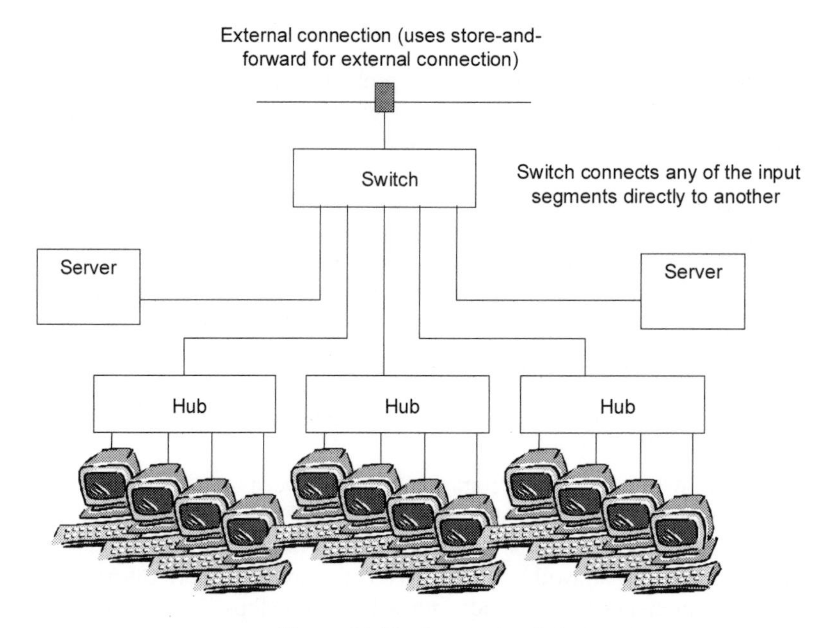

Figure 26.15 Segment switch

26.11.2 Desktop switch

A desktop switch can simply replace an existing 10BASET/100BASET hub. It has the advantage that any of the ports can connect directly to any other. In the network in Figure

26.14, any of the computers in the local workgroup can connect directly to any other, or to the printer, or the local disk drive. This type of switch works well if there is a lot of local traffic, typically between a local server and local peripherals.

26.11.3 Store-and-forward switching

Store-and-forwarding techniques have been used extensively in bridges and routers, and are now used with switches. It involves reading the entire Ethernet frame, before forwarding it, with the required protocol and at the correct speed, to the destination port. This has the advantages of:

- Improved error check – Bad frames are blocked from entering a network segment.
- Protocol filtering – Allows the switch to convert from one protocol to another.
- Speed matching – Typically, for Ethernet, reading at 10 Mbps or 100 Mbps and transmitting at 100 Mbps or 10 Mbps. Also, can be used for matching between ATM (155 Mbps), FDDI (100Mbps), Token Ring (4/16 Mbps) and Ethernet (10/100 Mbps).

The main disadvantage is:

- System delay – As the frame must be totally read before it is transmitted there is a delay in the transmission. The improvement in error checking normally overcomes this disadvantage.

26.12 Network interface card design

When receiving data, the network interface card (NIC) copies all data transmitted on the network, decodes it and transfers it to the computer. An Ethernet NIC contains three parts:

- Physical medium interface – the physical medium interface corresponds to the PLS and PMA in the standard and is responsible for the electrical transmission and reception of data. It consists of two parts: the transceiver, which receives and transmits data from or onto the transmission media; and a code converter that encodes/decodes the data. It also recognises a collision on the media.
- Data link controller – the controller corresponds to the MAC layer.
- Computer interface.

It can be split into four main functional blocks:

- Network interface.
- Manchester decoder.
- Memory buffer.
- Computer interface.

26.12.1 Network interface

The network interface must listen, recreate the waveform transmitted on the cable into a digital signal and transfer the digital signal to the Manchester decoder. The network interface consists of three parts:

- BNC/RJ-45 connector.
- Reception hardware – the reception hardware translates the waveforms transmitted on the cable to digital signals then copies them to the Manchester decoder.
- Isolator – the isolator is connected directly between the reception hardware and the rest of the Manchester decoder; it guarantees that no noise from the network affects the computer, and vice versa (as it isolates ground levels).

The reception hardware is called a receiver and is the main component in the network interface. It acts as an earphone, listening and copying the traffic on the cable. Unfortunately, the Ether and transceiver electronics are not perfect. The transmission line contains resistance and capacitance which distort the shape of the bit stream transmitted onto the Ether. Distortion in the system causes pulse spreading, which leads to intersymbol interference. There is also a possibility of noise affecting the digital pulse as it propagates through the cable. Therefore, the receiver also needs to recreate the digital signal and filter noise.

Figure 26.16 shows a block diagram of an Ethernet receiver. The received signal goes through a buffer with high input impedance and low capacitance to reduce the effects of loading on the coaxial cable. An equaliser passes high frequencies and attenuates low frequencies from the network, flattening the network passband. A 4-pole Bessel low-pass filter provides the average dc level from the received signal. The quench circuit activates the line driver only when it detects a true signal. This prevents noise activating the receiver.

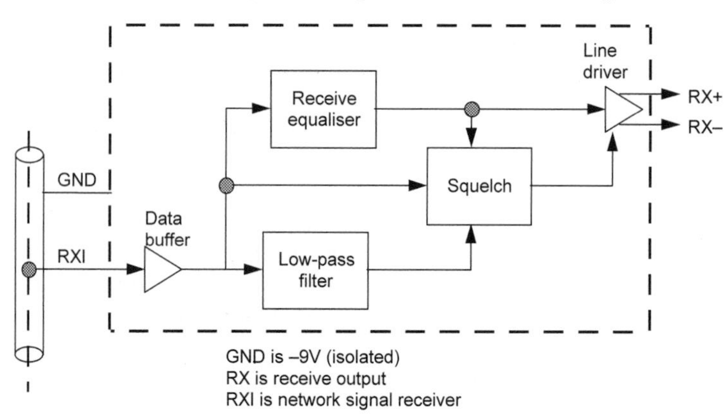

Figure 26.16 Ethernet receiver block diagram

26.12.2 Manchester decoder

Manchester coding has the advantage of embedding timing (clock) information within the transmitted bits. A positively edged pulse (low to high) represents a 1 and a negatively edged pulse (high to low) a 0, as shown in Figure 26.17. Another advantage of this coding method is that the average voltage is always zero when used with equal positive and negative voltage levels.

Figure 26.18 is an example of transmitted bits using Manchester encoding. The receiver passes the received Manchester-encoded bits through a low-pass filter. This extracts the lowest frequency in the received bit stream, i.e. the clock frequency. With this clock the receiver can then determine the transmitted bit pattern.

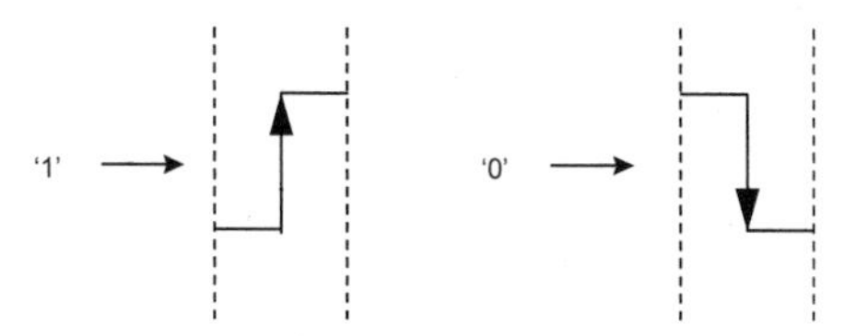

Figure 26.17 Manchester encoding

For Manchester decoding, the Manchester-encoded signal is first synchronised to the receiver (called bit synchronisation). A transition in the middle of each bit cell is used by a clock recovery circuit to produce a clock pulse in the center of the second half of the bit cell. In Ethernet the bit synchronisation is achieved by deriving the clock from the preamble field of the frame using a clock and data recovery circuit. Many Ethernet decoders use the SEEQ 8020 Manchester code converter, which uses a phase-locked loop (PLL) to recover the clock. The PLL is designed to lock onto the preamble of the incoming signal within 12-bit cells. Figure 26.19 shows a circuit schematic of bit synchronisation using Manchester decoding and a PLL.

The PLL is a feedback circuit which is commonly used for the synchronisation of digital signals. It consists of a phase detector (such as an EXOR gate) and a voltage-controlled oscillator (VCO) which uses a crystal oscillator as a clock source.

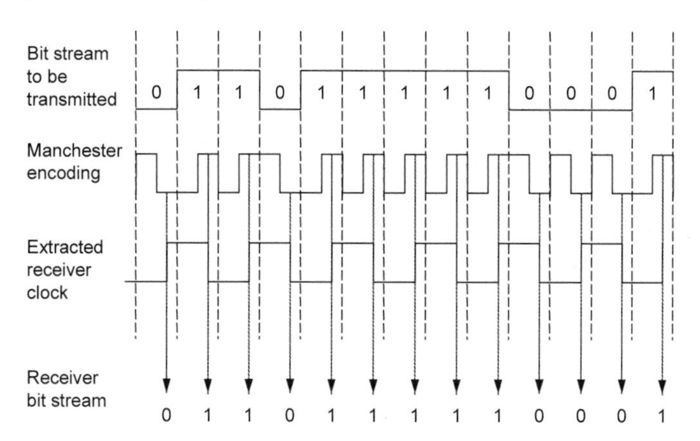

Figure 26.18 Example of Manchester coding

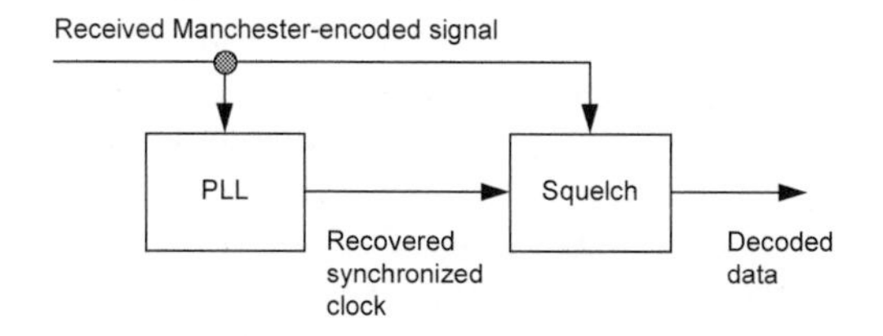

Figure 26.19 Manchester decoding with bit synchronization

The frequency of the crystal is twice the frequency of the received signal. It is so constant that it only needs irregular and small adjustments to be synchronised to the received signal. The function of the phase detector is to find irregularities between the two signals and adjust the VCO to minimise the error. This is accomplished by comparing the received signals and the output from the VCO. When the signals have the same frequency and phase the PLL is locked. Figure 26.20 shows the PLL components and the function of the EXOR.

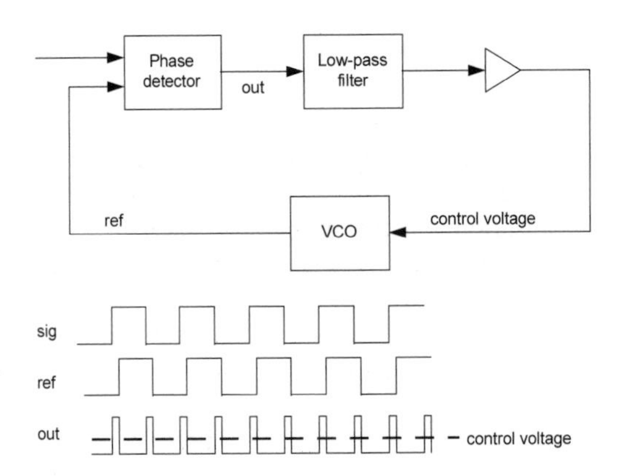

Figure 26.20 PLL and example waveform for the phase detector

26.12.3 Memory buffer

The rate at which data is transmitted on the cable differs from the data rate used by the receiving computer, and the data appears in bursts. To compensate for the difference between the data rate, a first-in first-out (FIFO) memory buffer is used to produce a constant data rate. An important condition is that the average data input rate should not exceed the frequency of the output clock; if this is not the case the buffer will be filled up regardless of its size.

A FIFO is a RAM that uses a queuing technique where the output data appears in the same order that it went in. The input and output are controlled by separate clocks, and the FIFO keeps track of the data that has been written and the data that has been read and can thus be overwritten. This is achieved with a pointer. Figure 26.21 shows a block diagram of the FIFO configuration. The FIFO status is indicated by flags, the empty flag (EF) and the full flag (FF), which show whether the FIFO is either empty or full.

26.12.4 Ethernet implementation

The completed circuit for the Ethernet receiver is given in Section 26.15 and is outlined in Figure 26.22. It uses the SEEQ Technologies 82C93A Ethernet transceiver as the receiver and the SEEQ 8020 Manchester code converter which decodes the Manchester code. A transformer and a dc-to-dc converter isolate the SEEQ 82C92A and the network cable from the rest of the circuit (and the computer). The isolated dc-to-dc converter converts a 5 V supply to the -9 V needed by the transceiver.

The memory buffer used is the AMD AM7204 FIFO which has 4096 data words with 9-bit words (but only eight bits are actually used). The output of the circuit is eight data lines, the control lines $\overline{FF}$, $\overline{EF}$, $\overline{RS}$, $\overline{R}$ and $\overline{W}$, and the +5 V and GND supply rails.

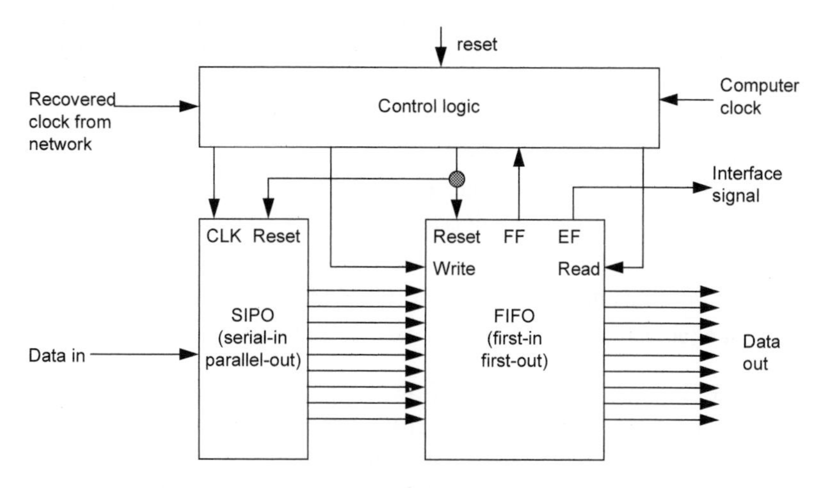

Figure 26.21 Memory buffering

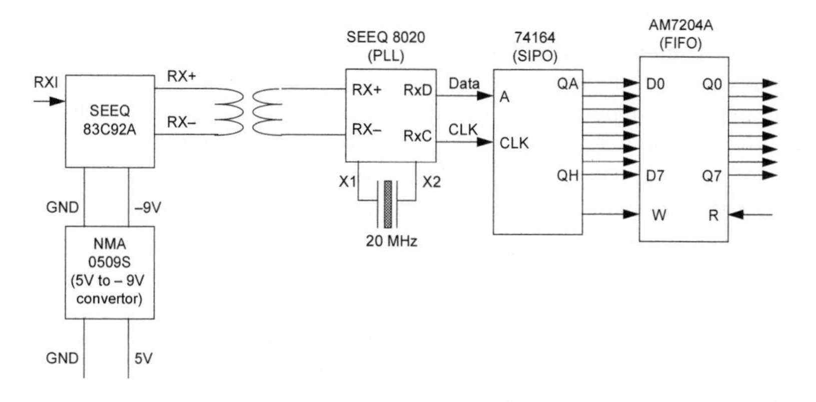

Figure 26.22 Ethernet receiver

26.13 Gigabit Ethernet

The IEEE 802.3 working group initiated the 802.3z gigabit Ethernet task force to create the gigabit Ethernet standard (which was finally defined in 1998). The Gigabit Ethernet Alliance (GEA) was founded in May 1996 and promotes gigabit Ethernet collaboration between organisations. Companies, which were initially involved in the GEA include: 3Com, Bay Networks, Cisco Systems, Compaq, Intel, LSI Logic, Sun and VLSI.

The amount of available bandwidth for a single segment is massive. For example, almost 125 million characters (125 MB) can be sent in a single second. A large reference book with over 1000 pages could be send over a network, 10 times in a single second. Compare it also with a ×24, CD-ROM drive which transmits at a maximum rate of 3.6 MB/s (24×150 kB/s). Gigabit Ethernet operates almost 35 times faster than this drive. With network switches, this bandwidth can be multiplied by a given factor, as they allow multiple simultaneous connections.

Gigabit Ethernet is an excellent challenger for network backbones as it interconnects 10/100BASE-T switches, and also provides high-bandwidth to high-performance servers. Initial aims were:

- Half/full-duplex operation at 1000 Mbps.
- Standard 802.3 Ethernet frame format. Gigabit Ethernet uses the same variable-length frame (64–1514-byte packets), and thus allows for easy upgrades.
- Standard CSMA/CD access method.
- Compatibility with existing 10BASE-T and 100BASE-T technologies.
- Development of an optional gigabit media independent interface (GMII).

The compatibility with existing 10/100BASE standards make the upgrading to Gigabit Ethernet much easier, and considerably less risky than changing to other networking types, such as FDDI and ATM. It will happily interconnect with, and autosense, existing slower rated Ethernet devices. Figure 26.23 illustrates the functional elements of Gigabit Ethernet. Its main characteristics are:

- **Full-duplex communication**. As defined by the IEEE 802.3x specification, two nodes connected via a full-duplex, switched path can simultaneously send and receive frames. Gigabit Ethernet supports new full-duplex operating modes for switch-to-switch and switch-to-end-station connections, and half-duplex operating modes for shared connections using repeaters and the CSMA/CD access method.
- **Standard flow control**. Gigabit Ethernet uses standard Ethernet flow control to avoid congestion and overloading. When operating in half-duplex mode, gigabit Ethernet adopts the same fundamental CSMA/CD access method to resolve contention for the shared media.
- **Enhanced CSMA/CD method**. This maintains a 200 m collision diameter at gigabit speeds. Without this, small Ethernet packets could complete their transmission before the transmitting node could sense a collision, thereby violating the CSMA/CD method. To resolve this issue, both the minimum CSMA/CD carrier time and the Ethernet slot time (the time, measured in bits, required for a node to detect a collision) have been extended from 64 bytes (which is 51.2 μs for 10BASE and 5.12 μs for 100BASE) to 512 bytes (which is 4.1 μs for 1000BASE). The minimum frame length is still 64 bytes. Thus, frames smaller than 512 bytes have a new carrier extension field following the CRC field. Packets larger than 512 bytes are not extended.
- **Packet bursting**. The slot time changes affect the small-packet performance, but this has been offset by a new enhancement to the CSMA/CD algorithm, called packet bursting. This allows servers, switches and other devices to send bursts of small packets in order to fully utilize the bandwidth.

Devices operating in full-duplex mode (such as switches and buffered distributors) are not subject to the carrier extension, slot time extension or packet bursting changes. Full-duplex devices use the regular Ethernet 96-bit interframe gap (IFG) and 64-byte minimum frame size.

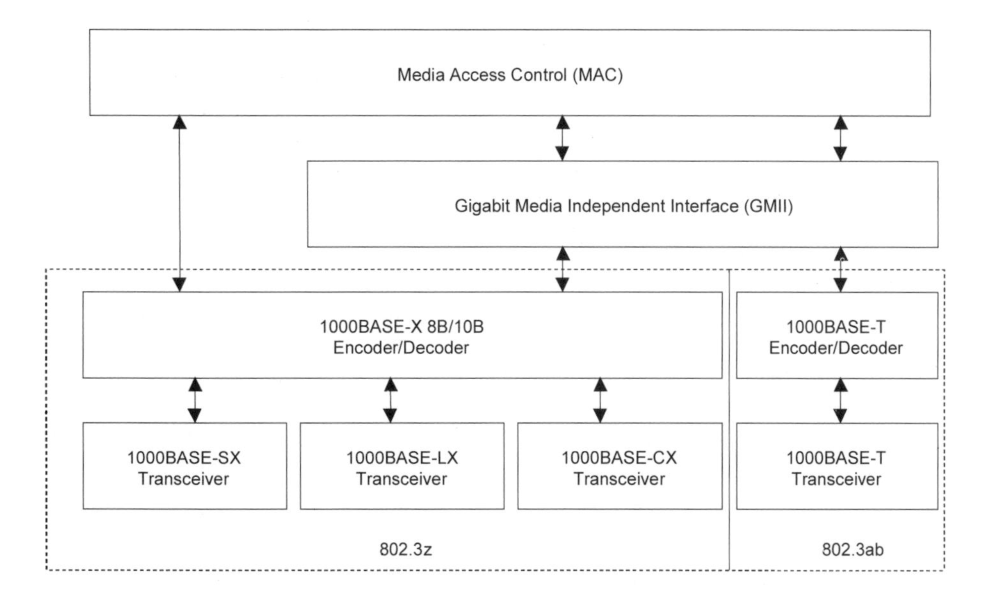

Figure 26.23 Gigabit Ethernet functional elements

26.13.1 Ethernet transceiver

The IEEE 802.3z task force spent much of their time defining the gigabit Ethernet standard for the transceiver (physical layer), which is responsible for the mechanical, electrical and procedural characteristics for establishing, maintaining and deactivating the physical link between network devices. The physical layers are:

- **1000BASE-SX** (low cost, multimode fibre cables). These can be used for short interconnections and short backbone networks. The IEEE 802.3z task force have tried to integrate the new standard with existing cabling, whether it be twisted-pair cable, coaxial cable or fibre optic cable. These tests involved firing lasers in long lengths of multimode fibre cables. Through these tests it was found that a jitter component results which is caused by a phenomenon known as differential mode delay (DMD). The 1000BASE-SX standard has resolved this by defining the launch of the laser signal, and enhanced conformance tests. Typical maximum lengths are: 62.5 μm, multimode fibre (up to 220 m) and 50 μm, multi-mode fibre (550 m).
- **1000BASE-LX** (multimode/singlemode fibre cables). These can be used for longer runs, such as on backbones and campus networks. Single-mode fibres are covered by the long-wavelength standard, and provide for greater distances. External patch cords are used to reduce DMD. Typical lengths are: 62.5 μm, multimode fibre (up to 550 m); 50 μm, multimode fibre (up to 550 m) and 50 μm, single-mode fibre (up to 5 km).
- **1000BASE-CX** (shielded balanced copper). This standard supports interconnection of equipment using a copper-based cable, typically up to 25 m. As with the 1000BASE-LX/SX standards, it uses the Fibre channel-based 8B/10B coding to give a serial line rate of 1.25 Gbps. The 1000BASE-T is likely to supersede this standard, but it has been relatively easy to define, and to implement.
- **1000BASE-T** (UTP). This is a useful standard for connecting directly to workstations.

The 802.3ab Task Force has been assigned the task of defining the 1000BASE-T physical layer standard for gigabit Ethernet over four pairs of Cat-5 UTP cable, for cable distances of up to 100 m, or networks with a diameter of 200 m. As it can be used with existing cabling, it allows easy upgrades. Unfortunately, it requires new technology and new coding schemes in order to meet the potentially difficult and demanding parameters set by the previous Ethernet and fast Ethernet standards.

26.13.2 Fibre Channel components

The IEEE 802.3 committee based much of the physical layer technology on the ANSI-backed X3.230 Fibre channel project. This allowed many manufacturers to re-use physical-layer Fibre channel components for new gigabit Ethernet designs, and has allowed a faster development time than is normal, and increased the volume production of the components. These include optical components and high-speed 8B/10B encoders.

The 1000BASE-T standard uses enhanced DSP (digital signal processing) and enhanced silicon technology to enable gigabit Ethernet over UTP cabling. As Figure 26.23 shows, it does not use the 8B/10B encoding.

26.13.3 Buffered distributors

Along with repeaters, bridges and switches, a new device called a buffered distributor (or full-duplex repeater) has been developed for gigabit Ethernet. It is a full-duplex, multiport, hub-like device that connects two or more gigabit Ethernet segments. Unlike a bridge, and like a repeater, it forwards all the Ethernet frames from one segment to the others, but unlike a standard repeater, a buffered distributor buffers one, or more, incoming frames on each link before forwarding them. This reduces collisions on connected segments. The maximum bandwidth for a buffered distributor will still only be 1 Gbps, as opposed to gigabit switches which allow multi-gigabit bandwidths.

26.13.4 Quality of service

Many, real-time, networked applications require a given quality of server (QoS), which might relate to bandwidth requirements, latency (network delays) and jitter. Unfortunately, there is nothing built into Ethernet that allows for a QoS, thus new techniques have been developed to overcome this. These include:

- RSVP – allows nodes to request and guarantee a QoS, and works at a higher level to Ethernet. For this, each network component in the chain must support RSVP and communicate appropriately. Unfortunately, this may require an extensive investment to totally support RSVP, thus many vendors have responded in implementing proprietary schemes, which may make parts of the network vendor-specific.
- IEEE 802.1p and IEEE 802.1Q – allows a QoS over Ethernet by tagging packets with an indication of the priority or class of service desired for the frames. These tags allow applications to communicate the priority of frames to internetworking devices. RSVP support can be achieved by mapping RSVP sessions into 802.1p service classes.
- Routing – implemented at a higher layer.

26.13.5 Gigabit Ethernet migration

The greatest advantage of gigabit Ethernet is that it is easy to upgrade existing Ethernet-based networks to higher bit rates. Typical migration might be:

- Switch-to-switch links – involves upgrading the connections between switches to 1 Gbps. As 1000BASE switches support both 100BASE and 1000BASE then not all the switches require to be upgraded at the same time; this allows for gradual migration.

- Switch-to-server links – involves upgrading the connection between a switch and the server to 1Gbps. The server requires an upgraded gigabit Ethernet interface card.
- Switched fast Ethernet backbone – involves upgrading a fast Ethernet backbone switch to a 100/1000BASE switch. It thus supports both 100BASE and 1000BASE switching, using existing cabling.
- Shared FDDI backbone – involves replacing FDDI attachments on the ring with gigabit Ethernet switches or repeaters. Gigabit Ethernet uses the existing fibre-optic cable, and provides a greatly increased segment bandwidth.
- Upgrade NICs on nodes to 1 Gbps. It is unlikely that users will require 1 Gbps connections, but this facility is possible.

26.13.6 1000BASE-T

One of the greatest challenges of gigabit Ethernet is to use existing Cat-5 cables, as this will allow fast upgrades. Two critical parameters, which are negligible at 10BASE speeds, are:

- Return loss – defines the amount of signal energy that is reflected back towards the transmitter due to impedance mismatches in the link (typically from connector and cable bends).
- Far-end crosstalk – noise that is leaked from another cable pair.

The 1000BASE-T task force estimates that less than 10% of the existing Cat-5 cable were improperly installed (as defined in ANSI/TIA/EIA568-A in 1995) and might not support 1000BASE-T (or even, 100BASE-TX). 100BASE-T uses two pairs, one for transmit and one for receive, and transmits at a symbol rate of 125 Mbaud with a 3-level code. 1000BASE-T uses:

- All four pairs with a symbol rate of 125 Mbaud (symbols/per second). One symbol contains two bits of information.
- Each transmitted pulse uses a five-level PAM (pulse amplitude modulation) line code, which allows two bits to be transmitted at a time.
- Simultaneous send and receive on each pair. Each connection uses a hybrid circuit to split the send and receive signals.
- Pulse shaping matches the characteristics of the transmitted signal to the channel so that the signal-to-noise ratio is minimised. It effectively reduces low frequency terms (which contain little data information, can cause distortion and cannot be passed over the transformer-coupled hybrid circuit), reduces high frequency terms (which increases crosstalk) and rejects any external high-frequency noise. It is thought that the transmitted signal spectrum for 1000BASE will be similar to 100BASE.
- Forward error correction (FEC) provides a second level of coding that helps to recover the transmitted symbols in the presence of high noise and crosstalk. The FEC bit uses the fifth level of the five-level PAM.

A five-level code (–2, –1, 0, +1, +2) allows two bits to be sent at a time, if all four pairs are used then eight bits are sent at a time. If each pair transmits at a rate of 125Mbaud (symbols/sec), the resulting bit rate will be 1 Gbps.

26.14 Exercises

26.14.1 The base bit rate of standard Ethernet is:

(a) 1 kbps (b) 1 Mbps (c) 10 Mbps (d) 100 Mbps

26.14.2 The base bit rate of fast Ethernet is:

(a) 1 kbps (b) 1 Mbps (c) 10 Mbps (d) 100 Mbps

26.14.3 Standard Ethernet (thick-wire Ethernet) is also known as:

(a) 10BASE2 (b) 10BASE5
(c) 10BASE-T (d) 10BASE-FL

26.14.4 Thin-wire Ethernet (Cheapernet) is also known as:

(a) 10BASE2 (b) 10BASE5
(c) 10BASE-T (d) 10BASE-FL

26.14.5 Standard Ethernet (thick-wire Ethernet) uses which type of cable:

(a) Twisted-pair cable (b) Coaxial cable
(c) Fibre optic cable (d) Radio link

26.14.6 Thin-wire Ethernet (Cheapernet) uses which type of cable:

(a) Twisted-pair cable (b) Coaxial cable
(c) Fibre optic cable (d) Radio link

26.14.7 The IEEE standard for Ethernet is:

(a) IEEE 802.1 (b) IEEE 802.2
(c) IEEE 802.3 (d) IEEE 802.4

26.14.8 The main disadvantage of Ethernet is that:

(a) Computers must contend for the network. (b) It does not network well.
(c) It is unreliable. (d) It is not secure.

26.14.9 A MAC address has how many bits:

(a) 8 bits (b) 24 bits
(c) 32 bits (d) 48 bits

26.14.10 Which bit pattern identifies the start of an Ethernet frame:

(a) 11001100…1100 (b) 00000000…0000
(c) 11111111…1111 (d) 10101010…1010

26.14.11 The main standards relating to Ethernet networks are:

(a)	IEEE 802.2 and IEEE 802.3	(b)	IEEE 802.3 and IEEE 802.4
(c)	ANSI X3T9.5 and IEEE 802.5	(d)	EIA RS-422 and IEEE 802.3

26.14.12 Which layer in the Ethernet standard communicates with the OSI network layer:

(a)	the MAC layer	(b)	the LLC layer
(c)	the physical layer	(d)	the protocol layer

26.14.13 Standard, or thick-wire, Ethernet is also known as:

(a)	10BASE2	(b)	10BASE5
(c)	10BASE-T	(d)	10BASE-F

26.14.14 Twisted-pair Ethernet is also known as:

(a)	10BASE2	(b)	10BASE5
(c)	10BASE-T	(d)	10BASE-FL

26.14.15 Fiber optic Ethernet is also known as:

(a)	10BASE2	(b)	10BASE5
(c)	10BASE-T	(d)	10BASE-F

26.14.16 Which type of connector does twisted-pair Ethernet use when connecting to a network hub:

(a)	N-type	(b)	BNC
(c)	RJ-45	(d)	SMA

26.14.17 Which type of connector does Cheapernet, or thin-wire Ethernet, use when connecting to the network backbone:

(a)	N-type	(b)	BNC
(c)	RJ-45	(d)	SMA

26.14.18 What is the function of a repeater in an Ethernet network:

(a)	It increases the bit rate	(b)	It isolates network segments
(c)	It prevents collisions	(d)	It boosts the electrical signal

26.14.19 Discuss the limitations of 10BASE5 and 10BASE2 Ethernet.

26.14.20 Discuss the main reasons for the preamble in an Ethernet frame.

26.14.21 Discuss 100 Mbps Ethernet technologies with respect to how they operate and their typical parameters.

26.14.22 Explain the usage of Ethernet SNAP.

26.14.23 State the main advantage of Manchester coding and show the line code for the bit sequence

01111010101101010001011010

26.14.24 Explain the main functional differences between 100BASE-T, 100BASE-4T and 100VG-AnyLAN.

26.14.25 Prove that the maximum length of segment that can be used with 10 Mbps Ethernet is 840 metres. Assume that the propagation speed is 1.5×10^8 m/s and the length of the preamble is 56 bits. Note, a collision must be detected by the end of the transmission of the preamble. Also, why might the maximum length of the segment be less than this?

26.14.26 What problem might be encountered with fast Ethernet, with respect to the maximum segment length?

26.15 Ethernet crossover connections

The standard connections for 10BASE and 100BASE are

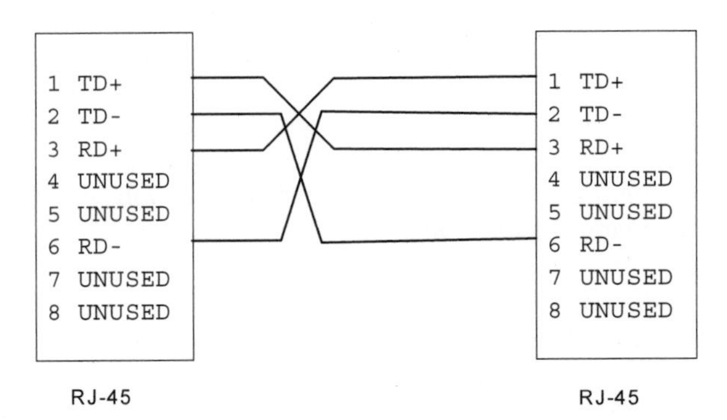

where RD is the receive signals (this is known as RECEIVE in 100BASE) and TD the transmit signals (TRANSMIT). These cable connections are difficult to set-up and most connections use a straight through connection (as given in Table 26.3). Ports which have the cross-over connection internal in the port are marked with an "X".

The standard connections for 100BASE-T4 is given next:

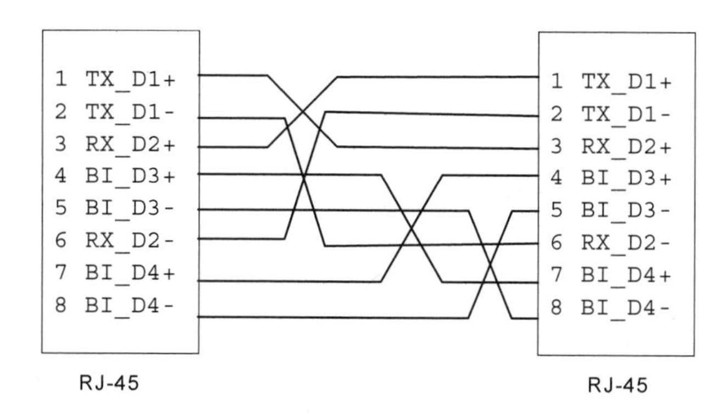

where BI represents the bi-directional transmission signals, TX the transmit signals and RX the receive signals. These cable connections are difficult to set-up and most connections use a straight through connection (as given in Table 26.3). Ports which have the cross-over connection internal in the port are marked with an "X".

26.16 Notes from the author

Until recently, it seemed unlikely that Ethernet would survive as a provider of network backbones and for campus networks, and its domain would stay, in the short-term, with connections to local computers. The world seemed distended for the global domination of ATM, the true integrator of real-time and non real-time data. This was due to Ethernet's lack of support for real-time traffic and that it does not cope well with traffic rates that approach the maximum bandwidth of a segment (as the number of collisions increases with the amount of traffic on a segment). ATM seemed to be the logical choice as it analyses the type of data being transmitted and reserves a route for the given quality of service. It looked as if ATM would migrate down from large-scale networks to the connection of computers, telephones, and all types analogue/digital communications equipment. But, remember, not always the best technological solution wins the battle for the market – a specialist is normally always trumped by a good all-rounder.

Ethernet also does not provide for quality of service and requires other higher-level protocols, such as IEEE 802.1p. These disadvantages are often outweighed by its simplicity, its upgradeability, its reliability and its compatibility. One way to overcome the contention problem is to provide a large enough bandwidth so that the network is not swamped by sources which burst data onto the network. For this, the gigabit Ethernet standard is likely to be the best solution for most networks.

RS-232 Programming using Visual Basic

27.1 Introduction

This chapter discusses how Visual Basic can be used to access serial communication functions. Windows hides much of the complexity of serial communications and automatically puts any received characters in a receive buffer and characters sent into a transmission buffer. The receive buffer can be read by the program whenever it has time and the transmit buffer is emptied when it is free to send characters.

27.2 Properties

The Comm component is added to a form whenever serial communications are required (as shown in left-hand side of Figure 27.1). The right-hand side of Figure 27.1 shows its properties. By default, the first created object is named MSComm1 (the second is named MSComm2, and so on). It can be seen that the main properties of the object are: CommPort, DTREnable, EOFEnable, Handshaking, InBufferSize, Index, InputLen, InputMode, Left, Name, NullDiscard, OutBufferSize, ParityReplace, RThreshold, RTSEnable, Settings, SThreshold, Tag and Top. The main properties are defined in Table 27.1.

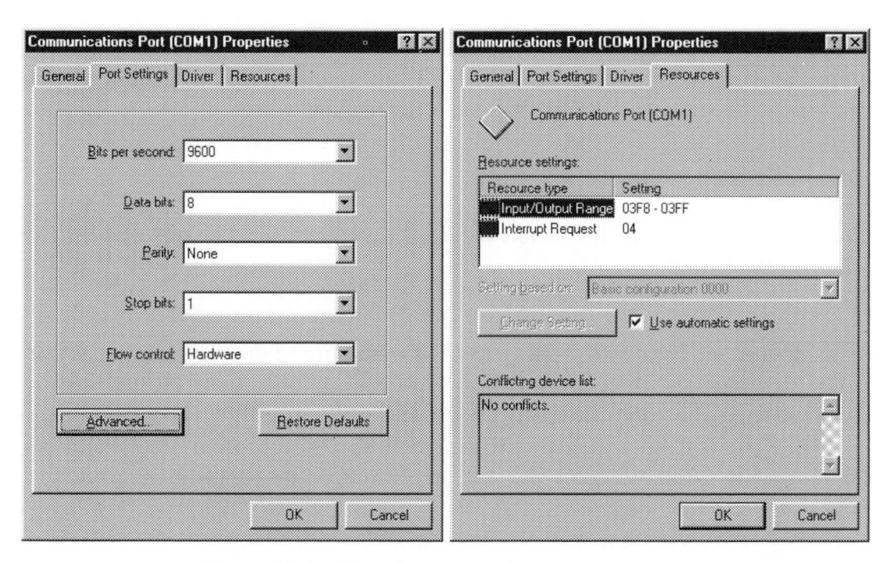

Figure 27.1 Changing port setting and parameters

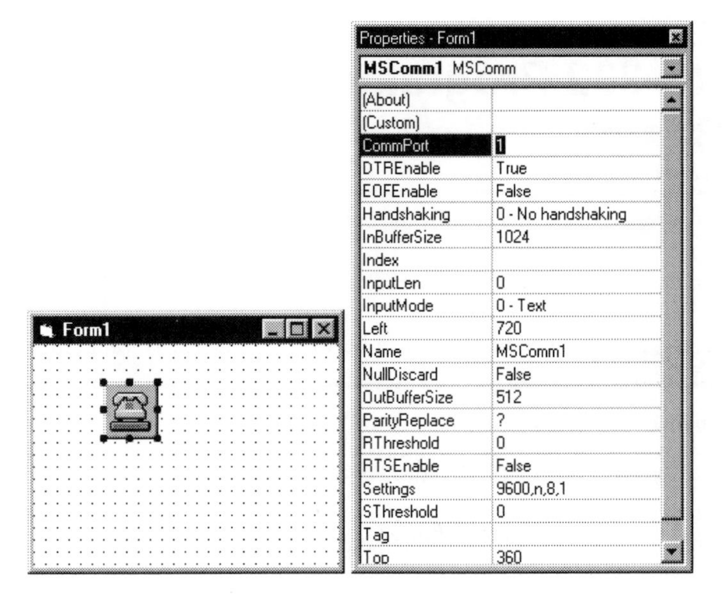

Figure 27.2 Communications control and MS Comm properties

Table 27.1 The main communications control properties

Properties	*Description*
CommPort	Sets and returns the communications port number.
Input	Returns and removes characters from the receive buffer.
Output	Writes a string of characters to the transmit buffer.
PortOpen	Opens and closes a port, and gets port settings
Settings	Sets and returns port parameters, such as bit rate, parity, number of data bits and so on.

27.2.1 Settings

The Settings property sets and returns the RS-232 parameters, such as baud rate, parity, the number of data bit, and the number of stop bits. Its syntax is

[*form.*]*MSComm*.`Settings`[= *setStr$*]

where the strStr is a string which contains the RS-232 settings. This string takes the form:

```
"BBBB,P,D,S"
```

where `BBBB` defines the baud rate, `P` the parity, `D` the number of data bits, and `S` the number of stop bits.

The following are the valid baud rates (default is 9600 baud):

110, 300, 600, 1200, 2400, 9600, 14 400, 19 200, 38 400, 56 000, 128 000, 256 000

The valid parity values are (default is N): E (even), M (mark), N (none), O (odd), S (space).
The valid data bit values are (default is 8): 4, 5, 6, 7 or 8.
The valid stop bit values are (default is 1). 1, 1.5 or 2.

An example of setting a control port to 4800 Baud, even parity, seven data bits and one stop bit is

```
Com1.Settings = "4800,E,7,1"
```

27.2.2 CommPort

The CommPort property sets and returns the communication port number. Its syntax is

[*form.*]*MSComm*.`CommPort`[= *portNumber%*]

which defines the portNumber from a value between 1 and 99. A value of 68 is returned if the port does not exist.

27.2.3 PortOpen

The PortOpen property sets and returns the state of the communications port. Its syntax is

[*form.*]*MSComm*.`PortOpen`[= {*True* | *False*}]

A True setting opens the port, whereas a False closes the port and clears the receive and transmit buffers (this automatically happens when an application is closed).

The following example opens communications port number 1 (COM1:) at 4800 baud with even parity, seven data bits and one stop bit:

```
Com1.Settings = "4800,E,7,1"
Com1.CommPort = 1
Com1.PortOpen = True
```

27.2.4 Inputting data

The three main properties used to read data from the receive buffer are Input, InBufferCount and InBufferSize.

Input

The Input property returns and removes a string of characters from the receive buffer. Its syntax is

[*form.*]*MSComm*.`Input`

To determine the number of characters in the buffer the InBufferCount property is tested (to be covered in the next section). Setting InputLen to 0 causes the Input property to read the entire contents of the receive buffer.

Program 27.1 shows an example of how to read data from the receiver buffer.

Program 27.1

```
' Check for characters in the buffer
If Com1.InBufferCount Then
      ' Read data in the buffer
      InStr$ = Com1.Input
End If
```

InBufferSize

The InBufferSize property sets and returns the maximum number of characters that can be received in the receive buffer (by default it is 1024 bytes). Its syntax is

[*form*.]*MSComm*.`InBufferSize`[= *numBytes*%]

The size of the buffer should be set so that it can store the maximum number of characters that will be received before the application program can read them from the buffer.

InBufferCount

The InBufferCount property returns the number of characters in the receive buffer. It can also be used to clear the buffer by setting the number of characters to 0. Its syntax is

[*form*.]*MSComm*.`InBufferCount`[= *count*%]

27.1.2 Outputting data

The three main properties used to write data to the transmit buffer are Output, OutBuffer-Count and OutBufferSize.

Output

The Output property writes a string of characters to the transmit buffer. Its syntax is

[*form*.]*MSComm*.`Output`[= *outString*$]

Program 27.2 uses the KeyPress event on a form to send the character to the serial port.

Program 27.2

```
Private Sub Form_KeyPress (KeyAscii As Integer)
      if (Com1.OutBufferCount < Com1.OutBufferSize)
         Com1.Output = Chr$(KeyAscii)
End Sub
```

OutBufferSize

The OutBufferSize property sets and returns the number of characters in the transmit buffer (default size is 512 characters). Its syntax is

[*form*.]*MSComm*.`OutBufferSize`[= *NumBytes*%]

OutBufferCount

The OutBufferCount property returns the number of characters in the transmit buffer. The transmit buffer can also be cleared by setting it to 0. Its syntax is

[*form*.]*MSComm*.`OutBufferCount`[= *0*]

27.1.3 Other properties

- Break – sets or clears the break signal. A True sets the break signal, whereas a False clears the break signal. When True, character transmission is suspended and a break level is set on the line. This continues until Break is set to False. Its syntax is

[form.]MSComm.`Break`[= {*True* | *False*}]

- CDTimeout – sets and returns the maximum amount of time that the control waits for a carried detect (CD) signal, in milliseconds, before a timeout. Its syntax is

  [form.]MSComm.`CDTimeout`[= *milliseconds&*]

- CTSHolding – determines whether the CTS line should be detected. CTS is typically used for hardware handshaking. Its syntax is

  [form.]MSComm.`CTSHolding`[= {*True* | *False*}]

- DSRHolding – determines the DSR line state. DSR is typically used to indicate the presence of a modem. If is a True then the DSR line is high, else it is low. Its syntax is

  [form.]MSComm.`DSRHolding`[= *setting*]

- DSRTimeout – sets and returns the number of milliseconds to wait for the DSR signal before an OnComm event occurs. Its syntax is

  [form.]MSComm.`DSRTimeout`[= *milliseconds&*]

- DTEEnable – determines whether the DTR signal is enabled. It is typically send from the computer to the modem to indicate that it is ready to receive data. A True setting enables the DTR line (output level high). It syntax is

  [form.]MSComm.`DTREnable`[= {*True* | *False*}]

- RTSEnable – determines whether the RTS signal is enabled. Normally used to handshake incoming data and is controlled by the computer. Its syntax is:

  [form.]MSComm.`RTSEnable`[= {*True* | *False*}]

- NullDiscard – determines whether null characters are read into the receive buffer. A True setting does not transfer the characters. Its syntax is

  [form.]MSComm.`NullDiscard`[= {*True* | *False*}]

- SThreshold – sets and returns the minimum number of characters allowable in the transmit buffer before the OnComm event. A 0 value disables generating the OnComm event for all transmission events, whereas a value of 1 causes the OnComm event to be called when the transmit buffer is empty. Its syntax is

  [form.]MSComm.`SThreshold`[= *numChars%*]

- Handshaking – sets and returns the handshaking protocol. It can be set to no handshaking, hardware handshaking (using RTS/CTS) or software handshaking (XON/XOFF).

Valid settings are given in Table 27.2. Its syntax is

[*form*.]*MSComm*.`Handshaking`[= *protocol%*]

- CommEvent – returns the most recent error message. Its syntax is

[*form*.]*MSComm*.`CommEvent`

Table 27.2 Settings for handshaking

Setting	*Value*	*Description*
`comNone`	0	no handshaking (default)
`comXOnXOff`	1	XON/XOFF handshaking
`comRTS`	2	RTS/CTS handshaking
`comRTSXOnXOff`	3	RTS/CTS and XON/XOFF handshaking

When a serial communication event (OnComm) occurs then the event (error or change) can be determined by testing the CommEvent property. Table 27.3 lists the error values and Table 27.4 lists the communications events.

Table 27.3 CommEvent property

Setting	*Value*	*Description*
`comEventBreak`	1001	Break signal received.
`comEventCTSTO`	1002	CTS timeout – occurs when transmitting a character and CTS was low for CTSTimeout milliseconds.
`comEventDSRTO`	1003	DSR timeout – occurs when transmitting a character and DTR was low for DTRTimeout milliseconds.
`comEventFrame`	1004	Framing error.
`comEventOverrun`	1006	Port overrun – the receive buffer is full and another character was written into the buffer, overwriting the previously received character.
`comEventCDTO`	1007	CD timeout – occurs when CD is low for CD timeout milliseconds, when transmitting a character.
`comEventRxOver`	1008	Receive buffer overflow.
`comEventRxParity`	1009	Parity error.
`comEventTxFull`	1010	Transmit buffer full.
`ComEventDCB`	1011	Unexpected error retrieving device control block (DCB) for the port.

Table 27.4 Communications events

Setting	*Value*	*Description*
`comEvSend`	1	Character has been sent.
`comEvReceive`	2	Character has been received.
`comEvCTS`	3	Change in CTS line.
`comEvDSR`	4	Change in DSR line from a high to a low.
`comEvCD`	5	Change in CD line.
`comEvRing`	6	Ring detected.
`comEvEOF`	7	EOF character received.

27.3 Events

The communication control generates an event (OnComm) when the value CommEvent property changes its value, and the CommEvent property is then tested to determine the source of the event. Figure 27.3 shows the event subroutine and Program 27.3 shows an example event routine which tests the CommEvent property. It also shows the property window, which is shown with a right click on the comms component.

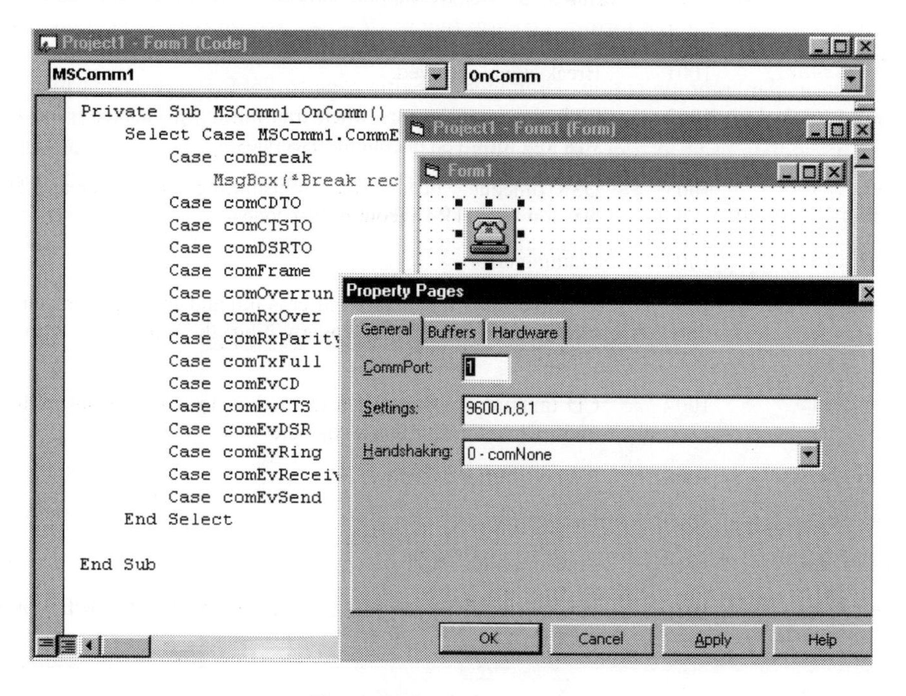

Figure 27.3 OnComm event

Program 27.3

```
Private Sub MSComm_OnComm ()
      Select Case MSComm1.CommEvent
         Case comEventBreak          ' A Break was received.
                                     MsgBox("Break received")
         Case comEventCDTO           ' CD (RLSD) Timeout.
         Case comEventCTSTO          ' CTS Timeout.
         Case comEventDSRTO          ' DSR Timeout.
         Case comEventFrame          ' Framing Error
         Case comEventOverrun        ' Data Lost.
         Case comEventRxOver         ' Receive buffer overflow.
         Case comEventRxParity       ' Parity Error.
         Case comEventTxFull         ' Transmit buffer full.
         Case comEventCD             ' Change in the CD.
         Case comEventCTS            ' Change in the CTS.
         Case comEventDSR            ' Change in the DSR.
         Case comEventRing           ' Change in the RI.
         Case comEventReceive
         Case comEventSend
      End Select
End Sub
```

27.4 Example program

Program 27.4 shows a simple transmit/receive program which uses COM1: to transmit and receive. A loopback connection which connects the transmit line to the receive line can be used to test the communications port. All the characters that are transmitted should be automatically received. A sample form is given in Figure 27.4.

The loading of the form (Form_Load) is called when the program is initially run. This is used to set-up the communication parameters (in this case to 9600 Baud, no parity, 8 data bits and 1 stop bit). When the user presses a key on the form the Form_Keypress event is called. This is then used to transmit the entered character and display it to the Transmit text window (Textl). When a character is received the OnComm event is called and the MSComml. CommEvent is set to 2 (comEvReceive) which identifies that a character has been received. This character is then displayed to the Receive text window (Text2). Figure 27.5 shows a sample run.

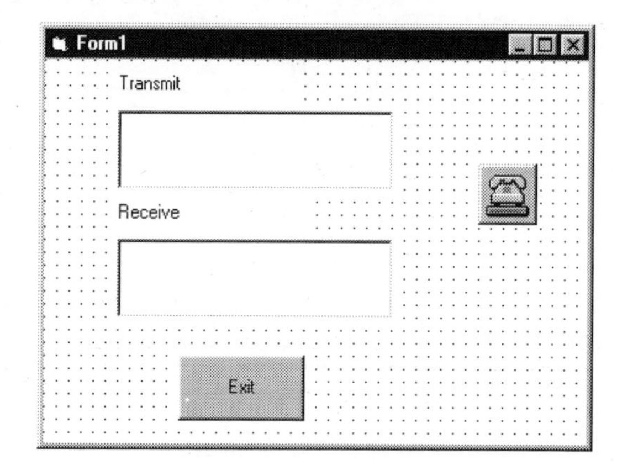

Figure 27.4 Simple serial communications transmit/receive form

Program 27.4

```
Private Sub Form_Load()

  MSComm1.CommPort = 1                ' Use COM1.
  MSComm1.Settings = "9600,N,8,1"     ' 9600 baud, no parity, 8 data,
                                      '  and 1 stop bit.
  MSComm1.InputLen = 0                ' Read entire buffer when Input
                                      ' is used
  MSComm1.PortOpen = True             ' Open port
End Sub

Private Sub Form_KeyPress(KeyAscii As Integer)
    MSComm1.Output = KeyAscii
    Text1.Text = KeyAscii
End Sub

Private Sub MSComm1_OnComm()
    If (MSComm1.CommEvent = comEvReceive) Then
        Text2.Text = MSComm1.Input
    End If
End Sub

Private Sub Command1_Click()
    End
End Sub
```

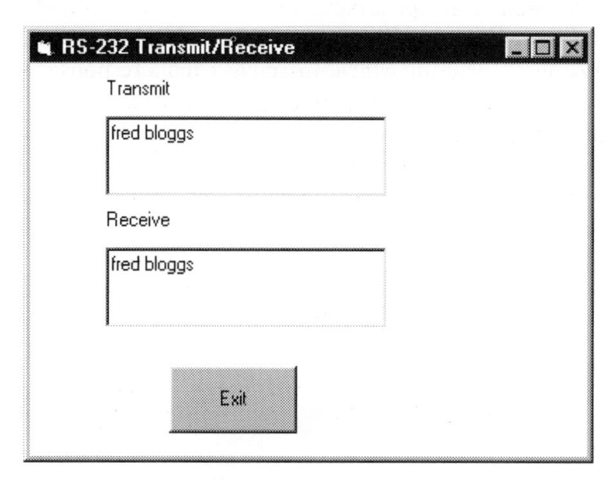

Figure 27.5 Sample run

27.5 Error messages

Table 27.5 identifies the run-time errors that can occur with the Communications control.

Table 27.5 Error messages

Error number	*Message explanation*	*Error number*	*Message explanation*
8000	Invalid operation on an opened port	8010	Hardware is not available
8001	Timeout value must be greater than zero	8011	Cannot allocate the queues
8002	Invalid port number	8012	Device is not open
8003	Property available only at run time	8013	Device is already open
8004	Property is read-only at run time	8014	Could not enable Comm notification
8005	Port already open	8015	Could not set Comm state
8006	Device identifier is invalid	8016	Could not set Comm event mask
8006	Device identifier is invalid	8018	Operation valid only when the port is open
8007	Unsupported baud rate	8019	Device busy
8008	Invalid byte size is invalid	8020	Error reading Comm device
8009	Error in default parameters		

27.6 RS-232 polling

The previous program used interrupt-driven RS-232. It is also possible to use polling to communicate over RS-232. Program 27.5 uses COM2 to send the message 'Hello' and then waits for a received string. It determines that there has been a response by continually testing the number of received characters in the receive buffer (InBufferCount). When there is more than one character in the input buffer it is read.

Program 27.5

```
Private Sub Form_Load()
  Dim Str As String                   ' String to hold input

  MSComm1.CommPort = 2                ' Use COM2
  MSComm1.Settings = "9600,N,8,1"     ' 9600 baud, no parity, 8 data,
                                      ' and 1 stop bit
  MSComm1.InputLen = 0                ' Read entire buffer when Input
                                      ' is used
  MSComm1.PortOpen = True             ' Open port

  Text1.Text = "Sending: Hello"
  MSComm1.Output = "Hello"            ' Send message

  Do  ' Wait for response from port
       DoEvents
  Loop Until MSComm1.InBufferCount >= 2
  Str = MSComm1.Input                 ' Read input buffer
  Text1.Text = "Received: " + Str
  MSComm1.PortOpen = False  ' Close serial port.
End Sub
```

27.6.1 Modems

Modems are programmed in a similar way to RS-232 communications. The program can get the attention of the modem by send '+++' and then can be program with AT codes, such as:

```
MSComm1.CommPort = 2              ' Use COM2
MSComm1.Settings = "9600,N,8,1"   ' 9600 baud, no parity, 8 data,
                                  ' and 1 stop bit
MSComm1.InputLen = 0              ' Read entire buffer when Input
                                  ' is used
MSComm1.PortOpen = True           ' Open port

MSComm1.Output = "+++" + Chr$(13)
MSComm1.Output = "AT 01314554231" + Chr$(13)  ' Telephone 0134554231
```

where `Chr$(13)` gives a carriage return character, and is required at the end of each command string. If successful, the modem returns the OK code, which can be read with:

```
Do
   DoEvents
Loop Until MSComm1.InBufferCount >= 2

' Read the "OK" response data in the serial port.
Instring = MSComm1.Input
```

27.7 Exercises

27.7.1 List the properties of the MSComm control and outline their uses.

27.7.2 Write a Visual Basic program that continuously sends the character 'A' to the serial line. If possible, observe the output on an oscilloscope and identify the bit pattern and the baud rate.

27.7.3 Write a program that continuously sends the characters from 'A' to 'Z' to the serial line. If possible, observe the output on an oscilloscope.

27.7.4 Write a Visual Basic program that prompts the user for the main RS-232 parameters, such as bit rate, parity, and so on. The user should then be able to transmit and receive with those parameters.

27.7.5 If possible, connect two computers together with a serial link and write a program which uses full-duplex communications.

27.7.6 If possible, connect two computers together with a serial link and write a program which uses full-duplex communications.

27.7.7 Write a program which tests some of the run-time errors given in Table 27.5.

27.7.8 Investigate the handshaking property of the MSComm control. Its settings are:

0 - comNone	1 - comXOnXoff
2 - comRTS	3 - comRTSXOnXoff

28 Interrupt-driven RS-232

28.1 Interrupt-driven RS-232

Interrupt-driven devices are efficient on processor time as they allow the processor to run a program without having to poll the devices. This allows fast devices almost instant access to the processor and stops slow devices from 'hogging' the processor. For example, a line printer tends to be slow in printing characters. If the printer only interrupted the processor when it was ready for data then the processor can do other things while the printer is printing the character. Another example can be found in serial communications. Characters sent over an RS-232 link are transmitted and received relatively slowly. In a non-interrupt-driven system the computer must poll the status register to determine if a character has been received, which is inefficient in processor time. But, if the amount of time spent polling the status register is reduced, there is a possibility of the computer missing the received character as another could be sent before the first is read from the receiver buffer. If the serial communications port was set up to interrupt the processor when a new character arrived then it is guaranteed that the processor will always process the receiver buffer.

A major disadvantage with non-interrupt-driven software is when the processor is involved in a 'heavy processing' task such as graphics or mathematical calculations. This can have the effect of reducing the amount of time that can be spent in polling and/or reading data.

28.2 DOS-based RS-232 program

Program 28.1 is a simple interrupt-driven DOS-based RS-232 program which is written for Turbo/Borland C/C++. If possible, connect two PCs together with a cable which swaps the TX and RX lines, as shown in Figure 13.18. Each of the computers should be able to transmit and receive concurrently. A description of this program is given in the next section. The header file associated with this program is **serial.h**.

Program 28.1

```
#include <dos.h>
#include <conio.h>
#include <stdio.h>
#include <bios.h>
#include "serial.h"

void  interrupt rs_interrupt(void);
void  setup_serial(void);
void  send_character(int ch);
```

```
int    get_character(void);
int    get_buffer(void);
void   set_vectors(void);
void   reset_vectors(void);
void   enable_interrupts(void);
void   disable_interrupts(void);

void   interrupt(*oldvect)();

char   buffer[RSBUFSIZE];

unsigned int   startbuf=0,endbuf = 0;

int    main(void)
{
int    ch, done  = FALSE;
       setup_serial();
       set_vectors(); /* set new interrupt vectors and store old ones       */
       enable_interrupts();
       printf("Terminal emulator, press [ESC] to quit\n");
       do
       {
          if (kbhit())
          {
             ch=getche();
             if (ch==ESC) break;
             send_character(ch);
          }
          /* empty RS232 buffer   */
          do
          {
             if ((ch=get_buffer()) != -1) putch(ch);
          } while (ch!=-1);
       } while (!done);
       disable_interrupts();
       reset_vectors();
       return(0);
}

void   interrupt rs_interrupt(void)
{
       disable();
       if ((inportb(IIR) & RX_MASK) == RX_ID)
       {
          buffer[endbuf] = inportb(RXR);
          endbuf++;
          if (endbuf == RSBUFSIZE) endbuf=0;
       }
       /* Set end of interrupt flag */
       outportb(ICR, EOI);
       enable();
}

void      setup_serial(void)
{
int       RS232_setting;
       RS232_setting=BAUD1200 | STOPBIT1 | NOPARITY | DATABITS7;
       bioscom(0,RS232_setting,COM1);
}

void   send_character(int ch)
{
```

```
char  status;
      do
      {
         status = inportb(LSR) & 0x40;
      } while (status!=0x40);
      /*repeat until Tx buffer empty ie bit 6 set*/
      outportb(TXDATA,(char) ch);
}

int   get_character(void)
{
int   status;
      do
      {
         status = inportb(LSR) & 0x01;
      } while (status!=0x01);
      /* Repeat until bit 1 in LSR is set */
      return( (int)inportb(TXDATA));
}

int   get_buffer(void)
{
int   ch;
      if (startbuf == endbuf) return (-1);
      ch = (int) buffer[startbuf];
      startbuf++;
      if (startbuf == RSBUFSIZE) startbuf = 0;
      return (ch);
}

void   set_vectors(void)
{
    oldvect = getvect(0x0C);
    setvect(0x0C, rs_interrupt);
}

/* Uninstall interrupt vectors before exiting the program */
void   reset_vectors(void)
{
    setvect(0x0C, oldvect);
}

void     disable_interrupts(void)
{
int      ch;
      disable();
      ch = inportb(IMR) | ~IRQ4; /* disable IRQ4 interrupt */
      outportb(IMR, ch);
      outportb(IER, 0);
      enable();
}

void     enable_interrupts(void)
{
int   ch;
      disable();
      /* initialize rs232 port   */
      ch = inportb(MCR) | MC_INT;
      outportb(MCR, ch);
      /* enable interrupts for IRQ4 */
      outportb(IER, 0x01);
      ch = inportb(IMR) & IRQ4;
```

```
	outportb(IMR, ch);
	enable();
}
```

Header file 28.1: serial.h

```
#define  FALSE       0
/* RS232 set up parameters */
#define COM1         0
#define COM2         1

#define DATABITS7    0x02
#define DATABITS8    0x03

#define STOPBIT1     0x00
#define STOPBIT2     0x04

#define NOPARITY     0x00
#define ODDPARITY    0x08
#define EVENPARITY   0x18

#define BAUD110      0x00
#define BAUD150      0x20
#define BAUD300      0x40
#define BAUD600      0x60
#define BAUD1200     0x80
#define BAUD2400     0xA0
#define BAUD4800     0xC0
#define BAUD9600     0xE0

#define ESC          0x1B         /* ASCII Escape character     */
#define RSBUFSIZE    10000        /* RS232 buffer size          */

#define COM1BASE     0x3F8        /* Base port address for COM1 */

#define TXDATA       COM1BASE           /* Transmit register    */
#define RXR          COM1BASE           /* Receive register     */
#define IER          (COM1BASE+1)       /* Interrupt Enable     */
#define IIR          (COM1BASE+2)       /* Interrupt ID         */
#define LCR          (COM1BASE+3)       /* Line control         */
#define MCR          (COM1BASE+4)       /* Line control         */
#define LSR          (COM1BASE+5)       /* Line Status          */

#define RX_ID        0x04
#define RX_MASK      0x07
#define MC_INT       0x08

/*    Addresses of the 8259 Programmable Interrupt Controller (PIC).*/

#define IMR          0x21  /* Interrupt Mask Register port     */
#define ICR          0x20  /* Interrupt Control Port           */

/*    An end of interrupt needs to be sent to the Control Port of         */
/*    the 8259 when a hardware interrupt ends.                 */
#define EOI          0x20  /* End Of Interrupt                 */

#define IRQ4         0xEF  /* COM1                             */
```

28.2.1 Description of program

The initial part of the program sets up the required RS-232 parameters. It uses `bioscom()` to set COM1: with the parameters of 1200 bps, 1 stop bit, no parity and 7 data bits.

```
void     setup_serial(void)
{
int      RS232_setting;
      RS232_setting=BAUD1200 | STOPBIT1 | NOPARITY | DATABITS7;
      bioscom(0,RS232_setting,COM1);
}
```

After the serial port has been initialized the interrupt service routine for the IRQ4 line is set to point to a new 'user-defined' service routine. The primary serial port COM1: sets the IRQ4 line active when it receives a character. The interrupt associated with IRQ4 is 0Ch (12). The getvect() function gets the ISR address for this interrupt, which is then stored in the variable oldvect so that at the end of the program it can be restored. Finally, in the set_vectors() function, the interrupt assigns a new 'user-defined' ISR (in this case it is the function rs_interrupt()).

```
void     set_vectors(void)
{
      oldvect = getvect(0x0C);      /* store IRQ4 interrupt vector         */
      setvect(0x0C, rs_interrupt); /* set ISR to rs_interrupt()           */
}
```

At the end of the program the ISR is restored with the following code.

```
void    reset_vectors(void)
{
      setvect(0x0C, oldvect);       /* reset IRQ4 interrupt vector         */
}
```

The COM1: port is initialized for interrupts with the code given next. The statement

```
      ch = inportb ( MCR ) | 0x08;
```

resets the RS-232 port by setting bit 3 for the modem control register (MCR) to a 1. Some RS-232 ports require this bit to be set. The interrupt enable register (IER) enables interrupts on a port. Its address is offset by 1 from the base address of the port (that is, 0x3F9 for COM1:). If the least significant bit of this register is set to a 1 then interrupts are enabled, else they are disabled.

To enable the IRQ4 line on the PIC, bit 5 of the IMR (interrupt mask register) is to be set to a 0 (zero). The statement:

```
      ch = inportb(IMR) & 0xEF;
```

achieves this as it bitwise ANDs all the bits, except for bit 4, with a 1. This is because any bit which is ANDed with a 0 results in a 0. The bit mask 0xEF has been defined with the macro IRQ4.

```
void     enable_interrupts(void)
{
int    ch;
      disable();
      ch = inportb(MCR) | MC_INT; /* initialize rs232 port */
      outportb(MCR, ch);
      outportb(IER, 0x01);
      ch = inportb(IMR) & IRQ4;
```

```
      outportb(IMR, ch);   /* enable interrupts for IRQ4 */

      enable();
}
```

At the end of the program the function `disable_interrupts()` sets the IER register to all 0s. This disables interrupts on the `COM1:` port. Bit 4 of the IMR is also set to a 1 which disables `IRQ4` interrupts.

```
void     disable_interrupts(void)
{
int      ch;
      disable();
      ch = inportb(IMR) | ~IRQ4; /* disable IRQ4 interrupt */
      outportb(IMR, ch);
      outportb(IER, 0);
      enable();
}
```

The ISR for the `IRQ4` function is set to `rs_interrupt()`. When it is called, the Interrupt Status Register (this is named IIR to avoid confusion with the interrupt service routine) is tested to determine if a character has been received. Its address is offset by 2 from the base address of the port (that is, 0x3FA for `COM1:`). The first 3 bits give the status of the interrupt. A `000b` indicates that there are no interrupts pending, a `100b` that data has been received, or a `111b` that an error or break has occurred. The statement `if ((inportb(IIR) & 0x7) == 0x4)` tests if data has been received. If this statement is true then data has been received and the character is then read from the receiver buffer array with the statement `buffer[endbuf] = inportb(RXR);`. The end of the buffer variable (`endbuf`) is then incremented by 1.

At the end of this ISR the end of interrupt flag is set in the interrupt control register with the statement `outportb(ICR, 0x20);`. The `startbuf` and `endbuf` variables are global, thus all parts of the program have access to them.

Turbo/Borland functions `enable()` and `disable()` in `rs_interrupt()` are used to enable and disable interrupts, respectively.

```
void  interrupt rs_interrupt(void)
{
      disable();
      if ((inportb(IIR) & RX_MASK) == RX_ID)
      {
         buffer[endbuf] = inportb(RXR);
         endbuf++;
         if (endbuf == RSBUFSIZE) endbuf=0;
      }
      /* Set end of interrupt flag  */
      outportb(ICR, EOI);
      enable();
}
```

The `get_buffer()` function is given next. It is called from the main program and it tests the variables `startbuf` and `endbuf`. If they are equal then it returns –1 to the `main()`. This indicates that there are no characters in the buffer. If there are characters in the buffer then the function returns, the character pointed to by the `startbuf` variable. This variable is then incremented. The difference between `startbuf` and `endbuf` gives the number of characters in the buffer. Note that when `startbuf` or `endbuf` reach the end of the buffer (`RSBUFSIZE`) they are set back to the first character, that is, element 0.

```
int    get_buffer(void)
{
int    ch;

       if (startbuf == endbuf) return (-1);
       ch = (int) buffer[startbuf];
       startbuf++;
       if (startbuf == RSBUFSIZE) startbuf = 0;
       return (ch);
}
```

The `get_character()` and `send_character()` functions are similar to those developed in Chapter 13. For completeness, these are listed next.

```
void   send_character(int ch)
{
char   status;
       do
       {
          status = inportb(LSR) & 0x40;
       } while (status!=0x40);

       /*repeat until Tx buffer empty ie bit 6 set*/
       outportb(TXDATA,(char) ch);
}

int       get_character(void)
{
int       status;

       do
       {
          status = inportb(LSR) & 0x01;
       } while (status!=0x01);
       /* Repeat until bit 1 in LSR is set */
       return( (int)inportb(TXDATA));
}
```

The `main()` function calls the initialization and the de-initialization functions. It also contains a loop, which continues until the Esc key is pressed. Within this loop, the keyboard is tested to determine if a key has been pressed. If it has then the `getche()` function is called. This function returns a key from the keyboard and displays it to the screen. Once read into the variable `ch` it is tested to determine if it is the Esc key. If it is then the program exits the loop, else it transmits the entered character using the `send_character()` function. Next the `get_buffer()` function is called. If there are no characters in the buffer then a –1 value is returned, else the character at the start of the buffer is returned and displayed to the screen using `putch()`.

```
int       main(void)
{
int       ch, done  = FALSE;

       setup_serial();
       /* set new interrupt vectors and store old ones*/
       set_vectors();
       enable_interrupts();
       printf("Terminal emulator, press [ESC] to quit\n");
       do
```

```
    {
       if (kbhit())
       {
          ch=getche();
          if (ch==ESC) break;
          send_character(ch);
       }
       /* empty RS232 buffer   */
       do
       {
          if ((ch=get_buffer()) != -1) putch(ch);
       } while (ch!=-1);
    } while (!done);
    disable_interrupts();
    reset_vectors();
    return(0);
}
```

28.3 Exercises

28.3.1 Modify Program 28.1 so that a new-line character is displayed properly.

28.3.2 Prove that Program 28.1 is a true multitasking system by inserting a delay in the main loop, as shown next. The program should be able to buffer all received characters and display them to the screen when the sleep delay is over.

```
do
{
   sleep(10);
      /* go to sleep for 10 seconds, real-time system          */
      /* will  buffer all received characters          */
   if (kbhit())
   {
      ch=getche();
      if (ch==ESC) break;
      send_character(ch);
   }
   /* empty RS232 buffer   */
   do
   {
      if ((ch=get_buffer()) != -1) putch(ch);
   } while (ch!=-1);
} while (!done);
```

28.3.3 Modify Program 28.1 so that the transmitted characters are displayed in the top half of the screen and then received in the bottom half of the screen.

28.3.4 Modify Program 28.1 so that it communicates via `COM2:` (if the PC has one).

28.3.5 Using a loopback connection on a serial port, write a program, which sends out the complete ASCII table and checks it against the received characters.

28.3.6 Outline how a program could communicate with a serial port card with eight serial lines on it. Normally the IRQ lines for COM3 to COM8 can set to either IRQ3 to IRQ4 with jumpers on the board. Investigate typical base addresses for COM3 to COM8. Thus explain how multiple devices can be connected to a single interrupt line.

28.3.7 Implement a program which has a write buffer system. This should fill a buffer with characters and only send them once every 30 seconds. Modify the program so that the user can enter the delay time.

28.3.8 Write a program which sends the complete contents of a text file from one computer to another. The receiving program should wait for the sending program to send the file and the sending program should prompt the user on the name of the file. The sending program must thus identify the name of the file to the receiver. A typical protocol could be:

1. Send name of the file followed by an invalid file name character, such as:

   ```
   myfile.txt*
   ```

   which would identify that the name of the file to be sent is `myfile.txt`.

2. Send each of the characters in the file one by one. The end of the file is then identified by the end-of-file (EOF) character.

28.3.9 Modify the program in Exercise 28.3.8 so that a binary file can be sent. Note that in a binary file the EOF character can occur at a point in the file. There are two possible methods which can be used to implement this. These are:

- Implement a time-out on the characters received. When no characters have been received after, say, 1 second then it is assumed that the EOF has occurred. The receiver will then close the file.

- The sender sends some initial information on the filename, number of bytes in the file, date and time, and so on. This could be a fixed format header.

- Send a second EOF character whenever there is an EOF character which is not the EOF marker. Thus when the receiver receives two consecutive EOF characters, it simply deletes one of them. When it receives a single EOF character (within a given time) then it knows it is at the end of the file.

28.3.10 Modify the program in Exercise 28.3.9 so that a user on the receiving computer can specify the filename which is to be sent from the sending program.

A PC Processors

A.1 Introduction

Intel marketed the first microprocessor, named the 4004. This device caused a revolution in the electronics industry because previous electronic systems had a fixed functionality. With this processor the functionality could be programmed by software. It could handle four bits of data at a time (a nibble), contained 2000 transistors, had 46 instructions and allowed 4 KB of program code and 1 KB of data. The PC has since evolved using Intel microprocessors (Intel is a contraction of *Int*egrated *El*ectronics).

The second generation of Intel microprocessors began in 1974. These could handle 8 bits (a byte) of data at a time and were named the 8008, 8080 and the 8085. They were much more powerful than the previous 4-bit devices and were used in many early microcomputers and applications such as electronic instruments and printers. The 8008 has a 14-bit address bus and can thus address up to 16 kB of memory (the 8080 has a 16-bit address bus giving it a 64 kB limit).

The third generation of microprocessors began with the launch of the 16-bit processors. Intel released the 8086 microprocessor which was mainly an extension to the original 8080 processor and thus retained a degree of software compatibility. IBM's designers realised the power of the 8086 and used it in the original IBM PC and IBM XT (eXtended Technology). It has a 16-bit data bus and a 20-bit address bus, and thus has a maximum addressable capacity of 1 MB.

A stripped-down 8-bit external data bus version called the 8088 is also available. This stripped down processor allowed designers to produce less complex (and cheaper) computer systems. The 8086 could handle either 8 or 16 bits of data at a time (although in a messy way). An improved architecture version, called the 80286, was launched in 1982, and was used in the IBM AT (Advanced Technology).

In 1985, Intel introduced its first 32-bit microprocessor, the 80386DX. This device was compatible with the previous 8088/8086/80286 (80X86) processors and gave excellent performance handling 8, 16 or 32 bits at a time. It has full 32-bit data and address buses and can thus address up to 4 GB of physical memory. A stripped-down 16-bit external data bus and 24-bit address bus version called the 80386SX was released in 1988. This processor can thus only access up to 16 MB of physical memory.

In 1989, Intel introduced the 80486DX which is basically an improved 80386DX with a memory cache and math co-processor integrated onto the chip. It had an improved internal structure making it around 50% faster than a comparable 80386. The 80486SX was also introduced, which is merely a 80486DX with the link to the math co-processor broken. Clock doubler/ trebler 80486 processors were also released. In these devices the processor runs at a higher speed than the system clock. Typically, systems with clock doubler processors are around 75% faster than the comparable non-doubled processors. Typical clock doubler processors are DX2-66 and DX2-50 which run from 33 MHz and 25 MHz clocks, respectively. Intel also produced a new range of microprocessors which run at three or four times the system clock speed and are referred to as DX4 processors. These include the Intel DX4-100 (25 MHz clock) and Intel DX4-75 (25 MHz clock).

The Pentium (or P-5) is a 64-bit 'superscalar' processor. It can execute more than one instruction at a time and has a full 64-bit (8-byte) data bus and a 32-bit address bus. In terms of performance, it operates almost twice as fast as the equivalent 80486. It also has improved floating-point operations (roughly three times faster) and is fully compatible with previous 80x86 processors.

The Pentium II/III is an enhancement of the P-5 and has a bus which supports up to four processors on the same bus without extra supporting logic. With clock multiplying speeds of over 500 MHz are possible. It also has major savings of electrical power and the minimisation of electromagnetic interference (EMI). A great enhancement of the Pentium II/III bus is that it detects and corrects all single bit data bus errors and also detects multiple bit errors on the data bus.

A.2 8086/88

A.2.1 Introduction

The great revolution in processing power arrived with the 16-bit 8086 processor. This had a 20-bit address bus and a 16-bit address bus, whereas the 8088 has an 8-bit external data bus. Figure A.1 shows the pin connections of the 8086 and also the main connections to the processor. Many of the 40 pins of the 8086 have dual functions. The lines AD0–AD7 act either a the lower eight bits of the address bus (A0–A7) or as the lower eight bits of the data bus (D0–D7). The lines A16/S3-A19/S6 also have a dual function, S3–S6 are normally not used by the PC thus they are used as the four upper bits of the address bus. The latching of the address is achieved when the ALE (address latch enable) goes from a high to a low.

The bus controller (8288) generates the required control signals form the 8088 status lines $\overline{S0}$ – $\overline{S2}$. For example, if $\overline{S0}$ is high, $\overline{S1}$ is low and $\overline{S2}$ is low then the $\overline{MEMR}$ line goes low. The main control signals are:

- $\overline{IOR}$ (I/O read) which means that the processor is reading from the contents of the address which is on the I/O bus.
- $\overline{IOW}$ (I/O write) which means that the processor is writing the contents of the data bus to the address which is on the I/O bus.
- $\overline{MEMR}$ (memory read) which means that the processor is reading from the contents of the address which is on the address bus.
- $\overline{MEMW}$ (memory write) which means that the processor is writing the contents of the data bus to the address which is on the address bus.
- $\overline{INTA}$ (interrupt acknowledgement) which is used by the processor to acknowledge an interrupt ($\overline{S0}$, $\overline{S1}$ and $\overline{S2}$ all go low). When a peripheral wants the attention of the processor it sends an interrupt request to the 8259 which, if it is allowed, sets the INTR high.

The processor either communicates directly with memory (with $\overline{MEMW}$ and $\overline{MEMR}$) or communicates with peripherals through isolated I/O ports (with $\overline{IOR}$ and $\overline{IOW}$).

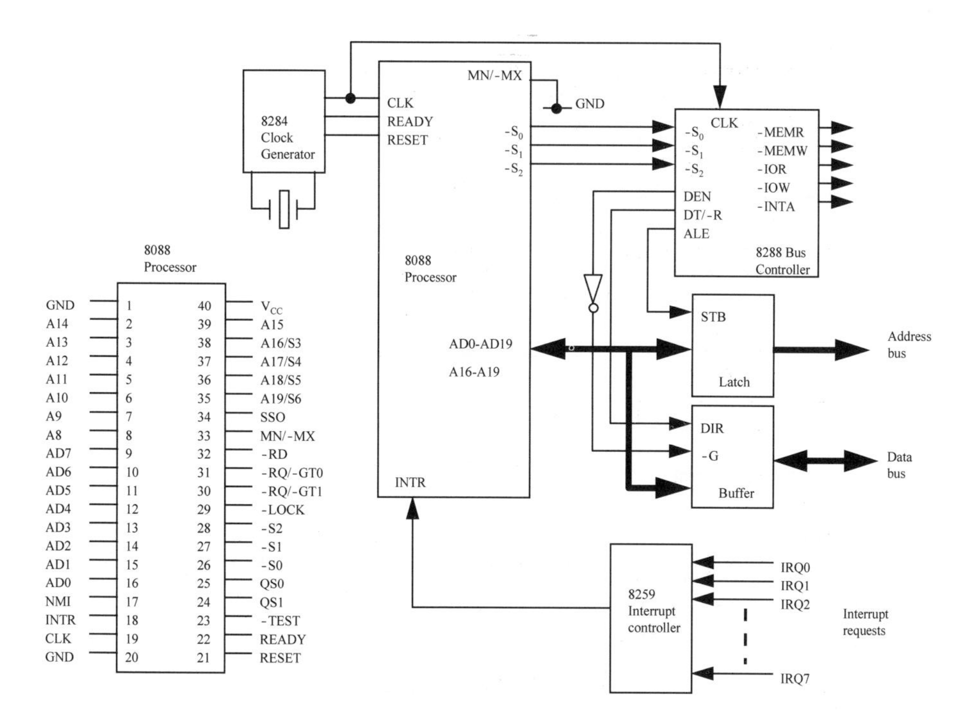

Figure A.1 8088 connections

Registers

Each of the PC-based Intel microprocessors is compatible with the original 8086 processor and are normally backwardly compatible. Thus, for example, a Pentium can run 8086 and 80386 code. Microprocessors use registers to perform their operations. These registers are basically special memory locations in that they are given names. The 8086/88 has 14 registers which are grouped into four categories, as illustrated in Figure A.2.

General-purpose registers

There are four general-purpose registers which are AX, BX, CX and DX. Each can be used to manipulate a whole 16-bit word or with two separate 8-bit bytes. These bytes are called the lower and upper order bytes. Each of these registers can be used as two 8-bit registers; for example, AL represents an 8-bit register which is the lower half of AX and AH represents the upper half of AX.

The AX register is the most general purpose of the four registers and is usually used for all types of operations. Each of other registers have one or more implied extra functions:

- AX is the accumulator. It is used for all input/output operations and some arithmetic operations. For example, multiply, divide and translate instructions assume the use of AX.
- BX is the base register. It can be used as an address register
- CX is the count register. It is used by instructions which require to count. Typically is it is used for controlling the number of times a loop is repeated and in bit shift operations.
- DX is the data register. It is used for some input/output and also when multiplying and dividing.

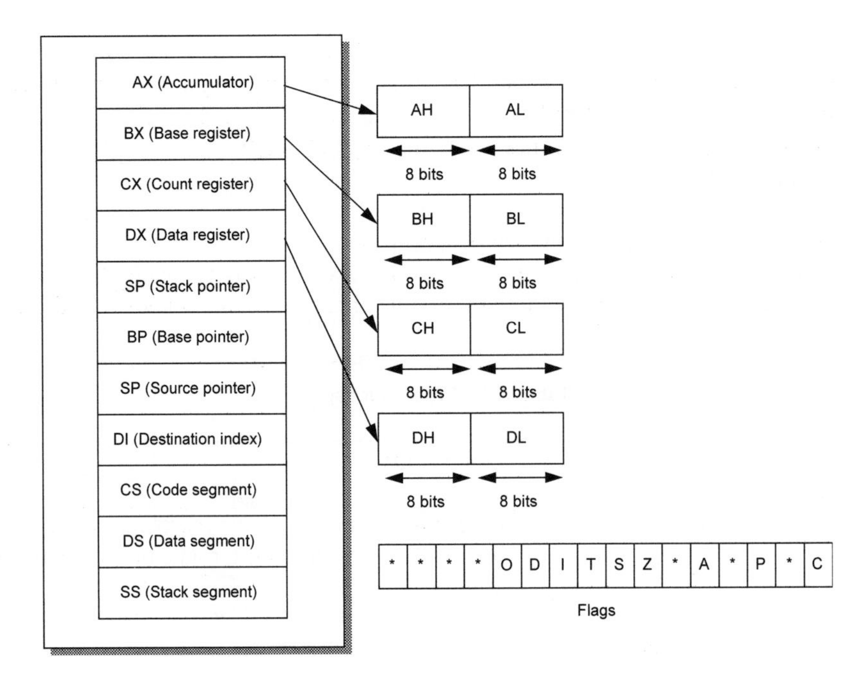

Figure A.2 8086/88 registers

Addressing registers
The addressing registers are used in memory addressing operations, such as the holding the source address of the memory and the destination address. These address registers are named BP, SP, SI and DI, which are:

- SI is the source index. This is used with extended addressing commands.
- DI is the destination index. The destination is used in some addressing modes.
- BP is the base pointer.
- SP is the stack pointer.

Status registers
Status registers are used to test for various conditions in an operation, such as 'is the result negative', 'is the result zero', and so on. The two status registers have 16 bits and are called the instruction pointer (IP) and the flag register (F):

- IP is the instruction pointer. The IP register contains the address of the next instruction of the program.
- Flag register. The flag register holds a collection of 16 different conditions. Table A.1 outlines the most-used flags.

Segments registers
There are four areas of memory called segments, each of which are 16 bits and can thus address up to 64 KB (from `0000h` to `FFFFh`). These segments are:

- Code segment (CS register) – defines the memory location where the program code (or instructions) is stored.

- Data segment (DS register) – defines where data from the program will be stored (DS stands for data segment register).
- Stack segment (SS register) – defines where the stack is stored.
- Extra segment (ES).

All addresses are with reference to the segment registers.

The 8086 has a segmented memory, these registers are used to manipulate these segments. Each segment provides 64K of memory, this area of memory is known as the current segment.

Table A.1 Processor flags

Bit	*Flag position*	*Name*	*Description*
C	0	Set on carry	Contains the carry from the most significant bit (left hand bit) following a shift, rotate or arithmetic operation.
A	4	Set on 1/2 carry	
S	7	Set on negative result	Contains the sign of an arithmetic operation (0 for positive, 1 for negative).
Z	6	Set on zero result	Contains results of last arithmetic or compare result (0 for nonzero, 1 for zero).
O	11	Set on overflow	Indicates that an overflow has occurred in the most significant bit from an arithmetic operation.
P	2	Set on even parity	
D	10	Direction	
I	9	Interrupt enable	Indicates whether the interrupt has been disabled.
T	8	Trap	

Memory Addressing

There are several methods of accessing memory locations, these are:

- Implied addressing which uses an instruction in which it is known the which registers are used.
- Immediate (or literal) addressing uses a simple constant number to define the address location.
- Register addressing which uses the address registers for the addressing (such as AX, BX , and so on).
- Memory addressing which is used to read or write to a specified memory location.

A.2.2 Memory segmentation

The 80386, 80486 and Pentium processors run in one of two modes, either virtual or real. When using the virtual mode they act as a pseudo-8086 16-bit processor, known as the protected mode. In the real mode they can use the full capabilities of their address and data bus. The mode and their addressing capabilities depend on the software and thus all DOS-based programs use the virtual mode.

The 8086 has a 20-bit address bus so that when the PC is running 8086-compatible code it can only address up to 1 MB of memory. It also has a segmented memory architecture and can only directly address 64 kB of data at a time. A chunk of memory is known as a segment and hence the phrase 'segmented memory architecture'.

Memory addresses are normally defined by their hexadecimal address. A 4-bit address bus can address 16 locations from `0000b` to `1111b`. This can be represented in hexadecimal as `0h` to `Fh`. An 8-bit bus can address up to 256 locations from `00h` to `FFh`.

Two important addressing capabilities for the PC relate to a 16- and a 20-bit address bus. A 16-bit address bus addresses up to 64 kB of memory from `0000h` to `FFFFh` and a 20-bit address bus addresses a total of 1 MB from `00000h` to `FFFFFh`. The 80386/80486/Pentium processors have a 32-bit address bus and can address from `00000000h` to `FFFFFFFFh`.

A memory location is identified with a segment and an offset address and the standard notation is `segment:offset`. A `segment` address is a 4-digit hexadecimal address which points to the start of a 64 kB chunk of data. The `offset` is also a 4-digit hexadecimal address which defines the address offset from the segment base pointer. This is illustrated in Figure A.3.

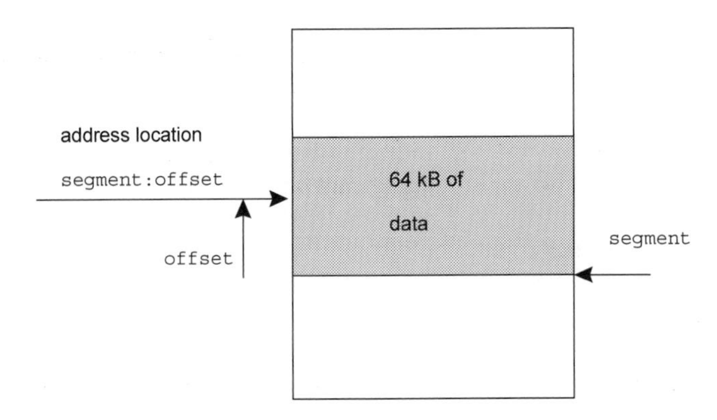

Figure A.3 Memory addressing

The `segment:offset` address is defined as the logical address, the actual physical address is calculated by shifting the segment address; 4 bits to the left and adding the offset. The example given next shows that the actual address of `2F84:0532` is `2FD72h`:

Segment (`2F84`):	0010	1111	1000	0100	0000
Offset (`0532`):		0000	0101	0011	0010
Actual address:	0010	1111	1101	0111	0010

A.3 80386/80486

A.3.1 Introduction

The 32-bit 80386 processor was a great leap in processing power and for the first time many PCs could properly run graphical user interface software (such as Microsoft Windows). A key to its success was that it was fully compatible with the previous 8088/8086/80286 processors. The DX version has full 32-bit data and address buses and can thus address up to 4 GB of physical memory. An SX version with a stripped-down 16-bit external data bus and 24-bit address bus version can access only up to 16 MB of physical memory (at its time of release this has a large amount of memory).

The 80486DX basically consists of an improved 80386 with a memory cache and math co-processor integrated onto the chip. An SX version had the link to the math co-processor broken. At the time a limiting factor was the speed of the system clock (which was limited to around 25 MHz or 33 MHz). Thus clock doublers, treblers or quadrupers allows the processor to multiply the system clock frequency to a high speed. Thus internal operations within the processor could be carried out at much higher speeds. Then accesses the external devices would slow down to the system clock. As most of the operations within the computer involve the processor then the overall speed of the computer is improved (roughly by about 75% for a clock doubler). 80486 processors were also released. In these devices the processor runs at a higher speed than the system clock. Typically, systems with clock doubler processors are around 75% faster than the comparable non-doubled processors.

A.3.2 80486 pin out

To allow for easy upgrades and to save space the 80486 and Pentium processors are available in pin-grid array (PGA) form. The 80486DX processor is available as a 168 pin PGA, as illustrated in Figure A.4. The PGA chip is inserted into a zero-insertion force (ZIF) socket on the motherboard of the PC.

It can be seen that the 486 processor has a 32-bit address bus (A0–A31) and a 32-bit data bus (D0–D31). The pin definitions are defined in Table A.2.
Table A.3 defines the how the control signals are interpreted. For the STOP/special bus cycle, the byte enable signals ($\overline{BE0} - \overline{BE3}$) are used to further define the cycle. These are:

- Write-back cycle $\overline{BE0}=1$, $\overline{BE1}=1$, $\overline{BE2}=1$, $\overline{BE3}=0$.
- Halt cycle $\overline{BE0}=1$, $\overline{BE1}=1$, $\overline{BE2}=0$, $\overline{BE3}=1$.
- Flush cycle $\overline{BE0}=1$, $\overline{BE1}=0$, $\overline{BE2}=1$, $\overline{BE3}=1$.
- Shut-down cycle $\overline{BE0}=0$, $\overline{BE1}=1$, $\overline{BE2}=1$, $\overline{BE3}=1$.

The 486 integrates a processor, cache and a math co-processor onto a single IC, its pin connections are:

A2–A31 (I/O) — The 30 most significant bits of the address bus.

$\overline{A20M}$ (I) — When active low, the processor internally masks the address bit A20 before every memory access.

$\overline{ADS}$ (O) — Indicates that the processor has valid control signals and a valid address signals.

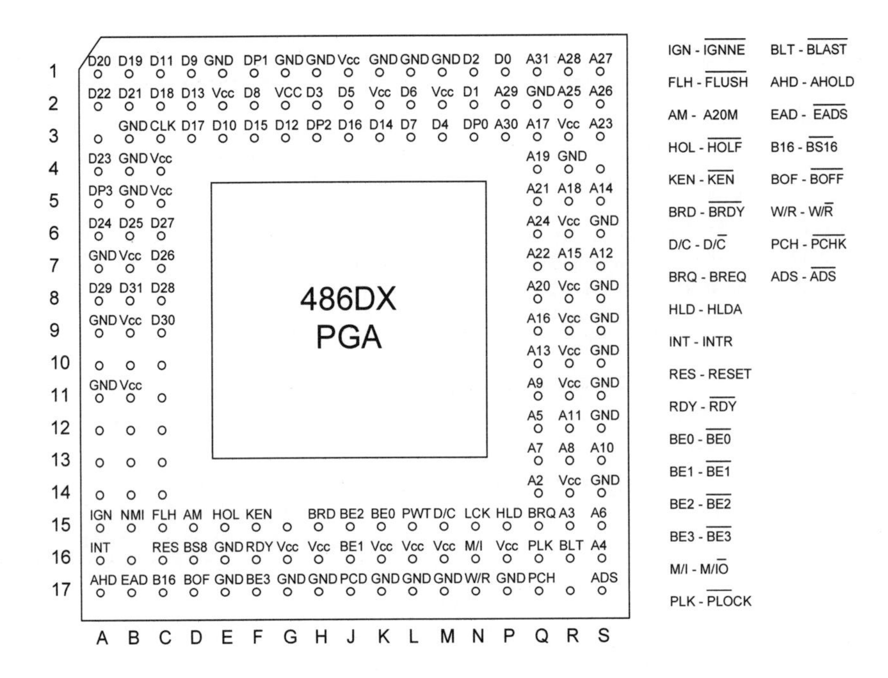

Figure A.4 i486DX processor

AHOLD (I) — When active a different bus controller can access have access to the address bus. This is typically used in a multi-processor system.

$\overline{BE0}$ – $\overline{BE3}$ (O) — The byte enable lines indicates which of the bytes of the 32-bit data bus are active.

$\overline{BLAST}$ (O) — Indicates that the current burst cycle will end after the next $\overline{BRDY}$ signal.

$\overline{BOFF}$ (I) — The backoff signal informs the processor to deactivate the bus on the next clock cycle.

$\overline{BRDY}$ (I) — The burst ready signal is used by an addressed system has sent data on the data bus or read data from the bus.

BREQ (O) — Indicates that the processor has internally requested the bus.

$\overline{BS16}$, $\overline{BS8}$ (I) — The $\overline{BS16}$ signal indicates that a 16-bit data bus is used, the $\overline{BS8}$ signal indicates that a 8-bit data bus is used. If both are high then a 32-bit data is used.

DP0-DP3 (I/O) — The data parity bits gives a parity check for each byte of the 32-bit data bus. The parity bits are always even parity.

$\overline{EADS}$ (I) — Indicates that an external bus controller has put a valid address on the address bus.

$\overline{\text{FERR}}$ (O) — Indicates that the processor has detected an error in the internal floating-point unit.

$\overline{\text{FLUSH}}$ (I) — When active the processor writes the complete contents of the cache to memory.

HOLD, HLDA (I/O) — The bus hold (HOLD) and acknowledge (HLDA) are used for bus arbitration and allow other bus controllers to take control of the buses.

$\overline{\text{IGNNE}}$ (I) — When active the processor ignores any numeric errors.

INTR (I) — The interrupt request line is used by external devices to interrupt the processor.

$\overline{\text{KEN}}$ (I) — This signal stops caching of a specific address.

$\overline{\text{LOCK}}$ (O) — If active the processor will not pass control to an external bus controller, when it receives a HOLD signal.

M/$\overline{\text{IO}}$, D/$\overline{\text{C}}$, W/$\overline{\text{R}}$ (O) — See Table A.2.

NMI (I) — The non- maskable interrupt signal causes an interrupt 2.

$\overline{\text{PCHK}}$ (O) — If it is set active then a data parity error has occurred.

$\overline{\text{PLOCK}}$ (O) — The active pseudolock signal identifies that the current data transfer requires more than one bus cycle.

PWT, PCD (O) — The page write through (PWT) and page cache disable (PCD) are used with cache control.

$\overline{\text{RDY}}$ (I) — When active the addressed system has sent data on the data bus or read data from the bus.

RESET (I) — If the reset signal is high for more than 15 clock cycles then the processor will reset itself.

Table A.2 Control signals

M/$\overline{\text{IO}}$	D/$\overline{\text{C}}$	W/$\overline{\text{R}}$	*Description*
0	0	0	Interrupt acknowledge sequence
0	0	1	STOP/special bus cycle
0	1	0	Reading from an I/O port
0	1	1	Writing to an I/O port
1	0	0	Reading an instruction form memory
1	0	1	Reserved
1	1	0	Reading data from memory
1	1	1	Writing data to memory

Figure A.5 shows the main 80386/80486 processor connections. The Pentium processor connections are similar but it has a 64-bit data bus. There are three main interface connections: the memory/IO interface, interrupt interface and DMA interface.

The write/read ($W/\overline{R}$) line determines whether data is written to (W) or read from ($\overline{R}$) memory. PCs can interface directly with memory or can interface to isolated memory. Signal line $M/\overline{IO}$ differentiates between the two types. If it is high then the direct memory is addressed, else if it is low then the isolated memory is accessed.

The 80386DX and 80486 have an external 32-bit data bus (D_0–D_{31}) and a 32-bit address bus ranging from A_2 to A_{31}. The two lower address lines, A_0 and A_1, are decoded to produce the byte enable signals $\overline{BE0}$, $\overline{BE1}$, $\overline{BE2}$ and $\overline{BE3}$. The $\overline{BE0}$ line activates when A_1A_0 is 00, $\overline{BE1}$ activates when A_1A_0 is 01, $\overline{BE2}$ activates when A_1A_0, $\overline{BE3}$ actives when A_1A_0 is 11. Figure A.6 illustrates this addressing.

The byte enable lines are also used to access 8, 16, 24 or 32 bits of data at a time. When addressing a single byte, only the $\overline{BE0}$ line will be active (D_0-D_7), if 16 bits of data are to be accessed then $\overline{BE0}$ and $\overline{BE1}$ will be active (D_0-D_{15}), if 32 bits are to be accessed then $\overline{BE0}$, $\overline{BE1}$, $\overline{BE2}$ and $\overline{BE3}$ are active (D_0-D_{31}).

The $D/\overline{C}$ line differentiates between data and control signals. When it is high then data is read from or written to memory, else if it is low then a control operation is indicated, such as a shutdown command.

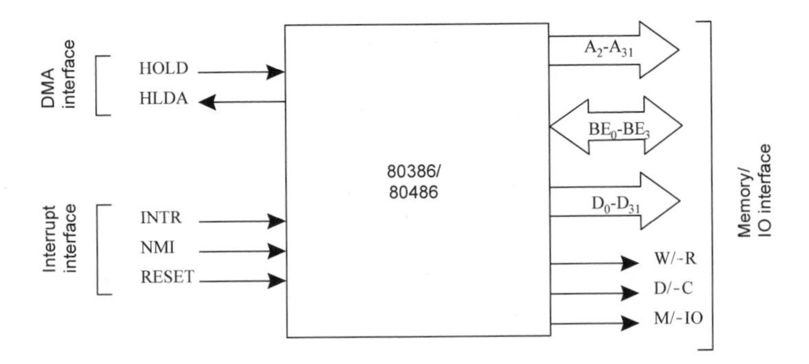

Figure A.5 Some of the 80386/80486 signal connections

The interrupt lines are interrupt request (INTR), nonmaskable interrupt request (NMI) and system reset (RESET), all of which are active high signals. The INTR line is activated when an external device, such as a hard disk or a serial port, wishes to communicate with the processor. This interrupt is maskable and the processor can ignore the interrupt if it wants. The NMI is a non-maskable interrupt and is always acted on. When it becomes active the processor calls the nonmaskable interrupt service routine. The RESET signal causes a hardware reset and is normally made active when the processor is powered up.

A.3.3 80386/80486 registers

The 80386 and 80486 are 32-bit processors and can thus operate on 32-bits at a time. It thus has expanded 32-bit registers, which can also be used as either 16-bit or 8-bit registers. The general purpose registers, such as AX, BX, CX, DX, SI, DI and BP have been expanded and are named EAX, EBX, ECX, EDX, ESI, EDI and EBP, respectively, as illustrated in Figure A.7. The CS, SS and DS registers are still 16 bits, but the flag register has been expanded to 32 bits and is named EFLAG.

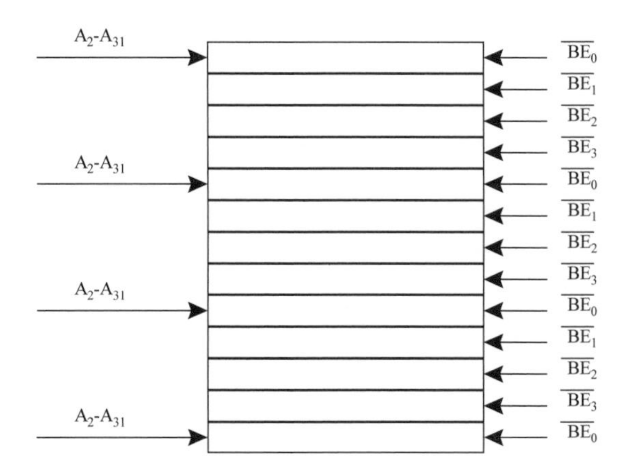

Figure A.6 Memory addressing

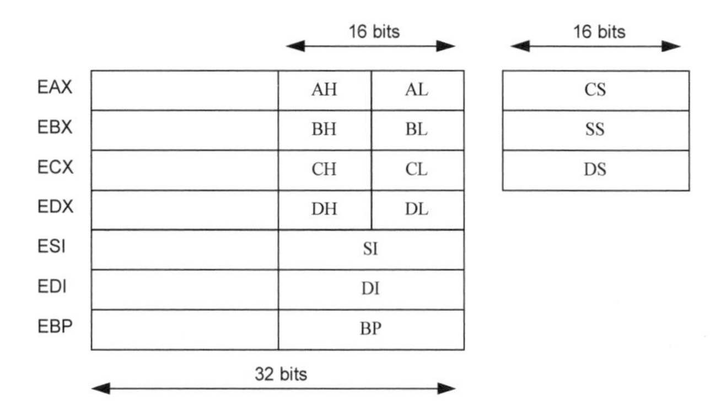

Figure A.7 80386/80486 registers

A.3.4 Memory cache

DRAM is a relatively slow type of memory compared with SRAM. A cache memory can be used to overcome this problem. This is a bank of fast memory (SRAM) that loads data from main memory (typically DRAM). The cache controller guesses the data the processor requires and loads this into the cache memory. If the controller guesses correctly then it is a cache hit, else if it is wrong it is a cache miss (as illustrated in Figure A.8). A miss causes the processor to access the memory in the normal way (that is, there may be wait states). Typical cache memory sizes are 16KB, 32KB and 64KB. This should be compared with the size of the RAM on a typical PC which is typically at least 64MB.

Many modern systems have extra cache memory added to improve the hit rate. Typically an 8KB cache memory gives 70% hit rate, a 16KB cache memory 85%, a 32K cache 93% and a 64KB cache 95%. Cache sizes above this do not significantly effect the hit rate and can actually slow the process down as they take so long to fill the cache memory. The Intel 80486 and Pentium have built-in cache controllers and, at least, 8KB of SRAM cache memory. Intel claim that this has a 96% hit rate, which is an extremely high hit rate for such a small amount of cache memory.

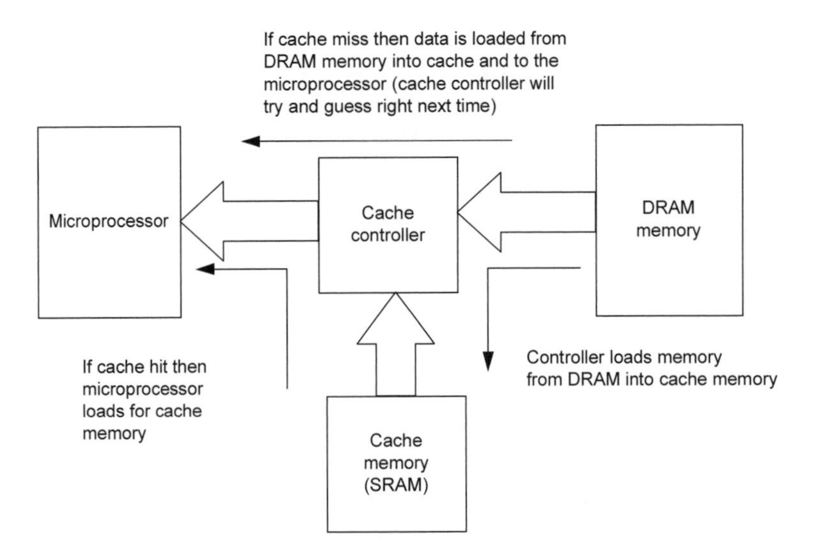

Figure A.8 Cache operation

Cache architecture

The main cache architectures are:

- **Look-through cache**. In a look-through cache the system memory is isolated from the processor address and control busses. In this case the processor directly sends a memory request to the cache controller which then determines whether it should forward the request to its own memory or the system memory. Figure A.9 illustrates this type of cache. It can be seen that the cache controls whether the processor address contents are latched through to the DRAM memory and it also controls whether the contents of the DRAMs memory is loaded onto the processor data bus (through the data transceiver). The operation is described as bus cycle forwarding.
- **Look-aside cache**. A look-aside cache is where the cache and system memory connect to a common bus. System memory and the cache controller see the beginning of the processor bus cycle at the same time. If the cache controller detects a cache hit then it must inform the system memory before it tries to find the data. If a cache miss is found then the memory access is allowed to continue.
- **Write-through cache**. With a write-through cache all memory address accesses are seen by the system memory when the processor performs a bus cycle.
- **Write-back cache**. With a write-back cache all system writes are controlled by the cache controller. It thus does not write the system memory unless it has to.

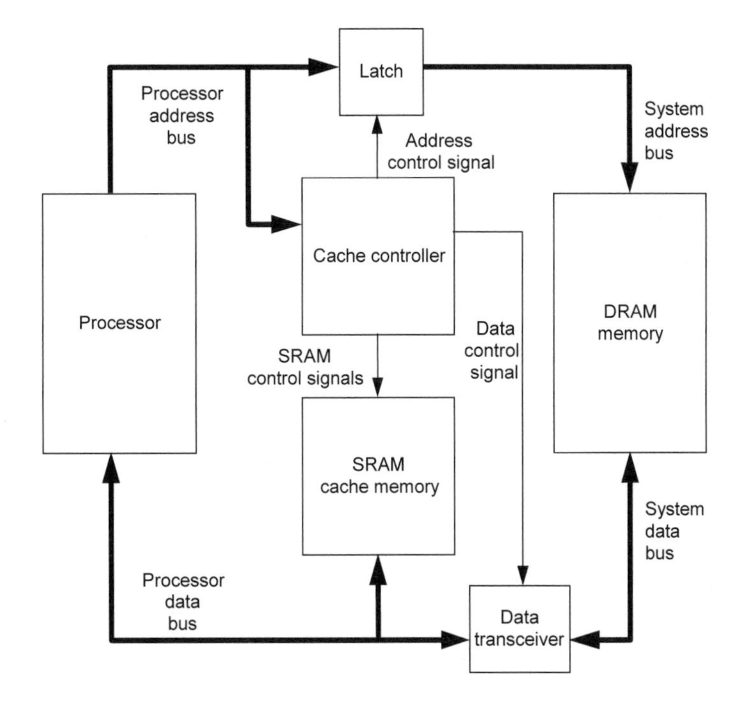

Figure A.9 Look-through cache

Second-level caches

A L1-cache (first-level cache) provides a relatively small on-chip cache, where a L2-cache (second-level cache) provides an external, on-board, cache which provides a cache memory of between 128 and 512 KB. The processor looks in its own L1-cache for a cache hit, if none is found then it searches in the on-board L2-cache. A cache hit in the L1-cache will obviously be faster than the off-chip cache.

An L2-cache for the 486 has a maximum 512 KB memory size and is typically available as 128 KB, 256 KB or 512 KB.

A.4 Pentium/Pentium Pro

Intel has gradually developed their range of processors from the original 16-bit 8086 processor to the 64-bit Pentium III processor. The original 8086 had just 29 000 transistors and operated at a clock speed of 8 MHz. It had an external 20-bit bus and could thus only access up to 1 MB of memory. Compare this with the Pentium III which can operate at 500 MHz, contains over 8,000,000 transistors and can access up to 64 GB of physical memory. Table A.3 contrasts the Intel processor range. It can also be seen from the table that the Pentium II processor is nearly a thousand times more powerful than an 8086 processor.

Table A.3 Processor comparison

Processor	Clock (when released)	Register size	External data bus	Maximum external memory	Cache	Performance (MIPs)
8086	8 MHz	16	16	1 MB		0.8
286	12.5 MHz	16	16	16 MB		2.7
386DX	20 MHz	32	32	4 GB		6.0
486DX	25 MHz	32	32	4 GB	8KB L1	20
Pentium	60 MHz	32	64	4 GB	16KB L1	100
Pentium Pro	200 MHz	32	64	64 GB	16KB L1 256KB L2	440
Pentium II/III	300 MHz	32	64	64 GB	16KB L1 512KB L2	700

A.4.1 Intel processor development

The 80386 processor was a great leap in processing power over the 8086 and 80286, but it required an on-board maths co-processor to be added to enhance its mathematical operations. It could also only execute one instruction at a time. The 486 brought many enhancements, such as:

- The addition of parallel execution with the expansion of the instruction decode and execution units into five pipelined stages. Each of these stages operate in parallel with the others on up to five instructions in different stages of execution. This allows up to five instructions to be completed at a time.
- The addition of an 8 KB on-chip cache to greatly reduce the time taken to access data and code.
- The addition of an integrated floating-point unit.
- Support for more complex and powerful systems, such as off-board L2 cache support and multiprocessor operation.

With the increase in notebook and palmtop computers, the 486 was also enhanced to support many energy and system management capabilities. These processors were named the 486SL processors. The new enhancements included:

- System management mode – this mode is triggered by the processor's own interrupt pin and allows complex system management features to be added to a system transparently to the operating system and application programs.
- Stop clock and auto halt powerdown – these allow the processor to either shut itself down (and preserve its current state) or run at a reduced clock rate.

The Intel Pentium processor added many enhancements to the previous processors, including:

- The addition of a second execution pipeline. These two pipelines, named u and v, can execute two instructions per clock. This is known as superscalar operation.
- Increased on-chip L1 cache 8 KB for code and another 8 KB for data. It uses the MESI protocol to support write-back mode, as well as the write-through mode (which is used by

the 486 processor).

- Branch prediction with an on-chip branch table which improves looping characteristics.
- Enhancement to the virtual-8086 mode to allow for 4 MB as well as 4 KB pages.
- 128-bit and 256-bit data paths are possible (although the main registers are still 32 bits).
- Burstable 64-bit external data bus.
- Addition of advanced programmable interrupt controller (APIC) to support multiple Pentium processors.
- New dual processing mode to support dual processor systems.

The Pentium processor has been extremely successful and has helped support enhanced multitasking operating systems such as Microsoft Windows. The Intel Pentium Pro enhanced the Pentium processor with the following:

- Incorporation of a three-way superscalar architecture, as apposed to a 2-way for the Pentium. This allows three instructions to be executed for every clock cycle.
- Uses enhanced prediction of parallel code (called dynamic execution microarchitecture) for the superscalar operation. This includes methods such as microdata flow analysis, out-of-order execution, enhanced branch prediction and speculative execution. The three instruction decode units work in parallel to decode object code into smaller operations called micro-ops. These micro-ops then go into an instruction pool, and, when there are no interdependencies they can be executed out-of-order by the five parallel execution units (two integer units, two for floating-point operations and one for memory operations). A retirement unit retires completed micro-ops in their original program order, taking account of any branches. This recovers the original program flow.
- Addition of register renaming. Multiple instructions not dependent on each other, using the same registers, allow the source and destination registers to be temporarily renamed. The original register names are used when instructions are retired and program flow is maintained.
- Addition of a, closely coupled, on-package, 256 KB L2 cache which has a dedicated 64-bit full clock speed bus. The L2 cache also supports up to four concurrent accesses through a 64-bit external data bus. Each of these accesses is transaction-oriented where each access is handled as a separate request and response. This allows for numerous requests while awaiting a response.
- Expanded 36-bit address bus to give a physical address size of 64 GB.

The Pentium II/III processor is a further enhancement to the processor range. Apart from increasing the clock speed it has several enhancements over the Pentium Pro, including:

- Integration of MMX technology. MMX instructions support high-speed multimedia operations and include the addition of eight new registers (MM0 to MM7), four MMX data types and an MMX instruction set.
- Single edge contact (SEC) cartridge packaging. This gives improved handling performance and socketability. It uses surface mount components and has a thermal plate (which accepts a standard heat sink), a cover and a substrate with an edge finger connection.
- Integrated on-chip L1 cache 16 KB for code and another 16 KB for data. This has since been increased to 512 KB cache.
- Increased size, on-package, 512 KB L2 cache.
- Enhanced low-power states, such as AutoHALT, Stop-Grant, Sleep and Deep Sleep.

A.4.2 Terms

Before giving an introduction to the Pentium Pro various terms have to be defined. These are given in Table A.4.

Table A.4 Pentium terms

Term	*Description*
Transaction	Used to define a bus cycle. It consists of a set of phases, which are related to a single bus request.
Bus agent	Devices that reside on the processor bus, that is, the processor, PCI bridge and memory controller.
Priority agent	The device handling reset, configuration, initialization, error detection and handling; generally the processor-to-PCI bridge.
Requesting agent	The device driving the transaction, that is, busmaster.
Addressed agent	The slave device addressed by the transaction, that is, target agent.
Responding agent	The device that provides the transaction response on RS<2:0># signals.
Snooping agent	A caching device that snoops on the transactions to maintain cache coherency.
Implicit write-back	When a hit to a modified line is detected during the snoop phase, an implicit write-back occurs. This is the mechanism used to write-back the cache line.

A.4.3 Pentium II/III and Pentium Pro

A major objective of electronic systems design is the saving of electrical power and the minimisation of electromagnetic interference (EMI). Thus gunning transceiver logic (GTL) has been used to reduce both power consumption and EMI as it has a low voltage swing. GTL requires a 1 V reference signal and signals which use GTL logic are terminated to 1.5 V. If a signal is 0.2 V above the reference voltage, that is, 1.2 V, then it is considered HIGH. If a signal is 0.2 V below the reference voltage, that is, 0.8 V, then it is considered LOW.

The Pentium Pro and II support up to four processors on the same bus without extra supporting logic. Integrated into the bus structure are cache coherency signals, advanced programmable interrupt control signals and bus arbitration.

A great enhancement of the Pentium Pro bus is data error detection and correction. The Pentium Pro bus detects and corrects all single-bit data bus errors and also detects multiple-bit errors on the data bus. Address and control bus signals have basic parity protection.

The Pentium Pro bus has a modified line write-back performed without backing off the current bus owner, where the processor must perform a write-back to memory when it detects a hit to a modified line. The following mechanism eliminates the need to back-off the current busmaster. If a memory write is being performed by the current bus owner then two writes will be seen on the bus, that is, the original one followed by the write-back. The memory controller latches, and merges the data from the two cycles, and performs one write to DRAM. If a memory read is being performed by the current bus owner then it accepts the data when it is being written to memory.

Other enhanced features are:

- Deferred reply transactions stop the processor from having to wait for slow devices; transactions that require a long time can be completed later, that is, deferred.
- Deeply pipelined bus transactions where the bus supports up to eight outstanding pipelined transactions.

The Pentium III processor integrates the best features of the Pentium microarchitecture processors, such as dynamic execution performance, a multitransaction system bus, and Intel MMX media enhancement technology. In addition, the Pentium III processor offers Streaming SIMD Extensions with 70 new instructions enabling advanced imaging, 3D, streaming audio and video, and speech recognition applications.

A.4.4 System overview

Figure A.10 outlines the main components of a Pentium system. A major upgrade is the support for up to four processors. The memory control and data path control logic provides the memory control signals, that is, memory address, $\overline{\text{RAS}}$ and $\overline{\text{CAS}}$ signals. The data path logic moves the data between the processor bus and the memory data bus. The memory interface component interfaces the memory data bus with the DRAM devices. Both interleaved and non-interleaved methods are generally supported. The memory consists of dual in line memory modules, that is, DIMMs. A DIMM module supports 64 bits of data, and eight parity or ECC bits.

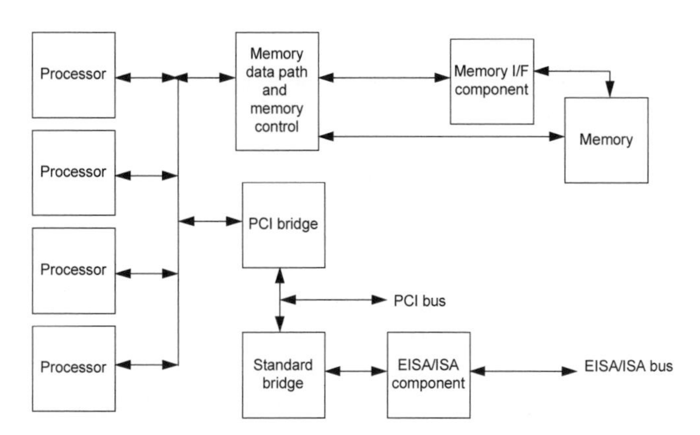

Figure A.10 Pentium architecture

The PCI bridge provides the interface between the processor bus and the PCI bus. The standard bridge provides an interface between the PCI bus and the EISA/ISA bus. The EISA/ISA support component provides the EISA/ISA bus support functions, for example, timers, interrupt control, flash ROM, keyboard interface, LA/SA translation and XD bus control.

A.5 Exercises

A.5.1 Which was one of the first 8-bit Intel processors:

(a) 8000 (b) 8080
(c) 8888 (d) 8280

A.5.2 The 8086 is classified as which type of processor:

(a) 8-bit (b) 16-bit
(c) 24-bit (d) 32-bit

A.5.3 The 80386 is classified as which type of processor:

(a) 8-bit (b) 16-bit
(c) 24-bit (d) 32-bit

A.5.4 How many bits does the address bus on the 8086 have:

(a) 8 (b) 16
(c) 20 (d) 32

A.5.5 How many bits does the address bus on the 8088 have:

(a) 8 (b) 16
(c) 20 (d) 32

A.5.6 How many bits does the data bus on the 8086 have:

(a) 8 (b) 16
(c) 20 (d) 32

A.5.7 How many bits does the data bus on the 8088 have:

(a) 8 (b) 16
(c) 20 (d) 32

A.5.8 What is the number of the general-purpose register in the 8086:

(a) AX (b) BX
(c) CX (d) DX

A.5.9 What register is used in counting operations register in the 8086:

(a) AX (b) BX
(c) CX (d) DX

A.5.10 What is the physical address for the segment address of `4444:0333h`:

(a) 4777h (b) 7774h
(c) 7777h (d) 44773h

A.5.11 How does a second-level cache differ from a primary cache:

(a) It is on the motherboard, and not located beside the processor
(b) It is located beside the processor, and not located on the motherboard
(c) It is faster
(d) It is made from DRAM memory rather than SRAM

A.5.12 What happens when a cache miss occurs in the second-level cache controller:

(a) It takes data from the nearest cached address
(b) The cache controller makes a guess about the data
(c) An access is made to the main DRAM memory
(d) The addressed memory is ignored

A.5.13 Explain how the byte enable lines (BE_0–BE_3) are used to address one or more bytes at a time. Outline how these lines are used with the other address lines (A_2–A_{31}).

A.5.14 Explain the method that the 80386/80486 uses to support 8-bit, 16-bit and 32-bit registers.

A.5.15 Outline the main cache control architectures.

A.5.16 Outline major enhancements that have occurred with the Pentium (if possible, access the www.intel.com WWW site and determine the most up-to-date information on the latest processors).

B VESA VL-Local Bus

Table B.1 lists the pin connections for the 32-bit VL-local bus and it shows that, in addition to the standard ISA connector, there are two sides of connections, the A and the B side. Each side has 58 connections giving a total of 116 connections. It has a full 32-bit data and address bus. The 32 data lines are labelled DAT00–DAT31 and 32 address lines are labelled from ADR00-ADR31. Note that although the data and address lines are contained within the extra VL-local bus extension, some of the standard ISA lines are used, such as the IRQ lines.

The VL-local bus uses a standard ISA connector and an extra connector to tap into the 32-bit data and address busses. Table B.1 lists the additional 32-bit VESA VL-local bus connections. It has a 32-bit data bus (D0–D31) and a full 32-bit address bus (A0–A31). The VL-Local bus is an extension to the standard ISA bus and can thus use the interrupt lines on the ISA bus connector, that is, IRQ3–IRQ7 and IRQ10–IRQ14. It also has the memory addressing lines ($M/\overline{IO}$ and $R/\overline{W}$).

Table B.1 32-bit VESA VL-local bus connections

Pin	*Side-A*	*Side-B*	*Pin*	*Side-A*	*Side-B*
1	D0	D1	30	A17	A16
2	D2	D3	31	A15	A14
3	D4	GND	32	VCC	A12
4	D6	D5	33	A13	A10
5	D8	D7	34	A11	A8
6	GND	D9	35	A9	GND
7	D10	D11	36	A7	A6
8	D12	D13	37	A5	A4
9	VCC	D15	38	GND	$\overline{WBACK}$
10	D14	GND	39	A3	$\overline{BE0}$
11	D16	D17	40	A2	VCC
12	D18	VCC	41	NC	$\overline{BE1}$
13	D20	D19	42	$\overline{RESET}$	$\overline{BE2}$
14	GND	D21	43	$D/\overline{C}$	GND
15	D22	D23	44	$M/\overline{IO}$	$\overline{BE3}$
16	D24	D25	45	$W/\overline{R}$	$\overline{ADS}$
17	D26	GND	46	KEY	KEY
18	D28	D27	47	KEY	KEY
19	D30	D29	48	$\overline{RDYRTN}$	$\overline{LRDY}$
20	VCC	D31	49	GND	$\overline{LDEV}$
21	A31	A30	50	IRQ9	$\overline{LREQ}$
22	GND	A28	51	$\overline{BRDY}$	GND
23	A29	A26	52	$\overline{BLAST}$	$\overline{LGNT}$
24	A27	GND	53	IDO VCC	
25	A25	A24	54	ID1	ID2
26	A23	A22	55	GND	ID3
27	A21	VCC	56	LCLK	ID4
28	A19	A20	57	VCC	NC
29	GND	A18	58	$\overline{LBS16}$	$\overline{LEADS}$

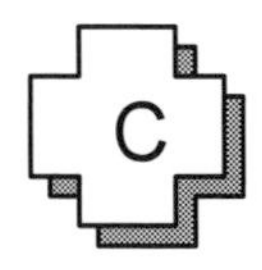

Modem Codes

C.1 AT commands

The AT commands are preceded by the attention code AT. They are:

A — **Go on-line in answer mode**
Instructs the modem to go off-hook immediately and then make a connection with a remote modem

Bn — **Select protocol to 300 bps to 1200 bps**
B0 Selects CCITT operation at 300 bps or 1200 bps
B1 Selects BELL operation at 300 bps or 1200 bps

D — **Go on-line in originate mode**
Instructs the modem to go off-hook and automatically dials the number contained in the dial string which follows the D command

En — **Command echo**
E0 Disable command echo
E1 Enables command echo (default)

Fn — **Select line modulation**
F0 Select auto-detect mode
F1 Select V.21 or Bell 103
F4 Select V.22 or Bell 212A 1200 bps
F5 Select V.22bis line modulation.
F6 Select V.32bis or V.32 4800 bps line modulation
F7 Select V.32bis or V.32 7200 bps line modulation
F8 Select V.32bis or V.32 9600 bps line modulation
F9 Select V.32bis 12000 line modulation
F10 Select V.32bis 14400 line modulation

Hn — **Hang-up**
H0 Go on-hook (hang-up connection)
H1 Goes off-hook

In — **Request product code or ROM checksum**
I0 Reports the product code
I1/I2 Reports the hardware ROM checksum
I3 Reports the product revision code
I4 Reports response programmed by an OEM
I5 Reports the country code number

Ln — **Control speaker volume**
L0 Low volume
L1 Low volume
L2 Medium volume (default)
L3 High volume

Mn — **Monitor speaker on/off**
M0/M Speaker is always off
M1 Speaker is off while receiving carrier (default)
M2 Speaker is always on
M3 Speaker is on when dialing but is off at any other time

Nn — **Automode enable**
N0 Automode detection is disabled
N1 Automode detection is enabled

On — **Return to the on-line state**
O0 Enters on-line data mode with a retrain
O1 Enters on-line data mode without a retrain

P — **Set pulse dial as default**

Q — **Result code display**
Q0 Send result codes to the computer
Q1 No return codes

Sn — **Reading and writing to S registers**
Sn? Reads the Sn register

	Sn=val	Writes the value of val to the Sn register
T	**Set tone dial as default**	
Vn	**Select word or digit result code**	
	V0	Display result codes in a numeric form
	V1	Display result code in a long form (default)
Wn	**Error correction message control**	
	W0	When connected report computer connection speed
	W1	When connected report computer connection speed, error correcting protocol and line speed
	W2	When connected report modem connection speed
Xn	**Select result code**	
	X0	Partial connect message, dial-tone monitor off, busy tone monitor off
	X1	Full connect message, dial-tone monitor off, busy tone monitor off
	X2	Full connect message, dial-tone monitor on, busy tone monitor off
	X3	Full connect message, dial-tone monitor off, busy tone monitor on
	X4	Full connect message, dial-tone monitor on, busy tone monitor on
Yn	**Enables or disables long space disconnection**	
	Y0	Disables long space disconnect (default)
	Y1	Enables long space disconnect
Zn	**Reset**	
	Z0	Resets modem and load stored profile 0
	Z1	Resets modem and load stored profile 1
&Cn	**Select DCD options**	
	&C0	Sets DCD permanently on
	&C1	Use state of carrier to set DCD (default)
&Dn	**DTR option**	

This is used with the &Qn setting to determine the operation of the DTR signal

	&D0	&D1	&D2	&D3
&Q0	a	c	d	e
&Q1	b	c	d	e
&Q2	d	d	d	d
&Q3	d	d	d	d
&Q4	b	c	d	e
&Q5	a	c	d	e
&Q6	a	c	d	e

where
a – modem ignore DTR signal
b – modem disconnects and sends OK result code
c – modem goes into command mode and sends OK result code
d – modem disconnects and sends OK result code.

&F	**Restore factory configuration**	
&Gn	**Set guard tone**	
	&G0	Disables guard tone (default)
	&G1	Disables guard tone
	&G2	Selects 1800 Hz guard tone
&Kn	**DTE/modem flow control**	
	&K0	Disables DTE/DCE flow control
	&K3	Enables RTS/CTS handshaking flow control (default)
	&K4	Enables XON/XOFF flow control
	&K5	Enables transparent XON/XOFF flow control
	&K6	Enables RTS/CTS and XON/XOFF flow control
&L	**Line selection**	
	&L0	Selects dial-up line operation (default)
	&L1	Selects leased line operation
&Mn	**Communications mode**	
&Pn	**Select pulse dialing make/break ratio**	
	&P0	Sets a 39/61 make-break ratio at 10 pps (default)
	&P1	Sets a 33/67 make-break ratio at 10 pps (default)
	&P2	Sets a 39/61 make-break ratio at 20 pps (default)
	&P3	Sets a 33/67 make-break ratio at 20 pps (default)

Command	Option	Description
&Qn	**Asynchronous/synchronous mode selection**	
	&Q0	Set direct asynchronous operation
	&Q1	Set synchronous operation with asynchronous off-line
	&Q2	Set synchronous connect mode with asynchronous off-line
	&Q3	Set synchronous connect mode
	&Q5	Modem negotiation for error-corrected link
	&Q6	Set asynchronous operation in normal mode
&Rn	**RTS/CTS option**	
	&R0	In synchronous mode, CTS changes with RTS (the delay is defined by the S26 register)
	&R1	In synchronous mode, CTS is always ON
&Sn	**DSR option**	
	&S0	DSR is always ON (default)
	&S1	DSR is active after the answer tone has been detected
&Tn	**Testing and diagnostics**	
	&T0	Terminates any current test
	&T1	Local analogue loopback test
	&T2	Local digital loopback test
&V	**View configuration profiles**	
&Wn	**Store the current configuration in non-volatile RAM**	
	&W0	Writes current settings to profile 0 in nonvolatile RAM
	&W1	Writes current settings to profile 1 in nonvolatile RAM
&Xn	**Clock source selection**	
	&X0	Selects internal timing, where the modem uses its own clock for transmitted data
	&X1	Selects external timing, where the modem gets its timing from the DTE (computer)
	&X2	Selects slave receive timing, where the modem gets its timing from the received signal
&Yn	**Select default profile**	
	&Y0	Use profile 0 on power-up (default)
	&Y1	Use profile 1 on power-up
&Zn	**Store telephone numbers**	
	&Z0	Store telephone number 1 &Z1 Store telephone number 2
	&Z2	Store telephone number 3 &Z3 Store telephone number 4
\An	**Maximum MNP block size**	
	\A0	64 characters \A1 128 characters
	\A2	192 characters \A3 256 characters
\Bn	**Transmit break**	
	\B1	Break length 100 ms \B2 Break length 200 ms
	\B3	Break length 300 ms (Default) *and so on.*
\Gn	**Modem/modem flow control**	
	\G0	Disable (Default) \G1 Enable
\Jn	**Enable/disable DTE auto rate adjustment**	
	\J0	Disable \J1 Enable
\Kn	**Break control**	
	\K0	Enter on-line command mode with no break signal
	\K1	Clear data buffers and send a break to the remote modem
	\K3	Send a break to the remote modem immediately
	\K5	Send a break to the remote modem with transmitted data
\Ln	**MNP block transfer control**	
	\L0	Use stream mode for MNP connection (default)
	\L1	Use interactive MNP block mode.

C.2 Result codes

After the modem has received an AT command it responds with a return code. A complete set of return codes are given in Table C.1.

Table C.1 Modem return codes

Message	*Digit*	*Description*
OK	0	Command executed without errors
CONNECT	1	A connection has been made
RING	2	An incoming call has been detected
NO CARRIER	3	No carrier detected
ERROR	4	Invalid command
CONNECT 1200	5	Connected to a 1200 bps modem
NO DIAL-TONE	6	Dial-tone not detected
BUSY	7	Remote line is busy
NO ANSWER	8	No answer from remote line
CONNECT 600	9	Connected to a 600 bps modem
CONNECT 2400	10	Connected to a 2400 bps modem
CONNECT 4800	11	Connected to a 4800 bps modem
CONNECT 9600	13	Connected to a 9600 bps modem
CONNECT 14400	15	Connected to a 14 400 bps modem
CONNECT 19200	16	Connected to a 19200 bps modem
CONNECT 28400	17	Connected to a 28400 bps modem
CONNECT 38400	18	Connected to a 38400 bps modem
CONNECT 115200	19	Connected to a 115200 bps modem
FAX	33	Connected to a FAX modem in FAX mode
DATA	35	Connected to a data modem in FAX mode
CARRIER 300	40	Connected to V.21 or Bell 103 modem
CARRIER 1200/75	44	Connected to V.23 backward channel carrier modem
CARRIER 75/1200	45	Connected to V.23 forwards channel carrier modem
CARRIER 1200	46	Connected to V.22 or Bell 212 modem
CARRIER 2400	47	Connected to V.22 modem
CARRIER 4800	48	Connected to V.32bis 4800 bps modem
CONNECT 7200	49	Connected to V.32bis 7200 bps modem
CONNECT 9600	50	Connected to V.32bis 9600 bps modem
CONNECT 12000	51	Connected to V.32bis 12000 bps modem
CONNECT 14400	52	Connected to V.32bis 14400 bps modem
CONNECT 19200	61	Connected to a 19 200 bps modem
CONNECT 28800	65	Connected to a 28 800 bps modem
COMPRESSION: CLASS 5	66	Connected to modem with MNP Class 5 compression
COMPRESSION: V.42bis	67	Connected to a V.42bis modem with compression
COMPRESSION: NONE	69	Connection to a modem with no data compression
PROTOCOL: NONE	70	
PROTOCOL: LAPM	77	
PROTOCOL: ALT	80	

C.3 S-registers

The modem contains various status registers called the S-registers which store modem settings. Table C.2 lists these registers.

S14 **Bitmapped options**

	0	1
Bit 1	E0	**E1**
Bit 2	**Q0**	Q1
Bit 3	V0	**V1**
Bit 4	Reserved	
Bit 5	**T** (tone dial)	P (pulse dial)
Bit 6	Reserved	
Bit 7	Answer mode	**Originate mode**

S16 **Modem test mode register**

	0	1
Bit 0	Local analogue loopback	Local analogue loopback

Register	Bit	0	1
		terminated	test in progress
	Bit 2	Local digital loopback terminated	Local digital loopback test in progress
	Bit 3	Remote modem analogue loopback test terminated	Remote modem analogue loopback test in progress
	Bit 4	Remote modem digital loopback test terminated	Remote modem digital loopback test in progress
	Bit 5	Remote modem digital self-test terminated	Remote modem digital self-test in progress
	Bit 6	Remote modem analogue self-test terminated	Remote modem analogue self-test in progress
	Bit 7	Unused	
S21	**Bitmapped options**		
		0	1
	Bit 0	**&J0**	&J1
	Bit 1		
	Bit 2	&R0	**&R1**
	Bit 5	&C0	**&C1**
	Bit 6	**&S0**	&S1
	Bit 7	**Y0**	Y1
	Bit 4, 3 = 00	&D0	
	Bit 4, 3 = 01	&D1	
	Bit 4, 3 = 10	**&D2**	
	Bit 4, 3 = 11	&D3	
S22	**Speaker/results bitmapped options**		
	Bit 1, 0 = 00	L0	
	Bit 1, 0 = 01	**L1**	
	Bit 1, 0 = 10	L2	
	Bit 1, 0 = 11	L3	
	Bit 3, 2 = 00	M0	
	Bit 3, 2 = 01	**M1**	
	Bit 3, 2 = 10	M2	
	Bit 3, 2 = 11	M3	
	Bit 6, 5, 4 = 000	X0	
	Bit 6, 5, 4 = 001	Reserved	
	Bit 6, 5, 4 = 010	Reserved	
	Bit 6, 5, 4 = 011	Reserved	
	Bit 6, 5, 4 = 100	X1	
	Bit 6, 5, 4 = 101	X2	
	Bit 6, 5, 4 = 110	X3	
	Bit 6, 5, 4 = 111	**X4**	
	Bit 7 Reserved		
S23	**Bitmapped options**		
		0	1
	Bit 0	&T5	**&T4**
	Bit 3, 2, 1 = 000	300 bps communications rate	
	Bit 3, 2, 1 = 001	600 bps communications rate	
	Bit 3, 2, 1 = 010	1200 bps communications rate	
	Bit 3, 2, 1 = 011	**2400 bps communications rate**	
	Bit 3, 2, 1 = 100	4800 bps communications rate	
	Bit 3, 2, 1 = 101	9600 bps communications rate	
	Bit 3, 2, 1 = 110	19 200 bps communications rate	
	Bit 3, 2, 1 = 111	Reserved	
	Bit 5, 4 = 00	Even parity	
	Bit 5, 4 = 01	**Not used**	
	Bit 5, 4 = 10	Odd parity	
	Bit 5, 4 = 11	No parity	
	Bit 7, 6 = 00	**G0**	
	Bit 7, 6 = 01	G1	
	Bit 7, 6 = 10	G2	
	Bit 7, 6 = 11	G3	

S23 **Bitmapped options**
Bit 3, 1, 0 = 000 &M0 or &Q0
Bit 3, 1, 0 = 001 &M1 or &Q1
Bit 3, 1, 0 = 010 &M2 or &Q2
Bit 3, 1, 0 = 011 &M3 or &Q3
Bit 3, 1, 0 = 100 &Q3
Bit 3, 1, 0 = 101 &Q4
Bit 3, 1, 0 = 110 **&Q5**
Bit 3, 1, 0 = 111 &Q6

	0	1
Bit 2	**&L0**	&L1
Bit 6	B0	**B1**

Bit 5, 4 = 00 **X0**
Bit 5, 4 = 01 X1
Bit 5, 4 = 10 X2

S28 **Bitmapped options**
Bit 4, 3 = 00 **&P0**
Bit 4, 3 = 01 &P1
Bit 4, 3 = 10 &P2
Bit 4, 3 = 11 &P3

S31 **Bitmapped options**

	0	1
Bit 1	**N0**	N1

Bit 3, 2 = 00 **W0**
Bit 3, 2 = 01 W1
Bit 3, 2 = 10 W2

S36 **LAPM failure control**
Bit 2, 1, 0 = 000 Modem disconnect
Bit 2, 1, 0 = 001 Modem stays on line and a direct mode connection
Bit 2, 1, 0 = 010 Reserved
Bit 2, 1, 0 = 011 Modem stays on line and normal mode connection is established
Bit 2, 1, 0 = 100 An MNP connection is made, if it fails then the modem disconnects
Bit 2, 1, 0 = 101 An MNP connection is made, if it fails then the modem makes a direct connection
Bit 2, 1, 0 = 110 Reserved
Bit 2, 1, 0 = 111 An MNP connection is made, if it fails then the modem makes a normal mode connection

S37 **Desired line connection speed**
Bit 3, 2, 1, 0 = 0000 **Auto mode connection (F0)**
Bit 3, 2, 1, 0 = 0001 Modem connects at 300 bps (F1)
Bit 3, 2, 1, 0 = 0010 Modem connects at 300 bps (F1)
Bit 3, 2, 1, 0 = 0011 Modem connects at 300 bps (F1)
Bit 3, 2, 1, 0 = 0100 Reserved
Bit 3, 2, 1, 0 = 0101 Modem connects at 1200 bps (F4)
Bit 3, 2, 1, 0 = 0110 Modem connects at 2400 bps (F5)
Bit 3, 2, 1, 0 = 0111 Modem connects at V.23 (F3)
Bit 3, 2, 1, 0 = 1000 Modem connects at 4800 bps (F6)
Bit 3, 2, 1, 0 = 1001 Modem connects at 9600 bps (F8)
Bit 3, 2, 1, 0 = 1010 Modem connects at 12 000 bps (F9)
Bit 3, 2, 1, 0 = 1011 Modem connects at 144 000 bps (F10)
Bit 3, 2, 1, 0 = 1100 Modem connects at 7200 bps (F7)

S39 **Flow control**
Bit 2, 1, 0 = 000 No flow control
Bit 2, 1, 0 = 011 **RTS/CTS (&K3)**
Bit 2, 1, 0 = 100 XON/XOFF (&K4)
Bit 2, 1, 0 = 101 Transparent XON (&K5)
Bit 2, 1, 0 = 110 RTS/CTS and XON/XOFF (&K6)

S39 **General bitmapped options**
Bit 5, 4, 3 = 000 \K0
Bit 5, 4, 3 = 001 \K1
Bit 5, 4, 3 = 010 \K2

Bit 5, 4, 3 = 011 \K3
Bit 5, 4, 3 = 100 \K4
Bit 5, 4, 3 = 101 **\K5**
Bit 7, 6 = 00 MNP 64 character block size (\A0)
Bit 7, 6 = 01 **MNP 128 character block size (\A1)**
Bit 7, 6 = 10 MNP 192 character block size (\A2)
Bit 7, 6 = 11 MNP 256 character block size (\A3)

Table C.2 Modem registers

Register	*Function*	*Range [typical default]*
S0	Rings to Auto-answer	0–255 rings [0 rings]
S1	Ring counter	0–255 rings [0 rings]
S2	Escape character	[43]
S3	Carriage return character	[13]
S6	Wait time for dial-tone	2–255 s [2 s]
S7	Wait time for carrier	1–255 s [50 s]
S8	Pause time for automatic dialing	0–255 s [2 s]
S9	Carrier detect response time	1–255 in 0.1 s units [6]
S10	Carrier loss disconnection time	1–255 in 0.1 s units [14]
S11	DTMF tone duration	50–255 in 0.001 s units [95]
S12	Escape code guard time	0–255 in 0.02 s units [50]
S13	Reserved	
S14	General bitmapped options	[8Ah (1000 1010b)]
S15	Reserved	
S16	Test mode bitmapped options (&T)	[0]
S17	Reserved	
S18	Test timer	0–255 s [0]
S19–S20	Reserved	
S21	V.24/General bitmapped options	[04h (0000 0100b)]
S22	Speak/results bitmapped options	[75h (0111 0101b)]
S23	General bitmapped options	[37h (0011 0111b)]
S24	Sleep activity timer	0–255 s [0]
S25	Delay to DSR off	0–255 s [5]
S26	RTS–CTS delay	0–255 in 0.01 s [1]
S27	General bitmapped options	[49h (0100 1001b)]
S28	General bitmapped options	[00h]
S29	Flash dial modifier time	0–255 in 10 ms [0]
S30	Disconnect inactivity timer	0–255 in 10 s [0]
S31	General bitmapped options	[02h (0000 0010b)]
S32	XON character	[Cntrl–Q, 11h (0001 0001b)]
S33	XOFF character	[Cntrl–S, 13h (0001 0011b)]
S34–S35	Reserved	
S36	LAMP failure control	[7]
S37	Line connection speed	[0]
S38	Delay before forced hang-up	0–255 s [20]
S39	Flow control	[3]
S40	General bitmapped options	[69h (0110 1001b)]
S41	General bitmapped options	[3]
S42–S45	Reserved	
S46	Data compression control	[8Ah (1000 1010b)]
S48	V.42 negotiation control	[07h (0000 0111b)]
S80	Soft-switch functions	[0]
S82	LAPM break control	[40h (0100 0000b)]
S86	Call failure reason code	0–255
S91	PSTN transmit attenuation level	0–15 dBm [10]
S92	Fax transmit attenuation level	0–15 dBm [10]
S95	Result code message control	[0]
S99	Leased line transmit level	0–15 dBm [10]

D Redundancy checking

D.1 Cyclic redundancy check (CRC)

The CRC is one of the most reliable error detection schemes and can detect up to 95.5% of all errors. The most commonly used code is the CRC-16 standard code which is defined by the CCITT.

The basic idea of a CRC can be illustrated using an example. Suppose the transmitter and receiver were both to agree that the numerical value sent by the transmitter would always be divisible by 9. Then should the receiver get a value which was not divisible by 9 then it would know that there was an error. For example, if a value of 32 were to be transmitted it could be changed to 320 so that the transmitter would be able to add to the least significant digit, making it divisible by 9. In this case the transmitter would add 4, making 324. If this transmitted value were to be corrupted in transmission then there would only be a 10% chance that an error would not be detected.

In CRC-CCITT, the error correction code is 16 bits long and is the remainder of the data message polynomial $G(x)$ divided by the generator polynomial $P(x)$ ($x^{16}+x^{12}+x^5+1$, i.e. 10001000000100001). The quotient is discarded and the remainder is truncated to 16 bits. This is then appended to the message as the coded word.

The division does not use standard arithmetic division. Instead of the subtraction operation an exclusive-OR operation is employed. This is a great advantage as the CRC only requires a shift register and a few XOR gates to perform the division.

The receiver and the transmitter both use the same generating function $P(x)$. If there are no transmission errors then the remainder will be zero.

The method used is as follows:

1. Let $P(x)$ be the generator polynomial and $M(x)$ the message polynomial.
2. Let n be the number of bits in $P(x)$.
3. Append n zero bits onto the right-hand side of the message so that it contains $m+n$ bits.
4. Using modulo-2 division, divide the modified bit pattern by $P(x)$. Modulo-2 arithmetic involves exclusive-OR operations, i.e. `0 - 1 = 1`, `1 - 1 = 0`, `1 - 0 = 1` and `0 - 0 = 0`.
5. The final remainder is added to the modified bit pattern.

Example: For a 7-bit data code 1001100 determine the encoded bit pattern using a CRC generating polynomial of $P(x)=x^3+x^2+x^0$. Show that the receiver will not detect an error if there are no bits in error.

Answer

$P(x)=x^3+x^2+x^0$ (1101)

$G(x)=x^6+x^3+x^2$ (1001100)

Multiply by the number of bits in the CRC polynomial:

$$x^3(x^6+x^3+x^2)$$
$$x^9+x^6+x^5 \qquad (1001100000)$$

Figure D.1 shows the operations at the transmitter. The transmitted message is 1001100001 and Figure D.2 shows the operations at the receiver. It can be seen that the remainder is zero, so there have been no errors in the transmission.

```
           1111101
1101 | 1001100000
       1101
        1001
        1101
         1000
         1101
          1010
          1101
           1110
           1101
             1100
             1101
              001
```

Figure D.1 CRC coding example

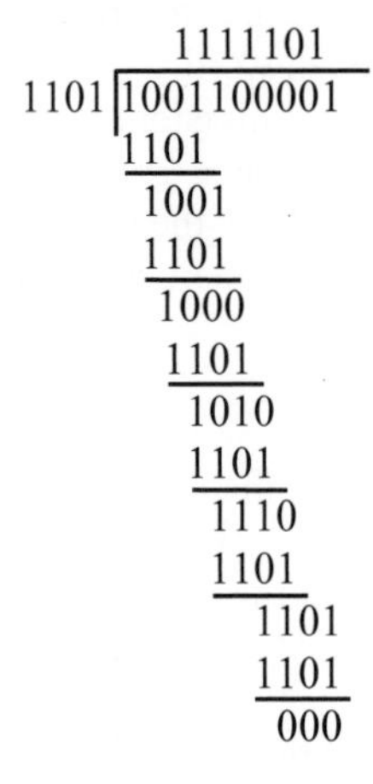

Figure D.2 CRC decoding example

The CRC-CCITT is a standard polynomial for data communications systems and can detect:

- All single and double bit errors.
- All errors with an odd number of bit.
- All burst errors of length 16 or less.
- 99.997% of 17-bit error bursts.
- 99.998% of 18-bit and longer bursts.

Table D.1 lists some typical CRC codes. CRC-32 is used in Ethernet, Token Ring and FDDI networks, whereas ATM uses CRC-8 and CRC-10.

Table D.1 Typical schemes

Type	*Polynomial*	*Polynomial binary equivalent*
CRC-8	$x^8+x^2+x^1+1$	100000111
CRC-10	$x^{10}+x^9+x^5+x^4+x^1+1$	11000110011
CRC-12	$x^{12}+x^{11}+x^3+x^2+1$	1100000001101
CRC-16	$x^{16}+x^{15}+x^2+1$	11000000000000101
CRC-CCITT	$x^{16}+x^{12}+x^5+1$	10001000000100001
CRC-32	$x^{32}+x^{26}+x^{23}+x^{16}+x^{12}+x^{11}$ $+x^{10}+x^8+x^7+x^5+x^4+x^2+x+1$	100000100100000010001110110110111

D.1.1 Mathematical representation of the CRC

The main steps to CRC implementation are:

1. Prescale the input polynomial of $M'(x)$ by the highest order of the generator polynomial $P(x)$.

$$M'(x) = x^n M(x)$$

2. Next divide $M'(x)$ by the generator polynomial to give:

$$\frac{M'(x)}{G(x)} = \frac{x^n M(x)}{G(x)} = Q(x) + \frac{R(x)}{G(x)}$$

which yields

$$x^n M(x) = G(x)Q(x) + R(x)$$

and rearranging gives

$$x^n M(x) + R(x) = G(x)Q(x)$$

This means that the transmitted message ($x^n M(x) + R(x)$) is now exactly divisible by $G(x)$.

D.1.2 CRC example

Question A

A CRC system uses a message of $1+x^2+x^4+x^5$. Design a cyclic encoder circuit with generator polynomial $G(x)=1+x^2+x^3$ and having appropriate gating circuitry to enable/disable the shift out of the CRC remainder.

Answer A

The generator polynomial is $G(x)=1+x^2+x^3$, the circuit is given in Figure D.3.

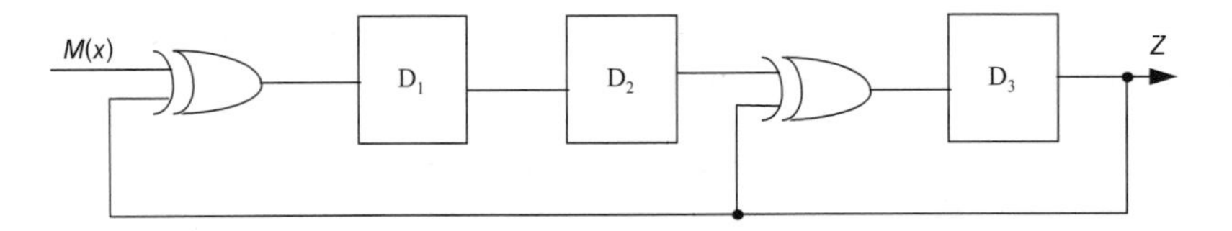

Figure D.3 CRC coder

Now to prove that this circuit does generate the polynomial. The output $Z(x)$ will be

$$Z(x) = Z(x)x^{-1} + \left[M(x)x^{-2} + Z(x)x^{-2} \right]x^{-1}$$
$$= Z(x)\left(x^{-3} + x^{-1}\right) + M(x)x^{-3}$$

Thus

$$M(x) = \frac{Z(x)\left(1 + x^{-1} + x^{-3}\right)}{x^{-3}}$$

giving

$$P(x) = \frac{M(x)}{Z(x)} = x^3 + x^2 + 1$$

Question B
If the previous CRC system uses a message of $1+x^2+x^4+x^5$ then determine the sequence of events that occur and hence determine the encoded message as a polynomial $T(x)$. Synthesise the same code algebraically using modulo-2 division.

Answer B
First prescale the input polynomial of $M(x)$ by x^3, the highest power of $G(x)$:

$$M'(x)=x^3M(x)= x^3+x^5+x^7+x^8$$

The input is thus $x^3+x^5+x^7+x^8$ (000101011), and the generated states are:

Time	$M'(x)$	D_1	D_2	D_3	D_4	
1	000101011	0	0	0	0	
2	00010101	1	0	0	0	← MSD
3	0001010	1	1	0	0	
4	000101	0	1	1	1	
5	00010	0	0	0	0	
6	0001	0	0	0	0	
7	000	1	0	0	0	
8	00	0	1	0	0	
9	0	0	0	1	1	← LSD
10		**1**	**0**	**1**		

The remainder is thus 101, so $R(x)$ is x^2+1. The transmitted polynomial will be

$$T(x)=x^3\,M(x)+R(x)=x^8+x^7+x^5+x^3+x^2+1 \;(110101101)$$

To check this, use modulo-2 division to give

$$\begin{array}{r|l} & x^5 \qquad\quad +1 \\ \hline x^3+x^2+1 & x^8+x^7+x^5+x^3 \\ & x^8+x^7+x^5 \\ \cline{2-2} & \qquad\qquad x^3 \\ & \qquad\qquad x^3+x^2+1 \\ \cline{2-2} \text{Remainder} \rightarrow & \qquad\qquad \boxed{x^2+1} \end{array}$$

This gives the same answer as the state table, i.e. x^2+1.

Prove that the transmitted message does not generate a remainder when divided by $P(x)$. The transmitted polynomial, $T(x)$, is $x^8+x^7+x^5+x^3+x^2+1$ (110101101) and the generator polynomial, $G(x)$, is $1+x^2+x^3$. Thus

$$\begin{array}{r|l} & x^5 \qquad\quad +1 \\ \hline x^3+x^2+1 & x^8+x^7+x^5+x^3+x^2+1 \\ & x^8+x^7+x^5 \\ \cline{2-2} & \qquad\qquad x^3+x^2+1 \\ & \qquad\qquad x^3+x^2+1 \\ \cline{2-2} \text{Remainder} \rightarrow & \qquad\qquad \boxed{0} \end{array}$$

As there is a zero remainder, there is no error.

D.2 Longitudinal/vertical redundancy checks (LRC/VRC)

RS-232 uses vertical redundancy checks (VRC) when it adds a parity bit to the transmitted character. Longitudinal (or horizontal) redundancy checks (LRC) adds a parity bit for all bits in the message at the same bit position. Vertical coding operates on a single character and is known as character error coding. Horizontal checks operate on groups of characters and described as message coding. LRC always uses even parity and the parity bit for the LRC character has the parity of the VRC code.

In the example given next, the character sent for LRC is thus `10101000` (28h) or a '`(`'. The message sent is '`F`', '`r`', '`e`', '`d`', '`d`', '`y`' and '`(`'.

Without VR checking, LR checking detects most errors but does not detect errors where an even number of characters have an error in the same bit position. In the previous example if bit 2 of the '`F`' and '`r`' were in error then LRC would be valid.

This problem is overcome if LRC and VRC are used together. With VRC/LRC the only time an error goes undetected is when an even number of bits, in an even number of characters, in the same bit positions of each character are in error. This is of course very unlikely.

On systems where only single-bit errors occur, the LRC/VRC method can be used to detect and correct the single-bit error. For systems where more than one error can occur it is not possible to locate the bits in error, so the receiver prompts the transmitter to retransmit the message.

Example

A communications channel uses ASCII character coding and LRC/VRC bits are added to each word sent. Encode the word 'Freddy' and, using odd parity for the VRC and even parity for the LRC; determine the LRC character.

Answer

	F	r	e	d	d	y	LRC
b0	0	0	1	0	0	1	**0**
b1	1	1	0	0	0	0	**0**
b2	1	0	1	1	1	0	**0**
b3	0	0	0	0	0	1	**1**
b4	0	1	0	0	0	1	**0**
b5	0	1	1	1	1	1	**1**
b6	1	1	1	1	1	1	**0**
VRC	0	1	1	0	0	0	**1**

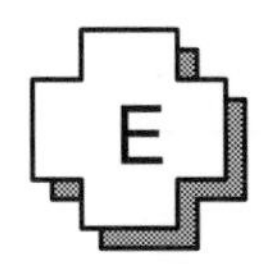

ASCII Character Code

E.1 Standard ASCII

ANSI defined a standard alphabet known as ASCII. This has since been adopted by the CCITT as a standard, known as IA5 (International Alphabet No. 5). The following tables define this alphabet in binary, as a decimal value, as a hexadecimal value and as a character.

Binary	*Decimal*	*Hex*	*Character*	*Binary*	*Decimal*	*Hex*	*Character*
00000000	0	00	NUL	00010000	16	10	DLE
00000001	1	01	SOH	00010001	17	11	DC1
00000010	2	02	STX	00010010	18	12	DC2
00000011	3	03	ETX	00010011	19	13	DC3
00000100	4	04	EOT	00010100	20	14	DC4
00000101	5	05	ENQ	00010101	21	15	NAK
00000110	6	06	ACK	00010110	22	16	SYN
00000111	7	07	BEL	00010111	23	17	ETB
00001000	8	08	BS	00011000	24	18	CAN
00001001	9	09	HT	00011001	25	19	EM
00001010	10	0A	LF	00011010	26	1A	SUB
00001011	11	0B	VT	00011011	27	1B	ESC
00001100	12	0C	FF	00011100	28	1C	FS
00001101	13	0D	CR	00011101	29	1D	GS
00001110	14	0E	SO	00011110	30	1E	RS
00001111	15	0F	SI	00011111	31	1F	US

Binary	*Decimal*	*Hex*	*Character*	*Binary*	*Decimal*	*Hex*	*Character*
00100000	32	20	SPACE	00110000	48	30	0
00100001	33	21	!	00110001	49	31	1
00100010	34	22	"	00110010	50	32	2
00100011	35	23	#	00110011	51	33	3
00100100	36	24	$	00110100	52	34	4
00100101	37	25	%	00110101	53	35	5
00100110	38	26	&	00110110	54	36	6
00100111	39	27	/	00110111	55	37	7
00101000	40	28	(	00111000	56	38	8
00101001	41	29	)	00111001	57	39	9
00101010	42	2A	*	00111010	58	3A	:
00101011	43	2B	+	00111011	59	3B	;
00101100	44	2C	,	00111100	60	3C	<
00101101	45	2D	-	00111101	61	3D	=
00101110	46	2E	.	00111110	62	3E	>
00101111	47	2F	/	00111111	63	3F	?

Binary	*Decimal*	*Hex*	*Character*	*Binary*	*Decimal*	*Hex*	*Character*
01000000	64	40	@	01010000	80	50	P
01000001	65	41	A	01010001	81	51	Q
01000010	66	42	B	01010010	82	52	R
01000011	67	43	C	01010011	83	53	S
01000100	68	44	D	01010100	84	54	T
01000101	69	45	E	01010101	85	55	U
01000110	70	46	F	01010110	86	56	V
01000111	71	47	G	01010111	87	57	W
01001000	72	48	H	01011000	88	58	X
01001001	73	49	I	01011001	89	59	Y
01001010	74	4A	J	01011010	90	5A	Z
01001011	75	4B	K	01011011	91	5B	[
01001100	76	4C	L	01011100	92	5C	\
01001101	77	4D	M	01011101	93	5D	]
01001110	78	4E	N	01011110	94	5E	`
01001111	79	4F	O	01011111	95	5F	_

Binary	*Decimal*	*Hex*	*Character*	*Binary*	*Decimal*	*Hex*	*Character*
01100000	96	60		01110000	112	70	p
01100001	97	61	a	01110001	113	71	q
01100010	98	62	b	01110010	114	72	r
01100011	99	63	c	01110011	115	73	s
01100100	100	64	d	01110100	116	74	t
01100101	101	65	e	01110101	117	75	u
01100110	102	66	f	01110110	118	76	v
01100111	103	67	g	01110111	119	77	w
01101000	104	68	h	01111000	120	78	x
01101001	105	69	i	01111001	121	79	y
01101010	106	6A	j	01111010	122	7A	z
01101011	107	6B	k	01111011	123	7B	{
01101100	108	6C	l	01111100	124	7C	:
01101101	109	6D	m	01111101	125	7D	}
01101110	110	6E	n	01111110	126	7E	~
01101111	111	6F	o	01111111	127	7F	DEL

E.2 Extended ASCII code

The standard ASCII character has 7 bits and the basic set ranges from 0 to 127. This code is rather limited as it does not contains symbols such as Greek letters, lines, and so on. For this purpose the extended ASCII code has been defined. This fits into character numbers 128 to 255. The following four tables define a typical extended ASCII character set.

Binary	*Decimal*	*Hex*	*Character*	*Binary*	*Decimal*	*Hex*	*Character*
10000000	128	80	Ç	10010000	144	90	É
10000001	129	81	ü	10010001	145	91	æ
10000010	130	82	é	10010010	146	92	Æ
10000011	131	83	â	10010011	147	93	ô
10000100	132	84	ä	10010100	148	94	ö
10000101	133	85	à	10010101	149	95	ò
10000110	134	86	å	10010110	150	96	û
10000111	135	87	ç	10010111	151	97	ù
10001000	136	88	ê	10011000	152	98	ÿ
10001001	137	89	ë	10011001	153	99	Ö
10001010	138	8A	è	10011010	154	9A	Ü
10001011	139	8B	ï	10011011	155	9B	¢
10001100	140	8C	î	10011100	156	9C	£
10001101	141	8D	ì	10011101	157	9D	¥
10001110	142	8E	Ä	10011110	158	9E	₧
10001111	143	8F	Å	10011111	159	9F	ƒ

Binary	*Decimal*	*Hex*	*Character*	*Binary*	*Decimal*	*Hex*	*Character*
10100000	160	A0	á	10110000	176	B0	░
10100001	161	A1	í	10110001	177	B1	▒
10100010	162	A2	ó	10110010	178	B2	▓
10100011	163	A3	ú	10110011	179	B3	│
10100100	164	A4	ñ	10110100	180	B4	┤
10100101	165	A5	Ñ	10110101	181	B5	╡
10100110	166	A6	ª	10110110	182	B6	╢
10100111	167	A7	º	10110111	183	B7	╖
10101000	168	A8	¿	10111000	184	B8	╕
10101001	169	A9	⌐	10111001	185	B9	╣
10101010	170	AA	¬	10111010	186	BA	║
10101011	171	AB	½	10111011	187	BB	╗
10101100	172	AC	¼	10111100	188	BC	╝
10101101	173	AD	¡	10111101	189	BD	╜
10101110	174	AE	«	10111110	190	BE	╛
10101111	175	AF	»	10111111	191	BF	┐

Binary	*Decimal*	*Hex*	*Character*	*Binary*	*Decimal*	*Hex*	*Character*
11000000	192	C0	└	11010000	208	D0	╨
11000001	193	C1	┴	11010001	209	D1	╤
11000010	194	C2	┬	11010010	210	D2	╥
11000011	195	C3	├	11010011	211	D3	╙
11000100	196	C4	─	11010100	212	D4	╘
11000101	197	C5	┼	11010101	213	D5	╒
11000110	198	C6	╞	11010110	214	D6	╓
11000111	199	C7	╟	11010111	215	D7	╫
11001000	200	C8	╚	11011000	216	D8	╪
11001001	201	C9	╔	11011001	217	D9	┘
11001010	202	CA	╩	11011010	218	DA	┌
11001011	203	CB	╦	11011011	219	DB	█
11001100	204	CC	╠	11011100	220	DC	▄
11001101	205	CD	═	11011101	221	DD	▌
11001110	206	CE	╬	11011110	222	DE	▐
11001111	207	CF	╧	11011111	223	DF	▀

Binary	*Decimal*	*Hex*	*Character*	*Binary*	*Decimal*	*Hex*	*Character*
11100000	224	E0	α	11110000	240	F0	Ξ
11100001	225	E1	ß	11110001	241	F1	±
11100010	226	E2	Γ	11110010	242	F2	≥
11100011	227	E3	π	11110011	243	F3	≤
11100100	228	E4	Σ	11110100	244	F4	⌠
11100101	229	E5	σ	11110101	245	F5	⌡
11100110	230	E6	µ	11110110	246	F6	÷
11100111	231	E7	τ	11110111	247	F7	≈
11101000	232	E8	Φ	11111000	248	F8	°
11101001	233	E9	Θ	11111001	249	F9	•
11101010	234	EA	Ω	11111010	250	FA	·
11101011	235	EB	δ	11111011	251	FB	√
11101100	236	EC	φ	11111100	252	FC	^{n}
11101101	237	ED	ϕ	11111101	253	FD	^{2}
11101110	238	EE	E	11111110	254	FE	■
11101111	239	EF	Λ	11111111	255	FF	

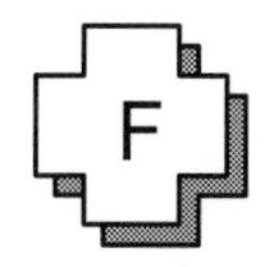

Quick Reference

Parallel port

Pin	Name	Pin	Name
1	Strobe	14	GND
2	Auto Feed	15	D6
3	D0	16	GND
4	Error	17	D7
5	D1	18	GND
6	INIT	19	ACK
7	D2	20	GND
8	SLCT IN	21	BUSY
9	D3	22	GND
10	GND	23	PE
11	D4	24	GND
12	GND	25	SLCT
13	D5		

Serial port

Pin	Name	Pin	Name
1	DCD	6	CTS
2	DSR	7	DTR
3	RX	8	RI
4	RTS	9	GND
5	TX		

IDE

Pin	Name	Pin	Name
1	Reset	2	GND
3	D7	4	D8
5	D6	6	D9
7	D5	8	D10
9	D4	10	D11
11	D3	12	D12
13	D2	14	D13
15	D1	16	D14
17	D0	18	D15
19	GND	20	Key
21	DRQ3	22	GND
23	-I/O W	24	GND
25	-I/O R	26	GND
27	IOCHRDY	28	BALE
29	-DACK3	30	GND
31	IRQ14	32	IOCS16
33	ADD 1	34	GND
35	ADD 0	36	ADD 2
37	-CS 0	38	CS 1
39	ACTIVITY	40	GND

Floppy disk

Pin	Name	Pin	Name
1	GND	2	FDHDIN
3	GND	4	Reserved
5	Key	6	FDEDIN
7	RTS	8	-Index
9	GND	10	Motor En A
11	GND	12	Drive Sel B
13	GND	14	Drive Sel A
15	GND	16	Motor En B
17	GND	18	DIR
19	GND	20	STEP
21	GND	22	Write Data
23	GND	24	Write Gate
25	GND	26	Track 00
27	GND	28	Write Protect
29	GND	30	Read Data
31	GND	32	Side 1 Sel
33	GND	34	Diskette

Bus connections	*Page*	*Bus connections*	*Page*
AGP	205	Cache	126
CAN bus	336	Games port	192
IDE/ISA	144	IEEE-1394	347
IEEE-488 bus	190	ISA bus	88
Keyboard	196	Parallel Port	272

Typical IRQs

IRQ	Use	IRQ	Use
0	Internal timer	1	Keyboard
2	Cascaded interrupt	3	COM2
4	COM1	5	(Soundcard)
6	Floppy disk	7	LPT1
8	Real-time clock	9	User available
10	User available	11	(PCI steering)
12	Serial bus mouse (if any)	13	Math coprocessor
14	Primary IDE	15	Secondary IDE

Typical DMA channels

Channel	Use
0	Any
1	Any
2	Floppy disk
3	Parallel port
4	Cascaded DMA
5	Any
6	Any
7	Any

Example I/O map

Address	Device	Address	Device
0000–000F	Slave DMA controller	0020–0021	Master PIC
0040–0043	System timer	0060	Keyboard
0061	Speaker	0064	Keyboard
0070–0071	Real-time clock	0080–008F	DMA
00A0–00A1	Slave PIC	00F0–00FF	Numeric processor
0170–0177	Secondary H/D	0200–020F	Game port
0220–022F	Soundcard	0294–0297	PCI bus
02F8–02FF	COM2	0330–0331	Soundcard
0370–0371	Soundcard	0376	Secondary IDE
0378–037A	LPT1	0388–03B8	Soundcard
03B0–03BB	VGA	03C0–03DF	VGA
03F6	Primary IDE	03F8–03FF	COM1
0480–048F	PCI bus	04D0–04D1	PCI bus
0530–0537	Soundcard	0778–077A	ECP Port (LPT1)
0CF8–0CFF	PCI bus	4000–403F	PCI bus
5000–5018	PCI bus	D000–DFFF	AGP controller
E000–E01F	USB controller	E400–E4FF	VGA

Bus specification

Bus	*Max. throughput*	*Data bus (bits)*	*Address bus (bits)*	*Notes*
AGP	500 MB/s	64	32	
EISA	32 MB/s	32	32	4 GB max address, 8 MHz clock
Ethernet	1.25 MB/s	1	N/A	10 Mbps (10BASE)
Fibre Channel	132.5 MB/s	1	N/A	1.06 Gbps
Firewire	50 MB/s	1	N/A	400 Mbps (S400)
IDE	16.6 MB/s	16	N/A	Mode 4, EIDE, Maximum 4 devices
IEEE-488	1 MB/s	8	N/A	
ISA	16 MB/s	16	24	16 MB max address, 8 MHz clock
ISDN	16 kB/s	1	N/A	2×64 kbps

MCA	100 MB/s	32	32	
Modem	7 kB/s	1	N/A	56 kbps
Parallel port	150 kB/s	8	N/A	150 kB/s is equivalent to 1.2 Mbps which is the required transfer rate for stereo, 44.1kHz, 16-bit sampled audio
Parallel port	1.2 MB/s	8	N/A	×8
PC	8 MB/s	8	20	1 MB max address, 8 MHz clock
PCI	132 MB/s	32	32	33 MHz clock
PCI (32-bit)	132 MB/s	32	32	33 MHz clock
PCI (64-bit)	264 MB/s	64	32	33 MHz clock
PCMCIA	16 MB/s	16	26	64 MB max address
RS-232	14.4 kB/s	1	N/A	115.2 kbps
RS-485	1.25 MB/s	1	N/A	10 Mbps
SCSI (Fast/wide)	40 MB/s	16	N/A	20 MHz clock
SCSI-I	5 MB/s	8	N/A	
SCSI-II (Wide)	20 MB/s	16	N/A	10 MHz clock
SCSI-II (Fast)	10 MB/s	8	N/A	10 MHz clock
USB	1.5 MB/s	1	N/A	12 Mbps
VL	132 MB/s	32	32	33 MHz clock

E.1 Notes from the author

Well, the book in nearly finished, so as a last little bit of fun here is a final league table for the busses. They are graded on usefulness (how useful it is in its application, and how well it can be used on other systems), availability (the ease that it can be purchased, and the number of applications that it has), data throughput (the speed of data throughput), cost (how expensive it is to purchase applications which use the bus) and configuration (how well and how easy it is to configure the bus).

From the table it can be seen that the winners are the PCI bus and Ethernet (100BASE). Over the past years, the PCI has rapidly moved up the table and takes away the top position from the ISA bus, and as it does everything well, and beats the ISA bus, for its ease of configuration. Its only problem is that it costs a bit more than the ISA bus, but it's worth it. The one to watch for is the USB bus. It has came straight into the forth position, and is sure to rise further as more applications use it, and as it bit rate increases. Busses such as the keyboard, joypad and PS/2 mouse port are not included as they are too specific to a single application (but they wouldn't do that well, as they are very slow, and can cause configuration problems). A special mention should go the Ethernet bus system. It is one of the oldest of the busses given here, but has withstood a lot of pressure from other busses that would like to take control of the networking applications, but it has beaten of all of them. Its main strength is its cheapness, and it general usage. In networking, more any other application area, standardness counts more than virtually anything else. If a company were to adopt a new network bus for their network, and within five years that technology was either too expensive to maintain, or was not even available, it would take a major investment to redesign the network. So, Ethernet wins because it has a virtual monopoly on the connection of computers to corporate networks. Its shortcomings have been overcome with gradual migration. Its slowness has been overcome with new standard such as 100BASE (100MBps) and 1000BASE (1Gbps). Its connection and grounding problems have been solved with hubs, twisted-pair cable and fibre optic cable.

Special mentions should go to RS-232 (the only bus to score three top scores) and the Parallel Port, who do some things extremely well, but their glory days have passed, and are hoping for glorious retirement as USB mops-up their main application areas. But, who knows, will RS-232 and Parallel Port connectors still be standard on the PCs in the year 2010? It's an even money bet at the present.

		Usefulness	*Availability*	*Data throughput*	*Cost*	*Configuration*	*Total*
=1	**PCI**	9	9	8	6	9	41
=1	**Ethernet (100BASE)**	10	9	5	10	7	41
=3	**ISA**	10	9	5	10	5	39
=3	**Ethernet (10BASE)**	10	9	3	10	7	39
=5	**IDE**	5	9	6	9	8	37
=5	**USB**	10	7	4	8	8	37
7	**RS-232**	10	10	2	10	3	35
8	**Parallel port (ECP/EPP)**	8	8	5	8	4	33
=9	**Parallel port**	7	8	3	8	5	31
=9	**SCSI-I**	8	6	5	5	7	31
=9	**AGP**	5	6	9	3	8	31
12	**SCSI-II**	8	4	7	4	7	30
13	**PC**	5	9	3	7	3	27
14	**IEEE-488**	7	5	3	6	5	26
15	**ISDN**	3	6	5	2	5	21
=16	**Modem**	3	9	1	3	4	20
=16	**RS-485**	4	5	4	3	4	20
=18	**Firewire**	3	3	7	2	4	19
=18	**PCMCIA**	4	5	5	1	4	19
=20	**Fibre Channel**	2	2	8	2	4	18
=20	**MCA**	4	1	6	2	5	18
=20	**VL**	5	1	6	1	5	18
23	**EISA**	1	1	3	2	5	12

Relegation zone

G ISDN

G.1 Introduction

A major problem in data communications and networks is the integration of real-time sampled data with non-real-time (normal) computer data. Sampled data tends to create a constant traffic flow whereas computer-type data has bursts of traffic. In addition, sampled data normally needs to be delivered at a given time but computer-type data needs a reliable path where delays are relatively unimportant.

The basic rate for real-time data is speech. It is normally sampled at a rate of 8 kHz and each sample is coded with eight bits. This leads to a transmission bit rate of 64 kbps. ISDN uses this transmission rate for its base transmission rate. Computer-type data can then be transmitted using this rate or can be split to transmit over several 64 kbps channels. The basic rate ISDN service uses two 64 kbps data lines and a 16 kbps control line, as illustrated in Figure G.1. Table G.1 summarizes the I series CCITT standards.

Typically, modems are used in the home for the transmission of computer-type data. Unfortunately, modems have a maximum bit rate of 56 kbps. With ISDN, this is automatically increased, on a single channel, to 64 kbps. The connections made by a modem and by ISDN are circuit switched.

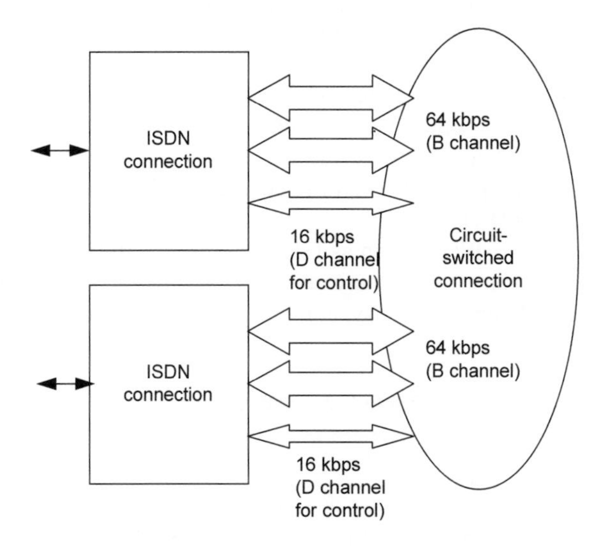

Figure G.1 Basic rate ISDN services

Table G.1 CCITT standards on ISDN

CCITT standard number	*Description*
I.1*XX*	ISDN terms and technology
I.2*XX*	ISDN services
I.3*XX*	ISDN addressing
I.430 and I.431	ISDN physical layer interface
I.440 and I.441	ISDN data layer interface
I.450 and I.451	ISDN network layer interface
I.5*XX*	ISDN internetworking
I.6*XX*	ISDN maintenance

The great advantage of an ISDN connection is that the type of data transmitted is irrelevant to the transmission and switching circuitry. Thus, it can carry other types of digital data, such as facsimile, teletex, videotex and computer data. This reduces the need for modems, which convert digital data into an analogue form, only for the public telephone network to convert the analogue signal back into a digital form for transmission over a digital link. It is also possible to multiplex the basic rate of 64 kbps to give even higher data rates. This multiplexing is known as N × 64 kbps or broadband ISDN (B-ISDN).

Another advantage of ISDN is that it is a circuit-switched connection where a permanent connection is established between two nodes. This connection is guaranteed for the length of the connection. It also has a dependable delay time and is thus suited to real-time data.

G.2 ISDN channels

ISDN uses channels to identify the data rate, each based on the 64 kbps provision. Typical channels are B, D, H0, H11 and H12. The B-channel has a data rate of 64 kbps and provides a circuit-switched connection between endpoints. A D-channel operates at 16 kbps and it controls the data transfers over the B channels. The other channels provide B-ISDN for much higher data rates. Table G.2 outlines the basic data rates for these channels.

The two main types of interface are the basic rate access and the primary rate access. Both are based around groupings of B- and D-channels. The basic rate access allows two B-channels and one 16 kbps D-channel.

Primary rate provides B-ISDN, such as H12 which gives 30 B-channels and a 64 kbps D-channel. For basic and primary rates, all channels multiplex onto a single line by combining channels into frames and adding extra synchronisation bits. Figure G.2 gives examples of the basic rate and primary rate.

Table G.2 ISDN channels

Channel	*Description*
B	64 kbps
D	16 kbps signaling for channel B (ISDN)
	64 kbps signaling for channel B (B-ISDN)
H0	384 kbps (6 × 64 kbps) for B-ISDN
H11	1.536 Mbps (24 × 64 kbps) for B-ISDN
H12	1.920 Mbps (30 × 64 kbps) for B-ISDN

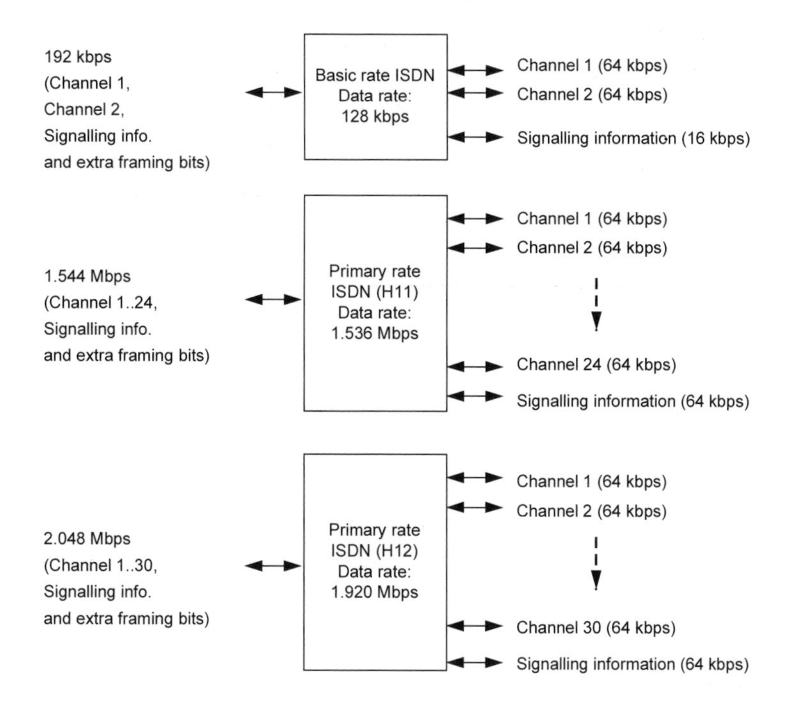

Figure G.2 Basic rate, H11 and H12 ISDN services

The basic rate ISDN gives two B-channels at 64 kbps and a signalling channel at 16 kbps. These multiplex into a frame and, after adding extra framing bits, the total output data rate is 192 kbps. The total data rate for the basic rate service is thus 128 kbps. One or many devices may multiplex their data, such as two devices transmitting at 64 kbps, a single device multiplexing its 128 kbps data over two channels (giving 128 kbps), or by several devices transmitting a sub-64 kbps data rate over the two channels. For example, four 32 kbps devices could simultaneously transmit their data, eight 16 kbps devices, and so on.

For H12, 30 × 64 kbps channels multiplex with a 64 kbps-signalling channel, and with extra framing bits, the resulting data rate is 2.048 Mbps (compatible with European PCM-TDM systems). This means the actual data rate is 1.920 Mbps. As with the basic service this could contain a number of devices with a data rate of less than or greater than a multiple of 64 kbps.

For H11, 24 × 64 kbps channels multiplex with a 64 kbps-signalling channel, and with extra framing bits, it produces a data rate of 1.544 Mbps (compatible with USA PCM-TDM systems). The actual data rate is 1.536 Mbps.

G.3 ISDN physical layer interfacing

The physical layer corresponds to layer 1 of the OSI seven-layer model and is defined in CCITT specifications I.430 and I.431. Pulses on the line are not coded as pure binary, they use a technique called alternate mark inversion (AMI).

G.3.1 Alternative mark inversion (AMI) line code

AMI line codes use three voltage levels. In pure AMI, 0 V represents a '0', and the voltage amplitude for each '1' is the inverse of the previous '1' bit. ISDN uses the inverse of this, i.e. 0 V for a '1' and an inverse in voltage for a '0', as shown in Figure G.3. Normally the pulse amplitude is 0.75 V.

Inversion of the AMI signal (i.e. inverting a '0' rather than a '1') allows for timing information to be recovered when there are long runs of zeros, which is typical in the idle state. AMI line code also automatically balances the signal voltage, and the average voltage will be approximately zero even when there are long runs of zeros (this is a requirement as the connection to the network is transformer coupled).

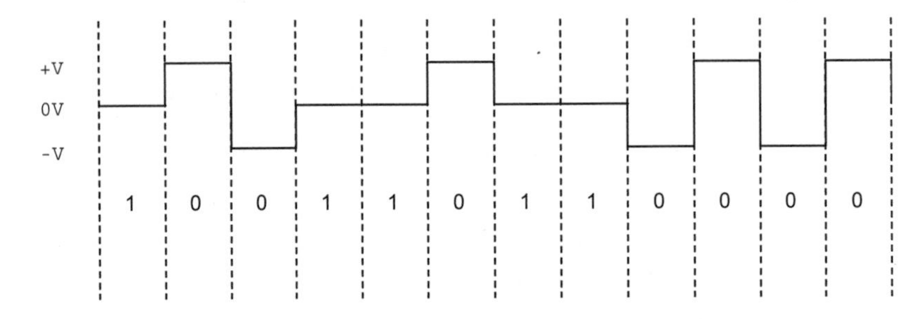

Figure G.3 AMI used in ISDN

G.3.2 System connections

In basic rate connections, up to eight devices, or items of termination equipment (TE), can connect to the network termination (NT). They connect over a common four-wire bus using two sets of twisted-pair cables. The transmit output (T_X) on each TE connects to the transmit output on the other TEs, and the receive input (R_X) on each TE connects to all other TEs. On the NT the receive input connects to the transmit of the TEs, and the transmit output of the NT connects to the receive input of the TEs. A contention protocol allows only one TE to communicate at a time.

An 8-pin ISO 8877 connector connects a TE to the NT; this is similar to the RJ-45 connector. Figure G.4 shows the pin connections. Pins 3 and 6 carry the T_X signal from the TE, pins 4 and 5 provide the R_X to the TEs. Pins 7 and 8 are the secondary power supply from the NT and pins 1 and 2 the power supply from the TE (if used). The T_X/R_X lines connect via transformers, thus only the AC part of the bitstream transfers into the PCM circuitry of the TE and the NT. This produces a need for a balanced DC line code such as AMI, as the DC component in the bitstream will not pass through the transformers.

G.3.3 Frame format

Figures G.5 and G.6 show the ISDN frame formats. Each frame is 250 μs long and contains 48 bits; this give a total bit rate of 192 kbps ($48/250 \times 10^{-6}$) made up of two 64 kbps B channels, one 16 kbps D-channel and extra framing, DC balancing and synchronisation bits.

The F/L pair of bits identify the start of each transmitted frame. When transmitting from a TE to an NT there is a 10-bit offset in the return of the frame back to the TE. The E bits echo the D-channel bits back to the TE.

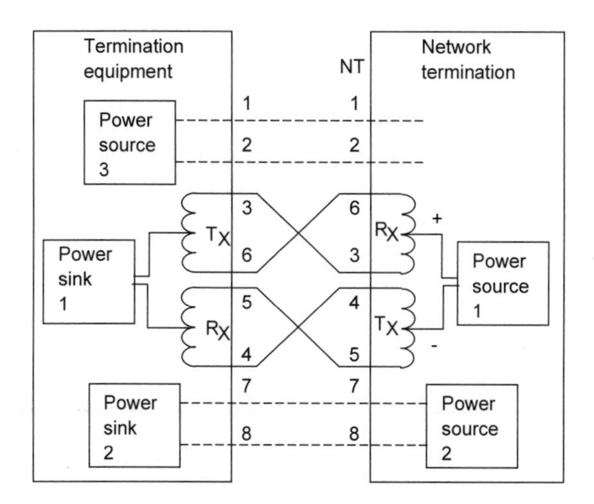

Figure G.4 Power supplies between NT and TE

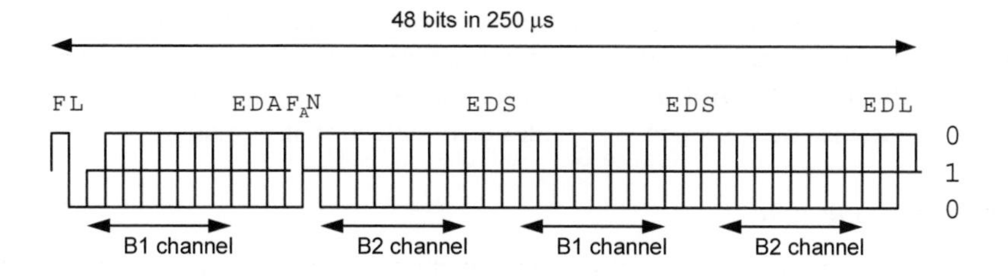

Figure G.5 ISDN frame format for NT to TE

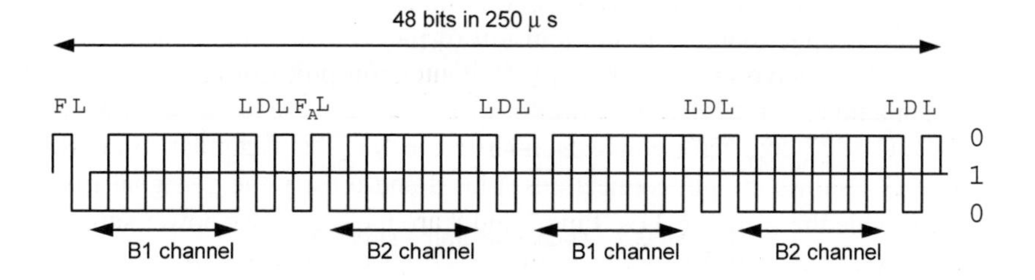

Figure G.6 ISDN frame format for TE to NT

where

F –	framing bit	N –	set to a 1
L –	DC balancing bit	D –	D-channel bit
E –	D-echo channel bit	F_A –	auxiliary framing bit (= 0)
S –	reserved for future use	A –	activation bit
M –	multiframing bit	B1 –	bits for channel 1
B2 –	bits for channel 2		

When transmitting from the NT to the TE, the bits after the F/L bits, in the B-channel, have a volition in the first 0. If any of these bits is a 0 then a volition will occur, but if they are 1s then no volition can occur. To overcome this the F_A bit forces a volition. As it is followed by

0 (the N bit) it will not be confused with the F/L pair. The start of the frame can thus be traced backwards to find the F/L pair.

There are 16 bits for each B-channel, giving a basic data rate of 64 kbps ($16/250 \times 10^{-6}$) and there are 4 bits in the frame for the D-channel, giving a bit rate of 16 kbps ($4/250 \times 10^{-6}$).

The L bit balances the DC level on the line. If the number of zeros following the last balancing bit is odd then the balancing bit is a 0, else it is a 1. When synchronised the NT informs the TEs by setting the A bit.

G.4 ISDN data link layer

The data link layer uses a protocol known as the link access procedure for the D-channel (LAPD). Figure G.7 shows the frame format. The unique bit sequence 01111110 identifies the start and end of the frame. This bit pattern cannot occur in the rest of the frame due to zero bit-stuffing.

The address field contains information on the type of data contained in the frame (the service access point identifier) and the physical address of the ISDN device (the terminal endpoint identifier). The control field contains a supervisory, an unnumbered or an information frame. The frame check sequence provides error detection information.

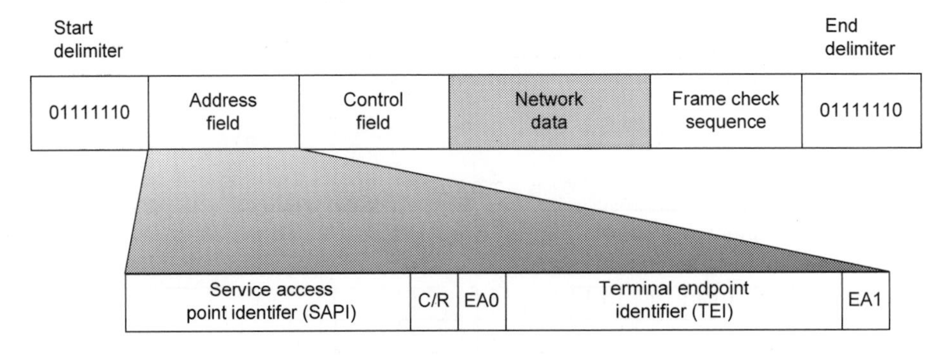

Figure G.7 D-channel frame structure

G.4.1 Address field

The data link address only contains addressing information to connect the TE to the NT and does not have network addresses. Figure G.7 shows the address field format. The SAPI identifies the type of ISDN service. For example, a frame from a telephone would be identified as such, and only telephones would read the frame.

All TEs connect to a single multiplexed bus, thus each has a unique data link address, known as a terminal endpoint identifier (TEI). The user or the network sets this; the ranges of available addresses are:

0–63	non-automatic assignment TEIs
64–126	automatic assignment TEIs
127	global TEI

The non-automatic assignment involves the user setting the address of each of the devices connected to the network. When a device transmits data it inserts its own TEI address and only receives data which has its TEI address. In most cases devices should not have the same TEI address, as this would cause all devices with the same TEI address, and the SAPI, to receive the same data (although, in some cases, this may be a requirement).

The network allocates addresses to devices requiring automatic assignment before they can communicate with any other devices. The global TEI address is used to broadcast messages to all connected devices. A typical example is when a telephone call is incoming to a group on a shared line where all the telephones would ring until one was answered.

The `C/R` bit is the command/response bit and `EA0`/`EA1` are extended address field bits.

G.4.2 Bit stuffing

With zero bit stuffing the transmitter inserts a zero into the bitstream when transmitting five consecutive 1s. When the receiver receives five consecutive 1s it deletes the next bit if it is a zero. This stops the unique `01111110` sequence occurring within the frame. For example if the bits to be transmitted are

`101000101011111100001010001010000111110101010`

then with the start and end delimiter this would be

01111110101000101*01111110*0001010001010000111110101010**01111110**

It can be seen from this bitstream that the stream to be transmitted contains the delimiter within the frame. This zero bit insertion is applied to give

01111110101000101*011111***0***10*000101000101000011111**0**0101010**01111110**

Notice that the transmitter has inserted a zero when five consecutive 1s occur. Thus the bit pattern 01111110 cannot occur anywhere in the bitstream. When the receiver receives five consecutive 1s it deletes the next bit if it is a zero. If it is a 1 then it is a valid delimiter. In the example the received stream will be

01111110101000101011111100001010001010000111110101010**01111110**

G.4.3 Control field

ISDN uses a 16-bit control field for information and supervisory frames and an 8-bit field for unnumbered frames, as illustrated in Figure G.8. Information frames contain sequenced data. The format is `0SSSSSSSXRRRRRRR`, where `SSSSSSS` is the send sequence number and `RRRRRRR` is the frame sequence number that the sender expects to receive next (`X` is the poll/final bit). As the extended mode uses a 7-bit sequence field then information frames are numbered from 0 to 127.

Supervisory frames contain flow control data. Table G.3 lists the supervisory frame types and the control field bit settings. The `RRRRRRR` value represent the 7-bit receive sequence number.

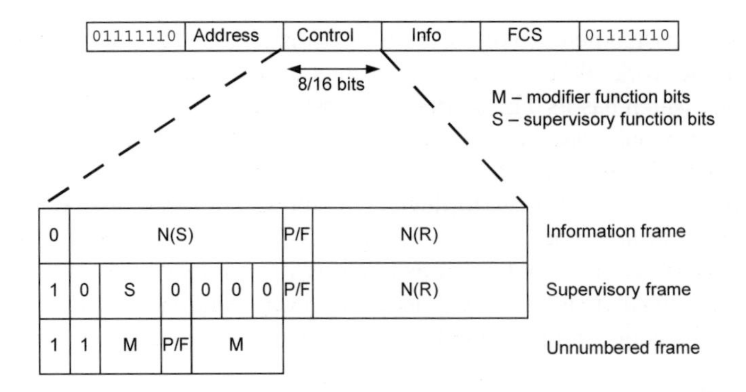

Figure G.8 ISDN control field

Table G.3 Supervisory frame types and control field settings

Type	*Control field setting*
Receiver ready (RR)	10000000PRRRRRRR
Receiver not ready (RNR)	10100000PRRRRRRR
Reject (REJ)	10010000PRRRRRRR

Unnumbered frames set up and clear connections between a node and the network. Table G.4 lists the unnumbered frame commands and Table G.5 lists the unnumbered frame responses.

Table G.4 Unnumbered frame commands and control field settings

Type	*Control field setting*
Set asynchronous balance mode extended (SABME)	1111P110
Unnumbered information (UI)	1100F000
Disconnect mode (DISC)	1100P010

Table G.5 Unnumbered frame responses and control field settings

Type	*Control field setting*
Disconnect mode (DM)	1111P110
Unnumbered acknowledgment (UA)	1100F000
Frame reject (FRMR)	1110P001

In ISDN all connected nodes and the network connection can send commands and receive responses. Figure G.9 shows a sample connection of an incoming call to an ISDN node (address `TEI_1`). The SABME mode is set up initially using the SABME command (`U[SABME,TEI_1,P=1]`), followed by an acknowledgement from the ISDN node (`U[UA,TEI_1,F=1]`). At any time, either the network or the node can disconnect the connection. In this case the ISDN node disconnects the connection with the command `U[DISC,TEI_1,P=1]`. The network connection acknowledges this with an unnumbered acknowledgement (`U[UA,TEI_1,F=1]`).

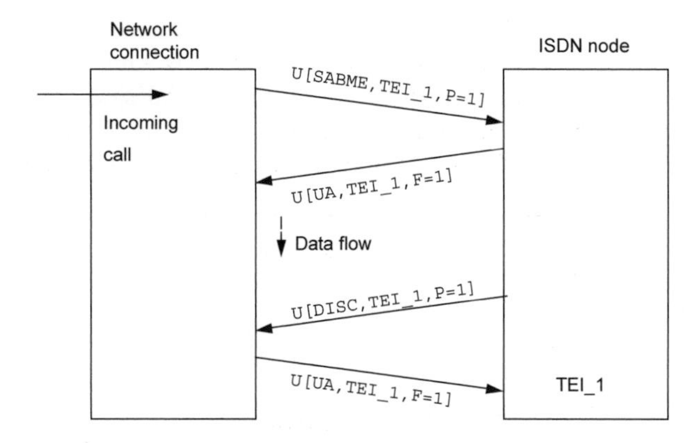

Figure G.9 Example connection between a primary/secondary

G.4.4 D-channel contention

The D-channel contention protocol ensures that only one terminal can transmit its data at a time. This happens because the start and the end of the D-channel bits have the bitstream `01111110`, as shown below:

```
1111101111110XXXXXXXXX...XXXXXXXX011111101111
```

When idle, each TE floats to a high-impedance state, which is taken as a binary 1. To transmit, a TE counts the number of 1s in the D-channel. A 0 resets this count. After a predetermined number, greater than a predetermined number of consecutive 1s, the TE transmits its data and monitors the return from the NT. If it does not receive the correct D-channel bitstream returned through the E bits then a collision has occurred. When a TE detects a collision it immediately stops transmitting and monitors the line.

When a TE has finished transmitting data it increases its count value for the number of consecutive 1s by 1. This gives other TEs an opportunity to transmit their data.

G.4.5 Frame check sequence

The frame check sequence (FCS) field contains an error detection code based on cyclic redundancy check (CRC) polynomials. It uses the CCITT V.41 polynomial, which is $G(x) = x^{16} + x^{12} + x^{5} + x^{1}$.

G.5 ISDN network layer

The D-channel carriers network layer information within the LAPD frame. This information establishes and controls a connection. The LAPD frames contain no true data as this is carried in the B-channel. Its function is to set up and manage calls and to provide flow control between connections over the network.

Figure G.10 shows the format of the layer-three signalling message frame. The first byte is the protocol discriminator. In the future, this byte will define different communications protocols. At present it is normally set to `0001000`. After the second byte the call reference length value is defined. This is used to identify particular calls with a reference number. The length of the call reference value is defined within the second byte. As it contains a 4-bit value, up to 16 bytes can be contained in the call reference value field. The next byte gives the message type and this type defines the information contained in the proceeding field.

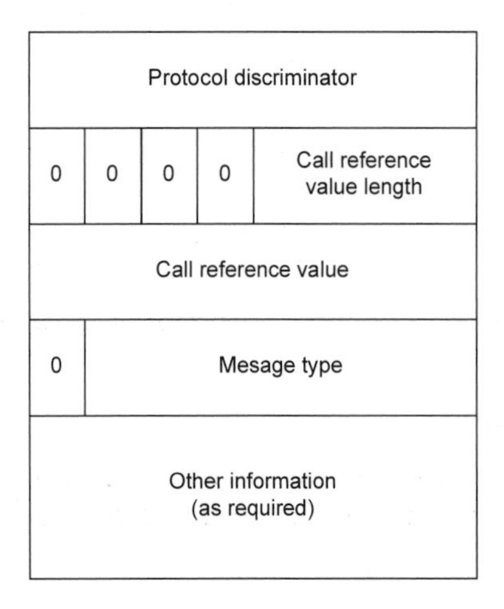

Figure G.10 Signalling message structure

There are four main types of message: call establish, call information, call clearing and miscellaneous messages. Table G.6 outlines the main messages. Figure G.11 shows an example connection procedure. The initial message sent is the setup. This may contain some of the following:

- Channel identification – identifies a channel with an ISDN interface.
- Calling party number.
- Calling party subaddress.
- Called party number.
- Called party subnumber.
- Extra data (2–131 bytes).

After the calling TE has sent the setup message, the network then returns the setup ACK message. If there is insufficient information in the setup message then other information needs to flow between the called TE and the network. After this the network sends back a call proceeding message and it also sends a setup message to the called TE. When the called TE detects its TEI address and SAPI, it sends back an alerting message. This informs the network that the node is alerting the user to answer the call. When it is answered, the called TE sends a connect message to the network. The network then acknowledges this with a connect ACK message, at the same time it sends a connect message to the calling TE. The calling TE then acknowledges this with a connect ACK. The connection is then established between the two nodes and data can be transferred.

To disconnect the connection the disconnect, release and release complete messages are used.

Table G.6 ISDN network messages

Call establish	*Information messages*	*Call clearing*
ALERTING	RESUME	DISCONNECT
CALL PROCEEDING	RESUME ACKNOWLEDGE	RELEASE
CONNECT	RESUME REJECT	RELEASE COMPLETE
CONNECT ACKNOWLEDGE	SUSPEND	RESTART
PROGRESS	SUSPEND ACKNOWLEDGE	RESTART ACKNOWLEDGE
SETUP	SUSPEND REJECT	
SETUP ACKNOWLEDGE	USER INFORMATION	

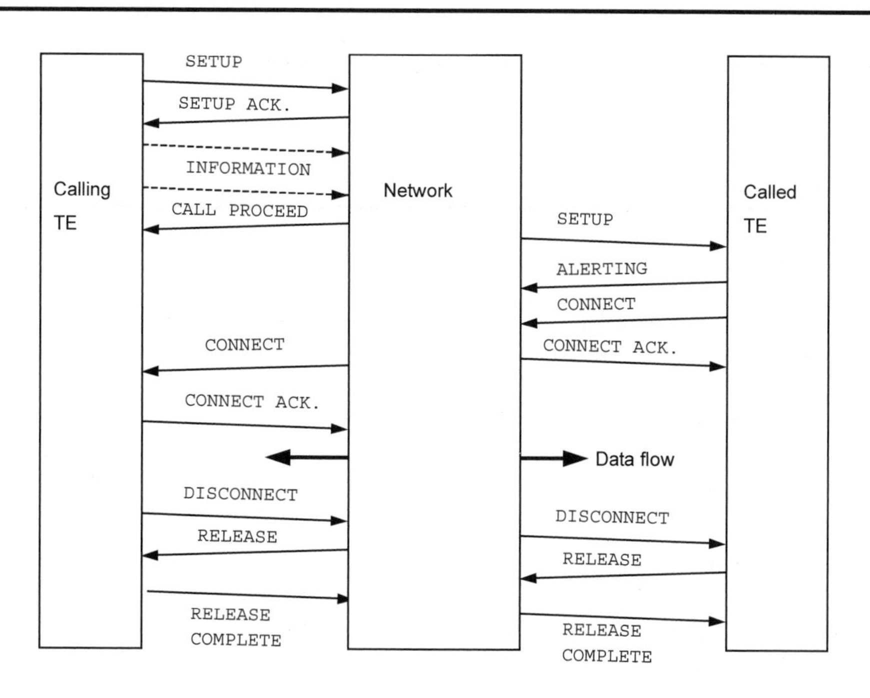

Figure G.11 Call establishment and clearing

G.6 Speech sampling

With telephone-quality speech the signal bandwidth is normally limited to 4 kHz, thus it is sampled at 8 kHz. If each sample is coded with eight bits then the basic bit rate will be:

Digitised speech signal rate = 8×8 kbps = 64 kbps

Table G.7 outlines the main compression techniques for speech. The G.722 standard allows the best-quality signal, as the maximum speech frequency is 7 kHz rather than 4 kHz in normal coding systems; and has the equivalent of 14 coding bits. The G.728 allows extremely low bit rates (16 kbps).

Table G.7 Speech compression standards

ITU standard	*Technology*	*Bit rate*	*Description*
G.711	PCM	64 kbps	Standard PCM
G.721	ADPCM	32 kbps	Adaptive delta PCM where each value is coded with four bits
G.722	SB-ADPCM	48, 56 and 64 kbps	Subband ADPCM allows for higher-quality audio signals with a sampling rate of 16 kHz
G.728	LD-CELP	16 kbps	Low-delay code excited linear prediction for low bit rates

G.7 Exercises

G.7.1 What is the function of a B-channel in ISDN:

(a) It transmits data (b) It sends control information
(c) It creates a network (d) It emulates a modem

G.7.2 What is the function of a D-channel in ISDN:

(a) It transmits data (b) It sends control information
(c) It creates a network (d) It emulates a modem

G.7.3 What is the bit rate of a single ISDN B-channel:

(a) 16 kbps (b) 64 kbps
(c) 128 kbps (d) 256 kbps

G.7.4 What is the bit rate of the an ISDN D-channel:

(a) 16 kbps (b) 64 kbps
(c) 128 kbps (d) 256 kbps

G.7.5 What is the maximum bit rate of an ISDN connection:

(a) 16 kbps (b) 64 kbps
(c) 128 kbps (d) 256 kbps

G.7.6 Which series of CCITT (ITU-T) specify ISDN specifications:

(a)	I-series	(b)	X-series
(c)	T-series	(d)	R-series

G.7.7 What is the base bit rate for a USA PCM-TDM system:

(a)	1.544 Mbps	(b)	64 kbps
(c)	2.048 Mbps	(d)	2 Gbps

G.7.8 What is the base bit rate for a UK PCM-TDM system:

(a)	1.544 Mbps	(b)	64 kbps
(c)	2.048 Mbps	(d)	2 Gbps

G.7.9 How many ISDN channels does the H11 (derived from USA PCM-TDM system) support:

(a)	20	(b)	24
(c)	30	(d)	32

G.7.10 How many ISDN channels does the H12 (derived from UK PCM-TDM system) support:

(a)	20	(b)	24
(c)	30	(d)	32

G.7.11 How does ISDN AMI (alternative mark inversion) operate:

(a) Every bit that is sent is inverted
(b) Every one that is sent has an alternating voltage level, but the zeros are sent as a zero voltage level
(c) Every zero that is sent has an alternating voltage level, but the ones are sent as a zero voltage level
(d) Ones are sent as negative voltages, and zeros as positive voltages

G.7.12 Why does the 01111110 bit sequence only occur at the start and end of a frame:

(a) Bit stuffing is used whenever it appears within the data frame
(b) It is a special code that can never occur within the data
(c) It is coded with special voltage levels
(d) It will hardly ever occur, so the occasional error is acceptable

G.7.13 Show why speech requires to be transmitted at 64 kbps.

G.7.14 If the bandwidth of hi-fi audio is 20 kHz and 16 bits are used to code each sample, determine the required bit rate for single-channel transmission. Hence prove that the bit rate required for professional hi-fi, which is sampled at 44.1 kHz is 1.4112 Mbps.

G.7.15 Using the rates determined in Question G.7.14, shows that the basic rate for a CD-ROM drive is 150 kB/s.

G.7.16 Explain the format of the ISDN frame.

G.7.17 Suppose that an ISDN frame has 48 bits and takes 250 μs to transmit. Show that the bit rate on each D-channel is 16 kbps and that the bit rate of the B-channel is 64 kbps.

G.7.18 Explain the different types of frames and show how a connection is made between ISDN nodes.

G.7.19 Show how supervisory frames are used to control the flow of data.

G.7.20 Discuss the format of the ISDN network layer packet.

G.7.21 How does an ISDN node set up and disconnect a network connection.

H Microsoft Windows

H.1 Introduction

DOS has long been the Achilles heel of the PC and has limited its development. It has also been its strength in that it provides a common platform for all packages. DOS and Windows 3.*x* operated in a 16-bit mode and had limited memory accessing. Windows 3.0 provided a great leap in PC systems as it provided an excellent graphical user interface to DOS. It suffered from the fact that it still used DOS as the core operating system. Windows 95/98 and Windows NT have finally moved away from DOS and operate as full 32-bit protected-mode operating systems. Their main features are:

- Run both 16-bit and 32-bit application programs.
- Allow access to a large virtual memory (up to 4 GB).
- Support for pre-emptive multitasking and multithreading of Windows-based and MS-DOS-based applications.
- Support for multiple file systems, including 32-bit installable file systems such as VFAT, CDFS (CD-ROM) and network redirectors. These allow better performance, use of long file names, and are an open architecture to support future growth.
- Support for 32-bit device drivers which give improved performance and intelligent memory usage.
- A 32-bit kernel which includes memory management, process scheduling and process management.
- Enhanced robustness and clean-up when an application ends or crashes.
- Enhanced dynamic environment configuration.

The three most widely used operating systems are MS-DOS, Microsoft Windows and UNIX. Microsoft Windows comes in many flavours; the main versions are outlined below and Table H.1 lists some of their attributes.

- Microsoft Windows 3.*x* – 16-bit PC-based operating system with limited multitasking. It runs from MS-DOS and thus still uses MS-DOS functionality and file system structure.
- Microsoft Windows 95/98 – robust 32-bit multitasking operating system (although there are some 16-bit parts in it) which can run MS-DOS applications, Microsoft Windows 3.*x* applications and 32-bit applications.
- Microsoft Windows NT Version 4 – robust 32-bit multitasking operating system with integrated networking. Networks are around NT servers and clients. As with Microsoft Windows 95/98 it can run MS-DOS, Microsoft Windows 3.*x* applications and 32-bit applications.
- Windows NT Version 5/2000 – available as Workstation, Server and SMP Server (multiprocessor). It runs on Alphas, Intel *x*86, Intel IA32, Intel IA64 and AMD K7 (which is similar to an Alpha).

Windows NT/2000 and 95/98 provide excellent network support as they can communicate directly with many different types of networks, protocols and computer architectures. They can create networks to make peer-to-peer connections and also connection to servers for access to file systems and print servers.

Windows NT/2000 Server has more security in running programs than Windows 95/98 as programs and data are insulated from the operation of other programs. The operating system parts of Windows NT/2000 and Windows 95/98 run at the most trusted level of privilege of the Intel processor, which is ring zero. Application programs run at the least trusted level of privilege, which is ring three. These programs can use either a 32-bit flat mode or any of the memory models, such as large, medium, compact or small.

There was a great leap in performance between the 16-bit Windows 3.*x* operating system (which was built on DOS) to Windows 95/98 and Window NT. Apart from running in a dual 16-bit and 32-bit mode, they also allow for application robustness. Figure H.1 outlines the internal architecture of Windows 95/98.

Table H.1 Windows comparisons

Facility	*Windows 3.1*	*Windows 95/98*	*Windows NT*
Pre-emptive multitasking		✓	✓
32-bit operating system		✓	✓
Long file names		✓	✓
TCP/IP	✓	✓	✓
32-bit applications		✓	✓
Flat memory model		✓	✓
32-bit disk access	✓	✓	✓
32-bit file access	✓	✓	✓
Centralised configuration storage		✓	✓
OpenGL 3D graphics		✓	✓

H.2 Windows registry

On DOS-based systems, the main configuration files were AUTOEXEC.BAT, CONFIG.SYS and INI files. INI files were a major problem in that each application program and device driver configuration required one or more of these files to store default settings (such as IRQ, I/O addresses, default directories, and so on). Several important INI files are:

- WIN.INI – information about the appearance of the Windows environment.
- SYSTEM.INI – system-specific information on the hardware and device driver configuration of the system.

Windows 95/98/NT/2000 use a central database called the Registry, which stores user-specific and configuration-specific information at a single location. This location could be on the local computer or stored on a networked computer. It thus allows network managers to standardise the configuration of networked PCs.

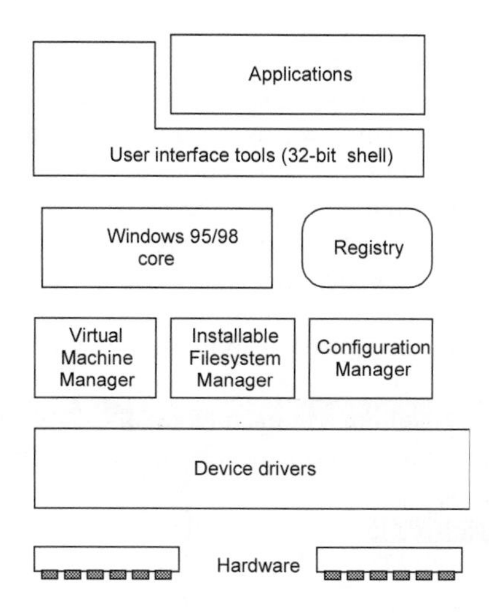

Figure H.1 Windows 95/98 architecture

When a computer is initially upgraded from Windows 3.*x* to Windows 95/98 the upgrade program reads the SYSTEM.INI file and system-specific information which it then puts into the Registry. Many INI files are still retained on the system as many Win16-based applications use them. For example, Microsoft Word Version 6 uses the WINWORD6.INI to store package information, such as location of filters, location of spell checker, location of grammar checker, and so on. An example is

```
[Microsoft Word]
WPHelp=0
USER-DOT-PATH=C:\MSOFFICE\WINWORD\TEMPLATE
PICTURE-PATH=C:\MSOFFICE\WINWORD
PROGRAMDIR=C:\MSOFFICE\WINWORD
TOOLS-PATH=C:\MSOFFICE\WINWORD
STARTUP-PATH=C:\DOCS\NOTES\
INI-PATH=C:\MSOFFICE\WINWORD
DOC-PATH=C:\DOCS\NOTES\
```

An important role for the registry is to store hardware-specific information which can be used by hardware detection and plug-and-play programs. The configuration manager determines the configuration of installed hardware (such as, IRQs, I/O addresses, and so on) and it uses this information to update the registry. This allows new devices to be installed and checked to see if they conflict with existing devices. If they are plug-and-play devices then the system assigns hardware parameters that do not conflict with existing devices.

The advantages of the registry over INI files include:

- No limit to size and data type – the Registry has no size restriction and can include binary and text values (INI files are text based and are limited to 64 KB).
- Hierarchical information – the registry is hierarchically arranged, whereas INI files are non-hierarchical and support only two levels of information.
- Standardised set-up – the registry provides a standardized method of setting up programs, whereas many INI files contain a whole host of switches and entries, and are complicated to configure.

- Support for user-specific information – the registry allows the storage of user-specific information, using the `Hkey_Users` key. This allows each user of a specific computer (or a networked computer) to have their own user-specific information. INI files do not support this.
- Remote administration and system policies – the registry can be used to remotely administer and set system policies (which are stored as registry values). These can be downloaded from a central server each time a new user logs on.

Figure H.2 shows an example of the registry in Windows 95/98.

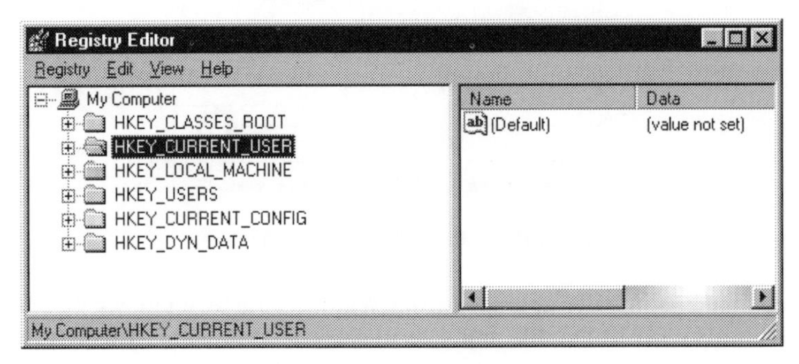

Figure H.2 Example registry

H.3 Device drivers

In Windows 3.*x*, device drivers were complex entities and were, in part, static and unchanging. Windows 95/98/NT now provide enhanced support for hardware devices and peripherals including disk devices. Windows NT will be discussed in Section H.10. Windows 95/98 uses a universal driver/mini-driver architecture that makes writing device-specific code much easier.

The universal driver provides for most of the code for a specific class of device (such as for printers or mice) and the mini-driver is a relatively small and simple driver that provides additional information for the hardware.

The actual system interface to the hardware (or some software parts) is through a virtual device driver (VxD), which is a 32-bit, protected-mode driver. These keep track of the state of the device for each application and ensure that the device is in the correct state whenever an application continues. This allows for multitasking programming and also for multi-access for a single device. VxD files also support hardware emulation, such as in the case of the MS-DOS device driver, where any calls to the PC hardware can be handled by the device driver and not by the physical hardware. Typical VxD drivers are:

EISA.VXD	EISA bus driver	ISAPNP.VXD	ISA plug-and-play
SERIAL.VXD	Serial port	LPTENUM.VXD	Parallel port
MSMOUSE.VXD	MS Mouse	PARALINK.VXD	Parallel port
PCI.VXD	PCI	QC117.VXD	Tape backup
IRCOMM.VXD	Infra-red comms	UNIMODEM.VXD	Modem
WSOCK.VXD	WinSock	LPT.VXD	LPT

VMM32.VXD	Memory management	JAVASUP.VXD	JavaScript
PPPMAC.VXD	PPP connection	NDIS.VXD	NDIS
NDIS2SUP.VXD	NDIS 2.0	NETBEUI.VXD	Net BEUI
NWREDIR.VXD	NetWare Redirect	VNETBIOS.VXD	Net BIOS
WSIPX.VXD	IPX	WSHTCP.VXD	TCP

In Windows 95/98, VxD files are loaded dynamically and are thus only loaded when they are required, whereas in Windows 3.*x* they were loaded statically (and thus took up a lot of memory). In Window 3.*x* these virtual device drivers have a 386 file extension.

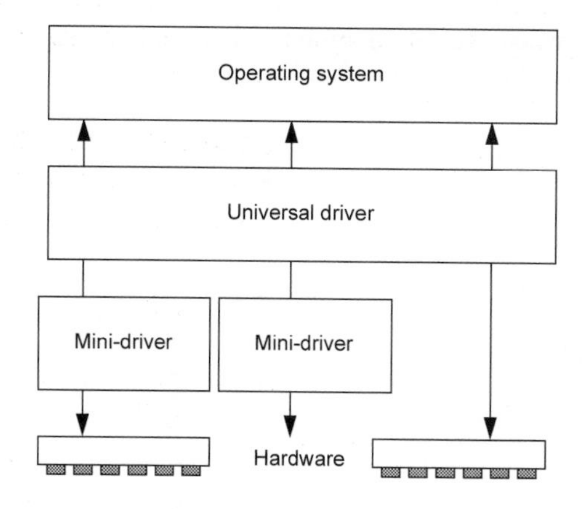

Figure H.3 Device drivers

H.4 Configuration manager

A major drawback with Windows 3.*x* and DOS is that they did not automate PC configuration. For this purpose, Windows 95/98 has a configuration manager. The left-hand side of Figure H.4 shows how it integrates into the system and the right side of Figure H.4 shows an example device connection of a PC. Its aims are:

- Determine, with the aid of several subcomponents, each bus and each device on the system, and their configuration settings. This is used to ensure that each device has unique IRQs and I/O port addresses and that there are no conflicts with other devices. With plug-and-play, devices can be configured so that they do not conflict with other devices.
- Monitor the PC for any changes to the number of devices connected and also the device types. If it detects any changes then it manages the reconfiguration of the devices.

The operation is as follows:

1. The configuration manager communicates with each of the bus enumerators and asks them to identify all the devices on the buses and their respective resource requirements. A bus enumerator is a driver that is responsible for creating a hardware tree, which is a hierarchical representation of all the buses and devices on a computer. Figure H.5 shows an example tree.

2. The bus enumerator locates and gathers information from either the device drivers or the BIOS services for that particular device type. For example, the CD-ROM bus enumerator calls the CD-ROM drivers to gather information.
3. Each of the drivers is then loaded and they wait for the configuration manager to assign their specific resources (such as IRQs, I/O addresses, and so on).
4. The configuration manager calls on resource arbitrators to allocate resources for each device.
5. Resource arbitrators identify any devices which are conflicting and tries to resolve them.
6. The configuration manager informs all device drivers of their device configuration. This process is repeated when the BIOS or one of the other bus enumerators informs the configuration manager about a system configuration change.

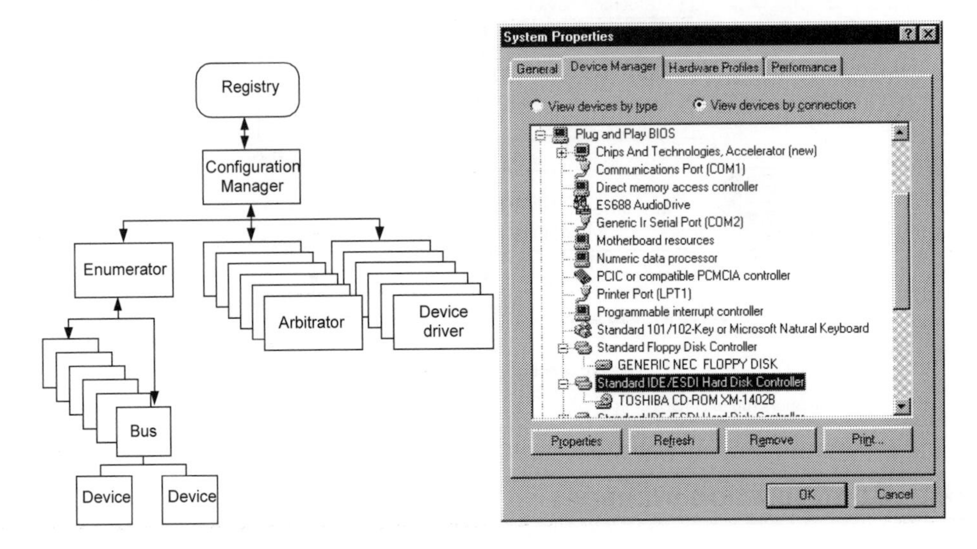

Figure H.4 Configuration manager and example connection of devices

H.5 Virtual machine manager (VMM)

The perfect environment for a program is to run on a stand-alone, dedicated computer, which does not have any interference from any other programs and can have access to any device when it wants. This is the concept of the virtual machine. In Windows 95/98 the virtual machine manager (VMM) provides each application with the system resources when it needs them. It creates and maintains the virtual machine environments in which applications and system processes run (in Windows 3.*x* the VMM was called WIN386.EXE).

The VMM is responsible for three areas:

- Process scheduling – responsible for scheduling processes. It allows for multiple applications to run concurrently and also for providing system resources to the applications and other processes that run. This allows multiple applications and other processes to run concurrently, using either co-operative multitasking or pre-emptive multitasking.

- Memory paging – Windows 95/98/NT uses a demand-paged virtual memory system, which is based on a flat, linear address space accessed using 32-bit addresses. The system allocates each process a unique virtual address space of 4 GB. The upper 2 GB is shared, while the lower 2 GB is private to the application. This virtual address space is divided into equal blocks (or pages).
- MS-DOS Mode support – provides support for MS-DOS-based applications which must have exclusive access to the hardware. When an MS-DOS-based application runs in this mode then no other applications or processes are allowed to compete for system resources. The application thus has sole access to the resources, as illustrated in Figure H.5.

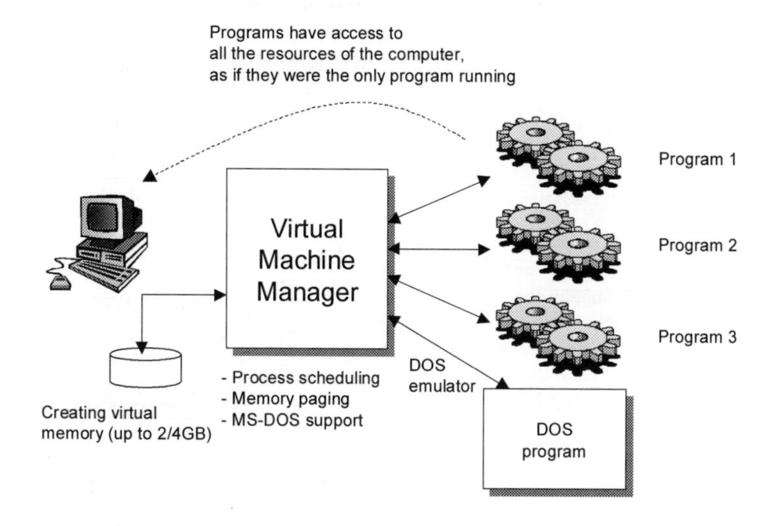

Figure H.5 Virtual Machine Manager

Windows 95/98 has a single VMM (named System VMM) in which all system processes run. Win32-based and Win16-based applications run within this VMM. Each MS-DOS-based application runs in its own VM.

H.5.1 Process scheduling and multitasking

This allows multiple applications and other processes to run concurrently, using either co-operative multitasking or pre-emptive multitasking. In Windows 3.*x*, applications ran using co-operative multitasking. This method requires that applications check the message queue periodically and give up control of the system to other applications. Unfortunately, applications that do not check the message queue at frequent intervals can effectively 'hog' the processor and prevent other applications from running. As this does not provide effective multi-processing, Windows 95/98/NT uses pre-emptive multitasking for Win32-based applications (but also supports co-operative multitasking for computability reasons). Thus, the operating system takes direct control away from the application tasks.

Win16 programs need to yield to other tasks in order to multitask properly, whereas Win32-based programs do not need to yield to share resources. This is because Win32-based applications (called processes) use multithreading, which provides for multi-processing. A thread in a program is a unit of code that can get a time slice from the operating system to run concurrently with other code units. Each process consists of one or more execution threads that identify the code path flow as it is run on the operating system. A Win32-based application can have multiple threads for a given process. This enhances the running of an application by improving throughput and responsiveness. It allows processes for smooth background processing.

H.5.2 Memory paging

Windows 95/98/NT use a demand-paged virtual memory system, which is based on a flat, linear address space using 32-bit addresses. The system allocates each process a unique virtual address space of 4 GB (which should be enough for most applications). The upper 2 GB is shared, while the lower 2 GB is private to the application. This virtual address space divides into equal blocks (or pages), as illustrated in Figure H.6.

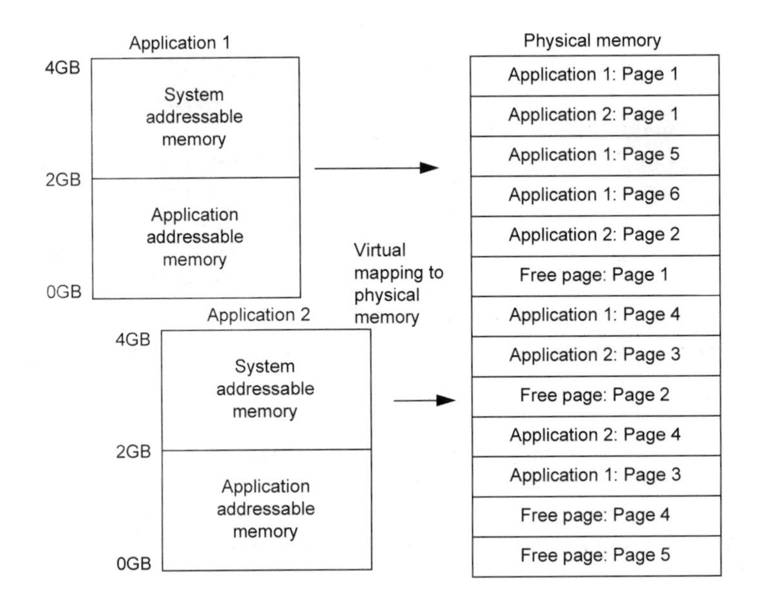

Figure H.6 Memory paging

Demand paging is a method by which code and data are moved in pages from physical memory to a temporary paging file on disk. When required, information is then paged back into physical memory.

The functions of the memory pager are:

- To map virtual addresses from the process's address space to physical pages in memory. This then hides the physical organisation of memory from the process's threads and ensures that the thread can access the required memory when required. It also stops other processes from writing to another memory location.
- To support a 16-bit segmented memory model for Windows 3.*x* and MS-DOS applications. In this addressing scheme the addresses are made from a 16-bit segment address and a 16-bit offset address.

Windows 95/98/NT use the full addressing capabilities of the 80x86/Pentium processors by supporting a flat, linear memory model for 32-bit operating system functionality and Win32-based applications. This linear addressing model simplifies the development process for application vendors, and removes the performance penalties of a segmented memory architecture.

H.6 Multiple file systems

Windows 95/98/NT supports a layered file system architecture that directly supports multiple file systems (such as FAT and CDFS). Windows 95/98/NT have great performance improvements over Windows 3.*x*, for example:

- Support for 32-bit protected-mode code when reading and writing information to and from a file system.
- Support for 32-bit dynamically allocated cache size.
- Support for an open file system architecture to enhance future system support.

Figure H.7 shows the file system architecture used by Windows 95/98. It has the following components:

- **IFS (installable file system) manager**. This is the arbiter for the access to different file system components. On MS-DOS and Windows 3.*x* it was provided by interrupt 21h. Unfortunately, some add-on components did not run correctly and interfered with other installed drivers. It also did not directly support multiple network redirections (the IFS manager can have an unlimited number of 32-bit redirectors).
- **File system drivers**. These provide support file systems, such as FAT-based disk devices, CD-ROM file systems and redirected network devices. They are ring 0 components, whereas Windows 3.*x* supported them through MS-DOS. The two enhanced file systems are:
  - 32-bit VFAT – the 'legacy' 16-bit FAT file system suffers from many problems, such as the 8.3 file format. The 32-bit VFAT format is an enhanced form which works directly in the protected mode, and thus provides smooth multitasking as it is re-entrant and multithreaded (a non re-entrant system does not allow an interrupt within an interrupt). It uses the VFAT.VXD driver and uses 32-bit code for all file accesses. Another advantage is that it provides for real-mode disk caching (VCACHE), where cache memory is automatically allocated or deallocated when it is required (in Windows 3.*x* this was provided by the SMARTDRV.EXE program).
  - 32-bit CDFS – the 32-bit, protected-mode CDFS format (as defined in the ISO 9660 standard) gives improved CD-ROM access and support for a dynamic cache (in Windows 3.*x* the MSCDEX driver provided to access CD-ROMs).
- **Block I/O subsystem**. This is responsible for the actual physical access to the disk drive. Its components are:

  - Input/Output Supervisor (IOS) – this component provides for an interface between the file systems and drivers. It is responsible for the queuing of file service requests and for routing the requests to the appropriate file system driver.

- Port driver – this component is a 32-bit, protected-mode driver that communicates with a specific IDE disk device. It implements the functionality of the SCSI manager and miniport driver.
- SCSI layer – this component is a 32-bit, protected-mode, universal driver model architecture for communicating with SCSI devices. It provides all the high-level SCSI functionality, and then uses a miniport driver to handle device-specific I/O calls.
- Miniport driver– in Windows 95/98 these miniport driver models are used to write device-specific code. The Windows 95/98 miniport driver is a 32-bit protected-mode code, and is binary-compatible with Windows NT miniport drivers.

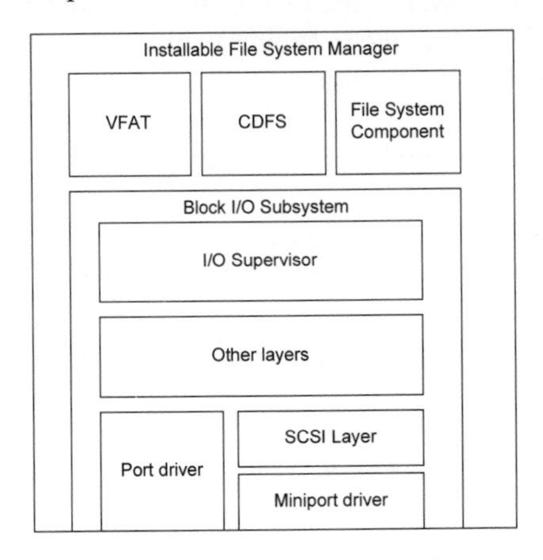

Figure H.7 File system architecture

In Windows 95/98, the I/O Supervisor (IOS) is a VxD that controls and manages all protected-mode file system and block device drivers. It loads and initialises protected-mode device drivers and provides services needed for I/O operations. In Windows 3.*x* the I/O Supervisor was *BLOCKDEV. Other responsibilities of the IOS include:

- Registering drivers.
- Routing and queuing I/O requests, and sending asynchronous notifications to drivers as needed.
- Providing services that drivers can use to allocate memory and complete I/O requests.

On Windows 95/98, the IOS stores port drivers, miniport and VxD drivers in the SYSTEM\IOSUBSYS directory. The PDR file extension identifies the port drivers, MPD identifies miniport drivers and VxD (or 386) identifies the VxD drivers. Other clients or virtual device drivers should be stored in other directories and explicitly loaded using device= entries in SYSTEM.INI. A sample listing of the IOSUBSYS directory is

```
Directory of C:\WINDOWS\SYSTEM\IOSUBSYS
AIC78XX.MPD        AMSINT.MPD      APIX.VXD         ATAPCHNG.VXD
BIGMEM.DRV         CDFS.VXD        CDTSD.VXD        CDVSD.VXD
DISKTSD.VXD        DISKVSD.VXD     DRVSPACX.VXD     ESDI_506.PDR
HSFLOP.PDR         NCRC710.MPD     NCRC810.MPD      NECATAPI.VXD
RMM.PDR            SCSI1HLP.VXD    SCSIPORT.PDR     VOLTRACK.VXD
```

H.7 Core system components

The core of Windows 95/98 has three components: user, kernel, and GDI (graphical device interface), each of which has a pair of DLLs (one for 32-bit accesses; the other for 16-bit accesses). The 16-bit DLLs (dynamic link libraries) allow for Win16 and MS-DOS computability.

Figure H.8 shows that the lowest-level services provided by the Windows 95/98 kernel are implemented as 32-bit code. In Windows 95/98 the names of the files are GDI32.DLL, KERNEL32.DLL and USER32.DLL; these are contained in the \WINDOWS\SYSTEM directory.

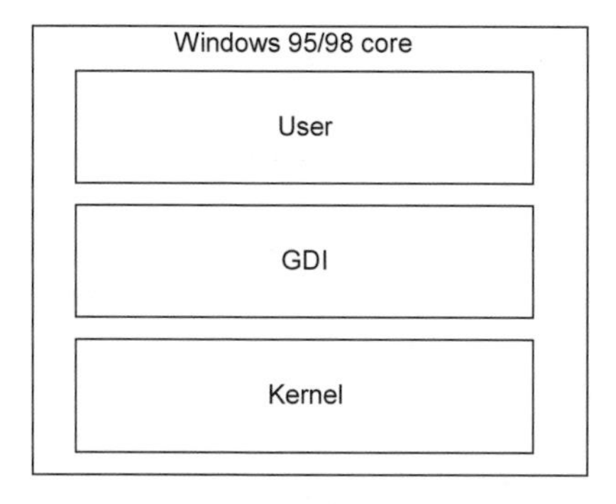

Figure H.8 Core components

H.7.1 User

The user component provides input and output to and from the user interface. Input is from the keyboard, mouse, and any other input device and the output is to the user interface. It also manages interaction with the sound driver, timer, and communications ports.

Win32 applications and Windows 95/98 use an asynchronous input model for system input. With this, devices have an associated interrupt handler (for example, the keyboard interrupts with IRQ1) which converts the interrupt into a message. This message is then sent to a raw input thread area, which then passes the message to the appropriate message queue. Each Win32 application can have its own message queue, whereas all Win16 applications share a common message queue.

H.7.2 Kernel

The kernel provides for core operating system components including file I/O services, virtual memory management, task scheduling and exception handling, such as:

- File I/O services.
- Exceptions – these are events that occur as a program runs and calls additional software which is outside of the normal flow of control. For example, if an application generates an exception, the Kernel is able to communicate that exception to the application to perform the necessary functions to resolve the problem. A typical exception is caused by a divide-by-zero error in a mathematical calculation, an exception routine can be designed so that it handles the error and does not crash the program.
- Virtual memory management – this resolves import references and supports demand paging for the application.

- Task scheduling – the Kernel schedules and runs threads of each process associated with an application.
- Provides services to both 16-bit and 32-bit applications by using a thunking process which is the translation process between 16-bit and 32-bit formats. It is typically used by a Win16 program to communicate with the 32-bit operating system core.

Virtual memory allows processes to allocate more memory than can be physically allocated. The operating system allocates each process a unique virtual address space, which is a set of addresses available for the process's threads. This virtual address space appears to be 4 GB in size, where 2 GB are reserved for program storage and 2 GB for system storage.

Figure H.9 illustrates where the system components and applications reside in virtual memory. Its contents are:

- 3 GB–4 GB – all Ring 0 components.
- 2 GB–3 GB – operating system core components and shared DLLs. These are available to all applications.
- 4 MB–2 GB – Win32-based applications, where each has its own address space. This memory is protected so that other programs cannot corrupt or otherwise hinder the application.
- 0–640 KB – real-mode device drivers and TSRs.

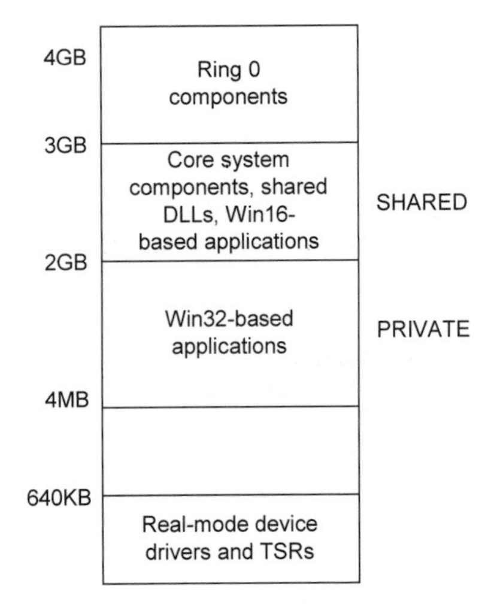

Figure H.9 System memory usage

H.7.3 GDI

The graphical device interface (GDI) is the graphical system that:

- Manages information that appears on the screen.
- Draws graphic primitives and manipulates bitmaps.
- Interacts with device-independent graphics drivers, such as display and printer drivers.

The graphics subsystem provides input and output graphics support. Windows uses a 32-bit graphics engine (known as DIB, device-independent bitmaps) which:

- Directly controls the graphics output on the screen.
- Provides a set of optimized generic drawing functions for monochrome, 16-colour, 16-bit high colour, 256-colour, and 24-bit true colour graphic devices. It also supports Bézier curves and paths.
- Support for image Colour Matching for better color matching between display and colour output devices.

The Windows graphics subsystem is included as a universal driver with a 32-bit mini-driver. The mini-driver provides only for the hardware-specific instructions.

The 32-bit Windows 95/98 printing subsystem has several enhancements over Windows 3.*x*. These include:

- They use a background thread processing to allow for smooth background printing.
- Smooth printing where the operating system only passes data to the printer when it is ready to receive more information.
- They send enhanced metafile (EMF) format files, rather than raw printer data. This EMF information is interpreted in the background and the results are then sent to the printer.
- Support for deferred printing, where a print job can be sent to a printer and then stored until the printer becomes available.
- Support for bi-directional communication protocols for printers using the extended communication port (ECP) printer communication standard. ECP mode allows printers to send messages to the user or to application programs. Typical messages are: 'Paper Jam', 'Out-of-paper', 'Out-of-Memory', 'Toner Low', and so on.
- Plug-and-play.

H.8 Multitasking and threading

Multitasking involves running several tasks at the same time. It normally involves running a process for a given amount of time, before releasing it and allowing another process a given amount of time. There are two forms of multitasking:

- Pre-emptive multitasking – this involves the operating system controlling how long a process stays on the processor. This allows for smooth multitasking and is used in Windows NT/95/98 32-bit programs.
- Co-operative multitasking – this relies on a process giving up the processor. It is used with Windows 3.*x* programs and suffers from processor hogging, where a process can stay on a processor and the operating system cannot kick it off.

The logical extension to multitasking programs is to split a program into a number a parts (threads) and run each of these on the multitasking system (multithreading). A program which is running more than one thread at a time is known as a multithreaded program. Multithreaded programs have many advantages over non-multithreaded programs, including:

- They make better use of the processor, where different threads can be run when one or more threads are waiting for data. For example, a thread could be waiting for keyboard input, while another thread could be reading data from the disk.
- They are easier to test, as each thread can be tested independently of other threads.
- They can use standard threads, which are optimised for given hardware.

They also have disadvantages, including:

- The program has to be planned properly so that threads know on which threads they depend.
- A thread may wait indefinitely for another thread which has crashed or terminated.

The main difference between multiple processes and multiple threads is that each process has independent variables and data, while multiple threads share data from the main program.

H.8.1 Scheduling

Scheduling involves determining which thread should be run on the process at a given time. This element is named a time slice, and its actual value depends on the system configuration.

Each thread currently running has a base priority. The programmer who created the program sets this base priority level of the thread. This value defines how the thread is executed in relation to other system threads. The thread with the highest priority gets use of the processor.

NT and 95/98 have 32 priority levels. The lowest priority is 0 and the highest is 31. A scheduler can change a threads base priority by increasing or decreasing it by two levels. This changes the threads priority.

The scheduler is made up from two main parts:

- **Primary scheduler**. This scheduler determines the priority numbers of the threads which are currently running. It then compares their priority and assigns resources to them depending on their priority. Threads with the highest priority are executed for the current time slice. When two or more threads have the same priority then the threads are put on a stack. One thread is run and then put to the bottom of the stack, then the next is run and it is put to the bottom, and so on. This continues until all threads with the same priority have been run for a given time slice.
- **Secondary scheduler**. The primary scheduler runs threads with the highest priority, whereas the secondary scheduler is responsible for increasing the priority of non-executing threads (which are all other threads apart from the currently executed thread). It is thus important for giving low priority threads a chance to run on the operating system. Threads which are given a higher or lower priority are:

  - A thread which is waiting for user input has its priority increased.
  - A thread that has completed a voluntary wait also has its priority increased.
  - Threads with a computation-bound thread get their priorities reduced. This prevents the blocking of I/O operations.

  Apart from these, all threads get a periodic increase. This prevents lower-priority threads hogging shared resources that are required by higher-priority threads.

H.8.2 Priority inheritance boosting

One problem that can occur is when a low priority thread accesses resources which are required by a higher priority thread. For example, an RS-232 program could be loading data into memory while another program requires to access the memory. One method which can be used to overcome this is priority inheritance boosting. In this case, low priority threads gets a boost so that they can quickly release resources. For example, suppose a system has three threads: Thread A, Thread B and Thread C. If Thread A has the highest priority and it requires a resource from Thread C then Thread C gets a boost in its priority. Thread A remains blocked until Thread C releases the required resource. When it does release it then Thread C goes back to its normal priority and Thread A then gets access to the resource.

H.9 Plug-and-play process

Plug-and-play allows the operating system to configure hardware as required. On system start-up, the configuration manager scans the system hardware. When it finds a new plug-and-play device it does the following:

- Sets the device into configuration mode – this is achieved by using three I/O ports. Some data (the initiation key) is written to one of the ports and enables the plug-and-play logic.
- Isolate and identify each device – each device is isolated, one at a time. The method used is to assign each device a unique number, which is a unique handle for the device. This number is made from a device ID and a serial number.
- Determine device specifications – each device sends its functionality to the operating system, such as how many joysticks it supports, its audio functions, its networking modes, and so on.
- Allocate resources – the operating system then allocates resources to the device depending on its functionality and the plug-and-play device is informed of the allocated resources (such as IRQs, I/O addresses, DMA channels, and so on). It also checks for conflicts on these resources.
- Activate device – when the above have been completed the device is enabled. Only the initiation key can re-initialise the device.

H.10 Windows NT architecture

Windows NT uses two modes:

- User mode – this is a lower privileged mode than kernel mode. It has no direct access to the hardware or to memory. It interfaces to the operating system through well-defined API (application program interface) calls.
- Kernel mode – this is a privileged mode of operation and allows all code direct access to the hardware and memory, including memory allocated to user mode processes. Kernel mode processes also have a higher priority over user mode processes.

Figure H.10 shows an outline of the architecture of NT. It can be seen that only the kernel mode has access to the hardware. This kernel includes executive services which include managers (for I/O, interprocess communications, and so on) and device drivers (which control the hardware). Its parts include:

- Microkernel – controls basic operating system services, such as interrupt handling and scheduling.
- HAL (hardware abstraction layer) – this is a library of hardware-specific programs which give a standard interface between the hardware and software. This can either be Microsoft written or manufacturer provided. They have the advantage of allowing for transportability of programs across different hardware platforms.
- Win32 window manager – supports Win32, MS-DOS and Windows 3.*x* applications.

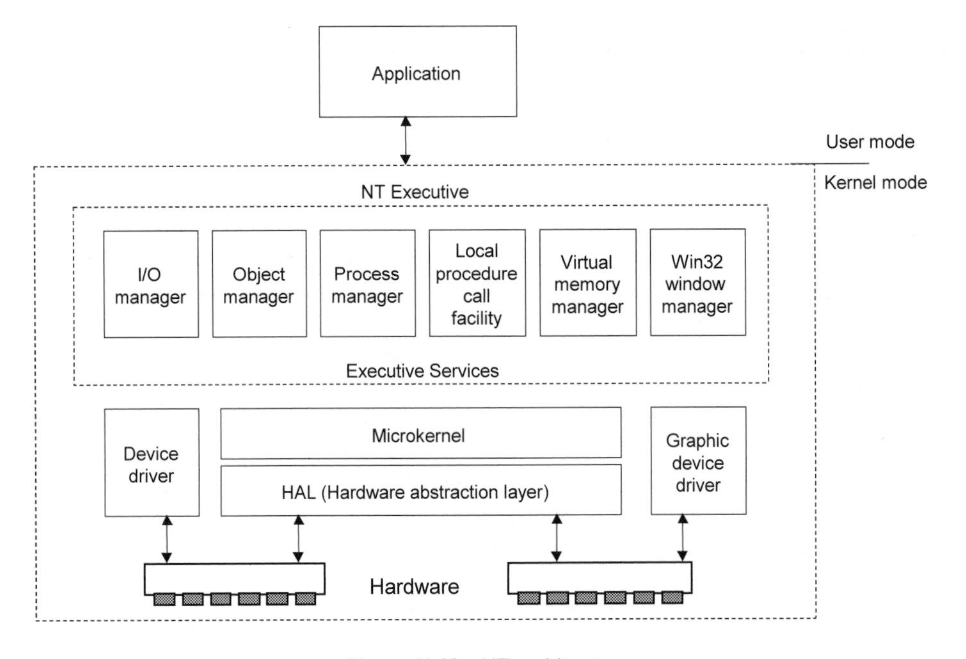

Figure H.10 NT architecture

H.10.1 MS-DOS support

Windows NT supports MS-DOS-based applications with an NT Virtual DOS Machine (NTVDM), where each MS-DOS application has its own NTVDM. The NTVDM is started by the application Ntvdm.exe and when this has started the application communicates with two system files Ntio.sys (equivalent to IO.SYS) and Ntdos.sys (equivalent to MSDOS.SYS). Note that the AUTOEXEC.BAT and CONFIG.SYS files have also been replaced by Autoexec.nt and Config.nt (which are normally located in `\WINNT\System32`).

Multiple NTVDMs have the advantage of being reliable because if one NTVDM fails then it does not affect any others. It also allows MS-DOS-based applications to be multitasked. Unfortunately, each NTVDM needs at least 1 MB of physical memory.

Some MS-DOS applications require direct access to the hardware. NT supports this by providing virtual device drivers (VDDs). These detect a call to hardware and communicate with the NT 32-bit device driver.

Windows NT communicates with hardware through device drivers. These drivers have a `.sys` file extension. An example listing of these is

```
Directory of C:\WINNT\system32\drivers
afd.sys        atapi.sys      atdisk.sys       beep.sys
cdaudio.sys    cdfs.sys       cdrom.sys        changer.sys
cirrus.sys     disk.sys       diskdump.sys     diskperf.sys
fastfat.sys    floppy.sys     ftdisk.sys       hpscan16.sys
i8042prt.sys   kbdclass.sys   ksecdd.sys       modem.sys
mouclass.sys   msfs.sys       mup.sys          ndis.sys
netdtect.sys   npfs.sys       ntfs.sys         null.sys
parallel.sys   parport.sys    parvdm.sys       pcmcia.sys
scsiport.sys   scsiprnt.sys   scsiscan.sys     serial.sys
sfloppy.sys    streams.sys    tape.sys         tdi.sys
vga.sys        videoprt.sys
```

With this, virtual memory applications can have access to the full available memory but NT then maps this to a private memory range (called a virtual memory space). It maps physical memory to virtual memory in 4 KB blocks (called pages). This was previously illustrated in Figure H.6. The driver used to perform the page file access is Pagefile.sys (which is normally found in the top-level directory).

Windows NT has 32 levels of priority (0 to 31). Levels 0 to 15 are used for dynamic applications (such as non-critical operations) and 16 to 31 are used for real-time applications (such as Kernel operations). NT provides a virtual memory by paging file(s) onto the hard disk. Priority levels 0 to 15 can be paged, but levels 16 to 31 cannot.

A summary of the `system32` directory is shown below. The `wowdeb.exe` and `wowexec.exe` files allow Windows 3.*x* programs to run in a 32-bit environment.

```
Directory of C:\winnt\system32
ansi.sys       append.exe     at.exe         atsvc.exe
attrib.exe     autoexec.nt    backup.exe     bootok.exe
bootvrfy.exe   cacls.exe      chcp.com       chkdsk.exe
clipsrv.exe    comm.drv       command.com    comp.exe
compact.exe    config.nt      control.exe    convert.exe
country.sys    csrss.exe      dcomcnfg.exe   ddeshare.exe
ddhelp.exe     ebug.exe       diskcomp.com   diskcopy.com
diskperf.exe   doskey.exe     dosx.exe       DRIVERS
edit.com       exe2bin.exe    expand.exe     fastopen.exe
fc.exe         find.exe       findstr.exe    finger.exe
fontview.exe   forcedos.exe   format.com     ftp.exe
gdi.exe        graftabl.com   graphics.com   grpconv.exe
help.exe       himem.sys      inetins.exe    internat.exe
kb16.com       keyb.com       keyboard.drv   keyboard.sys
krnl386.exe    label.exe      lights.exe     lodctr.exe
mem.exe        mode.com       more.com       mpnotify.exe
mscdexnt.exe   nddeagnt.exe   nddeapir.exe   net.exe
nlsfunc.exe    notepad.exe    ntdos.sys      ntio.sys
ntvdm.exe      os2ss.exe      pax.exe        pentnt.exe
ping.exe       portuas.exe    posix.exe      print.exe
psxss.exe      rdisk.exe      recover.exe    redir.exe
replace.exe    restore.exe    rpcss.exe      rundll32.exe
runonce.exe    savedump.exe   setup.exe      setver.exe
share.exe      shmgrate.exe   skeys.exe      smss.exe
sort.exe       SPOOL          sprestrt.exe   subst.exe
syncapp.exe    sysedit.exe    systray.exe    taskman.exe
```

```
taskmgr.exe    telnet.exe     tree.com       unlodctr.exe
ups.exe        user.exe       userinit.exe   VIEWERS
win.com        winhlp32.exe   winspool.exe   winver.exe
wowdeb.exe     wowexec.exe
```

H.11 Windows 95 and Windows 98

Windows 98 was really an upgrade to Windows 95 OSR2 (which includes NetMeeting, ScanDisk and Disk Defragmenter) and inherits many of the programs that were released with Windows 95 OSR2. The new features include:

- **Advanced plug and play**. USB devices can be added to the computer without rebooting it.
- **Automatic hardware detection**.
- **Enhanced power management**.
- **Increased WWW integration**. WWW page creation, integrated email, channels, and so on.
- **Windows updates**. This facility allows for a single source to update system drivers, system files and operating system programs, such as service packs.
- **System file checker**. The facility checks for system files and recovers old system files. It also checks the integrity of the operating system files and if necessary restores them or extracts them from the installation disks.
- **Maintenance wizard**. This facility allows tasks to be run at given time intervals.
- **Multiple monitors**. The facility allows the computer to display to multiple monitors. Different parts of the screen can be sent to the connected monitors, and thus expand the physical size of the desktop area.
- **NetShow**. The facility allows for the reception of streamed multimedia from a WWW server. It synchronises video, audio and graphics data.
- **WWW TV**. This facility combines broadcast TV and Internet-based content into a single program. With an Internet connection, television program listings are included giving lists of scheduled television shows. Broadcast TV requires a TV tuner card.
- **Support for new devices**. Windows 98 supports many new hardware devices, such as: universal serial bus (USB), IEEE 1394, accelerated graphics port (AGP) and DVD. IEEE 1394 defines a class of hardware that makes it easy to add serial devices to your computer. The AGP is an enhanced video card interface which give enhanced support for 3-D animation. DVD drives play software, videos and music CDs.
- **Subscription**. This facility allows Internet Explorer to check a WWW site for new content, at given time intervals. This new content can be automatically downloaded (or prompted for a download).
- **Channels**. This facility allows content from WWW sites to be automatically downloaded. It is similar to subscription, but the content provider can suggest a schedule for the subscription and it gives a rich map of the WWW site (rather than a single WWW page).

H.12 Fundamentals of Operating Systems

H.12.1Multitasking and threading

Multitasking involves running several tasks at the same time. It normally involves running a process for a given amount of time, before releasing it and allowing another process a given amount of time. The two forms of multitasking are illustrated in Figure H.11 and Figure H.13.

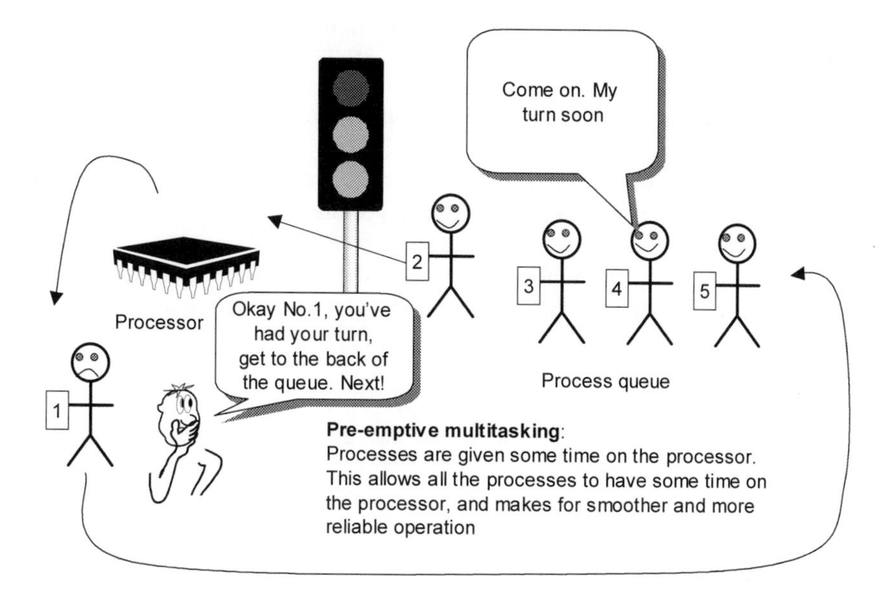

Figure H.11 Pre-emptive multitasking

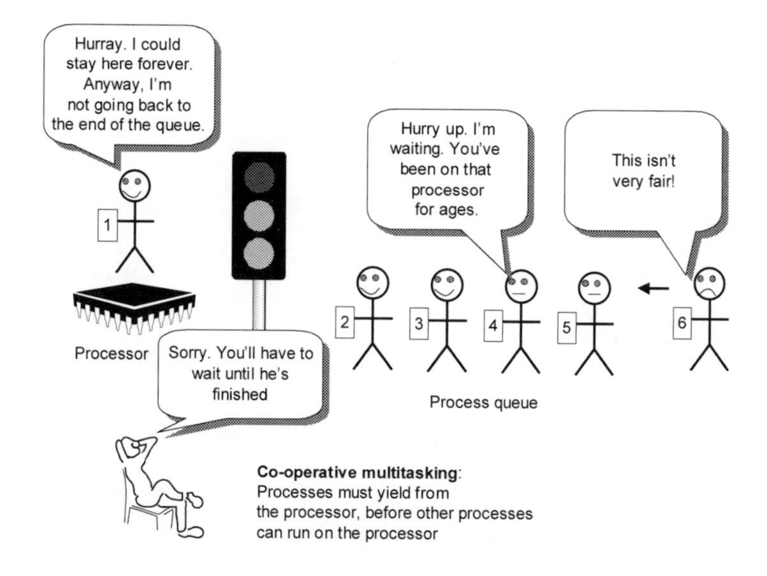

Figure H.12 Co-operative multitasking

The two types are defined as:

- **Pre-emptive multitasking**. This type of multitasking involves the operating system controlling how long a process stays on the processor. This allows for smooth multitasking and is used in 32-bit Microsoft Windows programs and the UNIX operating system.
- **Co-operative multitasking**. This type of multitasking relies on a process giving up the processor. It is used with Windows 3.*x* programs and suffers from processor hogging, where a process can stay on a processor and the operating system cannot kick it off.

The logical extension to multitasking programs is to split a program into a number of parts (threads) and run each of these on the multitasking system (multithreading). A program that is running more than one thread at a time is known as a multithreaded program. Multithreaded programs have many advantages over non-multithreaded programs, including:

- They make better use of the processor, where different threads can be run when one or more threads are waiting for data. For example, a thread could be waiting for keyboard input, while another thread could be reading data from the disk.
- They are easier to test, as each thread can be tested independently of other threads.
- They can use standard threads, which are optimised for given hardware.

They also have disadvantages, including:

- The program has to be planned properly so that threads know on which other threads they depend.
- A thread may wait indefinitely for another thread which has crashed or terminated.

The main difference between multiple processes and multiple threads is that each process has independent variables and data, while multiple threads share data from the main program, as illustrated in Figure H.13.

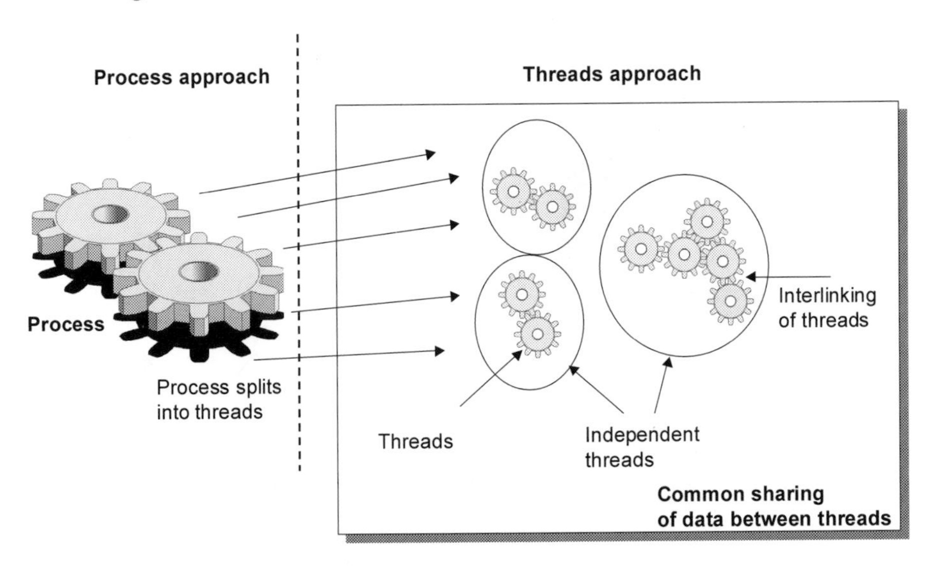

Figure H.13 Process splitting into threads

H.13 Exercises

The following questions are multiple choice. Please select from a to d.

H.13.1 Microsoft Windows 3.*x* in its standard form is:

(a) An 8-bit operating system (b) A 16-bit operating system
(c) A 32-bit operating system (d) A 64-bit operating system

H.13.2 Microsoft Windows NT in its standard form is:

(a) An 8-bit operating system
(b) A 16-bit operating system
(c) A 32-bit operating system
(d) A 64-bit operating system

H.13.3 How does the starting of Windows 95/98 differ from Windows 3.*x*?

(a) An 8-bit operating system
(b) A 16-bit operating system
(c) Windows 3.*x* boots from DOS where as Windows 95/98 has its own boot procedure
(d) A 64-bit operating system

H.13.4 Where do 16-bit application programs get their configuration data?

(a) INF files (b) INI files
(c) BAT files (d) The system registry

H.13.5 Where do 32-bit application programs get their configuration data?

(a) INF files (b) INI files
(c) BAT files (d) The system registry

H.13.6 Which file is used to set up hardware?

(a) SYSTEM.INI (b) WIN.INI
(c) SETUP.INI (d) AUTOEXEC.INI

H.13.7 Which files does Windows 3.*x* use for device drivers?

(a) `.386` (b) `.VXD`
(c) `.SYS` (d) `.DRV`

H.13.8 Which files does Windows 95/98 use for device drivers?

(a) `.386` (b) `.VXD`
(c) `.SYS` (d) `.DRV`

H.13.9 Which files does Windows NT use for device drivers?

(a) `.386` (b) `.VXD`
(c) `.SYS` (d) `.DRV`

H.13.10 What is the 32-bit FAT disk format known as?

(a) E-FAT (b) SFAT
(c) Extended FAT (d) VFAT

H.13.11 What the standard CD-ROM file system known as?

(a) CDFS (b) CD-FAT
(c) CD-RW (d) CD-R

H.13.12 Which of the following best describes pre-emptive multitasking?

(a) The operating system determines how long a process stays on the processor
(b) Processes yield to other processes
(c) Processes have a given time limit on the processor
(d) Processes have sole access to the processor

H.13.13 Which of the following best describes co-operative multitasking:

(a) The operating system determines how long a process stays on the processor
(b) Processes yield to other processes
(c) Processes have a given time limit on the processor
(d) Processes have sole access to the processor

H.13.14 Discuss the advantages that the registry has over INI files.

H.13.15 Discuss the architecture of the 32-bit Windows system.

H.13.16 Discuss how Windows 95/98 use VxD device drivers to interface to the hardware. How do equipment manufacturers develop drivers which use the VxD drivers? How do device drivers in NT differ from Windows 95/98?

H.13.17 Explain how the configuration manager is used to determine the devices which are connected to the system.

H.13.18 Explain the operation of the virtual machine manager.

H.13.19 Explain the main differences between pre-emptive and co-operative multitasking. Discuss also how multitasking and threading is implemented.

H.13.20 Discuss how Windows 95/98 use priority systems to schedule processes.

I HDLC

I.1 Introduction

The data link layer is the second layer in the OSI seven-layer model and its protocols define rules for the orderly exchange of data information between two adjacent nodes connected by a data link. Final framing, flow control between nodes, and error detection and correction are added at this layer. In previous chapters the data link layer was discussed in a practical manner. It is a use protocol as it provides a model for interfacing to a serial bus.

The two types of protocol are:

- Asynchronous protocol.
- Synchronous protocol.

Asynchronous communications uses start-stop method of communication where characters are sent between nodes, as illustrated in Figure I.1. Special characters are used to control the data flow. Typical flow control characters are End of Transmission (EOT), Acknowledgement (ACK), Start of Transmission (STX) and Negative Acknowledgement (NACK).

Synchronous communications involves the transmission of frames of bits with start and end bit characters to delimit the frame. The two of the most popular are IBM's synchronous data link communication (SDLC) and high-level data link control (HDLC). Many network data link layers are based upon these standards, examples include the LLC layer in IEE 802.*x* LAN standards and LAPB in the X.25 packet switching standard.

Synchronous communications normally uses a bit-oriented protocol (BOP), where data is sent one bit at a time. The data link control information is interpreted on a bit-by-bit basis rather than with unique data link control characters.

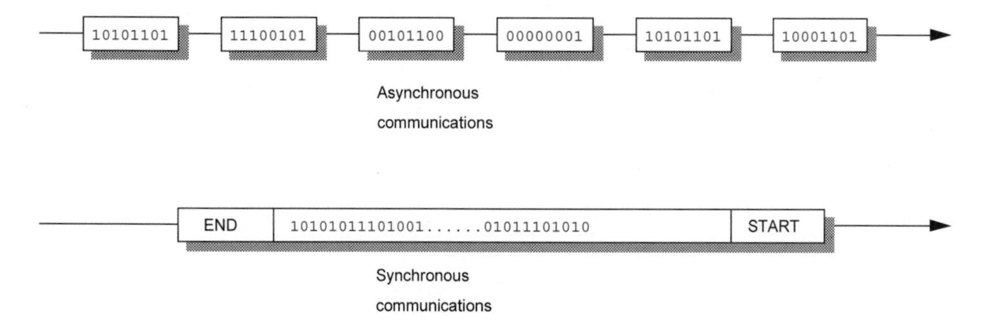

Figure I.1 Asynchronous and synchronous communications

HDLC is a standard developed by the ISO to provide a basis for the data link layer for point-to-point and multi-drop connections. It can transfer data either in a simplex, half-duplex, or full-duplex mode. Frames are generally limited to 256 bytes in length and a single control field performs most data link control functions.

I.2 HDLC protocol

In HDLC, a node is either defined as a primary station or a secondary station. A primary station controls the flow of information and issues commands to secondary stations. The secondary station then sends back responses to the primary. A primary station with one or more secondary stations is known as unbalanced configuration.

HDLC allows for point-to-point and multi-drop. In point-to-point communications a primary station communicates with a single secondary station. For multi-drop, one primary station communications with many secondary stations.

In point-to-point communications it is possible for a station be operate as a primary and a secondary station. At any time, one of the stations can be a primary and the other the secondary. Thus, commands and responses flow back and forth over the transmission link. This is known as a balanced configuration, or combined stations.

I.2.1 HDLC modes of operation

HDLC has three modes of operation. Unbalanced configurations can use the normal response mode (NRM). Secondary stations can only transmit when specifically instructed by the primary station. When used as a point-to-point or multi-drop configuration only one primary station is used. Figure I.2 shows a multi-drop NRM configuration.

Unbalanced configurations can also use the asynchronous response mode (ARM). It differs from NRM in that the secondary is allowed to communicate with the primary without receiving permission from the primary.

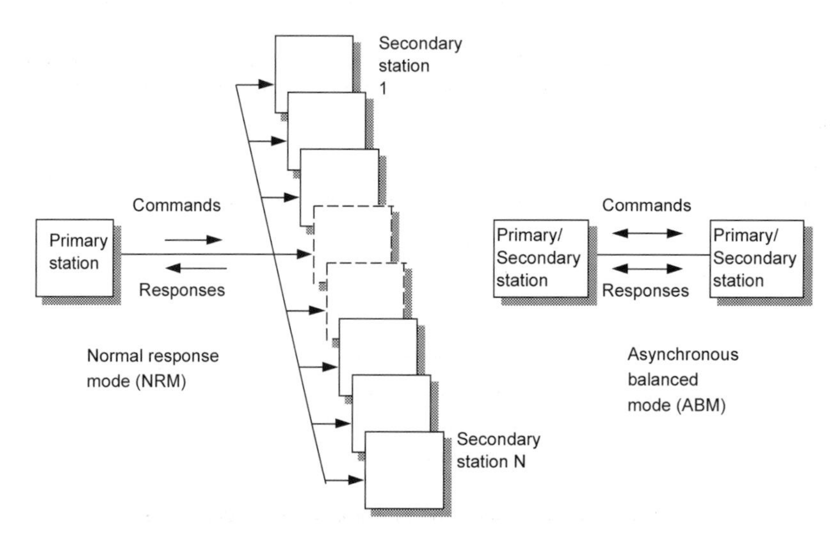

Figure I.2 NRM and ABM mode

In asynchronous balanced mode (ABM) all stations have the same priority and can perform the functions of a primary and secondary station.

I.2.2 HDLC frame format

HDLC frames are delimited by the bit sequence `01111110`. Figure I.3 shows the standard format of the HDLC frame, the 5 fields are the:

- Flag field.
- Address field.
- Control field.
- Information field.
- Frame check sequence (FCS) field.

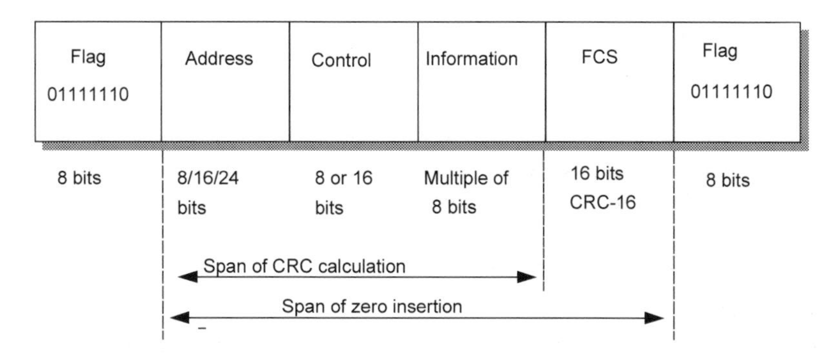

Figure I.3 HDLC frame structure

I.2.3 Information field

The information fields contain data, such as OSI level 3, and above, information. It contains an integer number of bytes and thus the number of bits contained is always a multiple of eight. The receiver determines the number of bytes in the data because it can detect the start and end flag. By this method, it also finds the FCS field. Note that the number of characters in the information can be zero as not all frames contain data.

I.2.4 Flag field

A unique flag sequence, `01111110` (or `7Eh`), delimits the start and end of the frame. As this sequence could occur anywhere within the frame a technique called bit-insertion is used to stop this happening except at the start and end of the frame.

I.2.5 Address field

The address field is used to address connected stations an, in basic addressing, it contains an 8-bit address. It can also be extended, using extended addressing, to give any multiple of 8 bits.

When it is 8 bits wide it can address up to 254 different nodes, as illustrated in Figure I.4. Two special addresses are `00000000` and `11111111`. The `00000000` address defines the null or void address and the `11111111` broadcasts a message to all secondaries. The other 254 addresses are used to address secondary nodes individually.

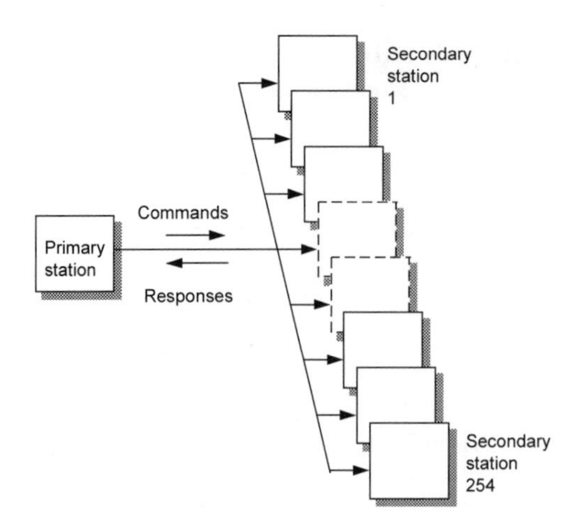

Figure I.4 HDLC addressing range

If there are a large number of secondary stations then extended address can be used to extend the address field indefinitely. A 0 in the first bit of the address field allows a continuation of the address, or a 1 ends it. For example:

```
XXXXXXX1 XXXXXXX0 XXXXXXX0 XXXXXXX0
```

I.2.6 Control field

The control field can either be 8 or 16 bits wide. It is used to identify the frame type and can also contain flow control information. The first two bits of the control field define the frame type, as shown in Figure I.5. There are three types of frames, these are:

- Information frames.
- Supervisory frames.
- Unnumbered frames.

When sent from the primary the `P/F` bit indicates that it is polling the secondary station. In an unbalanced mode, a secondary station cannot transmit frames unless the primary sets the poll bit.

When sending frames from the secondary, the `P/F` bit indicates whether the frame is the last of the message, or not. Thus if the `P/F` bit is set by the primary it is a poll bit (`P`), if it is set by the secondary it is a final bit (`F`).

The following sections describe 8-bit control fields. Sixteen-bit control fields are similar but reserve a 7-bit field for the frame counter variables `N(R)` and `N(S)`.

Information frame

An information frame contains sequenced data and is identified by a 0 in the first bit position of the control field. The 3-bit variable `N(R)` is used to confirm the number of transmitted frames received correctly and `N(S)` is used to number an information frame. The first frame transmitted is numbered 0 as (`000`), the next as 1 (`001`), until the eighth which is numbered `111`. The sequence then starts back at 0 again and this gives a sliding window of eight frames.

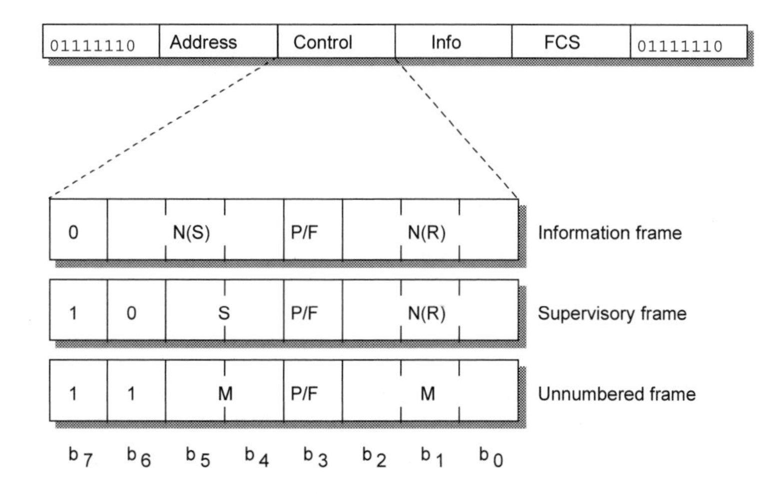

Figure I.5 Format of an 8-bit control field

Supervisory frame

Supervisory frames contain flow control data. They confirm or reject previously received information frames and also can indicate whether a station is ready to receive frames.

The N(S) field is used with the S bits to acknowledge, or reject, previously transmitted frames. Responses from the receiver are set in the S field, these are receiver ready (RR), ready not to receive (RNR), reject (REJ) and selectively reject (SREJ). Table I.1 gives the format of these bits.

RR informs the receiver that it acknowledges the frames sent up to N(R). RNR tells the transmitter that the receiver cannot receive any more frames at the present time (RR will cancel this). It also acknowledges frames up to N(R). The REJ control rejects all frames after N(R). The transmitter must then send frames starting at N(R).

Table I.1 Supervisory bits

b_5	b_4	*Receiver status*
0	0	Receiver ready (RR)
1	0	Receiver not ready (RNR)
0	1	Reject (REJ)
1	1	Selectively reject (SREJ)

Unnumbered frame

If the first two bits of the control field are 1's then it is an unnumbered frame. Apart from the P/F flag the other bits are used to send unnumbered commands. When sending commands, the P/F flag is a poll bit (asking for a response), and for responses it is a flag bit (end of response).

The available commands are SARM (set asynchronous response mode), SNRM (set normal response mode), SABM (set asynchronous balance mode), RSET (reset), FRMR (frame reject) and Disconnect (DISC). The available responses are UA (unnumbered acknowledge), CMDR (command reject), FRMR (frame reject) and DM (disconnect mode). Bit definitions for some of these are:

SABM	1111P110	DM	1111F000	DISC	1100P010
UA	1100F110	FRMR	1110F001		

I.2.7 Frame check sequence field

The frame check sequence (FCS) field contains an error detection code based on cyclic redundancy check (CRC) polynomials. It is used to check the address, control and information fields, as previously illustrated in Figure I.2. HDLC uses a polynomial specified by CCITT V.41, which is $G(x) = x^{16} + x^{12} + x^{5} + x^{1}$. This is also known as CRC-16 or CRC-CCITT.

I.3 Transparency

The flag sequence `01111110` can occur anywhere in the frame. To prevent this a transparency mechanism called zero-bit insertion or zero stuffing is used. There are two main rules that are applied, these are:

- In the transmitter, a 0 is automatically inserted after five consecutive 1's, except when the flag occurs.
- At the receiver, when five consecutive 1's are received and the next bit is a 0 then the 0 is deleted and removed. If it is a 1 then it must be a valid flag.

In the following example a flag sequence appears in the data stream where it is not supposed to (spaces have been inserted around it). Notice that the transmitter detects five 1's in a row and inserts a 0 to break them up.

```
Message:  00111000101000 01111110  01011111 1111010101
Sent:     00111000101000 011111010 0101111101111010101
```

I.4 Flow control

Supervisory frames (`S[]`) send flow control information to acknowledge the reception of data frames or to reject frames. Unnumbered frames (`U[]`) set up the link between a primary and a secondary, by the primary sending commands and the secondary replying with responses. Information frames (`I[]`) contain data.

I.4.1 Link connection

Figure I.6 shows how a primary station (node A) sets up a connection with a secondary station (node B) in NRM (normal response mode). In this mode one or many secondary stations can exist. First the primary station requests a link by sending an unnumbered frame with: node B's address (`ADDR_B`), the set normal response mode (`SNRM`) command and with poll flag set (`P=1`), that is, `U[SNRM,ABBR_B,P=1]`. If the addressed secondary wishes to make a connection then it replies back with an unnumbered frame containing: its own address

(ADDR_B), the unnumbered acknowledge (UA) response and the final bit set (F=1), i.e. U[UA,ABBR_B,F=1]. The secondary sends back its own address because many secondaries can exist and it thus identifies which station has responded. There is no need to send the primary station address as only one primary exists.

Once the link is set up data can flow between the nodes. To disconnect the link, the primary station sends an unnumbered frame with: node B's address (ADDR_B), the disconnect (DISC) command and the poll flag set (P=1), that is, U[DISC,ABBR_B,P=1]. If the addressed secondary accepts the disconnection then it replies back with an unnumbered frame containing: its own address (ADDR_B), the unnumbered acknowledge (UA) response and the final bit set (F=1), i.e. U[UA,ABBR_B,F=1].

When two stations act as both primaries and secondaries then they use the asynchronous balanced mode (ABM). Each station has the same priority and can perform the functions of a primary and secondary station. Figure I.7 shows a typical connection. The ABM mode is set up initially using the SABM command (U[SABM,ABBR_B,P=1]). The connection between node A and node B is then similar to the NRM but, as node B operates as a primary station, it can send a disconnect command to node A (U[DISC,ABBR_B,P=1]).

The SABM, SARM and SNRM modes set up communications using an 8-bit control field. Three other commands exist which set up a 16-bit control field, these are SABME (set asynchronous balanced mode extended), SARME and SNRME. The format of the 16-bit control field is given in Figure I.8.

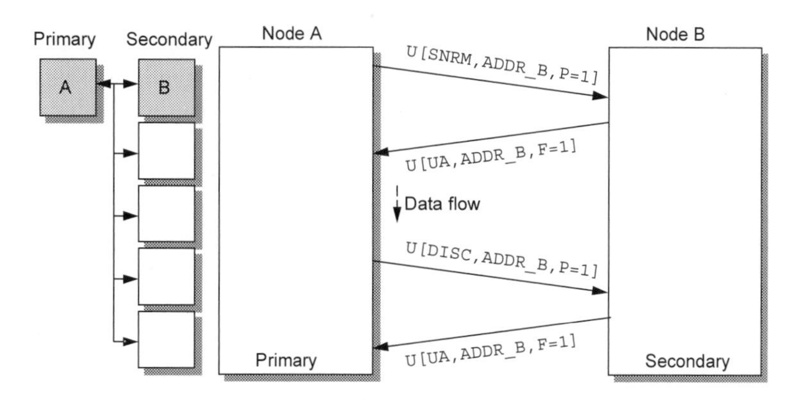

Figure I.6 Connection between a primary and secondary in NRM

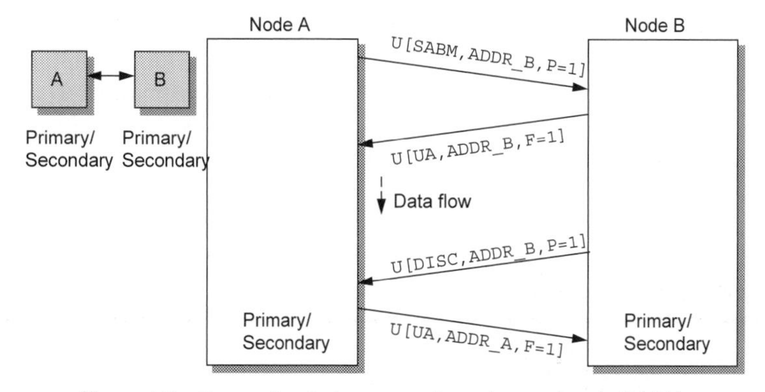

Figure I.7 Connection between a primary/secondary in SABM

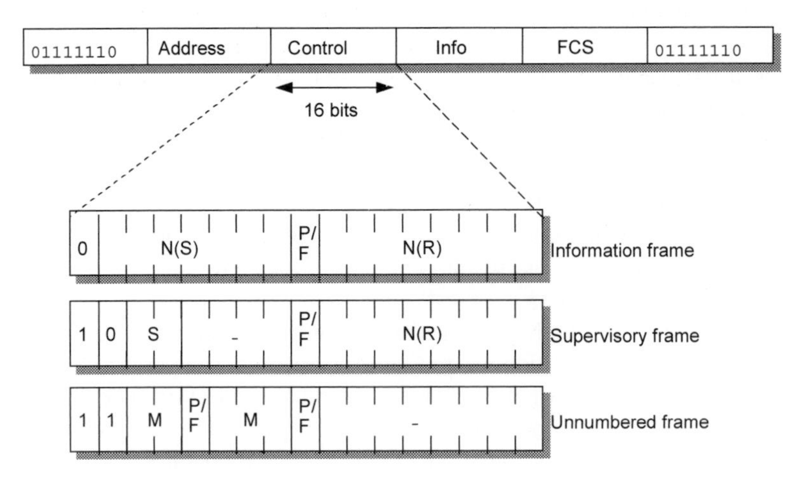

Figure I.8 Extended control field.

Figure I.9 shows an example conversation between a sending station (node A) and a receiving station (node B). Initially three information frames are sent numbered 2, 3 and 4 (`I[N(S)=2]`, `I[N(S)=3]` and `I[N(S)=4, P=1]`). The last of these frames has the poll bit set, which indicates to node B that node A wishes it to respond, either to acknowledge or reject previously unacknowledged frames. Node B does this by sending back a supervisory frame (`S[RR, N(R)=5]`) with the receiver ready (RR) acknowledgement. This informs node A that node B expects to receive frame number 5 next. Thus it has acknowledged all frames up to and including frame 4.

In the example in Figure I.9 an error has occurred in the reception of frame 5. The recipient informs the sender by sending a supervisory frame with a reject flow command (`S[REJ, N(R)=5]`). After the sender receives this it resends each frame after and including frame 5.

If the receiver does not want to communicate, at the present, it sends a receiver not ready flow command. For example `S[RNR, N(R)=5]` tells the transmitter to stop sending data, at the present. It also informs the sender that all frames up to frame 5 have been accepted. The sender will transmit frames once it has received a receiver ready frame from the receiver.

Figure I.9 shows an example of data flow in only the one direction. With ABM both stations can transmit and receive data. Thus each frame sent contains receive and send counter values. When stations send information frames the previously received frames can be acknowledged, or rejected, by piggy-backing the receive counter value. In Figure I.10, node A sends three information frames with `I[N(S)=0,N(R)=0]`, `I[N(S)=1, N(R)=0]`, and `I[N(S)=2,N(R)=0]`. The last frame informs node B that node A expects to receive frame 0 next. Node B then sends frame 0 and acknowledges the reception of all frames up to, and including frame 2 with `I[N(S)=0,N(R)=3]`, and so on.

I.5 Derivatives of HDLC

There are many derivatives of HDLC, including:

- LAPB (link access procedure balanced) is used in X.25 packet switched networks;
- LAPM(link access procedure for modems) is used in error correction modems;
- LLC (logical link control) is used in Ethernet and Token Ring networks;
- LAPD (link access procedure D-channel) is used in Integrated Services Digital Networks (ISDNs).

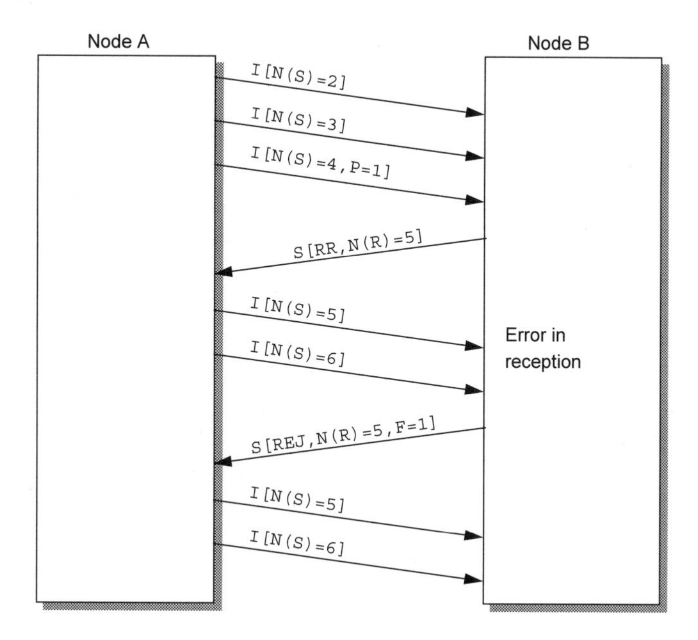

Figure I.9 Example flow

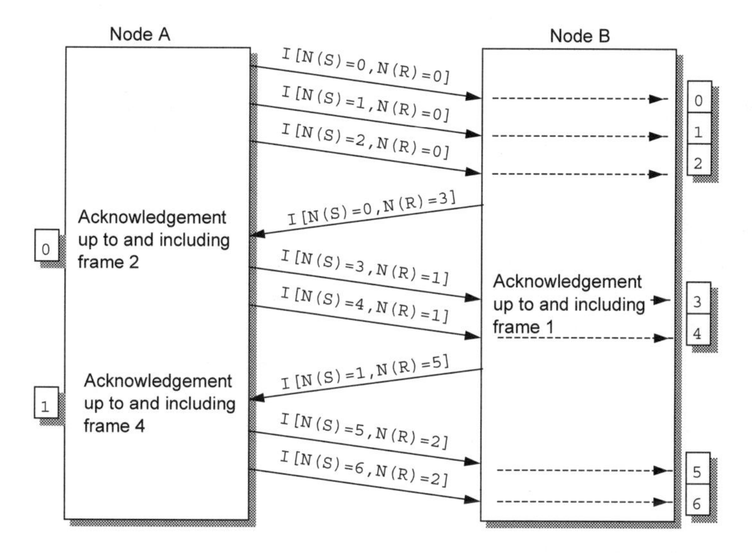

Figure I.10 Example flow with piggy-backed acknowledgement

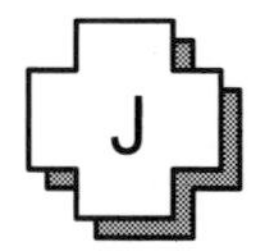

Example WinSock Code for Visual Basic

J.1 My client (myClient.frm)

```
VERSION 5.00
Object = "{248DD890-BB45-11CF-9ABC-0080C7E7B78D}#1.0#0"; "mswinsck.ocx"
Begin VB.Form myClient
   Caption         =   "Client Application"
   ClientHeight    =   3210
   ClientLeft      =   2250
   ClientTop       =   6195
   ClientWidth     =   8265
   LinkTopic       =   "Form1"
   MaxButton       =   0   'False
   ScaleHeight     =   3210
   ScaleWidth      =   8265
   StartUpPosition =   1  'CenterOwner
   Begin VB.CommandButton HelpClient
      Caption         =   "Example"
      Height          =   375
      Left            =   4800
      TabIndex        =   15
      Top             =   2640
      Width           =   1335
   End
   Begin VB.ListBox PortNameC
      Height          =   840
      ItemData        =   "myClient.frx":0000
      Left            =   4560
      List            =   "myClient.frx":0016
      TabIndex        =   1
      Top             =   1080
      Width           =   1815
   End
   Begin VB.TextBox AddressPort
      Height          =   285
      IMEMode         =   3  'DISABLE
      Left            =   6480
      TabIndex        =   14
      TabStop         =   0   'False
      Text            =   "Port Number"
      Top             =   1080
      Width           =   1575
   End
   Begin VB.TextBox EOTClient
      Enabled         =   0   'False
      Height          =   195
      Left            =   0
      TabIndex        =   11
```

```
      TabStop          =   0   'False
      Text             =   " EOT "
      Top              =   120
      Visible          =   0   'False
      Width            =   150
   End
   Begin VB.CommandButton CloseC
      Caption          =   "Finish"
      Height           =   375
      Left             =   6480
      TabIndex         =   5
      Top              =   2640
      Width            =   1335
   End
   Begin VB.ComboBox AddressIP
      Height           =   315
      ItemData         =   "myClient.frx":0050
      Left             =   4560
      List             =   "myClient.frx":005D
      TabIndex         =   0
      Text             =   "IP or DNS"
      Top              =   480
      Width            =   3495
   End
   Begin VB.CommandButton cmdDisConnect
      Caption          =   "Disconnect"
      Height           =   375
      Left             =   6480
      TabIndex         =   4
      Top              =   2160
      Width            =   1335
   End
   Begin VB.CommandButton cmdConnect
      Caption          =   "Connect"
      Height           =   375
      Left             =   4800
      TabIndex         =   2
      Top              =   2160
      Width            =   1335
   End
   Begin VB.TextBox ShowText
      Height           =   1965
      Left             =   240
      Locked           =   -1  'True
      MultiLine        =   -1  'True
      ScrollBars       =   2  'Vertical
      TabIndex         =   6
      TabStop          =   0   'False
      Top              =   1080
      Width            =   3975
   End
   Begin VB.TextBox SendTextData
      Height           =   285
      Left             =   240
      MultiLine        =   -1  'True
      TabIndex         =   3
      Top              =   480
      Width            =   3975
   End
   Begin MSWinsockLib.Winsock myTCPClient
      Left             =   7800
      Top              =   0
```

```
      _ExtentX          =   741
      _ExtentY          =   741
      _Version          =   327681
   End
   Begin VB.Label Label1
      Caption           =   "PortName"
      Height            =   255
      Left              =   4560
      TabIndex          =   13
      Top               =   840
      Width             =   735
   End
   Begin VB.Label CopyRight1
      Caption           =   "email_address"
      Enabled           =   0   'False
      BeginProperty Font
         Name              =   "Times New Roman"
         Size              =   8.25
         Charset           =   0
         Weight            =   400
         Underline         =   0   'False
         Italic            =   0   'False
         Strikethrough     =   0   'False
      EndProperty
      Height            =   255
      Left              =   6360
      TabIndex          =   12
      Top               =   3000
      Width             =   1935
   End
   Begin VB.Label Label4
      Caption           =   "AddressPort"
      Height            =   255
      Left              =   6480
      TabIndex          =   10
      Top               =   840
      Width             =   975
   End
   Begin VB.Label Label3
      Caption           =   "AddressIP"
      Height            =   255
      Left              =   4560
      TabIndex          =   9
      Top               =   240
      Width             =   855
   End
   Begin VB.Label ShowLabel
      Caption           =   "ShowText"
      Enabled           =   0   'False
      Height            =   255
      Left              =   240
      TabIndex          =   8
      Top               =   840
      Width             =   3495
   End
   Begin VB.Label SendLabel
      Caption           =   "SendTextData"
      Height            =   255
      Left              =   240
      TabIndex          =   7
      Top               =   240
      Width             =   3495
```

```
   End
End
Attribute VB_Name = "myClient"
Attribute VB_GlobalNameSpace = False
Attribute VB_Creatable = False
Attribute VB_PredeclaredId = True
Attribute VB_Exposed = False
'Last modification: 15/11/99

Private Sub Form_Load()
    Unload ChoiceSC     'Close the main menu properly
    SendTextData.Enabled = False     'Initialisation
    SendLabel.Enabled = False
    cmdDisConnect.Enabled = False
    cmdConnect.Enabled = True
    AddressIP.Enabled = True
    AddressPort.Enabled = False
End Sub

Private Sub cmdConnect_Click()
    'Connect to the server
    myTCPClient.Connect
    SendTextData.Enabled = True
    SendLabel.Enabled = True
    cmdConnect.Enabled = False
    cmdDisConnect.Enabled = True
    AddressIP.Enabled = False
    AddressPort.Enabled = False
    CloseC.Enabled = False
    PortNameC.Enabled = False
End Sub

Private Sub cmdDisConnect_Click()
    'Disconnect from the server
    myTCPClient.Close
    SendTextData.Enabled = False
    SendLabel.Enabled = False
    cmdConnect.Enabled = True
    cmdDisConnect.Enabled = False
    AddressIP.Enabled = True
    CloseC.Enabled = True
    PortNameC.Enabled = True
End Sub

Private Sub PortNameC_Click()
    'Choice of the port (name)
    If PortNameC.Text = "--Manual enter-- >" Then AddressPort.Enabled = True
Else AddressPort.Enabled = False
    If PortNameC.Text = "Test" Then AddressPort.Text = "1001"
    If PortNameC.Text = "Echo" Then AddressPort.Text = "7"
    If PortNameC.Text = "Daytime" Then AddressPort.Text = "13"
    If PortNameC.Text = "FTP" Then AddressPort.Text = "21"
    If PortNameC.Text = "Telnet" Then AddressPort.Text = "23"
    If PortNameC.Text = "SMTP" Then AddressPort.Text = "25"

End Sub

Private Sub myTCPClient_DataArrival(ByVal bytesTotal As Long)
    'Display incoming data
    Dim str1 As String, str2 As String, str As String  'declare old, new, to-
tal data
    str1 = ShowText.Text    'old data
```

```
    myTCPClient.GetData str2    'incoming data (new data)
    str = str1 + str2   'total data to display
    ShowText.Text = str 'display to ShowText
End Sub

Private Sub AddressIP_Click()
    'Choose IP Address
    myTCPClient.RemoteHost = AddressIP.Text
End Sub

Private Sub AddressIP_Change()
    'Enter IP or DNS address
    myTCPClient.RemoteHost = AddressIP.Text
End Sub

Private Sub AddressPort_Change()
    'Change port number directly in the AddressPort box (manually)
    myTCPClient.RemotePort = AddressPort.Text
End Sub

Private Sub CloseC_Click()
    'Return to main menu
    ChoiceSC.Show
End Sub

Private Sub SendTextData_KeyPress(KeyAscii As Integer)
    'When you press the ENTER key the contain of the top box is sent
    If KeyAscii = 13 Then
    myTCPClient.SendData SendTextData.Text
    SendTextData.Text = ""
    End If
End Sub
```

J.2 My server (myServer.frm)

```
VERSION 5.00
Object = "{248DD890-BB45-11CF-9ABC-0080C7E7B78D}#1.0#0"; "mswinsck.ocx"
Begin VB.Form myServer
   Caption         =   "Server Application"
   ClientHeight    =   3810
   ClientLeft      =   60
   ClientTop       =   345
   ClientWidth     =   3840
   LinkTopic       =   "Form1"
   MaxButton       =   0   'False
   ScaleHeight     =   3810
   ScaleWidth      =   3840
   StartUpPosition =   1  'CenterOwner
   Begin VB.CommandButton DisConnect
      Caption         =   "Disconnect"
      Height          =   375
      Left            =   240
      TabIndex        =   7
      Top             =   3120
      Width           =   1335
   End
   Begin VB.TextBox ConnectionState
```

```
      Height          =   375
      Left            =   1800
      Locked          =   -1  'True
      TabIndex        =   5
      TabStop         =   0   'False
      Top             =   2520
      Width           =   1695
   End
   Begin VB.CommandButton CloseS
      Caption         =   "Finish"
      Height          =   375
      Left            =   2280
      TabIndex        =   4
      Top             =   3120
      Width           =   1335
   End
   Begin VB.TextBox ShowText
      Enabled         =   0   'False
      Height          =   650
      Left            =   120
      MultiLine       =   -1  'True
      ScrollBars      =   2  'Vertical
      TabIndex        =   1
      Top             =   1440
      Width           =   3615
   End
   Begin VB.TextBox SendTextData
      Height          =   650
      Left            =   120
      MultiLine       =   -1  'True
      ScrollBars      =   2  'Vertical
      TabIndex        =   0
      Top             =   480
      Width           =   3615
   End
   Begin MSWinsockLib.Winsock myTCPServer
      Left            =   240
      Top             =   3120
      _ExtentX        =   741
      _ExtentY        =   741
      _Version        =   327681
   End
   Begin VB.Label CopyRight2
      Caption         =   "ncenciar@engineer.com"
      Enabled         =   0   'False
      BeginProperty Font
         Name            =   "Times New Roman"
         Size            =   8.25
         Charset         =   0
         Weight          =   400
         Underline       =   0   'False
         Italic          =   0   'False
         Strikethrough   =   0   'False
      EndProperty
      Height          =   255
      Left            =   1920
      TabIndex        =   8
      Top             =   3600
      Width           =   1815
   End
   Begin VB.Label Label3
      Caption         =   "Connection State:"
```

```
      Height          =   375
      Left            =   120
      TabIndex        =   6
      Top             =   2520
      Width           =   1575
   End
   Begin VB.Label Label2
      Caption         =   "ShowText"
      Height          =   255
      Left            =   120
      TabIndex        =   3
      Top             =   1200
      Width           =   3615
   End
   Begin VB.Label Label1
      Caption         =   "SendTextData"
      Height          =   255
      Left            =   120
      TabIndex        =   2
      Top             =   240
      Width           =   3615
   End
End
Attribute VB_Name = "myServer"
Attribute VB_GlobalNameSpace = False
Attribute VB_Creatable = False
Attribute VB_PredeclaredId = True
Attribute VB_Exposed = False
'Last modification: 15/11/99

Private Sub Form_Load()
    Unload ChoiceSC 'Close main menu properly
    'Set the local port to 1001 and listen for a connection
    myTCPServer.LocalPort = 1001
    myTCPServer.Listen
    ConnectionState.Text = "Disconnected"
    DisConnect.Enabled = False
End Sub

Private Sub myTCPServer_Close()
    'Close method
    myTCPServer.Close
    ConnectionState.Text = "Disconnected"
    SendTextData.Enabled = False
End Sub

Private Sub myTCPServer_ConnectionRequest(ByVal requestID As Long)
    'Check state of socket, if it is not closed then close it.
    If myTCPServer.State <> sckClosed Then myTCPServer.Close
    'Accept the request with the requestID parameter.
    myTCPServer.Accept requestID
    ConnectionState.Text = "Connected"
    DisConnect.Enabled = True
    CloseS.Enabled = False
    SendTextData.Enabled = True
End Sub

Private Sub SendTextData_Change()
    'SendTextData contains the data to be sent
    'This data is sent using the SendData method
    myTCPServer.SendData SendTextData.Text
End Sub
```

```
Private Sub myTCPServer_DataArrival(ByVal bytesTotal As Long)
    'Read incoming data into the str variable,
    'then display it to ShowText
    Dim str As String
    myTCPServer.GetData str
    ShowText.Text = str
End Sub

Private Sub CloseS_Click()
    'Return to main menu
    ChoiceSC.Show
End Sub

Private Sub DisConnect_Click()
    myTCPServer.Close
    ConnectionState.Text = "Disconnected"
    CloseS.Enabled = True
    DisConnect.Enabled = False
End Sub
```

J.3 Choice form (ChoiceSC.frm)

```
VERSION 5.00
Begin VB.Form ChoiceSC
   Caption         =   "Server or Client?"
   ClientHeight    =   1530
   ClientLeft      =   60
   ClientTop       =   345
   ClientWidth     =   4125
   LinkTopic       =   "Form1"
   MaxButton       =   0   'False
   Moveable        =   0   'False
   ScaleHeight     =   1530
   ScaleWidth      =   4125
   StartUpPosition =   1  'CenterOwner
   Begin VB.CommandButton Client1
      Caption         =   "Client"
      Height          =   375
      Left            =   2160
      TabIndex        =   2
      Top             =   720
      Width           =   1215
   End
   Begin VB.CommandButton Server1
      Caption         =   "Server"
      Height          =   375
      Left            =   840
      TabIndex        =   1
      Top             =   720
      Width           =   1215
   End
   Begin VB.Label CopyRight0
      Caption         =   "email_address"
      Enabled         =   0   'False
      BeginProperty Font
         Name            =   "Times New Roman"
```

```
         Size            =   8.25
         Charset         =   0
         Weight          =   400
         Underline       =   0   'False
         Italic          =   0   'False
         Strikethrough   =   0   'False
      EndProperty
      Height          =   255
      Left            =   2160
      TabIndex        =   3
      Top             =   1320
      Width           =   1815
   End
   Begin VB.Label Question1
      Caption         =   "Which mode do you want to use?"
      Height          =   375
      Left            =   840
      TabIndex        =   0
      Top             =   240
      Width           =   2535
   End
End
Attribute VB_Name = "ChoiceSC"
Attribute VB_GlobalNameSpace = False
Attribute VB_Creatable = False
Attribute VB_PredeclaredId = True
Attribute VB_Exposed = False
'Last modification: 15/11/99

Private Sub Form_Load()
    Unload myClient 'Close Client form properly
    Unload myServer 'Close Server form properly
End Sub

Private Sub Server1_Click()
    Unload myClient 'Close Client form properly
    Unload myServer 'Close Server form properly
    myServer.Show
End Sub

Private Sub Client1_Click()
    Unload myClient 'Close Client form properly
    Unload myServer 'Close Server form properly
    myClient.Show
End Sub
```

J.4 Error panel (ErrorPanel.frm)

```
VERSION 5.00
Begin VB.Form ErrorPanel
   BorderStyle     =   4  'Fixed ToolWindow
   Caption         =   "00 Type Error"
   ClientHeight    =   1545
   ClientLeft      =   45
   ClientTop       =   285
   ClientWidth     =   2385
   LinkTopic       =   "Form1"
```

```
   MaxButton       =   0   'False
   MinButton       =   0   'False
   Moveable        =   0   'False
   NegotiateMenus  =   0   'False
   ScaleHeight     =   1545
   ScaleWidth      =   2385
   ShowInTaskbar   =   0   'False
   StartUpPosition =   1  'CenterOwner
   Begin VB.CommandButton OkButton
      Caption         =   "OK"
      Height          =   375
      Left            =   960
      TabIndex        =   1
      Top             =   840
      Width           =   615
   End
   Begin VB.Label CopyRight3
      Caption         =   "email_address"
      BeginProperty Font
         Name            =   "Times New Roman"
         Size            =   8.25
         Charset         =   0
         Weight          =   400
         Underline       =   0   'False
         Italic          =   0   'False
         Strikethrough   =   0   'False
      EndProperty
      Height          =   255
      Left            =   600
      TabIndex        =   3
      Top             =   1320
      Width           =   1815
   End
   Begin VB.Label ErrorLabel
      Caption         =   "Click OK to continue"
      Height          =   255
      Left            =   480
      TabIndex        =   2
      Top             =   480
      Width           =   1575
   End
   Begin VB.Label ErrorMessage
      Caption         =   "Error"
      BeginProperty Font
         Name            =   "System"
         Size            =   9.75
         Charset         =   0
         Weight          =   700
         Underline       =   0   'False
         Italic          =   0   'False
         Strikethrough   =   0   'False
      EndProperty
      Height          =   255
      Left            =   1080
      TabIndex        =   0
      Top             =   120
      Width           =   615
   End
End
Attribute VB_Name = "ErrorPanel"
Attribute VB_GlobalNameSpace = False
Attribute VB_Creatable = False
```

```
Attribute VB_PredeclaredId = True
Attribute VB_Exposed = False
'Last modification 10/11/99

Private Sub Form_Load()
    ErrorPanel.Enabled = True
    ErrorPanel.Visible = True
End Sub

Private Sub OkButton_Click()
    ErrorPanel.Enabled = False
    ErrorPanel.Visible = False
End Sub
```

J.5 Help form (help.frm)

```
VERSION 5.00
Begin VB.Form HELP
   BorderStyle     =   4  'Fixed ToolWindow
   Caption         =   "Example - Help"
   ClientHeight    =   3195
   ClientLeft      =   45
   ClientTop       =   285
   ClientWidth     =   4680
   MaxButton       =   0   'False
   MinButton       =   0   'False
   ScaleHeight     =   3195
   ScaleWidth      =   4680
   ShowInTaskbar   =   0   'False
   Begin VB.Label Copyright4
      Caption         =   "email_address"
      BeginProperty Font
         Name            =   "Times New Roman"
         Size            =   8.25
         Charset         =   0
         Weight          =   400
         Underline       =   0   'False
         Italic          =   0   'False
         Strikethrough   =   0   'False
      EndProperty
      Height          =   255
      Left            =   2880
      TabIndex        =   0
      Top             =   2880
      Width           =   1695
   End
End
Attribute VB_Name = "HELP"
Attribute VB_GlobalNameSpace = False
Attribute VB_Creatable = False
Attribute VB_PredeclaredId = True
Attribute VB_Exposed = False
```

Index